This Changes Everything

NAOMI KLEIN

This Changes Everything

Capitalism vs. the Climate

ALLEN LANE
an imprint of
PENGUIN BOOKS

ALLEN LANE

Published by the Penguin Group
Penguin Books Ltd, 80 Strand, London WC2R 0RL, England
Penguin Group (USA) Inc., 375 Hudson Street, New York, New York 10014, USA
Penguin Group (Canada), 90 Eglinton Avenue East, Suite 700, Toronto, Ontario, Canada M4P 2Y3
(a division of Pearson Penguin Canada Inc.)
Penguin Ireland, 25 St Stephen's Green, Dublin 2, Ireland (a division of Penguin Books Ltd)
Penguin Group (Australia), 707 Collins Street, Melbourne, Victoria 3008, Australia
(a division of Pearson Australia Group Pty Ltd)
Penguin Books India Pvt Ltd, 11 Community Centre, Panchsheel Park, New Delhi – 110 017, India
Penguin Group (NZ), 67 Apollo Drive, Rosedale, Auckland 0632, New Zealand
(a division of Pearson New Zealand Ltd)
Penguin Books (South Africa) (Pty) Ltd, Block D, Rosebank Office Park,
181 Jan Smuts Avenue, Parktown North, Gauteng 2193, South Africa

Penguin Books Ltd, Registered Offices: 80 Strand, London WC2R 0RL, England

www.penguin.com

First published in the United States of America by Simon & Schuster 2014
First published in Great Britain by Allen Lane 2014
001

Copyright © Naomi Klein, 2014

The moral right of the author has been asserted

Printed in Great Britain by Clays Ltd, St Ives plc

A CIP catalogue record for this book is available from the British Library

Hardback ISBN: 978-1-846-14505-6
Trade Paperback ISBN: 978-1-846-14506-3

www.greenpenguin.co.uk

Penguin Books is committed to a sustainable
future for our business, our readers and our planet.
This book is made from Forest Stewardship
Council™ certified paper.

For Toma

"We need to remember that the work of our time is bigger than climate change. We need to be setting our sights higher and deeper. What we're really talking about, if we're honest with ourselves, is transforming everything about the way we live on this planet."

—Rebecca Tarbotton, Executive Director of the Rainforest Action Network, 1973–2012 [1]

"In my books I've imagined people salting the Gulf Stream, damming the glaciers sliding off the Greenland ice cap, pumping ocean water into the dry basins of the Sahara and Asia to create salt seas, pumping melted ice from Antarctica north to provide freshwater, genetically engineering bacteria to sequester more carbon in the roots of trees, raising Florida 30 feet to get it back above water, and (hardest of all) comprehensively changing capitalism."

—Science fiction writer Kim Stanley Robinson, 2012 [2]

CONTENTS

PART THREE

STARTING ANYWAY

ONE WAY OR ANOTHER, EVERYTHING CHANGES

"Most projections of climate change presume that future changes—greenhouse gas emissions, temperature increases and effects such as sea level rise—will happen incrementally. A given amount of emission will lead to a given amount of temperature increase that will lead to a given amount of smooth incremental sea level rise. However, the geological record for the climate reflects instances where a relatively small change in one element of climate led to abrupt changes in the system as a whole. In other words, pushing global temperatures past certain thresholds could trigger abrupt, unpredictable and potentially irreversible changes that have massively disruptive and large-scale impacts. At that point, even if we do not add any additional CO_2 to the atmosphere, potentially unstoppable processes are set in motion. We can think of this as sudden climate brake and steering failure where the problem and its consequences are no longer something we can control."

—Report by the American Association for the Advancement of Science, the world's largest general scientific society, 2014[1]

"I love that smell of the emissions."

—Sarah Palin, 2011[2]

A voice came over the intercom: would the passengers of Flight 3935, scheduled to depart Washington, D.C., for Charleston, South Carolina, kindly collect their carry-on luggage and get off the plane.

They went down the stairs and gathered on the hot tarmac. There they saw something unusual: the wheels of the US Airways jet had sunk into the black pavement as if it were wet cement. The wheels were lodged so deep, in fact, that the truck that came to tow the plane away couldn't pry it loose. The airline had hoped that without the added weight of the flight's thirty-five passengers, the aircraft would be light enough to pull. It wasn't. Someone posted a picture: "Why is my flight cancelled? Because DC is so damn hot that our plane sank 4" into the pavement."[3]

Eventually, a larger, more powerful vehicle was brought in to tow the plane and this time it worked; the plane finally took off, three hours behind schedule. A spokesperson for the airline blamed the incident on "very unusual temperatures."[4]

The temperatures in the summer of 2012 were indeed unusually hot. (As they were the year before and the year after.) And it's no mystery why this has been happening: the profligate burning of fossil fuels, the very thing that US Airways was bound and determined to do despite the inconvenience presented by a melting tarmac. This irony—the fact that the burning of fossil fuels is so radically changing our climate that it is getting in the way of our capacity to burn fossil fuels—did not stop the passengers of Flight 3935 from reembarking and continuing their journeys. Nor was climate change mentioned in any of the major news coverage of the incident.

I am in no position to judge these passengers. All of us who live high consumer lifestyles, wherever we happen to reside, are, metaphorically, passengers on Flight 3935. Faced with a crisis that threatens our survival as a species, our entire culture is continuing to do the very thing that caused the crisis, only with an extra dose of elbow grease behind it. Like the airline bringing in a truck with a more powerful engine to tow that plane, the global economy is upping the ante from conventional sources of fossil fuels to even dirtier and more dangerous versions—bitumen from the Alberta tar sands, oil from deepwater drilling, gas from hydraulic fracturing (fracking), coal from detonated mountains, and so on.

Meanwhile, each supercharged natural disaster produces new irony-laden snapshots of a climate increasingly inhospitable to the very industries most responsible for its warming. Like the 2013 historic floods in Calgary that forced the head offices of the oil companies mining the Alberta tar sands to go dark and send their employees home, while a train carrying flammable petroleum products teetered on the edge of a disintegrating rail bridge. Or the drought that hit the Mississippi River one year earlier, pushing water levels so low that barges loaded with oil and coal were unable to move for days, while they waited for the Army Corps of Engineers to dredge a channel (they had to appropriate funds allocated to rebuild from the previous year's historic flooding along the same waterway). Or the coal-fired power plants in other parts of the country that were temporarily shut down because the waterways that they draw on to cool their machinery were either too hot or too dry (or, in some cases, both).

Living with this kind of cognitive dissonance is simply part of being alive in this jarring moment in history, when a crisis we have been studiously ignoring is hitting us in the face—and yet we are doubling down on the stuff that is causing the crisis in the first place.

I denied climate change for longer than I care to admit. I knew it was happening, sure. Not like Donald Trump and the Tea Partiers going on about how the continued existence of winter proves it's all a hoax. But I stayed pretty hazy on the details and only skimmed most of the news stories, especially the really scary ones. I told myself the science was too complicated and that the environmentalists were dealing with it. And I continued to behave as if there was nothing wrong with the shiny card in my wallet attesting to my "elite" frequent flyer status.

A great many of us engage in this kind of climate change denial. We look for a split second and then we look away. Or we look but then turn it into a joke ("more signs of the Apocalypse!"). Which is another way of looking away.

Or we look but tell ourselves comforting stories about how humans are clever and will come up with a technological miracle that will safely suck the carbon out of the skies or magically turn down the heat of the sun. Which, I was to discover while researching this book, is yet another way of looking away.

Or we look but try to be hyper-rational about it ("dollar for dollar it's more efficient to focus on economic development than climate change, since wealth is the best protection from weather extremes")—as if having a few more dollars will make much difference when your city is underwater. Which is a way of looking away if you happen to be a policy wonk.

Or we look but tell ourselves we are too busy to care about something so distant and abstract—even though we saw the water in the subways in

New York City, and the people on their rooftops in New Orleans, and know that no one is safe, the most vulnerable least of all. And though perfectly understandable, this too is a way of looking away.

Or we look but tell ourselves that all we can do is focus on ourselves. Meditate and shop at farmers' markets and stop driving—but forget trying to actually change the systems that are making the crisis inevitable because that's too much "bad energy" and it will never work. And at first it may appear as if we are looking, because many of these lifestyle changes are indeed part of the solution, but we still have one eye tightly shut.

Or maybe we do look—really look—but then, inevitably, we seem to forget. Remember and then forget again. Climate change is like that; it's hard to keep it in your head for very long. We engage in this odd form of on-again-off-again ecological amnesia for perfectly rational reasons. We deny because we fear that letting in the full reality of this crisis will change everything. And we are right.[5]

We know that if we continue on our current path of allowing emissions to rise year after year, climate change will change everything about our world. Major cities will very likely drown, ancient cultures will be swallowed by the seas, and there is a very high chance that our children will spend a great deal of their lives fleeing and recovering from vicious storms and extreme droughts. And we don't have to do anything to bring about this future. All we have to do is nothing. Just continue to do what we are doing now, whether it's counting on a techno-fix or tending to our gardens or telling ourselves we're unfortunately too busy to deal with it.

All we have to do is *not* react as if this is a full-blown crisis. All we have to do is keep on denying how frightened we actually are. And then, bit by bit, we will have arrived at the place we most fear, the thing from which we have been averting our eyes. No additional effort required.

There are ways of preventing this grim future, or at least making it a lot less dire. But the catch is that these also involve changing everything. For us high consumers, it involves changing how we live, how our economies function, even the stories we tell about our place on earth. The good news is that many of these changes are distinctly un-catastrophic. Many are downright exciting. But I didn't discover this for a long while.

I remember the precise moment when I stopped averting my eyes to the reality of climate change, or at least when I first allowed my eyes to rest there for a good while. It was in Geneva, in April 2009, and I was meeting with Bolivia's ambassador to the World Trade Organization (WTO), who was then a surprisingly young woman named Angélica Navarro Llanos. Bolivia being a poor country with a small international budget, Navarro Llanos had recently taken on the climate portfolio in addition to her trade responsibilities. Over lunch in an empty Chinese restaurant, she explained to me (using chopsticks as props to make a graph of the global emission trajectory) that she saw climate change both as a terrible threat to her people—but also an opportunity.

A threat for the obvious reasons: Bolivia is extraordinarily dependent on glaciers for its drinking and irrigation water and those white-capped mountains that tower over its capital were turning gray and brown at an alarming rate. The opportunity, Navarro Llanos said, was that since countries like hers had done almost nothing to send emissions soaring, they were in a position to declare themselves "climate creditors," owed money and technology support from the large emitters to defray the hefty costs of coping with more climate-related disasters, as well as to help them develop on a green energy path.

She had recently given a speech at a United Nations climate conference in which she laid out the case for these kinds of wealth transfers, and she gave me a copy. "Millions of people," it read, "in small islands, least-developed countries, landlocked countries as well as vulnerable communities in Brazil, India and China, and all around the world—are suffering from the effects of a problem to which they did not contribute. . . . If we are to curb emissions in the next decade, we need a massive mobilization larger than any in history. We need a Marshall Plan for the Earth. This plan must mobilize financing and technology transfer on scales never seen before. It must get technology onto the ground in every country to ensure we reduce emissions while raising people's quality of life. We have only a decade."[6]

Of course a Marshall Plan for the Earth would be very costly—hundreds of billions if not trillions of dollars (Navarro Llanos was reluctant to name a figure). And one might have thought that the cost alone would make it a nonstarter—after all, this was 2009 and the global financial crisis was in full swing. Yet the grinding logic of austerity—passing on the bankers' bills to the people in the form of public sector layoffs, school closures, and the like—had not yet been normalized. So rather than making Navarro Llanos's ideas seem less plausible, the crisis had the opposite effect.

We had all just watched as trillions of dollars were marshaled in a moment when our elites decided to declare a crisis. If the banks were allowed to fail, we were told, the rest of the economy would collapse. It was a matter of collective survival, so the money had to be found. In the process, some rather large fictions at the heart of our economic system were exposed (Need more money? Print some!). A few years earlier, governments took a similar approach to public finances after the September 11 terrorist attacks. In many Western countries, when it came to constructing the security/surveillance state at home and waging war abroad, budgets never seemed to be an issue.

Climate change has never received the crisis treatment from our leaders, despite the fact that it carries the risk of destroying lives on a vastly greater scale than collapsed banks or collapsed buildings. The cuts to our greenhouse gas emissions that scientists tell us are necessary in order to greatly reduce the risk of catastrophe are treated as nothing more than gentle suggestions, actions that can be put off pretty much indefinitely. Clearly, what gets declared a crisis is an expression of power and priorities as much as hard facts. But we need not be spectators in all this: politicians aren't the only ones with the power to declare a crisis. Mass movements of regular people can declare one too.

Slavery wasn't a crisis for British and American elites until abolitionism turned it into one. Racial discrimination wasn't a crisis until the civil rights movement turned it into one. Sex discrimination wasn't a crisis until feminism turned it into one. Apartheid wasn't a crisis until the anti-apartheid movement turned it into one.

In the very same way, if enough of us stop looking away and decide that climate change is a crisis worthy of Marshall Plan levels of response, then it will become one, and the political class will have to respond, both by making resources available and by bending the free market rules that have proven so pliable when elite interests are in peril. We occasionally catch glimpses of this potential when a crisis puts climate change at the front of our minds for a while. "Money is no object in this relief effort. Whatever money is needed for it will be spent," declared British prime minister David Cameron—Mr. Austerity himself—when large parts of his country were underwater from historic flooding in February 2014 and the public was enraged that his government was not doing more to help.[7]

Listening to Navarro Llanos describe Bolivia's perspective, I began to understand how climate change—if treated as a true planetary emergency akin to those rising flood waters—could become a galvanizing force for humanity, leaving us all not just safer from extreme weather, but with societies that are safer and fairer in all kinds of other ways as well. The resources required to rapidly move away from fossil fuels and prepare for the coming heavy weather could pull huge swaths of humanity out of poverty, providing services now sorely lacking, from clean water to electricity. This is a vision of the future that goes beyond just surviving or enduring climate change, beyond "mitigating" and "adapting" to it in the grim language of the United Nations. It is a vision in which we collectively use the crisis to leap somewhere that seems, frankly, better than where we are right now.

After that conversation, I found that I no longer feared immersing myself in the scientific reality of the climate threat. I stopped avoiding the articles and the scientific studies and read everything I could find. I also stopped outsourcing the problem to the environmentalists, stopped telling myself this was somebody else's issue, somebody else's job. And through conversations with others in the growing climate justice movement, I began to see all kinds of ways that climate change could become a catalyzing force for positive change—how it could be the best argument progressives have ever had to demand the rebuilding and reviving of local economies; to reclaim our democracies from corrosive corporate influence; to block harmful new free trade deals and rewrite old ones; to invest in starving public infrastructure like mass transit and affordable housing; to take back ownership of essential services like energy and water; to remake our sick agricultural system into something much healthier; to open borders to migrants whose displacement is linked to climate impacts; to finally respect Indigenous land rights—all of which would help to end grotesque levels of inequality within our nations and between them.

And I started to see signs—new coalitions and fresh arguments—hinting at how, if these various connections were more widely understood, the urgency of the climate crisis could form the basis of a powerful mass movement, one that would weave all these seemingly disparate issues into a coherent narrative about how to protect humanity from the ravages of both a savagely unjust economic system and a destabilized climate system. I have written this book because I came to the conclusion that climate action could provide just such a rare catalyst.

A People's Shock

But I also wrote it because climate change can be a catalyst for a range of very different and far less desirable forms of social, political, and economic transformation.

I have spent the last fifteen years immersed in research about societies undergoing extreme shocks—caused by economic meltdowns, natural disasters, terrorist attacks, and wars. And I have looked deeply into how societies change in these periods of tremendous stress. How these events change the collective sense of what is possible, for better but mostly for worse. As I discussed in my last book, *The Shock Doctrine*, over the past four decades corporate interests have systematically exploited these various forms of crisis to ram through policies that enrich a small elite—by lifting regulations, cutting social spending, and forcing large-scale privatizations of the public sphere. They have also been the excuse for extreme crackdowns on civil liberties and chilling human rights violations.

And there are plenty of signs that climate change will be no exception—that, rather than sparking solutions that have a real chance of preventing catastrophic warming and protecting us from inevitable disasters, the crisis will once again be seized upon to hand over yet more resources to the 1 percent. You can see the early stages of this process already. Communal forests around the world are being turned into privatized tree farms and preserves so their owners can collect something called "carbon credits," a lucrative scam I'll explore later. There is a booming trade in "weather futures," allowing companies and banks to gamble on changes in the weather as if deadly disasters were a game on a Vegas craps table (between 2005 and 2006 the weather derivatives market jumped nearly fivefold, from $9.7 billion to $45.2 billion). Global reinsurance companies are making billions in profits, in part by selling new kinds of protection schemes to developing countries that have done almost nothing to create the climate crisis, but whose infrastructure is intensely vulnerable to its impacts.[8]

And in a moment of candor, the weapons giant Raytheon explained, "Expanded business opportunities are likely to arise as consumer behaviour and needs change in response to climate change." Those opportunities include not just more demand for the company's privatized disaster response services but also "demand for its military products and services as security concerns may arise as results of droughts, floods, and storm events occur as a result of climate change."[9] This is worth remembering whenever doubts creep in about the urgency of this crisis: the private militias are already mobilizing.

Droughts and floods create all kinds of business opportunities besides a growing demand for men with guns. Between 2008 and 2010, at least 261 patents were filed related to growing "climate-ready" crops—seeds supposedly able to withstand extreme weather conditions; of these patents close to 80 percent were controlled by six agribusiness giants, including Monsanto and Syngenta. Superstorm Sandy, meanwhile, has been a windfall for New Jersey real estate developers who have received millions for new construction in lightly damaged areas, while it continues to be a nightmare for those living in hard-hit public housing, much as the aftermath of Hurricane Katrina played out in New Orleans.[10]

None of this is surprising. Finding new ways to privatize the commons and profit from disaster is what our current system is built to do; left to its own devices, it is capable of nothing else. The shock doctrine, however, is not the only way societies respond to crises. We have all witnessed this in recent years as the financial meltdown that began on Wall Street in 2008 reverberated around the world. A sudden rise in food prices helped create the conditions for the Arab Spring. Austerity policies have inspired mass movements from Greece to Spain to Chile to the United States to Quebec. Many of us are getting a lot better at standing up to those who would cynically exploit crises to ransack the public sphere. And yet these protests have also shown that saying no is not enough. If opposition movements are to do more than burn bright and then burn out, they will need a comprehensive vision for what should emerge in the place of our failing system, as well as serious political strategies for how to achieve those goals.

Progressives used to know how to do this. There is a rich populist history of winning big victories for social and economic justice in the midst of large-scale crises. These include, most notably, the policies of the New Deal after the market crash of 1929 and the birth of countless social programs after World War II. These policies were so popular with voters that getting them passed into law did not require the kind of authoritarian trickery that I documented in *The Shock Doctrine*. What was essential was building muscular mass movements capable of standing up to those defending a failing status quo, and that demanded a significantly fairer share of the economic pie for everyone. A few of the lasting (though embattled) legacies of these exceptional historical moments include: public health insurance in many countries, old age pensions, subsidized housing, and public funding for the arts.

I am convinced that climate change represents a historic opportunity on an even greater scale. As part of the project of getting our emissions down to the levels many scientists recommend, we once again have the chance to advance policies that dramatically improve lives, close the gap between rich and poor, create huge numbers of good jobs, and reinvigorate democracy from the ground up. Rather than the ultimate expression of the shock doctrine—a frenzy of new resource grabs and repression—climate change can be a People's Shock, a blow from below. It can disperse power into the hands of the many rather than consolidating it in the hands of the few, and radically expand the commons, rather than auctioning it off in pieces. And where right-wing shock doctors exploit emergencies (both real and manufactured) in order to push through policies that make us even more crisis prone, the kinds of transformations discussed in these pages would do the exact opposite: they would get to the root of why we are facing serial crises in the first place, and would leave us with both a more habitable climate than the one we are headed for and a far more just economy than the one we have right now.

But before any of these changes can happen—before we can believe that climate change can change us—we first have to stop looking away.

———

"You have been negotiating all my life." So said Canadian college student Anjali Appadurai, as she stared down the assembled government negotiators at the 2011 United Nations climate conference in Durban, South Africa. She was not exaggerating. The world's governments have been talking about preventing climate change for more than two decades; they began negotiating the year that Anjali, then twenty-one years old, was born. And yet as she pointed out in her memorable speech on the convention floor, delivered on behalf of all of the assembled young people: "In that time, you've failed to meet pledges, you've missed targets, and you've broken promises."[11]

In truth, the intergovernmental body entrusted to prevent "dangerous" levels of climate change has not only failed to make progress over its twenty-odd years of work (and more than ninety official negotiation meetings since the agreement was adopted), it has overseen a process of virtually uninterrupted backsliding. Our governments wasted years fudging numbers and squabbling over start dates, perpetually trying to get extensions like undergrads with late term papers.

The catastrophic result of all this obfuscation and procrastination is now undeniable. Preliminary data shows that in 2013, global carbon dioxide emissions were 61 *percent* higher than they were in 1990, when negotiations toward a climate treaty began in earnest. As MIT economist John Reilly puts it: "The more we talk about the need to control emissions, the more they are growing." Indeed the only thing rising faster than our emissions is the output of words pledging to lower them. Meanwhile, the annual U.N. climate summit, which remains the best hope for a political breakthrough on climate action, has started to seem less like a forum for serious negotiation than a very costly and high-carbon group therapy session, a place for the representatives of the most vulnerable countries in the world to vent their grief and rage while low-level representatives of the nations largely responsible for their tragedies stare at their shoes.[12]

This has been the mood ever since the collapse of the much-hyped 2009 U.N. climate summit in Copenhagen. On the last night of that massive gathering, I found myself with a group of climate justice activists, including one of the most prominent campaigners in Britain. Throughout the summit, this young man had been the picture of confidence and composure, briefing dozens of journalists a day on what had gone on during each round of negotiations and what the various emission targets meant in the real world. Despite the challenges, his optimism about the summit's prospects never flagged. Once it was all over, however, and the pitiful deal was done, he fell apart before our eyes. Sitting in an overlit Italian restaurant, he began to sob uncontrollably. "I really thought Obama understood," he kept repeating.

I have come to think of that night as the climate movement's coming of age: it was the moment when the realization truly sank in that no one was coming to save us. The British psychoanalyst and climate specialist Sally Weintrobe describes this as the summit's "fundamental legacy"—the acute and painful realization that our "leaders are not looking after us . . . we are not cared for at the level of our very survival."[13] No matter how many times we have been disappointed by the failings of our politicians, this realization still comes as a blow. It really is the case that we are on our own and any credible source of hope in this crisis will have to come from below.

In Copenhagen, the major polluting governments—including the United States and China—signed a nonbinding agreement pledging to keep temperatures from increasing more than 2 degrees Celsius above where they were before we started powering our economies with coal. (That converts to an increase of 3.6 degrees Fahrenheit.) This well-known target, which supposedly represents the "safe" limit of climate change, has always been a highly political choice that has more to do with minimizing economic disruption than with protecting the greatest number of people. When the 2 degrees target was made official in Copenhagen, there were impassioned objections from many delegates who said the goal amounted to a "death sentence" for some low-lying island states, as well as for large parts of Sub-Saharan Africa. In fact it is a very risky target for all of us: so far, temperatures have increased by just .8 degree Celsius and we are already experiencing many alarming impacts, including the unprecedented melting of the Greenland ice sheet in the summer of 2012 and the acidification of oceans far more rapidly than expected. Allowing temperatures to warm by more than twice that amount will unquestionably have perilous consequences.[14]

In a 2012 report, the World Bank laid out the gamble implied by that target. "As global warming approaches and exceeds 2-degrees Celsius, there is a risk of triggering nonlinear tipping elements. Examples include the disintegration of the West Antarctic ice sheet leading to more rapid sea-level rise, or large-scale Amazon dieback drastically affecting ecosystems, rivers, agriculture, energy production, and livelihoods. This would further add to 21st-century global warming and impact entire continents."[15] In other words, once we allow temperatures to climb past a certain point, where the mercury stops is not in our control.

But the bigger problem—and the reason Copenhagen caused such great despair—is that because governments did not agree to binding targets, they are free to pretty much ignore their commitments. Which is precisely what is happening. Indeed, emissions are rising so rapidly that unless something radical changes within our economic structure, 2 degrees now looks like a utopian dream. And it's not just environmentalists who are raising the alarm. The World Bank also warned when it released its report that "we're on track for a 4°C warmer world [by century's end] marked by extreme heat waves, declining global food stocks, loss of ecosystems and biodiversity, and life-threatening sea level rise." And the report cautioned that, "there is also no certainty that adaptation to a 4°C world is possible." Kevin Anderson, former director (now deputy director) of the Tyndall Centre for Climate Change Research, which has quickly established itself as one of the U.K.'s premier climate research institutions, is even blunter; he says 4 degrees Celsius warming—7.2 degrees Fahrenheit—is "incompatible with any reasonable characterization of an organized, equitable and civilized global community."[16]

We don't know exactly what a 4 degrees Celsius world would look like, but even the best-case scenario is likely to be calamitous. Four degrees of warming could raise global sea levels by 1 or possibly even 2 meters by 2100 (and would lock in at least a few additional meters over future centuries). This would drown some island nations such as the Maldives and Tuvalu, and inundate many coastal areas from Ecuador and Brazil to the Netherlands to much of California and the northeastern United States, as well as huge swaths of South and Southeast Asia. Major cities likely in jeopardy include Boston, New York, greater Los Angeles, Vancouver, London, Mumbai, Hong Kong, and Shanghai.[17]

Meanwhile, brutal heat waves that can kill tens of thousands of people, even in wealthy countries, would become entirely unremarkable summer events on every continent but Antarctica. The heat would also cause staple crops to suffer dramatic yield losses across the globe (it is possible that Indian wheat and U.S. corn could plummet by as much as 60 percent), this at a time when demand will be surging due to population growth and a growing demand for meat. And since crops will be facing not just heat stress but also extreme events such as wide-ranging droughts, flooding, or pest outbreaks, the losses could easily turn out to be more severe than the models have predicted. When you add ruinous hurricanes, raging wildfires, fisheries collapses, widespread disruptions to water supplies, extinctions, and globe-trotting diseases to the mix, it indeed becomes difficult to imagine that a peaceful, ordered society could be sustained (that is, where such a thing exists in the first place).[18]

And keep in mind that these are the optimistic scenarios in which warming is more or less stabilized at 4 degrees Celsius and does not trigger tipping points beyond which runaway warming would occur. Based on the latest modeling, it is becoming safer to assume that 4 degrees could bring about a number of extremely dangerous feedback loops—an Arctic that is regularly ice-free in September, for instance, or, according to one recent study, global vegetation that is too saturated to act as a reliable "sink," leading to more carbon being emitted rather than stored. Once this happens, any hope of predicting impacts pretty much goes out the window. And this process may be starting sooner than anyone predicted. In May 2014, NASA and University of California, Irvine scientists revealed that glacier melt in a section of West Antarctica roughly the size of France now "appears unstoppable." This likely spells doom for the entire West Antarctic ice sheet, which according to lead study author Eric Rignot "comes with a sea level rise of between three and five metres. Such an event will displace millions of people worldwide." The disintegration, however, could unfold over centuries and there is still time for emission reductions to slow down the process and prevent the worst.[19]

Much more frightening than any of this is the fact that plenty of mainstream analysts think that on our current emissions trajectory, we are headed for even more than 4 degrees of warming. In 2011, the usually staid International Energy Agency (IEA) issued a report projecting that we are actually on track for 6 degrees Celsius—10.8 degrees Fahrenheit—of warming. And as the IEA's chief economist put it: "Everybody, even the school children, knows that this will have catastrophic implications for all of us." (The evidence indicates that 6 degrees of warming is likely to set in motion several major tipping points—not only slower ones such as the aforementioned breakdown of the West Antarctic ice sheet, but possibly more abrupt ones, like massive releases of methane from Arctic permafrost.) The accounting giant PricewaterhouseCoopers has also published a report warning businesses that we are headed for "4°C, or even 6°C" of warming.[20]

These various projections are the equivalent of every alarm in your house going off simultaneously. And then every alarm on your street going off as well, one by one by one. They mean, quite simply, that climate change has become an existential crisis for the human species. The only historical precedent for a crisis of this depth and scale was the Cold War fear that we were heading toward nuclear holocaust, which would have made much of the planet uninhabitable. But that was (and remains) a threat; a slim possibility, should geopolitics spiral out of control. The vast majority of nuclear scientists never told us that we were almost certainly going to put our civilization in peril if we kept going about our daily lives as usual, doing exactly what we were already doing, which is what the climate scientists have been telling us for years.

As the Ohio State University climatologist Lonnie G. Thompson, a world-renowned specialist on glacier melt, explained in 2010, "Climatologists, like other scientists, tend to be a stolid group. We are not given to theatrical rantings about falling skies. Most of us are far more comfortable in our laboratories or gathering data in the field than we are giving interviews to journalists or speaking before Congressional committees. Why then are climatologists speaking out about the dangers of global warming? The answer is that virtually all of us are now convinced that global warming poses a clear and present danger to civilization."[21]

It doesn't get much clearer than that. And yet rather than responding with alarm and doing everything in our power to change course, large parts of humanity are, quite consciously, continuing down the same road. Only, like the passengers aboard Flight 3935, aided by a more powerful, dirtier engine.

What is wrong with us?

Really Bad Timing

Many answers to that question have been offered, ranging from the extreme difficulty of getting all the governments in the world to agree on anything, to an absence of real technological solutions, to something deep in our human nature that keeps us from acting in the face of seemingly remote threats, to—more recently—the claim that we have blown it anyway and there is no point in even trying to do much more than enjoy the scenery on the way down.

Some of these explanations are valid, but all are ultimately inadequate. Take the claim that it's just too hard for so many countries to agree on a course of action. It is hard. But many times in the past, the United Nations has helped governments to come together to tackle tough cross-border challenges, from ozone depletion to nuclear proliferation. The deals produced weren't perfect, but they represented real progress. Moreover, during the same years that our governments failed to enact a tough and binding legal architecture requiring emission reductions, supposedly because cooperation was too complex, they managed to create the World Trade Organization—an intricate global system that regulates the flow of goods and services around the planet, under which the rules are clear and violations are harshly penalized.

The assertion that we have been held back by a lack of technological solutions is no more compelling. Power from renewable sources like wind and water predates the use of fossil fuels and is becoming cheaper, more efficient, and easier to store every year. The past two decades have seen an explosion of ingenious zero-waste design, as well as green urban planning. Not only do we have the technical tools to get off fossil fuels, we also have no end of small pockets where these low carbon lifestyles have been tested with tremendous success. And yet the kind of large-scale transition that would give us a collective chance of averting catastrophe eludes us.

Is it just human nature that holds us back then? In fact we humans have shown ourselves willing to collectively sacrifice in the face of threats many times, most famously in the embrace of rationing, victory gardens, and victory bonds during World Wars I and II. Indeed to support fuel conservation during World War II, pleasure driving was virtually eliminated in the U.K., and between 1938 and 1944, use of public transit went up by 87 percent in the U.S. and by 95 percent in Canada. Twenty million U.S. households—representing three fifths of the population—were growing victory gardens in 1943, and their yields accounted for 42 percent of the fresh vegetables consumed that year. Interestingly, all of these activities together dramatically reduce carbon emissions.[22]

Yes, the threat of war seemed immediate and concrete but so too is the threat posed by the climate crisis that has already likely been a substantial contributor to massive disasters in some of the world's major cities. Still, we've gone soft since those days of wartime sacrifice, haven't we? Contemporary humans are too self-centered, too addicted to gratification to live without the full freedom to satisfy our every whim—or so our culture tells us every day. And yet the truth is that we continue to make collective sacrifices in the name of an abstract greater good all the time. We sacrifice our pensions, our hard-won labor rights, our arts and after-school programs. We send our kids to learn in ever more crowded classrooms, led by ever more harried teachers. We accept that we have to pay dramatically more for the destructive energy sources that power our transportation and our lives. We accept that bus and subway fares go up and up while service fails to improve or degenerates. We accept that a public university education should result in a debt that will take half a lifetime to pay off when such a thing was unheard of a generation ago. In Canada, where I live, we are in the midst of accepting that our mail can no longer be delivered to our homes.

The past thirty years have been a steady process of getting less and less in the public sphere. This is all defended in the name of austerity, the current justification for these never-ending demands for collective sacrifice. In the past, other words and phrases, equally abstracted from daily life, have served a similar purpose: balanced budgets, increased efficiency, fostering economic growth.

It seems to me that if humans are capable of sacrificing this much collective benefit in the name of stabilizing an economic system that makes daily life so much more expensive and precarious, then surely humans should be capable of making some important lifestyle changes in the interest of stabilizing the physical systems upon which all of life depends. Especially because many of the changes that need to be made to dramatically cut emissions would also materially improve the quality of life for the majority of people on the planet—from allowing kids in Beijing to play outside without wearing pollution masks to creating good jobs in clean energy sectors for millions. There seems to be no shortage of both short-term and medium-term incentives to do the right thing for our climate.

Time is tight, to be sure. But we could commit ourselves, tomorrow, to radically cutting our fossil fuel emissions and beginning the shift to zero-carbon sources of energy based on renewable technology, with a full-blown transition underway within the decade. We have the tools to do that. And if we did, the seas would still rise and the storms would still come, but we would stand a much greater chance of preventing truly catastrophic warming. Indeed, entire nations could be saved from the waves. As Pablo Solón, Bolivia's former ambassador to the United Nations, puts it: "If I burned your house the least I can do is welcome you into my house . . . and if I'm burning it right now I should try to stop the fire now."[23]

But we are not stopping the fire. In fact we are dousing it with gasoline. After a rare decline in 2009 due to the financial crisis, global emissions surged by a whopping 5.9 percent in 2010—the largest absolute increase since the Industrial Revolution.[24]

So my mind keeps coming back to the question: what is wrong with us? What is really preventing us from putting out the fire that is threatening to burn down our collective house?

I think the answer is far more simple than many have led us to believe: we have not done the things that are necessary to lower emissions because those things fundamentally conflict with deregulated capitalism, the reigning ideology for the entire period we have been struggling to find a way out of this crisis. We are stuck because the actions that would give us the best chance of averting catastrophe—and would benefit the vast majority—are extremely threatening to an elite minority that has a stranglehold over our economy, our political process, and most of our major media outlets. That problem might not have been insurmountable had it presented itself at another point in our history. But it is our great collective misfortune that the scientific community made its decisive diagnosis of the climate threat at the precise moment when those elites were enjoying more unfettered political, cultural, and intellectual power than at any point since the 1920s. Indeed, governments and scientists began talking seriously about radical cuts to greenhouse gas emissions in 1988—the exact year that marked the dawning of what came to be called "globalization," with the signing of the agreement representing the world's largest bilateral trade relationship between Canada and the United States, later to be expanded into the North American Free Trade Agreement (NAFTA) with the inclusion of Mexico.[25]

When historians look back on the past quarter century of international negotiations, two defining processes will stand out. There will be the climate process: struggling, sputtering, failing utterly to achieve its goals. And there will be the corporate globalization process, zooming from victory to victory: from that first free trade deal to the creation of the World Trade Organization to the mass privatization of the former Soviet economies to the transformation of large parts of Asia into sprawling free-trade zones to the "structural adjusting" of Africa. There were setbacks to that process, to be sure—for example, popular pushback that stalled trade rounds and free trade deals. But what remained successful were the ideological underpinnings of the entire project, which was never really about trading goods across borders—selling French wine in Brazil, for instance, or U.S. software in China. It was always about using these sweeping deals, as well as a range of other tools, to lock in a global policy framework that provided maximum freedom to multinational corporations to produce their goods as cheaply as possible and sell them with as few regulations as possible—while paying as little in taxes as possible. Granting this corporate wishlist, we were told, would fuel economic growth, which would trickle down to the rest of us, eventually. The trade deals mattered only in so far as they stood in for, and plainly articulated, this far broader agenda.

The three policy pillars of this new era are familiar to us all: privatization of the public sphere, deregulation of the corporate sector, and lower corporate taxation, paid for with cuts to public spending. Much has been written about the real-world costs of these policies—the instability of financial markets, the excesses of the super-rich, and the desperation of the increasingly disposable poor, as well as the failing state of public infrastructure and services. Very little, however, has been written about how market fundamentalism has, from the very first moments, systematically sabotaged our collective response to climate change, a threat that came knocking just as this ideology was reaching its zenith.

The core problem was that the stranglehold that market logic secured over public life in this period made the most direct and obvious climate responses seem politically heretical. How, for instance, could societies invest massively in zero-carbon public services and infrastructure at a time when the public sphere was being systematically dismantled and auctioned off? How could governments heavily regulate, tax, and penalize fossil fuel companies when all such measures were being dismissed as relics of "command and control" communism? And how could the renewable energy sector receive the supports and protections it needed to replace fossil fuels when "protectionism" had been made a dirty word?

A different kind of climate movement would have tried to challenge the extreme ideology that was blocking so much sensible action, joining with other sectors to show how unfettered corporate power posed a grave threat to the habitability of the planet. Instead, large parts of the climate movement wasted precious decades attempting to make the square peg of the climate crisis fit into the round hole of deregulated capitalism, forever touting ways for the problem to be solved by the market itself. (Though it was only years into this project that I discovered the depths of collusion between big polluters and Big Green.)

But blocking strong climate action wasn't the only way that the triumph of market fundamentalism acted to deepen the crisis in this period. Even more directly, the policies that so successfully freed multinational corporations from virtually all constraints also contributed significantly to the underlying cause of global warming—rising greenhouse gas emissions. The numbers are striking: in the 1990s, as the market integration project ramped up, global emissions were going up an average of 1 percent a year; by the 2000s, with "emerging markets" like China now fully integrated into the world economy, emissions growth had sped up disastrously, with the annual rate of increase reaching 3.4 percent a year for much of the decade. That rapid growth rate continues to this day, interrupted only briefly in 2009 by the world financial crisis.[26]

With hindsight, it's hard to see how it could have turned out otherwise. The twin signatures of this era have been the mass export of products across vast distances (relentlessly burning carbon all the way), and the import of a uniquely wasteful model of production, consumption, and agriculture to every corner of the world (also based on the profligate burning of fossil fuels). Put differently, the liberation of world markets, a process powered by the liberation of unprecedented amounts of fossil fuels from the earth, has dramatically sped up the same process that is liberating Arctic ice from existence.

As a result, we now find ourselves in a very difficult and slighty ironic position. Because of those decades of hardcore emitting exactly when we were supposed to be cutting back, the things we must do to avoid catastrophic warming are no longer just in conflict with the particular strain of deregulated capitalism that triumphed in the 1980s. They are now in conflict with the fundamental imperative at the heart of our economic model: grow or die.

Once carbon has been emitted into the atmosphere, it sticks around for hundreds of years, some of it even longer, trapping heat. The effects are cumulative, growing more severe with time. And according to emissions specialists like the Tyndall Centre's Kevin Anderson (as well as others), so much carbon has been allowed to accumulate in the atmosphere over the past two decades that now our only hope of keeping warming below the internationally agreed-upon target of 2 degrees Celsius is for wealthy countries to cut their emissions by somewhere in the neighborhood of 8–10 percent a year.[27] The "free" market simply cannot accomplish this task. Indeed, this level of emission reduction has happened only in the context of economic collapse or deep depressions.

I'll be delving deeper into those numbers in Chapter 2, but the bottom line is what matters here: our economic system and our planetary system are now at war. Or, more accurately, our economy is at war with many forms of life on earth, including human life. What the climate needs to avoid collapse is a contraction in humanity's use of resources; what our economic model demands to avoid collapse is unfettered expansion. Only one of these sets of rules can be changed, and it's not the laws of nature.

Fortunately, it is eminently possible to transform our economy so that it is less resource-intensive, and to do it in ways that are equitable, with the most vulnerable protected and the most responsible bearing the bulk of the burden. Low-carbon sectors of our economies can be encouraged to expand and create jobs, while high-carbon sectors are encouraged to contract. The problem, however, is that this scale of economic planning and management is entirely outside the boundaries of our reigning ideology. The only kind of contraction our current system can manage is a brutal crash, in which the most vulnerable will suffer most of all.

So we are left with a stark choice: allow climate disruption to change everything about our world, or change pretty much everything about our economy to avoid that fate. But we need to be very clear: because of our decades of collective denial, no gradual, incremental options are now available to us. Gentle tweaks to the status quo stopped being a climate option when we supersized the American Dream in the 1990s, and then proceeded to take it global. And it's no longer just radicals who see the need for radical change. In 2012, twenty-one past winners of the prestigious Blue Planet Prize—a group that includes James Hansen, former director of NASA's Goddard Institute for Space Studies, and Gro Harlem Brundtland, former prime minister of Norway—authored a landmark report. It stated that, "In the face of an absolutely unprecedented emergency, society has no choice but to take dramatic action to avert a collapse of civilization. Either we will change our ways and build an entirely new kind of global society, or they will be changed for us."[28]

That's tough for a lot of people in important positions to accept, since it challenges something that might be even more powerful than capitalism, and that is the fetish of centrism—of reasonableness, seriousness, splitting the difference, and generally not getting overly excited about anything. This is the habit of thought that truly rules our era, far more among the liberals who concern themselves with matters of climate policy than among conservatives, many of whom simply deny the existence of the crisis. Climate change presents a profound challenge to this cautious centrism because half measures won't cut it: "all of the above energy" programs, as U.S. President Barack Obama describes his approach, has about as much chance of success as an all of the above diet, and the firm deadlines imposed by science require that we get very worked up indeed.

By posing climate change as a battle between capitalism and the planet, I am not saying anything that we don't already know. The battle is already under way, but right now capitalism is winning hands down. It wins every time the need for economic growth is used as the excuse for putting off climate action yet again, or for breaking emission reduction commitments already made. It wins when Greeks are told that their only path out of economic crisis is to open up their beautiful seas to high-risk oil and gas drilling. It wins when Canadians are told our only hope of not ending up like Greece is to allow our boreal forests to be flayed so we can access the semisolid bitumen from the Alberta tar sands. It wins when a park in Istanbul is slotted for demolition to make way for yet another shopping mall. It wins when parents in Beijing are told that sending their wheezing kids to school in pollution masks decorated to look like cute cartoon characters is an acceptable price for economic progress. It wins every time we accept that we have only bad choices available to us: austerity or extraction, poisoning or poverty.

The challenge, then, is not simply that we need to spend a lot of money and change a lot of policies; it's that we need to think differently, radically differently, for those changes to be remotely possible. Right now, the triumph of market logic, with its ethos of domination and fierce competition, is paralyzing almost all serious efforts to respond to climate change. Cutthroat competition between nations has deadlocked U.N. climate negotiations for decades: rich countries dig in their heels and declare that they won't cut emissions and risk losing their vaulted position in the global hierarchy; poorer countries declare that they won't give up their right to pollute as much as rich countries did on their way to wealth, even if that means deepening a disaster that hurts the poor most of all. For any of this to change, a worldview will need to rise to the fore that sees nature, other nations, and our own neighbors not as adversaries, but rather as partners in a grand project of mutual reinvention.

That's a big ask. But it gets bigger. Because of our endless delays, we also have to pull off this massive transformation without delay. The International Energy Agency warns that if we do not get our emissions under control by a rather terrifying 2017, our fossil fuel economy will "lock-in" extremely dangerous warming. "The energy-related infrastructure then in place will generate all the CO_2 emissions allowed" in our carbon budget for limiting warming to 2 degrees Celsius—"leaving no room for additional power plants, factories and other infrastructure unless they are zero-carbon, which would be extremely costly." This assumes, probably accurately, that governments would be unwilling to force the closure of still-profitable power plants and factories. As Fatih Birol, the IEA's chief economist, bluntly put it: "The door to reach two degrees is about to close. In 2017 it will be closed forever." In short, we have reached what some activists have started calling

"Decade Zero" of the climate crisis: we either change now or we lose our chance.[29]

All this means that the usual free market assurances—A techno-fix is around the corner! Dirty development is just a phase on the way to a clean environment, look at nineteenth-century London!—simply don't add up. We don't have a century to spare for China and India to move past their Dickensian phases. Because of our lost decades, it is time to turn this around now. Is it possible? Absolutely. Is it possible without challenging the fundamental logic of deregulated capitalism? Not a chance.

One of the people I met on this journey and who you will meet in these pages is Henry Red Cloud, a Lakota educator and entrepreneur who trains young Native people to become solar engineers. He tells his students that there are times when we must accept small steps forward—and there are other times "when you need to run like a buffalo."[30] Now is one of those times when we must run.

Power, Not Just Energy

I was struck recently by a mea culpa of sorts, written by Gary Stix, a senior editor of *Scientific American*. Back in 2006, he edited a special issue on responses to climate change and, like most such efforts, the articles were narrowly focused on showcasing exciting low-carbon technologies. But in 2012 Stix wrote that he had overlooked a much larger and more important part of the story—the need to create the social and political context in which these technological shifts stand a chance of displacing the all too profitable status quo. "If we are ever to cope with climate change in any fundamental way, radical solutions on the social side are where we must focus, though. The relative efficiency of the next generation of solar cells is trivial by comparison."[31]

This book is about those radical changes on the social side, as well as on the political, economic, and cultural sides. What concerns me is less the mechanics of the transition—the shift from brown to green energy, from sole-rider cars to mass transit, from sprawling exurbs to dense and walkable cities—than the power and ideological roadblocks that have so far prevented any of these long understood solutions from taking hold on anything close to the scale required.

It seems to me that our problem has a lot less to do with the mechanics of solar power than the politics of human power—specifically whether there can be a shift in who wields it, a shift away from corporations and toward communities, which in turn depends on whether or not the great many people who are getting a rotten deal under our current system can build a determined and diverse enough social force to change the balance of power. I have also come to understand, over the course of researching this book, that the shift will require rethinking the very nature of humanity's power—our right to extract ever more without facing consequences, our capacity to bend complex natural systems to our will. This is a shift that challenges not only capitalism, but also the building blocks of materialism that preceded modern capitalism, a mentality some call "extractivism."

Because, underneath all of this is the real truth we have been avoiding: climate change isn't an "issue" to add to the list of things to worry about, next to health care and taxes. It is a civilizational wake-up call. A powerful message—spoken in the language of fires, floods, droughts, and extinctions—telling us that we need an entirely new economic model and a new way of sharing this planet. Telling us that we need to evolve.

Coming Out of Denial

Some say there is no time for this transformation; the crisis is too pressing and the clock is ticking. I agree that it would be reckless to claim that the only solution to this crisis is to revolutionize our economy and revamp our worldview from the bottom up—and anything short of that is not worth doing. There are all kinds of measures that would lower emissions substantively that could and should be done right now. But we aren't taking those measures, are we? The reason is that by failing to fight these big battles that stand to shift our ideological direction and change the balance of who holds power in our societies, a context has been slowly created in which any muscular response to climate change seems politically impossible, especially during times of economic crisis (which lately seems to be all the time).

So this book proposes a different strategy: think big, go deep, and move the ideological pole far away from the stifling market fundamentalism that has become the greatest enemy to planetary health. If we can shift the cultural context even a little, then there will be some breathing room for those sensible reformist policies that will at least get the atmospheric carbon numbers moving in the right direction. And winning is contagious so, who knows? Maybe within a few years, some of the ideas highlighted in these pages that sound impossibly radical today—like a basic income for all, or a rewriting of trade law, or real recognition of the rights of Indigenous people to protect huge parts of the world from polluting extraction—will start to seem reasonable, even essential.

For a quarter of a century, we have tried the approach of polite incremental change, attempting to bend the physical needs of the planet to our economic model's need for constant growth and new profit-making opportunities. The results have been disastrous, leaving us all in a great deal more danger than when the experiment began.

There are, of course, no guarantees that a more systemic approach will be any more successful—though there are, as will be explored later on, historical precedents that are grounds for hope. The truth is that this is the hardest book I have ever written, precisely because the research has led me to search out such radical responses. I have no doubt of their necessity, but I question their political feasibility every day, especially given that climate change puts us on such a tight and unforgiving deadline.

———

It's been a harder book to write for personal reasons too.

What gets me most are not the scary scientific studies about melting glaciers, the ones I used to avoid. It's the books I read to my two-year-old. *Have You Ever Seen a Moose?* is one of his favorites. It's about a bunch of kids that really, really, really want to see a moose. They search high and low—through a forest, a swamp, in brambly bushes and up a mountain, for "a long legged, bulgy nosed, branchy antlered moose." The joke is that there are moose hiding on each page. In the end, the animals all come out of hiding and the ecstatic kids proclaim: "We've never ever seen so many moose!"

On about the seventy-fifth reading, it suddenly hit me: he might never see a moose. I tried to hold it together. I went back to my computer and began to write about my time in northern Alberta, tar sands country, where members of the Beaver Lake Cree Nation told me about how the moose had changed—one woman described killing a moose on a hunting trip only to find that the flesh had already turned green. I heard a lot about strange tumors too, which locals assumed had to do with the animals drinking water contaminated by tar sands toxins. But mostly I heard about how the moose were simply gone.

And not just in Alberta. "Rapid Climate Changes Turn North Woods into Moose Graveyard," reads a May 2012 headline in *Scientific American*. A year and a half later, *The New York Times* was reporting that one of Minnesota's two moose populations had declined from four thousand in the 1990s to just one hundred today.[32]

Will he ever see a moose?

Then, the other day, I was slain by a miniature board book called *Snuggle Wuggle*. It involves different animals cuddling, with each posture given a ridiculously silly name: "How does a bat hug?" it asks. "Topsy turvy, topsy turvy." For some reason my son reliably cracks up at this page. I explain that it means upside down, because that's the way bats sleep.

But all I could think about was the report of some 100,000 dead and dying bats raining down from the sky in the midst of record-breaking heat across part of Queensland, Australia. Whole colonies devastated.[33]

Will he ever see a bat?

I knew I was in trouble when the other day I found myself bargaining with starfish. Red and purple ones are ubiquitous on the rocky coast of British Columbia where my parents live, where my son was born, and where I have spent about half of my adult life. They are always the biggest kid pleasers, because you can gently pick one up and give it a really good look. "This is the best day of my life!" my seven-year-old niece Miriam, visiting from Chicago, proclaimed after a long afternoon spent in the tide pools.

But in the fall of 2013, stories began to appear about a strange wasting disease that was causing starfish along the Pacific Coast to die by the tens of thousands. Termed the "sea star wasting syndrome," multiple species were disintegrating alive, their vibrant bodies melting into distorted globs, with legs falling off and bodies caving in. Scientists were mystified.[34]

As I read these stories, I caught myself praying for the invertebrates to hang in for just one more year—long enough for my son to be amazed by them. Then I doubted myself: maybe it's better if he never sees a starfish at all—certainly not like this . . .

When fear like that used to creep through my armor of climate change denial, I would do my utmost to stuff it away, change the channel, click past it. Now I try to feel it. It seems to me that I owe it to my son, just as we all owe it to ourselves and one another.

But what should we do with this fear that comes from living on a planet that is dying, made less alive every day? First, accept that it won't go away. That it is a fully rational response to the unbearable reality that we are living in a dying world, a world that a great many of us are helping to kill, by doing things like making tea and driving to the grocery store and yes, okay, having kids.

Next, use it. Fear is a survival response. Fear makes us run, it makes us leap, it can make us act superhuman. But we need somewhere to run *to*. Without that, the fear is only paralyzing. So the real trick, the only hope, really, is to allow the terror of an unlivable future to be balanced and soothed by the prospect of building something much better than many of us have previously dared hope.

Yes, there will be things we will lose, luxuries some of us will have to give up, whole industries that will disappear. And it's too late to stop climate change from coming; it is already here, and increasingly brutal disasters are headed our way no matter what we do. But it's not too late to avert the worst, and there is still time to change ourselves so that we are far less brutal to one another when those disasters strike. And that, it seems to me, is worth a great deal.

Because the thing about a crisis this big, this all-encompassing, is that it changes everything. It changes what we can do, what we can hope for, what we can demand from ourselves and our leaders. It means there is a whole lot of stuff that we have been told is inevitable that simply cannot stand. And it means that a whole lot of stuff we have been told is impossible has to start happening right away.

Can we pull it off? All I know is that nothing is inevitable. Nothing except that climate change changes everything. And for a very brief time, the nature of that change is still up to us.

PART ONE

BAD TIMING

"Coal, in truth, stands not beside but entirely above all other commodities. It is the material energy of the country—the universal aid—the factor in everything we do."

—William Stanley Jevons, economist, 1865 [1]

"How sad to think that nature speaks and mankind doesn't listen."

—Victor Hugo, 1840 [2]

1

THE RIGHT IS RIGHT

The Revolutionary Power of Climate Change

"Climate scientists agree: climate change is happening here and now. Based on well-established evidence, about 97 percent of climate scientists have concluded that human-caused climate change is happening. This agreement is documented not just by a single study, but by a converging stream of evidence over the past two decades from surveys of scientists, content analyses of peer-reviewed studies, and public statements issued by virtually every membership organization of experts in this field."

—Report by the American Association for the Advancement of Science, 2014[1]

"There is no way this can be done without fundamentally changing the American way of life, choking off economic development, and putting large segments of our economy out of business."

—Thomas J. Donohue, President of the U.S. Chamber of Commerce,
on ambitious carbon reduction[2]

There is a question from a gentleman in the fourth row.

He introduces himself as Richard Rothschild. He tells the crowd that he ran for county commissioner in Maryland's Carroll County because he had come to the conclusion that policies to combat global warming were actually "an attack on middle-class American capitalism." His question for the panelists, gathered in a Washington, D.C., Marriott, is: "To what extent is this entire movement simply a green Trojan horse, whose belly is full with red Marxist socioeconomic doctrine?"[3]

At the Heartland Institute's Sixth International Conference on Climate Change, held in late June 2011, the premier gathering for those dedicated to denying the overwhelming scientific consensus that human activity is warming the planet, this qualifies as a rhetorical question. Like asking a meeting of German central bankers if Greeks are untrustworthy. Still, the panelists aren't going to pass up an opportunity to tell the questioner just how right he is.

First up is Marc Morano, editor of the denialists' go-to news site Climate Depot. "In America today we are regulated down to our shower heads, to our light bulbs, to our washing machines," he says. And "we're allowing the American SUV to die right before our eyes." If the greens have their way, Morano warns, we will be looking at "a CO_2 budget for every man, woman, and child on the planet, monitored by an international body."[4]

Next is Chris Horner, a senior fellow at the Competitive Enterprise Institute who specializes in harassing climate scientists with burdensome lawsuits and Freedom of Information Act fishing expeditions. He angles the table mic over to his mouth. "You can believe this is about the climate," he says darkly, "and many people do, but it's not a reasonable belief." Horner, whose prematurely silver hair makes him look like Anderson Cooper's frat boy doppelgänger, likes to invoke 1960s counterculture icon Saul Alinsky: "The issue isn't the issue." The issue, apparently, is that "no free society would do to itself what this agenda requires. . . . The first step to [doing] that is to remove these nagging freedoms that keep getting in the way."[5]

Claiming that climate change is a plot to steal American freedom is rather tame by Heartland standards. Over the course of this two-day conference, I will hear modern environmentalism compared to virtually every mass-murderous chapter in human history, from the Catholic Inquisition to Nazi Germany to Stalin's Russia. I will learn that Barack Obama's campaign promise to support locally owned biofuels refineries was akin to Chairman Mao's scheme to put "a pig iron furnace in everybody's backyard" (the Cato Institute's Patrick Michaels). That climate change is "a stalking horse for National Socialism" (former Republican senator and retired astronaut Harrison Schmitt, referencing the Nazis). And that environmentalists are like Aztec priests, sacrificing countless people to appease the gods and change the weather (Marc Morano again).[6]

But most of all, I will hear versions of the opinion expressed by the county commissioner in the fourth row: that climate change is a Trojan horse designed to abolish capitalism and replace it with some kind of "green communitarianism." As conference speaker Larry Bell succinctly puts it in his book *Climate of Corruption*, climate change "has little to do with the state of the environment and much to do with shackling capitalism and transforming the American way of life in the interests of global wealth re-distribution."[7]

Yes, there is a pretense that the delegates' rejection of climate science is rooted in serious disagreement about the data. And the organizers go to some lengths to mimic credible scientific conferences, calling the gathering "Restoring the Scientific Method" and even choosing a name, the International Conference on Climate Change, that produces an organizational acronym, ICCC, just one letter off from that of the world's leading authority on climate change, the United Nations' Intergovernmental Panel on Climate Change (IPCC), a collaboration of thousands of scientists and 195 governments. But the various contrarian theses presented at the Heartland conference—tree rings, sunspots, the Medieval Warm Period—are old news and were thoroughly debunked long ago. And most of the speakers are not even scientists but rather hobbyists: engineers, economists, and lawyers, mixed in with a weatherman, an astronaut, and a "space architect"—all convinced they have outsmarted 97 percent of the world's climate scientists with their back-of-the-envelope calculations.[8]

Australian geologist Bob Carter questions whether warming is happening at all, while astrophysicist Willie Soon acknowledges some warming has occurred, but says it has nothing to do with greenhouse emissions and is instead the result of natural fluctuations in the activity of the sun. Cato's Patrick Michaels contradicts them both by conceding that CO_2 is indeed increasing temperatures, but insists the impacts are so minor we should "do nothing" about it. Disagreement is the lifeblood of any intellectual gathering, but at the Heartland conference, this wildly contradictory material sparks absolutely no debate among the deniers—no one attempts to defend one position over another, or to sort out who is actually correct. Indeed as the temperature graphs are presented, several members of the mostly elderly audience seem to doze off.[9]

The entire room comes to life, however, when the rock stars of the movement take the stage—not the C-team scientists but the A-team ideological warriors like Morano and Horner. This is the true purpose of the gathering: providing a forum for die-hard denialists to collect the rhetorical cudgels with which they will attempt to club environmentalists and climate scientists in the weeks and months to come. The talking points tested here will jam the comment sections beneath every article and YouTube video that contains the phrase "climate change" or "global warming." They will also fly from the mouths of hundreds of right-wing commentators and politicians—from Republican presidential hopefuls all the way down to county commissioners like Richard Rothschild. In an interview outside the sessions, Joseph Bast, president of the Heartland Institute, takes credit for "thousands of articles and op-eds and speeches . . . that were informed by or motivated by somebody attending one of these conferences."[10]

More impressive, though left unspoken, are all the news stories that were never published and never aired. The years leading up to the gathering had seen a precipitous collapse of media coverage of climate change, despite a rise in extreme weather: in 2007, the three major U.S. networks—CBS, NBC, and ABC—ran 147 stories on climate change; in 2011 the networks ran just fourteen stories on the subject. That too is the denier strategy at work, because the goal was never just to spread doubt but also to spread fear—to send a clear message that saying anything at all about climate change was a surefire way to find your inbox and comment threads jammed with a toxic strain of vitriol.[11]

The Heartland Institute, a Chicago-based think tank devoted to "promoting free-market solutions," has been holding these confabs since 2008, sometimes twice a year. And at the time of the gathering, the strategy appeared to be working. In his address, Morano—whose claim to fame is having broken the Swift Boat Veterans for Truth story that helped sink John Kerry's 2004 presidential bid—led the audience through a series of victory laps. Climate legislation in the U.S. Senate: dead! The U.N. summit on climate change in Copenhagen: failure! The climate movement: suicidal! He even projected on a screen a couple of quotes from climate activists beating up on themselves (as progressives do so well) and exhorted the audience to "celebrate!"[12]

The only things missing were balloons and confetti descending from the rafters.

––––––––

When public opinion on the big social and political issues changes, the trends tend to be relatively gradual. Abrupt shifts, when they come, are usually precipitated by dramatic events. Which is why pollsters were so surprised by what had happened to perceptions about climate change in just four years. A 2007 Harris poll found that 71 percent of Americans believed that the continued burning of fossil fuels would alter the climate. By 2009 the figure had dropped to 51 percent. In June 2011 the number was down to 44 percent—well under half the population. Similar trends have been tracked in the U.K. and Australia. Scott Keeter, director of survey research at the Pew Research Center for People & the Press, described the statistics in the United States as "among the largest shifts over a short period of time seen in recent public opinion history."[13]

The overall belief in climate change has rebounded somewhat since its 2010–11 low in the United States. (Some have hypothesized that experience with extreme weather events could be contributing, though "the evidence is at best very sketchy at this point," says Riley Dunlap, a sociologist at Oklahoma State University who specializes in the politics of climate change.) But what remains striking is that on the right-wing side of the political spectrum, the numbers are still way down.[14]

It seems hard to believe today, but as recently as 2008, tackling climate change still had a veneer of bipartisan support, even in the United States. That year, Republican stalwart Newt Gingrich did a TV spot with Democratic congresswoman Nancy Pelosi, then Speaker of the House, in which they pledged to join forces and fight climate change together. And in 2007, Rupert Murdoch—whose Fox News channel relentlessly amplifies the climate change denial movement—launched an incentive program at Fox to encourage employees to buy hybrid cars (Murdoch announced he had purchased one himself).

Those days of bipartisanship are decidedly over. Today, more than 75 percent of self-identified Democrats and liberals believe humans are changing the climate—a level that, despite yearly fluctuations, has risen only slightly since 2001. In sharp contrast, Republicans have overwhelmingly chosen to reject the scientific consensus. In some regions, only about 20 percent of self-identified Republicans accept the science. This political rift can also be found in Canada. According to an October 2013 poll conducted by Environics, only 41 percent of respondents who identify with the ruling Conservative Party believe that climate change is real and human-caused, while 76 percent of supporters of the left-leaning New Democratic Party and 69 percent of supporters of the centrist Liberal Party believe it is real. And the same phenomenon has once again been documented in Australia and the U.K., as well as Western Europe.[15]

Ever since this political divide opened up over climate change, a great deal of social science research has been devoted to pinpointing precisely how and why political beliefs are shaping attitudes toward global warming. According to Yale's Cultural Cognition Project, for example, one's "cultural worldview"—that would be political leanings or ideological outlook to the rest of us—explains "individuals' beliefs about global warming more powerfully than any other individual characteristic."[16] More powerfully, that is, than age, ethnicity, education, or party affiliation.

The Yale researchers explain that people with strong "egalitarian" and "communitarian" worldviews (marked by an inclination toward collective action and social justice, concern about inequality, and suspicion of corporate power) overwhelmingly accept the scientific consensus on climate change. Conversely, those with strong "hierarchical" and "individualistic" worldviews (marked by opposition to government assistance for the poor and minorities, strong support for industry, and a belief that we all pretty much get what we deserve) overwhelmingly reject the scientific consensus.[17]

The evidence is striking. Among the segment of the U.S. population that displays the strongest "hierarchical" views, only 11 percent rate climate change as a "high risk," compared with 69 percent of the segment displaying the strongest "egalitarian" views.[18]

Yale law professor Dan Kahan, the lead author on this study, attributes the tight correlation between "worldview" and acceptance of climate science to "cultural cognition," the process by which all of us—regardless of political leanings—filter new information in ways that will protect our "preferred vision of the good society." If new information seems to confirm that vision, we welcome it and integrate it easily. If it poses a threat to our belief system, then our brain immediately gets to work producing intellectual antibodies designed to repel the unwelcome invasion.[19]

As Kahan explained in *Nature*, "People find it disconcerting to believe that behavior that they find noble is nevertheless detrimental to society, and behavior that they find base is beneficial to it. Because accepting such a claim could drive a wedge between them and their peers, they have a strong emotional predisposition to reject it." In other words, it is always easier to deny reality than to allow our worldview to be shattered, a fact that was as true of die-hard Stalinists at the height of the purges as it is of libertarian climate change deniers today. Furthermore, leftists are equally capable of denying inconvenient scientific evidence. If conservatives are inherent system justifiers, and therefore bridle before facts that call the dominant economic system into question, then most leftists are inherent system questioners, and therefore prone to skepticism about facts that come from corporations and government. This can lapse into the kind of fact resistance we see among those who are convinced that multinational drug companies have covered up the link between childhood vaccines and autism. No matter what evidence is marshaled to disprove their theories, it doesn't matter to these crusaders—it's just the system covering up for itself.

This kind of defensive reasoning helps explain the rise of emotional intensity that surrounds the climate issue today. As recently as 2007, climate change was something most everyone acknowledged was happening—they just didn't seem to care very much. (When Americans are asked to rank their political concerns in order of priority, climate change still consistently comes in last.)[21]

But today there is a significant cohort of voters in many countries who care passionately, even obsessively, about climate change. What they care about, however, is exposing it as a "hoax" being perpetrated by liberals to force them to change their light bulbs, live in Soviet-style tenements, and surrender their SUVs. For these right-wingers, opposition to climate change has become as central to their belief system as low taxes, gun ownership, and opposition to abortion. Which is why some climate scientists report receiving the kind of harassment that used to be reserved for doctors who perform abortions. In the Bay Area of California, local Tea Party activists have disrupted municipal meetings when minor sustainability strategies are being discussed, claiming they are part of a U.N.-sponsored plot to usher in world government. As Heather Gass of the East Bay Tea Party put it in an open letter after one such gathering: "One day (in 2035) you will wake up in subsidized government housing, eating government subsidized food, your kids will be whisked off by government buses to indoctrination training centers while you are working at your government assigned job on the bottom floor of your urban transit center village because you have no car and who knows where your aging parents will be but by then it will be too late! WAKE UP!!!!"[22]

Clearly there is something about climate change that has some people feeling very threatened indeed.

Unthinkable Truths

Walking past the lineup of tables set up by the Heartland conference's sponsors, it's not terribly hard to see what's going on. The Heritage Foundation is hawking reports, as are the Cato Institute and the Ayn Rand Institute. The climate change denial movement—far from an organic convergence of "skeptical" scientists—is entirely a creature of the ideological network on display here, the very one that deserves the bulk of the credit for redrawing the global ideological map over the last four decades. A 2013 study by Riley Dunlap and political scientist Peter Jacques found that a striking 72 percent of climate denial books, mostly published since the 1990s, were linked to right-wing think tanks, a figure that rises to 87 percent if self-published books (increasingly common) are excluded.[23]

Many of these institutions were created in the late 1960s and early 1970s, when U.S. business elites feared that public opinion was turning dangerously against capitalism and toward, if not socialism, then an aggressive Keynesianism. In response, they launched a counterrevolution, a richly funded intellectual movement that argued that greed and the limitless pursuit of profit were nothing to apologize for and offered the greatest hope for human emancipation that the world had ever known. Under this liberationist banner, they fought for such policies as tax cuts, free trade deals, for the auctioning off of core state assets from phones to energy to water—the package known in most of the world as "neoliberalism."

At the end of the 1980s, after a decade of Margaret Thatcher at the helm in the U.K. and Ronald Reagan in the United States, and with communism collapsing, these ideological warriors were ready to declare victory: history was officially over and there was, in Thatcher's often repeated words, "no alternative" to their market fundamentalism. Filled with confidence, the next task was to systematically lock in the corporate liberation project in every country that had previously held out, which was usually best accomplished in the midst of political turmoil and large-scale economic crises, and further entrenched through free trade agreements and membership in the World Trade Organization.

It had all been going so well. The project had even managed to survive, more or less, the 2008 financial collapse directly caused by a banking sector that had been liberated of so much burdensome regulation and oversight. But to those gathered here at the Heartland conference, climate change is a threat of a different sort. It isn't about the political preferences of Republicans versus Democrats; it's about the physical boundaries of the atmosphere and ocean. If the dire projections coming out of the IPCC are left unchallenged, and business as usual is indeed driving us straight toward civilization-threatening tipping points, then the implications are obvious: the ideological crusade incubated in think tanks like Heartland, Cato, and Heritage will have to come to a screeching halt. Nor have the various attempts to soft-pedal climate action as compatible with market logic (carbon trading, carbon offsets, monetizing nature's "services") fooled these true believers one bit. They know very well that ours is a global economy created by, and fully reliant upon, the burning of fossil fuels and that a dependency that foundational cannot be changed with a few gentle market mechanisms. It requires heavy-duty interventions: sweeping bans on polluting activities, deep subsidies for green alternatives, pricey penalties for violations, new taxes, new public works programs, reversals of privatizations—the list of ideological outrages goes on and on. Everything, in short, that these think tanks—which have always been public proxies for far more powerful corporate interests—have been busily attacking for decades.

And there is also the matter of "global equity" that keeps coming up in the climate negotiations. The equity debate is based on the simple scientific fact that global warming is caused by the accumulation of greenhouse gases in the atmosphere over two centuries. That means that the countries that got a large head start on industrialization have done a great deal more emitting than most others. And yet many of the countries that have emitted least are getting hit by the impacts of climate change first and worst (the result of geographical bad luck as well as the particular vulnerabilities created by poverty). To address this structural inequity sufficiently to persuade fast-growing countries like China and India not to destabilize the global climate system, earlier emitters, like North America and Europe, will have to take a greater share of the burden at first. And there will obviously need to be substantial transfers of resources and technology to help battle poverty using low carbon tools. This is what Bolivia's climate negotiator Angélica Navarro Llanos meant when she called for a Marshall Plan for the Earth. And it is this sort of wealth redistribution that represents the direst of thought crimes at a place like the Heartland Institute.

Even climate action at home looks suspiciously like socialism to them; all the calls for high-density affordable housing and brand-new public transit are obviously just ways to give backdoor subsidies to the undeserving poor. Never mind what this war on carbon means to the very premise of global free trade, with its insistence that geographical distance is a mere fiction to be collapsed by Walmart's diesel trucks and Maersk's container ships.

More fundamentally than any of this, though, is their deep fear that if the free market system really has set in motion physical and chemical processes that, if allowed to continue unchecked, threaten large parts of humanity at an existential level, then their entire crusade to morally redeem capitalism has been for naught. With stakes like these, clearly greed is not so very good after all. And that is what is behind the abrupt rise in climate change denial among hardcore conservatives: they have come to understand that as soon as they admit that climate change is real, they will lose the central ideological battle of our time—whether we need to plan and manage our societies to reflect our goals and values, or whether that task can be left to the magic of the market.

Imagine, for a moment, how all of this looks to a guy like Heartland president Joseph Bast, a genial bearded fellow who studied economics at the University of Chicago and who told me in a sit-down interview that his personal calling is "freeing people from the tyranny of other people."[24] To Bast, climate action looks like the end of the world. It's not, or at least it doesn't have to be, but, for all intents and purposes, robust, science-based emission reduction is the end of *his* world. Climate change detonates the ideological scaffolding on which contemporary conservatism rests. A belief system that vilifies collective action and declares war on all corporate regulation and all things public simply cannot be reconciled with a problem that demands collective action on an unprecedented scale and a dramatic reining in of the market forces that are largely responsible for creating and deepening the crisis.

And for many conservatives, particularly religious ones, the challenge goes deeper still, threatening not just faith in markets but core cultural narratives about what humans are doing here on earth. Are we masters, here to subdue and dominate, or are we one species among many, at the mercy of powers more complex and unpredictable than even our most powerful computers can model? As Robert Manne, a professor of politics at La Trobe University in Melbourne, puts it, climate science is for many conservatives "an affront to their deepest and most cherished basic faith: the capacity and indeed the right of 'mankind' to subdue the Earth and all its fruits and to establish a 'mastery' over Nature." For these conservatives, he notes, "such a thought is not merely mistaken. It is intolerable and deeply offensive. Those preaching this doctrine have to be resisted and indeed denounced."[25]

And denounce they do, the more personal, the better—whether it's former Vice President Al Gore for his mansions, or famed climate scientist James Hansen for his speaking fees. Then there is "Climategate," a manufactured scandal in which climate scientists' emails were hacked and their contents distorted by the Heartlanders and their allies, who claimed to find evidence of manipulated data (the scientists were repeatedly vindicated of wrongdoing). In 2012, the Heartland Institute even landed itself in hot water by running a billboard campaign that compared people who believe in climate change ("warmists" in denialist lingo) to murderous cult leader Charles Manson and Unabomber Ted Kaczynski. "I still believe in Global

Warming. Do you?" the first ad demanded in bold red letters under a picture of Kaczynski. For Heartlanders, denying climate science is part of a war, and they act like it.[26]

Many deniers are quite open about the fact that their distrust of the science grew out of a powerful fear that if climate change is real, the political implications would be catastrophic. As British blogger and regular Heartland speaker James Delingpole has pointed out, "Modern environmentalism successfully advances many of the causes dear to the left: redistribution of wealth, higher taxes, greater government intervention, regulation." Heartland president Joseph Bast puts it even more bluntly. For the left, "Climate change is the perfect thing. . . . It's the reason why we should do everything [the left] wanted to do anyway."[27]

Bast, who has little of the swagger common to so many denialists, is equally honest about the fact he and his colleagues did not become engaged with climate issues because they found flaws in the scientific facts. Rather, they became alarmed about the economic and political implications of those facts and set out to disprove them. "When we look at this issue, we say, This is a recipe for massive increase in government," Bast told me, concluding that, "Before we take this step, let's take another look at the science. So conservative and libertarian groups, I think, stopped and said, Let's not simply accept this as an article of faith; let's actually do our own research."[28]

Nigel Lawson, Margaret Thatcher's former chancellor of the exchequer who has taken to declaring that "green is the new red," has followed a similar intellectual trajectory. Lawson takes great pride in having privatized key British assets, lowered taxes on the wealthy, and broken the power of large unions. But climate change creates, in his words, "a new license to intrude, to interfere and to regulate." It must, he concludes, be a conspiracy—the classic teleological reversal of cause and effect.[29]

The climate change denial movement is littered with characters who are twisting themselves in similar intellectual knots. There are the old-timer physicists like S. Fred Singer, who used to develop rocket technologies for the U.S. military and who hears in emissions regulation a distorted echo of the communism he fought during the Cold War (as documented compellingly by Naomi Oreskes and Erik Conway in *Merchants of Doubt*). In a similar vein, there is former Czech president Václav Klaus, who spoke at a Heartland climate conference while still head of state. For Klaus, whose career began under communist rule, climate change appears to have induced a full-fledged Cold War flashback. He compares attempts to prevent global warming to "the ambitions of communist central planners to control the entire society" and says, "For someone who spent most of his life in the 'noble' era of communism this is impossible to accept."[30]

And you can understand that, from their perspective, the scientific reality of climate change must seem spectacularly unfair. After all, the people at the Heartland conference thought they had won these ideological wars—if not fairly, then certainly squarely. Now climate science is changing everything: how can you win an argument against government intervention if the very habitability of the planet depends on intervening? In the short term, you might be able to argue that the economic costs of taking action are greater than allowing climate change to play out for a few more decades (and some neoliberal economists, using cost-benefit calculations and future "discounting," are busily making those arguments). But most people don't actually like it when their children's lives are "discounted" in someone else's Excel sheet, and they tend to have a moral aversion to the idea of allowing countries to disappear because saving them would be too expensive.

Which is why the ideological warriors gathered at the Marriott have concluded that there is really only one way to beat a threat this big: by claiming that thousands upon thousands of scientists are lying and that climate change is an elaborate hoax. That the storms aren't really getting bigger, it's just our imagination. And if they are, it's not because of anything humans are doing—or could stop doing. They deny reality, in other words, because the implications of that reality are, quite simply, unthinkable.

So here's my inconvenient truth: I think these hard-core ideologues understand the real significance of climate change better than most of the "warmists" in the political center, the ones who are still insisting that the response can be gradual and painless and that we don't need to go to war with anybody, including the fossil fuel companies. Before I go any further, let me be absolutely clear: as 97 percent of the world's climate scientists attest, the Heartlanders are completely wrong about the science. But when it comes to the political and economic *consequences* of those scientific findings, specifically the kind of deep changes required not just to our energy consumption but to the underlying logic of our liberalized and profit-seeking economy, they have their eyes wide open. The deniers get plenty of the details wrong (no, it's not a communist plot; authoritarian state socialism, as we will see, was terrible for the environment and brutally extractivist), but when it comes to the scope and depth of change required to avert catastrophe, they are right on the money.

About That Money . . .

When powerful ideologies are challenged by hard evidence from the real world, they rarely die off completely. Rather, they become cultlike and marginal. A few of the faithful always remain to tell one another that the problem wasn't with the ideology; it was the weakness of leaders who did not apply the rules with sufficient rigor. (Lord knows there is still a smattering of such grouplets on the neo-Stalinist far left.) By this point in history—after the 2008 collapse of Wall Street and in the midst of layers of ecological crises—free market fundamentalists should, by all rights, be exiled to a similarly irrelevant status, left to fondle their copies of Milton Friedman's *Free to Choose* and Ayn Rand's *Atlas Shrugged* in obscurity. They are saved from this ignominious fate only because their ideas about corporate liberation, no matter how demonstrably at war with reality, remain so profitable to the world's billionaires that they are kept fed and clothed in think tanks by the likes of Charles and David Koch, owners of the diversified dirty energy giant Koch Industries, and ExxonMobil.

According to one recent study, for instance, the denial-espousing think tanks and other advocacy groups making up what sociologist Robert Brulle calls the "climate change counter-movement" are collectively pulling in more than $900 million per year for their work on a variety of right-wing causes, most of it in the form of "dark money"—funds from conservative foundations that cannot be fully traced.[31]

This points to the limits of theories like cultural cognition that focus exclusively on individual psychology. The deniers are doing more than protecting their personal worldviews—they are protecting powerful political and economic interests that have gained tremendously from the way Heartland and others have clouded the climate debate. The ties between the deniers and those interests are well known and well documented. Heartland has received more than $1 million from ExxonMobil together with foundations linked to the Koch brothers and the late conservative funder Richard Mellon Scaife. Just how much money the think tank receives from companies, foundations, and individuals linked to the fossil fuel industry remains unclear because Heartland does not publish the names of its donors, claiming the information would distract from the "merits of our positions." Indeed, leaked internal documents revealed that one of Heartland's largest donors is anonymous—a shadowy individual who has given more than $8.6 million specifically to support the think tank's attacks on climate science.[32]

Meanwhile, scientists who present at Heartland climate conferences are almost all so steeped in fossil fuel dollars that you can practically smell the fumes. To cite just two examples, the Cato Institute's Patrick Michaels, who gave the 2011 conference keynote, once told CNN that 40 percent of his consulting company's income comes from oil companies (Cato itself has received funding from ExxonMobil and Koch family foundations). A Greenpeace investigation into another conference speaker, astrophysicist Willie Soon, found that between 2002 and 2010, 100 percent of his new research grants had come from fossil fuel interests.[33]

The people paid to amplify the views of these scientists—in blogs, op-eds, and television appearances—are bankrolled by many of the same sources. Money from big oil funds the Committee for a Constructive Tomorrow, which houses Marc Morano's website, just as it funds the Competitive Enterprise Institute, one of Chris Horner's intellectual homes. A February 2013 report in The Guardian revealed that between 2002 and 2010, a network of anonymous U.S. billionaires had donated nearly $120 million to "groups casting doubt about the science behind climate change . . . the ready stream of cash set off a conservative backlash against Barack Obama's environmental agenda that wrecked any chance of Congress taking action on climate change."[34]

There is no way of knowing exactly how this money shapes the views of those who receive it or whether it does at all. We do know that having a significant economic stake in the fossil fuel economy makes one more prone to deny the reality of climate change, regardless of political affiliation. For example, the only parts of the U.S. where opinions about climate change are slightly less split along political lines are regions that are highly dependent on fossil fuel extraction—such as Appalachian coal country and the Gulf Coast. There, Republicans still overwhelmingly deny climate change, as they do across the country, but many of their Democratic neighbors do as well (in parts of Appalachia, just 49 percent of Democrats believe in human-created climate change, compared with 72–77 percent in other parts of the country). Canada has the same kinds of regional splits: in Alberta, where incomes are soaring thanks to the tar sands, only 41 percent of residents told pollsters that humans are contributing to climate change. In Atlantic Canada, which has seen far less extravagant benefits from fossil fuel extraction, 68 percent of respondents say that humans are warming the planet.[35]

A similar bias can be observed among scientists. While 97 percent of active climate scientists believe humans are a major cause of climate change, the numbers are radically different among "economic geologists"—scientists who study natural formations so that they can be commercially exploited by the extractive industries. Only 47 percent of these scientists believe in human-caused climate change. The bottom line is that we are all inclined to denial when the truth is too costly—whether emotionally, intellectually, or financially. As Upton Sinclair famously observed: "It is difficult to get a man to understand something, when his salary depends upon his not understanding it!"[36]

Plan B: Get Rich off a Warming World

One of the most interesting findings of the many recent studies on climate perceptions is the clear connection between a refusal to accept the science of climate change and social and economic privilege. Overwhelmingly, climate change deniers are not only conservative but also white and male, a group with higher than average incomes. And they are more likely than other adults to be highly confident in their views, no matter how demonstrably false. A much discussed paper on this topic by sociologists Aaron McCright and Riley Dunlap (memorably titled "Cool Dudes") found that as a group, conservative white men who expressed strong confidence in their understanding of global warming were almost six times as likely to believe climate change "will never happen" as the rest of the adults surveyed. Mc-

Cright and Dunlap offer a simple explanation for this discrepancy: "Conservative white males have disproportionately occupied positions of power within our economic system. Given the expansive challenge that climate change poses to the industrial capitalist economic system, it should not be surprising that conservative white males' strong system-justifying attitudes would be triggered to deny climate change."[37]

But deniers' relative economic and social privilege doesn't just give them more to lose from deep social and economic change; it gives them reason to be more sanguine about the risks of climate change should their contrarian views turn out to be false. This occurred to me as I listened to yet another speaker at the Heartland conference display what can only be described as an utter absence of empathy for the victims of climate change. Larry Bell (the space architect) drew plenty of laughs when he told the crowd that a little heat isn't so bad: "I moved to Houston intentionally!" (Houston was, at that time, in the midst of what would turn out to be Texas's worst single-year drought on record.) Australian geologist Bob Carter offered that "the world actually does better from our human perspective in warmer times." And Patrick Michaels said people worried about climate change should do what the French did after the devastating 2003 heat wave across Europe killed nearly fifteen thousand people in France alone: "they discovered Walmart and air-conditioning."[38]

I listened to these zingers as an estimated thirteen million people in the Horn of Africa faced starvation on parched land. What makes this callousness among deniers possible is their firm belief that if they're wrong about climate science, a few degrees of warming isn't something wealthy people in industrialized countries have to worry much about.* ("When it rains, we find shelter. When it's hot, we find shade," Texas congressman Joe Barton explained at an energy and environment subcommittee hearing.)[39]

As for everyone else, well, they should stop looking for handouts and get busy making money. (Never mind that the World Bank warned in a 2012

---

* Much of this confidence is based on fantasy. Though the ultra-rich may be able to buy a measure of protection for a while, even the wealthiest nation on the planet can fall apart in the face of a major shock (as Hurricane Katrina showed). And no society, no matter how well financed or managed, can truly adapt to massive natural disasters when one comes fast and furious on the heels of the last.

report that for poor countries, the increased cost of storms, droughts, and flooding is already so high that it "threatens to roll back decades of sustainable development.") When I asked Patrick Michaels whether rich countries have a responsibility to help poor ones pay for costly adaptations to a warmer climate, he scoffed: There is no reason to give resources to countries "because, for some reason, their political system is incapable of adapting." The real solution, he claimed, was more free trade.[40]

Michaels surely knows that free trade is hardly going to help islanders whose countries are disappearing, just as he is doubtlessly aware that most people on the planet who are hit hardest by heat and drought can't solve their problems by putting a new AC system on their credit cards. And this is where the intersection between extreme ideology and climate denial gets truly dangerous. It's not simply that these "cool dudes" deny climate science because it threatens to upend their dominance-based worldview. It is that their dominance-based worldview provides them with the intellectual tools to write off huge swaths of humanity, and indeed, to rationalize profiting from the meltdown.

Recognizing the threat posed by this empathy-exterminating mind-set—which the cultural theorists describe as "hierarchical" and "individualistic"—is a matter of great urgency because climate change will test our moral character like little before. The U.S. Chamber of Commerce, in its bid to prevent the Environmental Protection Agency from regulating carbon emissions, argued in a petition that in the event of global warming, "populations can acclimatize to warmer climates via a range of behavioral, physiological, and technological adaptations."[41]

It is these adaptations that worry me most of all. Unless our culture goes through some sort of fundamental shift in its governing values, how do we honestly think we will "adapt" to the people made homeless and jobless by increasingly intense and frequent natural disasters? How will we treat the climate refugees who arrive on our shores in leaky boats? How will we cope as freshwater and food become ever more scarce?

We know the answers because the process is already under way. The corporate quest for natural resources will become more rapacious, more violent. Arable land in Africa will continue to be seized to provide food and fuel to wealthier nations, unleashing a new stage of neocolonial plunder layered on top of the most plundered places on earth (as journalist Christian Parenti documents so well in *Tropic of Chaos*). When heat stress and vicious storms wipe out small farms and fishing villages, the land will be handed over to large developers for mega-ports, luxury resorts, and industrial farms. Once self-sufficient rural residents will lose their lands and be urged to move into increasingly crowded urban slums—for their own protection, they will be told. Drought and famine will continue to be used as pretexts to push genetically modified seeds, driving farmers further into debt.[42]

In the wealthier nations, we will protect our major cities with costly seawalls and storm barriers while leaving vast areas of coastline that are inhabited by poor and Indigenous people to the ravages of storms and rising seas. We may well do the same on the planetary scale, deploying techno-fixes to lower global temperatures that will pose far greater risks to those living in the tropics than in the Global North (more on this later). And rather than recognizing that we owe a debt to migrants forced to flee their lands as a result of our actions (and inactions), our governments will build ever more high-tech fortresses and adopt even more draconian anti-immigration laws. And, in the name of "national security," we will intervene in foreign conflicts over water, oil, and arable land, or start those conflicts ourselves. In short our culture will do what it is already doing, only with more brutality and barbarism, because that is what our system is built to do.

In recent years, quite a number of major multinational corporations have begun to speak openly about how climate change might impact their businesses, and insurance companies closely track and discuss the increased frequency of major disasters. The CEO of Swiss Re Americas admitted, for instance, that "What keeps us up at night is climate change," while companies like Starbucks and Chipotle have raised the alarm about how extreme weather may impact the availability of key ingredients. In June 2014, the Risky Business project, led by billionaire and former New York City mayor Michael Bloomberg, as well as former U.S. treasury secretary Henry Paulson and hedge fund founder and environmental philanthropist Tom Steyer, warned that climate change would cost the U.S. economy billions of dollars each year as a result of rising sea levels alone, and that the corporate world must take such climate costs seriously.[43]

This kind of talk is often equated with support for strong action to

warming. It shouldn't be. Just because companies are willing to
wledge the probable effects of climate change does not mean they
support the kinds of aggressive measures that would significantly reduce
those risks by keeping warming below 2 degrees. In the U.S., for instance,
the insurance lobby has been, by far, the corporate sector most vocal about
the mounting impacts, with the largest companies employing teams of climate scientists to help them prepare for the disasters to come. And yet the
industry hasn't done much to push more aggressive climate policy—on the
contrary, many companies and trade groups have provided substantial funding to the think tanks that created the climate change denial movement.[44]

For some time, this seemingly contradictory dynamic played out within
different divisions of the Heartland Institute itself. The world's premier climate denial institution houses something called the Center on Finance,
Insurance, and Real Estate. Up until May 2012, it was pretty much a mouth-
piece for the insurance industry, headed by conservative Washington insider Eli Lehrer. What made Lehrer different from his Heartland colleagues,
however, is that he is willing to state matter-of-factly, "Climate change is
obviously real and obviously caused to a significant extent by people. I don't
really think there's room for serious debate on either of those points."[45]

So even as his Heartland colleagues were organizing global conferences
designed specifically to manufacture the illusion of a serious scientific debate, Lehrer's division was working with the insurance lobby to protect their
bottom lines in a future of climate chaos. According to Lehrer, "In general
there was no enormous conflict, day-to-day" between his work and that of
his climate-change-denying colleagues.[46] That's because what many of the
insurance companies wanted from Heartland's advocacy was not action to
prevent climate chaos but rather policies that would safeguard or even increase their profits no matter the weather. That means pushing government
out of the subsidized insurance business, giving companies greater freedom
to raise rates and deductibles and to drop customers in high-risk areas, as
well as other "free market" measures.

Eventually, Lehrer split away from the Heartland Institute after the
think tank launched its billboard comparing people who believe in climate
change to mass murderers. Since climate change believers include the insurance companies that were generously funding the Heartland Institute,

that stunt didn't sit at all well. Still, in an interview, Eli Lehrer was quick to stress that the differences were over public relations, not policy. "The public policies that Heartland supported are generally ones I still favor," he said.[47] In truth the work was more or less compatible. Heartland's denier division did its best to cast so much doubt on the science that it helped to paralyze all serious attempts to regulate greenhouse emissions, while the insurance arm pushed policies that would allow corporations to stay profitable regardless of the real-world results of those emissions.

And this points to what really lies behind the casual attitude about climate change, whether it is being expressed as disaster denialism or disaster capitalism. Those involved feel free to engage in these high-stakes gambles because they believe that they and theirs will be protected from the ravages in question, at least for another generation or so.

On a large scale, many regional climate models do predict that wealthy countries—most of which are located at higher latitudes—may experience some economic benefits from a slightly warmer climate, from longer growing seasons to access to shorter trade routes through the melting Arctic ice. At the same time, the wealthy in these regions are already finding ever more elaborate ways to protect themselves from the coming weather extremes. Sparked by events like Superstorm Sandy, new luxury real estate developments are marketing their gold-plated private disaster infrastructure to would-be residents—everything from emergency lighting to natural-gas-powered pumps and generators to thirteen-foot floodgates and watertight rooms sealed "submarine-style," in the case of a new Manhattan condominium. As Stephen G. Kliegerman, the executive director of development marketing for Halstead Property, told *The New York Times*: "I think buyers would happily pay to be relatively reassured they wouldn't be terribly inconvenienced in case of a natural disaster."[48]

Many large corporations, meanwhile, have their own backup generators to keep their lights on through mass blackouts (as Goldman Sachs did during Sandy, despite the fact that its power never actually went out); the capacity to fortify themselves with their own sandbags (which Goldman also did ahead of Sandy); and their own special teams of meteorologists (FedEx). Insurance companies in the United States have even begun dispatching teams of private firefighters to their high-end customers when their mansions in California and Colorado are threatened by wildfires, a "concierge" service pioneered by AIG.[49]

Meanwhile, the public sector continues to crumble, thanks in large part to the hard work of the warriors here at the Heartland conference. These, after all, are the fervent dismantlers of the state, whose ideology has eroded so many parts of the public sphere, including disaster preparedness. These are the voices that have been happy to pass on the federal budget crisis to the states and municipalities, which in turn are coping with it by not repairing bridges or replacing fire trucks. The "freedom" agenda that they are desperately trying to protect from scientific evidence is one of the reasons that societies will be distinctly less prepared for disasters when they come.

For a long time, environmentalists spoke of climate change as a great equalizer, the one issue that affected everyone, rich or poor. It was supposed to bring us together. Yet all signs are that it is doing precisely the opposite, stratifying us further into a society of haves and have-nots, divided between those whose wealth offers them a not insignificant measure of protection from ferocious weather, at least for now, and those left to the mercy of increasingly dysfunctional states.

The Meaner Side of Denial

As the effects of climate change become impossible to ignore, the crueler side of the denial project—now lurking as subtext—will become explicit. It has already begun. At the end of August 2011, with large parts of the world still suffering under record high temperatures, the conservative blogger Jim Geraghty published a piece in *The Philadelphia Inquirer* arguing that climate change "will help the U.S. economy in several ways and enhance, not diminish, the United States' geopolitical power." He explained that since climate change will be hardest on developing countries, "many potentially threatening states will find themselves in much more dire circumstances." And this, he stressed, was a good thing: "Rather than our doom, climate change could be the centerpiece of ensuring a second consecutive American Century." Got that? Since people who scare Americans are unlucky enough to live in poor, hot places, climate change will cook them, leaving the United States to rise like a phoenix from the flames of global warming.*[50]

Expect more of this monstrousness. As the world warms, the ideology so threatened by climate science—the one that tells us it's everyone for themselves, that victims deserve their fate, that we can master nature—will take us to a very cold place indeed. And it will only get colder, as theories of racial superiority, barely under the surface in parts of the denial movement, make a raging comeback.†[51] In the grossly unequal world this ideology has done so much to intensify and lock in, these theories are not optional: they are necessary to justify the hardening of hearts to the largely blameless victims of climate change in the Global South and to the predominantly African American cities like New Orleans that are most vulnerable in the Global North.

In a 2007 report on the security implications of climate change, copublished by the Center for Strategic and International Studies, former CIA director R. James Woolsey predicted that on a much warmer planet "altruism and generosity would likely be blunted."[52] We can already see that emotional blunting on display from Arizona to Italy. Already, climate change is changing us, coarsening us. Each massive disaster seems to inspire less horror, fewer telethons. Media commentators speak of "compassion fatigue," as if empathy, and not fossil fuels, was the finite resource.

As if to prove the point, after Hurricane Sandy devastated large parts of New York and New Jersey, the Koch-backed organization Americans for Prosperity (AFP) launched a campaign to block the federal aid package going to these states. "We need to suck it up and be responsible for taking care of ourselves," said Steve Lonegan, then director of AFP's New Jersey chapter.[53]

---

* In early 2011, Joe Read, a newly elected representative to the Montana state legislature, made history by introducing the first bill to officially declare climate change a good thing. "Global warming is beneficial to the welfare and business climate of Montana," the bill stated. Read explained, "Even if it does get warmer, we're going to have a longer growing season. It could be very beneficial to the state of Montana. Why are we going to stop this progress?" The bill did not pass.

† In a telling development, the American Freedom Alliance hosted its own conference challenging the reality of climate change in Los Angeles in June 2011. Part of the Alliance's stated mission is "to identify threats to Western civilization," and it is known for its fearmongering about "the Islamic penetration of Europe" and similar supposed designs in the U.S. Meanwhile, one of the books on sale at the Heartland conference was *Going Green* by Chris Skates, a fictional "thriller" in which climate activists plot with Islamic terrorists to destroy America's electricity grid.

And then there is Britain's *Daily Mail* newspaper. In the midst of the extraordinary 2014 winter floods, the tabloid ran a front-page headline asking its readers to sign a petition calling on the government "to divert some of the £11 billion a year spent on overseas aid to ease the suffering of British flood victims."[54] Within days, more than 200,000 people had signed onto the demand to cut foreign aid in favor of local disaster relief. Of course Britain—the nation that invented the coal-fired steam engine—has been emitting industrial levels of carbon for longer than any nation on earth and therefore bears a particularly great responsibility to increase, as opposed to claw back, foreign aid. But never mind that. Screw the poor. Suck it up. Everyone for themselves.

Unless we radically change course, these are the values that will rule our stormy future, even more than they already rule our present.

Coddling Conservatives

Some climate activists have attempted to sway deniers away from their hardened positions, arguing that delaying climate action will only make the government interventions required more extreme. The popular climate blogger Joe Romm, for instance, writes that "if you hate government intrusion into people's lives, you'd better stop catastrophic global warming, because nothing drives a country more towards activist government than scarcity and deprivation. . . . Only Big Government—which conservatives say they don't want—can relocate millions of citizens, build massive levees, ration crucial resources like water and arable land, mandate harsh and rapid reductions in certain kinds of energy—all of which will be inevitable if we don't act now."[55]

It's true that catastrophic climate change would inflate the role of government to levels that would likely disturb most thinking people, whether left or right. And there are legitimate fears too of what some call "green fascism"—an environmental crisis so severe that it becomes the pretext for authoritarian forces to seize control in the name of restoring some kind of climate order. But it's also the case that there is no way to get cuts in emissions steep or rapid enough to avoid those catastrophic scenarios *without* levels of government intervention that will never be acceptable to right-wing ideologues.

This was not always so. If governments, including in the U.S., had started cutting emissions back when the scientific consensus first solidified, the measures for avoiding catastrophic warming would not have been nearly so jarring to the reigning economic model. For instance, the first major international gathering to set specific targets for emission reductions was the World Conference on the Changing Atmosphere, held in Toronto in 1988, with more than three hundred scientists and policymakers from forty-six countries represented. The conference, which set the groundwork for the Rio Earth Summit, was a breakthrough, recommending that governments cut emissions by 20 percent below 1988 levels by 2005. "If we choose to take on this challenge," remarked one scientist in attendance, "it appears that we can slow the rate of change substantially, giving us time to develop mechanisms so that the cost to society and the damage to ecosystems can be minimized. We could alternatively close our eyes, hope for the best, and pay the cost when the bill comes due."[56]

If we had heeded this advice and got serious about meeting that goal immediately after the 1992 signing of the U.N. climate convention in Rio, the world would have needed to reduce its carbon emissions by about 2 percent per year until 2005.[57] At that rate, wealthy countries could have much more comfortably started rolling out the technologies to replace fossil fuels, cutting carbon at home while helping to launch an ambitious green transition throughout the world. Since this was before the globalization juggernaut took hold, it would have created an opportunity for China and India and other fast-growing economies to battle poverty on low-carbon pathways. (Which was the stated goal of "sustainable development" as championed in Rio.)

Indeed this vision could have been built into the global trade architecture that would rise up in the early to mid-1990s. If we had continued to reduce our emissions at that pace we would have been on track for a completely de-carbonized global economy by mid-century.

But we didn't do any of those things. And as the famed climate scientist Michael Mann, director of the Penn State Earth System Science Center, puts it, "There's a huge procrastination penalty when it comes to emitting carbon into the atmosphere": the longer we wait, the more it builds up, the more dramatically we must change to reduce the risks of catastrophic warming. Kevin Anderson, deputy director of the Tyndall Centre for Climate Change Research, further explains: "Perhaps at the time of the 1992 Earth Summit, or even at the turn of the millennium, 2°C levels of mitigation could have been achieved through significant evolutionary changes within the political and economic hegemony. But climate change is a cumulative issue! Now, in 2013, we in high-emitting (post)industrial nations face a very different prospect. Our ongoing and collective carbon profligacy has squandered any opportunity for the 'evolutionary change' afforded by our earlier (and larger) 2°C carbon budget. Today, after two decades of bluff and lies, the remaining 2°C budget demands revolutionary change to the political and economic hegemony."[58]

Put a little more simply: for more than two decades, we kicked the can down the road. During that time, we also expanded the road from a two-lane carbon-spewing highway to a six-lane superhighway. That feat was accomplished in large part thanks to the radical and aggressive vision that called for the creation of a single global economy based on the rules of free market fundamentalism, the very rules incubated in the right-wing think tanks now at the forefront of climate change denial. There is a certain irony at work: it is the success of their own revolution that makes revolutionary levels of transformation to the market system now our best hope of avoiding climate chaos.

———

Some are advancing a different strategy to bring right-wingers back into the climate fold. Rather than trying to scare them with scenarios of interventionist governments if we procrastinate further, this camp argues that we need approaches to emission reduction that are less offensive to conservative values.

Yale's Dan Kahan points out that while those who poll as highly "hierarchical" and "individualist" bridle at any mention of regulation, they tend to like big, centralized technologies that do not challenge their belief that humans can dominate nature. In one of his studies, Kahan and his colleagues polled subjects on their views about climate change after showing some of them fake news stories. Some of the subjects were given a story about how global warming could be solved through "anti-pollution" measures. Others were given a story that held up nuclear power as the solution. Some were shown no story at all. The scientific facts about global warming were identical in all news stories. The researchers discovered that hard-core conservatives who received the nuclear solution story were more open to the scientific facts proving that humans are changing the climate. However, those who received the story about fighting pollution "were even more skeptical about these facts than were hierarchs and individualists in a control group that received no newspaper story."[59]

It's not hard to figure out why. Nuclear is a heavy industrial technology, based on extraction, run in a corporatist manner, with long ties to the military-industrial complex. And as renowned psychiatrist and author Robert Jay Lifton has noted, no technology does more to confirm the notion that man has tamed nature than the ability to split the atom.[60]

Based on this research, Kahan and others argue, environmentalists should sell climate action by playing up concerns about national security and emphasizing responses such as nuclear power and "geoengineering"—global-scale technological interventions that would attempt to reverse rapid warming by, for instance, blocking a portion of the sun's rays, or by "fertilizing" the oceans so that they trap more carbon, among other untested, extraordinarily high-risk schemes. Kahan reasons that since climate change is perceived by many on the right as a gateway to dreaded anti-industry policies, the solution is "to remove what makes it threatening." In a similar vein, Irina Feygina and John T. Jost, who have conducted parallel research at NYU, advise policymakers to package environmental action as being about protecting "our way of life" and a form of patriotism, something they revealingly call "system-sanctioned change."[61]

This kind of advice has been enormously influential. For instance, the Breakthrough Institute—a think tank that specialized in attacking grassroots environmentalism for its supposed lack of "modernity"—is forever charting this self-styled middle path, pushing nuclear power, fracked natural gas, and genetically modified crops as climate solutions, while attacking renewable energy programs. And as we will see later on, some greens are even warming up to geoengineering.[62] Moreover, in the name of reaching across the aisle, green groups are constantly "reframing" climate action so that it is about pretty much anything other than preventing catastrophic warming to protect life on earth. Instead climate action is about all the things conservatives are supposed to care about more than that, from cutting off revenues to Arab states to reasserting American economic dominance over China.

The first problem with this strategy is that it doesn't work: this has been the core messaging for many large U.S. green groups for five years ("Forget about climate change," counsels Jonathan Foley, director of the Institute on the Environment at the University of Minnesota. "Do you love America?"[63]) And as we have seen, conservative opposition to climate action has only hardened in this period.

The far more troubling problem with this approach is that rather than challenging the warped values fueling both disaster denialism and disaster capitalism, it actively reinforces those values. Nuclear power and geoengineering are not solutions to the ecological crisis; they are a doubling down on exactly the kind of reckless, short-term thinking that got us into this mess. Just as we spewed greenhouse gases into the atmosphere thinking that tomorrow would never come, both of these hugely high-risk technologies would create even more dangerous forms of waste, and neither has a discernible exit strategy (subjects that I will be exploring in greater depth later on). Hyper-patriotism, similarly, is an active barrier to coming up with any kind of global climate agreement, since it further pits countries against one another rather than encouraging them to cooperate. As for pitching climate action as a way to protect America's high-consumerist "way of life"—that is either dishonest or delusional because a way of life based on the promise of infinite growth cannot be protected, least of all exported to every corner of the globe.

The Battle of Worldviews

I am well aware that all of this raises the question of whether I am doing the same thing as the deniers—rejecting possible solutions because they threaten my ideological worldview. As I outlined earlier, I have long been greatly concerned about the science of global warming—but I was propelled into a deeper engagement with it partly because I realized it could be a catalyst for forms of social and economic justice in which I already believed.

But there are a few important differences to note. First, I am not asking anyone to take my word on the science; I think that all of us should take the word of 97 percent of climate scientists and their countless peer-reviewed articles, as well as every national academy of science in the world, not to mention establishment institutions like the World Bank and the International Energy Agency, all of which are telling us we are headed toward catastrophic levels of warming. Nor am I suggesting that the kind of equity-based responses to climate change that I favor are inevitable results of the science.

What I am saying is that the science forces us to *choose* how we want to respond. If we stay on the road we are on, we will get the big corporate, big military, big engineering responses to climate change—the world of a tiny group of big corporate winners and armies of locked-out losers that we have imagined in virtually every fictional account of our dystopic future, from *Mad Max* to *The Children of Men* to *The Hunger Games* to *Elysium*. Or we can choose to heed climate change's planetary wake-up call and change course, steer away not just from the emissions cliff but from the logic that brought us careening to that precipice. Because what the "moderates" constantly trying to reframe climate action as something more palatable are really asking is: How can we create change so that the people responsible for the crisis do not feel threatened by the solutions? How, they ask, do you reassure members of a panicked, megalomaniacal elite that they are still masters of the universe, despite the overwhelming evidence to the contrary?

The answer is: you don't. You make sure you have enough people on your side to change the balance of power and take on those responsible, knowing that true populist movements always draw from both the left and the right. And rather than twisting yourself in knots trying to appease a lethal worldview, you set out to deliberately strengthen those values ("egalitarian" and "communitarian" as the cultural cognition studies cited here describe them) that are currently being vindicated, rather than refuted, by the laws of nature.

Culture, after all, is fluid. It has changed many times before and can change again. The delegates at the Heartland conference understand this, which is why they are so determined to suppress the mountain of evidence proving that their worldview is a threat to life on earth. The task for the rest of us is to believe, based on that same evidence, that a very different worldview can be our salvation.

The Heartlanders understand that culture can shift quickly because they are part of a movement that did just that. "Economics are the method," Margaret Thatcher said, "the object is to change the heart and soul." It was a mission largely accomplished. To cite just one example, in 1966, a survey of U.S. college freshmen found that only about 44 percent of them said that making a lot of money was "very important" or "essential." By 2013, the figure had jumped to 82 percent.[64]

It's enormously telling that as far back as 1998, when the American Geophysical Union (AGU) convened a series of focus groups designed to gauge attitudes toward global warming, it discovered that "Many respondents in our focus groups were convinced that the underlying cause of environmental problems (such as pollution and toxic waste) is a pervasive climate of rampant selfishness and greed, and since they see this moral deterioration to be irreversible, they feel that environmental problems are unsolvable."[65]

Moreover, a growing body of psychological and sociological research shows that the AGU respondents were exactly right: there is a direct and compelling relationship between the dominance of the values that are intimately tied to triumphant capitalism and the presence of anti-environment views and behaviors. While a great deal of research has demonstrated that having politically conservative or "hierarchical" views and a pro-industry slant makes one particularly likely to deny climate change, there is an even larger number of studies connecting materialistic values (and even free market ideology) to carelessness not just about climate change, but to a great many environmental risks. At Knox College in Illinois, psychologist Tim Kasser has been at the forefront of this work. "To the extent people prioritize values and goals such as achievement, money, power, status and image, they tend to hold more negative attitudes towards the environment, are less likely to engage in positive environmental behaviors, and are more likely to use natural resources unsustainably," write Kasser and British environmental strategist Tom Crompton in their 2009 book, *Meeting Environmental Challenges: The Role of Human Identity.*[66]

In other words, the culture that triumphed in our corporate age pits us against the natural world. This could easily be a cause only for despair. But if there is a reason for social movements to exist, it is not to accept dominant values as fixed and unchangeable but to offer other ways to live—to wage, and win, a battle of cultural worldviews. That means laying out a vision of the world that competes directly with the one on harrowing display at the Heartland conference and in so many other parts of our culture, one that resonates with the majority of people on the planet because it is true: That we are not apart from nature but of it. That acting collectively for a greater good is not suspect, and that such common projects of mutual aid are responsible for our species' greatest accomplishments. That greed must be disciplined and tempered by both rule and example. That poverty amidst plenty is unconscionable.

It also means defending those parts of our societies that already express these values outside of capitalism, whether it's an embattled library, a public park, a student movement demanding free university tuition, or an immigrant rights movement fighting for dignity and more open borders. And most of all, it means continually drawing connections among these seemingly disparate struggles—asserting, for instance, that the logic that would cut pensions, food stamps, and health care before increasing taxes on the rich is the same logic that would blast the bedrock of the earth to get the last vapors of gas and the last drops of oil before making the shift to renewable energy.

Many are attempting to draw these connections and are expressing these alternative values in myriad ways. And yet a robust movement responding to the climate crisis is not emerging fast enough. Why? Why aren't we, as a species, rising to our historical moment? Why are we so far letting "decade zero" slip away?

It's rational for right-wing ideologues to deny climate change—to recognize it would be intellectually cataclysmic. But what is stopping so many who reject that ideology from demanding the kinds of powerful measures that the Heartlanders fear? Why aren't liberal and left political parties around the world calling for an end to extreme energy extraction and full transitions to renewal and regeneration-based economies? Why isn't climate change at the center of the progressive agenda, the burning basis for demanding a robust and reinvented commons, rather than an often forgotten footnote? Why do liberal media outlets still segregate stories about melting ice sheets in their "green" sections—next to viral videos of cuddly animals making unlikely friendships? Why are so many of us not doing the things that must be done to keep warming below catastrophic levels?

The short answer is that the deniers won, at least the first round. Not the battle over climate science—their influence in that arena is already waning. But the deniers, and the ideological movement from which they sprang, won the battle over which values would govern our societies. Their vision—that greed should guide us, that, to quote the late economist Milton Friedman, "the major error" was "to believe that it is possible to do good with other people's money"—has dramatically remade our world over the last four decades, decimating virtually every countervailing power.[67] Extreme free-market ideology was locked in through the harsh policy conditions attached to much-needed loans issued by the World Bank and the International Monetary Fund. It shaped the model of export-led development that dotted the developing world with free trade zones. It was written into countless trade agreements. Not everyone was convinced by these arguments, not by a long shot. But too many tacitly accepted Thatcher's dictum that there is no alternative.

Meanwhile, denigration of collective action and veneration of the profit motive have infiltrated virtually every government on the planet, every major media organization, every university, our very souls. As that American Geophysical Union survey indicated, somewhere inside each of us dwells a belief in their central lie—that we are nothing but selfish, greedy, self-gratification machines. And if we are that, then what hope do we have of taking on the grand, often difficult, collective work that will be required to save ourselves in time? This, without a doubt, is neoliberalism's single most damaging legacy: the realization of its bleak vision has isolated us enough from one another that it became possible to convince us that we are not just incapable of self-preservation but fundamentally *not worth saving*.

Yet at the same time, many of us know the mirror that has been held up to us is profoundly distorted—that we are, in fact, a mess of contradictions, with our desire for self-gratification coexisting with deep compassion, our greed with empathy and solidarity. And as Rebecca Solnit vividly documents in her 2009 book, *A Paradise Built in Hell*, it is precisely when humanitarian crises hit that these other, neglected values leap to the fore, whether it's the incredible displays of international generosity after a massive earthquake or tsunami, or the way New Yorkers gathered to spontaneously meet and comfort one another after the 9/11 attacks. Just as the Heartlanders fear, the existential crisis that is climate change has the power to release these suppressed values on a global and sustained scale, to provide us with a chance for a mass jailbreak from the house that their ideology built—a structure already showing significant cracks and fissures.[68]

But before that can happen, we need to take a much closer look at precisely how the legacy of market fundamentalism, and the much deeper cultural narratives on which it rests, still block critical, life-saving climate action on virtually every front. The green movement's mantra that climate is not about left and right but "right and wrong" has gotten us nowhere. The traditional political left does not hold all the answers to this crisis. But there can be no question that the contemporary political right, and the triumphant ideology it represents, is a formidable barrier to progress.

As the next four chapters will show, the real reason we are failing to rise to the climate moment is because the actions required directly challenge our reigning economic paradigm (deregulated capitalism combined with public austerity), the stories on which Western cultures are founded (that we stand apart from nature and can outsmart its limits), as well as many of the activities that form our identities and define our communities (shopping, living virtually, shopping some more). They also spell extinction for the richest and most powerful industry the world has ever known—the oil and gas industry, which cannot survive in anything like its current form if we humans are to avoid our own extinction. In short, we have not responded to this challenge because we are locked in—politically, physically, and culturally. Only when we identify these chains do we have a chance of breaking free.

HOT MONEY

How Free Market Fundamentalism Helped Overheat the Planet

> "We always had hope that next year was gonna be better. And even this year was gonna be better. We learned slowly, and what didn't work, you tried it harder the next time. You didn't try something different. You just tried harder, the same thing that didn't work."
>
> —Wayne Lewis, Dust Bowl survivor, 2012[1]

> "As leaders we have a responsibility to fully articulate the risks our people face. If the politics are not favorable to speaking truthfully, then clearly we must devote more energy to changing the politics."
>
> —Marlene Moses, Ambassador to the United Nations for Nauru, 2012[2]

During the globalization wars of the late nineties and early 2000s, I used to follow international trade law extremely closely. But I admit that as I immersed myself in the science and politics of climate change, I stopped paying attention to trade. I told myself that there was only so much abstract, bureaucratic jargon one person could be expected to absorb, and my quota was filled up with emission mitigation targets, feed-in tariffs, and the United Nations' alphabet soup of UNFCCCs and IPCCs.

Then about three years ago, I started to notice that green energy programs—the strong ones that are needed to lower global emissions fast— were increasingly being challenged under international trade agreements, particularly the World Trade Organization's rules.

In 2010, for instance, the United States challenged one of China's wind power subsidy programs on the grounds that it contained supports for local industry considered protectionist. China, in turn, filed a complaint in 2012 targeting various renewable energy programs in the European Union, singling out Italy and Greece (it has also threatened to bring a dispute against renewables subsidies in five U.S. states). Washington, meanwhile, has launched a World Trade Organization attack on India's ambitious Jawaharlal Nehru National Solar Mission, a large, multiphase solar support program—once again, for containing provisions, designed to encourage local industry, considered to be protectionist. As a result, brand-new factories that should be producing solar panels are now contemplating closure. Not to be outdone, India has signaled that it might take aim at state renewable energy programs in the U.S.[3]

This is distinctly bizarre behavior to exhibit in the midst of a climate emergency. Especially because these same governments can be counted upon to angrily denounce each other at United Nations climate summits for not doing enough to cut emissions, blaming their own failures on the other's lack of commitment. Yet rather than compete for the best, most effective supports for green energy, the biggest emitters in the world are rushing to the WTO to knock down each other's windmills.

As one case piled on top of another, it seemed to me that it was time to delve back into the trade wars. And as I explored the issue further, I discovered that one of the key, precedent-setting cases pitting "free trade" against climate action was playing out in Ontario, Canada—my own backyard. Suddenly, trade law became a whole lot less abstract.

———

Sitting at the long conference table overlooking his factory floor, Paolo Maccario, an elegant Italian businessman who moved to Toronto to open a solar factory, has the proud, resigned air of a captain determined to go down with his ship. He makes an effort to put on a brave face: True, "the Ontario market is pretty much gone," but the company will find new customers for its solar panels, he tells me, maybe in Europe, or the United States. Their products are good, best in class, and "the cost is competitive enough."[4]

As chief operating officer of Silfab Ontario, Maccario has to say these things; anything else would be a breach of fiduciary duty. But he is also frank that the last few months have been almost absurdly bad. Old customers are convinced the factory is going to close down and won't be able to honor the twenty-five-year warranty on the solar panels they purchased. New customers aren't placing orders over the same concerns, opting to go with Chinese companies that are selling less efficient but cheaper modules.* Suppliers who had been planning to set up their own factories nearby to cut down on transport costs are now keeping their distance.

Even his own board back home in Italy (Silfab is owned by Silfab SpA, whose founder was a pioneer in Italian photovoltaic manufacturing) seemed to be jumping ship. The parent company had committed to invest around $7 million on a custom piece of machinery that, according to Maccario, would have created solar modules that "have an efficiency that has not been reached by any manufacturer in China and in the Western world." But at the last minute, and after all the research and design for the machinery was complete, "It was decided that we cannot spend the money to bring the technology here," Maccario explains. We put on hair nets and lab coats and he shows me an empty rectangle in the middle of the factory floor, the space set aside for equipment that is not coming.

What are the chances he would choose to open this factory here today, given all that has happened, I ask. At this, all attempts at PR drop away and he replies, "I would say below zero if such a number exists."

With his finely tailored wool suit and trim salt and pepper goatee, Maccario looks as if he should be sipping espresso in a piazza in Turin, working for Fiat perhaps—not stuck in this concrete box with an unopened yogurt on his desk, across the street from Imperial Chilled Juice and down the road from the ass end of an AMC multiplex.

And yet in 2010, the decision to locate the company's first North American solar manufacturing plant in Ontario seemed to make a great deal of sense. Back then the mood in Ontario's renewable sector was positively giddy. One year earlier, at the peak of the Wall Street financial crisis, the

---

* China has of course emerged as the world's dominant supplier of inexpensive modules, and in that role has helped to drive dramatic drops in solar prices. It has also flooded the market with cheap panels in recent years, contributing to a global oversupply that has outpaced demand.

province had unveiled its climate action plan, the Green Energy and Green Economy Act, centered on a bold pledge to wean Canada's most populous province completely off coal by 2014.[5]

The plan was lauded by energy experts around the world, particularly in the U.S., where such ambition was lagging. On a visit to Toronto, Al Gore offered his highest blessing, proclaiming it "widely recognized now as the single best green energy [program] on the North American continent." And Michael T. Eckhart, then president of the American Council on Renewable Energy, described it as "the most comprehensive renewable energy policy entered anywhere around the world."[6]

The legislation created what is known as a feed-in tariff program, which allowed renewable energy providers to sell power back to the grid, offering long-term contracts with guaranteed premium prices. It also contained a variety of provisions to ensure that the developers weren't all big players but that local municipalities, co-ops, and Indigenous communities could all get into the renewable energy market and benefit from those premium rates. The catch was that in order for most of the energy providers to qualify, they had to ensure that a minimum percentage of their workforces and materials were local to Ontario. And the province set the bar high: solar energy developers had to source at least 40–60 percent of their content from within the province.[7]

The provision was an attempt to revive Ontario's moribund manufacturing sector, which had long been centered on the Big Three U.S. automakers (Chrysler, Ford, and General Motors) and was, at that time, reeling from the near bankruptcy of General Motors and Chrysler. Compounding these challenges was the fact that Alberta's tar sands oil boom had sent the Canadian dollar soaring, making Ontario a much costlier place to build anything.[8]

In the years that followed the announcement, Ontario's efforts to get off coal were plagued by political blunders. Large natural gas and wind developers ran roughshod over local communities, while the government wasted hundreds of millions (at least) trying to clean up the unnecessary messes. Yet even with all these screwups, the core of the program was an undeniable success. By 2012, Ontario was the largest solar producer in Canada and by 2013, it had only one working coal-fired power plant left. The local content requirements—as the "buy local" and "hire local" provisions are called—were also proving to be a significant boost to the ailing manufacturing sector: by 2014, more than 31,000 jobs had been created and a wave of solar and wind manufacturers had set up shop.[9]

Silfab is a great example of how it worked. The Italian owners had already decided to open a solar panel plant in North America. The company had considered Mexico but was leaning toward the United States. The obvious choices, Maccario told me, were California, Hawaii, and Texas, all of which offered lots of sunshine and corporate incentives, as well as large and growing markets for their product. Ontario—overcast and cold a lot of the year—wasn't "on the radar screen," he admitted. That changed when the province introduced the green energy plan with its local-content provisions, which Maccario described as a "very gutsy and very well intended program." The provisions meant that in communities that switched to renewable energy, companies like his could count on a stable market for their products, one that was protected from having to compete head-to-head with cheaper solar panels from China. So Silfab chose Toronto for its first North American solar plant.

Ontario's politicians loved Silfab. It helped that the building the company purchased to produce its panels was an abandoned auto parts factory, then sitting idle like so many others. And many of the workers the company hired also came from the auto sector—men and women from Chrysler and the autoparts giant Magna, who had years of experience working with the kind of robotic arms that are used to assemble Silfab's high-tech panels. When the plant opened, Wayne Wright, a laid-off autoworker who landed a job as a production operator on the Silfab line, spoke movingly about his seventeen-year-old son, who told him that "finally" his dad's new job would be "creating a better future for all the younger kids."[10]

And then things started to go very wrong. Just as the U.S. has acted against local renewable supports in China and India, so Japan and then the European Union let it be known that they considered Ontario's local-content requirement to be a violation of World Trade Organization rules. Specifically, they claimed that the requirement that a fixed percentage of renewable energy equipment be made in Ontario would "discriminate against equipment for renewable energy generation facilities produced outside Ontario."[11]

The WTO ruled against Canada, determining that Ontario's buy-local provisions were indeed illegal. And the province wasted little time in nixing the local-content rules that had been so central to its program.[12] It was this, Maccario said, that led his foreign investors to pull their support for factory expansion. "Seeing all those, for lack of a better term, mixed messages . . . was the straw that broke the camel's back."

It was also why many plants like his could well close, and others have decided not to open in the first place.

Trade Trumps Climate

From a climate perspective, the WTO ruling was an outrage: if there is to be any hope of meeting the agreed-upon 2 degree Celsius target, wealthy economies like Canada must make getting off fossil fuels their top priority. It is a moral duty, one that the federal government undertook when it signed the Kyoto Protocol in 1997. Ontario was putting real policies in place to honor that commitment (unlike the Canadian government as a whole, which has allowed emissions to balloon, leading it to withdraw from the Kyoto Protocol rather than face international censure). Most importantly, the program was working. How absurd, then, for the WTO to interfere with that success—to let trade trump the planet itself.

And yet from a strictly legal standpoint, Japan and the EU were perfectly correct. One of the key provisions in almost all free trade agreements involves something called "national treatment," which requires governments to make no distinction between goods produced by local companies and goods produced by foreign firms outside their borders. Indeed, favoring local industry constitutes illegal "discrimination." This was a flashpoint in the free trade wars back in the 1990s, precisely because these restrictions effectively prevent governments from doing what Ontario was trying to do: create jobs by requiring the sourcing of local goods as a condition of government support. This was just one of the many fateful battles that progressives lost in those years.

Defenders of these trade deals argue that protections like Ontario's buy-local provisions distort the free market and should be eliminated. Some green energy entrepreneurs (usually those that purchase their products from

China) have made similar arguments, insisting that it doesn't matter where solar panel and wind turbines are produced: the goal should be to get the cheapest products to the consumer so that the green transition can happen as quickly as possible.

The biggest problem with these arguments is the notion that there is any free market in energy to be protected from distortion. Not only do fossil fuel companies receive $775 billion to $1 trillion in annual global subsidies, but they pay nothing for the privilege of treating our shared atmosphere as a free waste dump—a fact that has been described by the *Stern Review on the Economics of Climate Change* as "the greatest market failure the world has ever seen." That freebie is the real distortion, that theft of the sky the real subsidy.[13]

In order to cope with these distortions (which the WTO has made no attempt to correct), governments need to take a range of aggressive steps—from price guarantees to straight subsidies—so that green energy has a fair shot at competing. We know from experience that this works: Denmark has among the most successful renewable energy programs in the world, with 40 percent of its electricity coming from renewables, mostly wind. But it's significant that the program was rolled out in the 1980s, before the free trade era began, when there was no one to argue with the Danish government's generous subsidies to the community-controlled energy projects putting up wind turbines (in 1980, new installations were subsidized by up to 30 percent).[14]

As Scott Sinclair of the Canadian Centre for Policy Alternatives has pointed out, "many of the policies Denmark used to launch its renewable energy industry would have been inconsistent with . . . international trade and investment agreements," since favoring "locally owned cooperatives would conflict with non-discrimination rules requiring that foreign companies be treated no less favourably than domestic suppliers."[15]

And Aaron Cosbey, a development economist and trade and climate expert who is generally supportive of the WTO, rightly notes that the promise of local job creation has been key to the political success of renewable energy programs. "In many cases the green jobs argument is the deciding factor that convinces governments to dole out support. And such requirements, if attached to subsidies or investment privileges, violate WTO obligations."[16]

Which is why governments adopting these tried-and-tested policies—of which there have been far too few—are the ones getting dragged into trade court, whether China, India, Ontario, or the European Union.

Worse, it's not only critical supports for renewable energy that are at risk of these attacks. Any attempt by a government to regulate the sale or extraction of particularly dirty kinds of fossil fuels is also vulnerable to similar trade challenges. The European Union, for instance, is considering new fuel quality standards that would effectively restrict the sales of oil derived from such high-carbon sources as the Alberta tar sands. It's excellent climate policy, of the kind we need much more, but the effort has been slowed down by Canada's not so subtle threats of trade retaliation. Meanwhile, the European Union is using bilateral trade talks to try to circumvent longstanding U.S. restrictions on oil and gas exports, including a decades-old export ban on crude oil. In July 2014, a leaked negotiating document revealed that Europe is pushing for a "legally binding commitment" that would guarantee its ability to import fracked gas and oil from North Dakota's Bakken formation and elsewhere.[17]

Almost a decade ago, a WTO official claimed that the organization enables challenges against "almost any measure to reduce greenhouse gas emissions"—there was little public reaction at the time, but clearly there should have been. And the WTO is far from the only trade weapon that can be used in such battles—so too can countless bilateral and regional free trade and investment agreements.[18]

As we will see later on, these trade deals may even give multinationals the power to overturn landmark grassroots victories against highly controversial extractive activities like natural gas fracking: in 2012, an oil company began taking steps to use NAFTA to challenge Quebec's hard-won fracking moratorium, claiming it robbed the company of its right to drill for gas in the province.[19] (The case is ongoing.) As more activist victories are won, more such legal challenges should be expected.

In some of these cases, governments may successfully defend their emission-reducing activities in trade court. But in too many others, they can be relied upon to cave in early, not wanting to appear anti–free trade (which is likely what is behind Ontario's quiet acceptance of the WTO's ruling against its green energy plan). These challenges aren't killing renewable energy; in the U.S. and China, for instance, the solar market continues to grow impressively. But it is not happening fast enough. And the legal uncertainty that now surrounds some of the most significant green energy programs in the world is bogging us down at the very moment when science is telling us we need to leap ahead. To allow arcane trade law, which has been negotiated with scant public scrutiny, to have this kind of power over an issue so critical to humanity's future is a special kind of madness. As Nobel Prize–winning economist Joseph Stiglitz puts it, "Should you let a group of foolish lawyers, who put together something before they understood these issues, interfere with saving the planet?"[20]

Clearly not. Steven Shrybman, an international trade and public interest lawyer who has worked with a broad range of civil society groups to defend against these trade challenges, says that the problem is structural. "If the trade rules don't permit all kinds of important measures to deal with climate change—and they don't—then the trade rules obviously have to be rewritten. Because there is no way in the world that we can have a sustainable economy and maintain international trade rules as they are. There's no way at all."[21]

This is exactly the sort of commonsense conclusion that has the Heartlanders so very scared of climate change. Because when people wake up to the fact that our governments have locked us into dozens of agreements that make important parts of a robust climate change response illegal, they will have an awfully powerful argument to oppose any such new deals until the small matter of our planet's habitability is satisfactorily resolved.

The same goes for all kinds of free market orthodoxies that threaten our capacity to respond boldly to this crisis, from the suffocating logic of austerity that prevents governments from making the necessary investments in low-carbon infrastructure (not to mention firefighting and flood response), to the auctioning off of electric utilities to private corporations that, in many cases, refuse to switch over to less profitable renewables.

Indeed the three policy pillars of the neoliberal age—privatization of the public sphere, deregulation of the corporate sector, and the lowering of income and corporate taxes, paid for with cuts to public spending—are each incompatible with many of the actions we must take to bring our emissions to safe levels. And together these pillars form an ideological wall that has blocked a serious response to climate change for decades. Before delving more deeply into the ways the climate crisis calls for dismantling that wall, it's helpful to look a little more closely at the epic case of bad timing that landed us where we are today.

A Wall Comes Down, Emissions Go Up

If the climate movement had a birthday, a moment when the issue pierced the public consciousness and could no longer be ignored, it would have to be June 23, 1988. Global warming had been on the political and scientific radar long before that, however. The basic insights central to our current understanding date back to the beginning of the second half of the nineteenth century, and the first scientific breakthroughs demonstrating that burning carbon could be warming the planet were made in the late 1950s. In 1965, the concept was so widely accepted among specialists that U.S. president Lyndon B. Johnson was given a report from his Science Advisory Committee warning that, "Through his worldwide industrial civilization, Man is unwittingly conducting a vast geophysical experiment. . . . The climatic changes that may be produced by the increased CO_2 content could be deleterious from the point of view of human beings."[22]

But it wasn't until James Hansen, then director of NASA's Goddard Institute for Space Studies, testified before a packed congressional hearing on June 23, 1988, that global warming became the stuff of chat shows and political speeches. With temperatures in Washington, D.C., a sweltering 98 degrees Fahrenheit (still a record for that day), and the building's air conditioning on the fritz, Hansen told a room filled with sweaty lawmakers that he had "99 percent confidence" in "a real warming trend" linked to human activity. In a comment to *The New York Times* he added that it was "time to stop waffling" about the science. Later that same month, hundreds of scientists and policymakers held the historic World Conference on the Changing Atmosphere in Toronto where the first emission reductions were discussed. The United Nations' Intergovernmental Panel on Climate Change (IPCC), the premier scientific body advising governments on the climate threat, held its first session that November. By the following year,

79 percent of Americans had heard of the greenhouse effect—a leap from just 38 percent in 1981.[23]

The issue was so prominent that when the editors of *Time* magazine announced their 1988 "Man of the Year," they went for an unconventional choice: "Planet of the Year: Endangered Earth," read the magazine's cover line, over an image of the globe held together with twine, the sun setting ominously in the background. "No single individual, no event, no movement captured imaginations or dominated headlines more," journalist Thomas Sancton explained, "than the clump of rock and soil and water and air that is our common home."[24]

More striking than the image was Sancton's accompanying essay. "This year the earth spoke, like God warning Noah of the deluge. Its message was loud and clear, and suddenly people began to listen, to ponder what portents the message held." That message was so profound, so fundamental, he argued, that it called into question the founding myths of modern Western culture. Here it is worth quoting Sancton at length as he described the roots of the crisis:

> In many pagan societies, the earth was seen as a mother, a fertile giver of life. Nature—the soil, forest, sea—was endowed with divinity, and mortals were subordinate to it. The Judeo-Christian tradition introduced a radically different concept. The earth was the creation of a monotheistic God, who, after shaping it, ordered its inhabitants, in the words of Genesis: "Be fruitful and multiply, and replenish the earth and subdue it: and have dominion over the fish of the sea and over the fowl of the air and over every living thing that moveth upon the earth." The idea of dominion could be interpreted as an invitation to use nature as a convenience.[25]

The diagnosis wasn't original—indeed it was a synthesis of the founding principles of ecological thought. But to read these words in America's most studiously centrist magazine was nothing short of remarkable. For this reason and others, the start of 1989 felt to many in the environmental movement like a momentous juncture, as if the thawing of the Cold War and the warming of the planet were together helping to birth a new consciousness, one in which cooperation would triumph over domination, and humility before nature's complexity would challenge technological hubris.

As governments came together to debate responses to climate change, strong voices from developing countries spoke up, insisting that the core of the problem was the high-consumption lifestyle that dominated in the West. In a speech in 1989, for instance, India's President R. Venkataraman argued that the global environmental crisis was the result of developed countries' "excessive consumption of all materials and through large-scale industrialization intended to support their styles of life."[26] If wealthy countries consumed less, then everyone would be safer.

But if that was the way 1989 began, it would end very differently. In the months that followed, popular uprisings would spread across the Soviet-controlled Eastern Bloc, from Poland to Hungary and finally to East Germany where, in November 1989, the Berlin Wall collapsed. Under the banner "the End of History," right-wing ideologues in Washington seized on this moment of global flux to crush all political competition, whether socialism, Keynesianism, or deep ecology. They waged a frontal attack on political experimentation, on the idea that there might be viable ways of organizing societies other than deregulated capitalism.

Within a decade, all that would be left standing would be their own extreme, pro-corporate ideology. Not only would the Western consumer lifestyle survive intact, it would grow significantly more lavish, with U.S. credit card debt per household increasing fourfold between 1980 and 2010.[27] Simultaneously, that voracious lifestyle would be exported to the middle and upper classes in every corner of the globe—including, despite earlier protestations, India, where it would wreak environmental damage on a scale difficult to fathom. The victories in the new era would be faster and bigger than almost anyone predicted; and the armies of losers would be left to pick through the ever-growing mountains of methane-spewing waste.

Trade and Climate: Two Solitudes

Throughout this period of rapid change, the climate and trade negotiations closely paralleled one another, each winning landmark agreements within a couple of years of each other. In 1992, governments met for the first United Nations Earth Summit in Rio, where they signed the United Nations Framework Convention on Climate Change (UNFCCC), the document that formed the basis for all future climate negotiations. That same year, the North American Free Trade Agreement was signed, going into effect two years later. Also in 1994, negotiations establishing the World Trade Organization concluded, and the new global trade body made its debut the next year. In 1997, the Kyoto Protocol was adopted, containing the first binding emission reduction targets. In 2001, China gained full membership in the WTO, the culmination of a trade and investment liberalization process that had begun decades earlier.

What is most remarkable about these parallel processes—trade on the one hand, climate on the other—is the extent to which they functioned as two solitudes. Indeed, each seemed to actively pretend that the other did not exist, ignoring the most glaring questions about how one would impact the other. Like, for example: How would the vastly increased distances that basic goods would now travel—by carbon-spewing container ships and jumbo jets, as well as diesel trucks—impact the carbon emissions that the climate negotiations were aiming to reduce? How would the aggressive protections for technology patents enshrined under the WTO impact the demands being made by developing nations in the climate negotiations for free transfers of green technologies to help them develop on a low-carbon path? And perhaps most critically, how would provisions that allowed private companies to sue national governments over laws that impinged on their profits dissuade governments from adopting tough antipollution regulations, for fear of getting sued?

These questions were not debated by government negotiators, nor was any attempt made to resolve their obvious contradictions. Not that there was ever any question about which side would win should any of the competing pledges to cut emissions and knock down commercial barriers ever come into direct conflict: the commitments made in the climate negotiations all effectively functioned on the honor system, with a weak and unthreatening mechanism to penalize countries that failed to keep their promises. The commitments made under trade agreements, however, were enforced by a dispute settlement system with real teeth, and failure to comply would land governments in trade court, often facing harsh penalties.

In fact, the hierarchy was so clear that the climate negotiators formally declared their subservience to the trading system from the start. When the U.N. climate agreement was signed at the Rio Earth Summit in 1992, it made clear that "measures taken to combat climate change, including unilateral ones, should not constitute . . . a disguised restriction on international trade." (Similar language appears in the Kyoto Protocol.) As Australian political scientist Robyn Eckersley puts it, this was "the pivotal moment that set the shape of the relationship between the climate and trade regimes" because, "Rather than push for the recalibration of the international trade rules to conform with the requirements of climate protection . . . the Parties to the climate regime have ensured that liberalized trade and an expanding global economy have been protected against trade-restrictive climate policies." This practically guaranteed that the negotiating process would be unable to reckon with the kinds of bold but "trade-restrictive" policy options that could have been coordinated internationally—from buy-local renewable energy programs to restrictions on trade in goods produced with particularly high carbon footprints.[28]

A few isolated voices were well aware that the modest gains being made in the negotiations over "sustainable development" were being actively unmade by the new trade and investment architecture. One of those voices belonged to Martin Khor, then director of the Third World Network, which has been a key advisor to developing country governments in both trade and climate talks. At the end of the 1992 Rio Earth Summit, Khor cautioned that there was a "general feeling among Southern country delegates . . . that events outside the [summit] process were threatening to weaken the South further and to endanger whatever positive elements exist in" the Rio agenda. The examples he cited were the austerity policies being pushed at the time by the World Bank and the International Monetary Fund, as well as the trade negotiations that would soon result in the creation of the WTO.[29]

Another early warning was sounded by Steven Shrybman, who observed a decade and a half ago that the global export of industrial agriculture had already dealt a devastating blow to any possible progress on emissions. In a paper published in 2000, Shrybman argued that "the globalization of agricultural systems over recent decades is likely to have been one of the most important causes of overall increases in greenhouse gas emissions."[30]

This had far less to do with current debates about the "food miles" associated with imported versus local produce than with the way in which the trade system, by granting companies like Monsanto and Cargill their regulatory wish list—from unfettered market access to aggressive patent protection to the maintenance of their rich subsidies—has helped to entrench and expand the energy-intensive, higher-emissions model of industrial agriculture around the world. This, in turn, is a major explanation for why the global food system now accounts for between 19 and 29 percent of world greenhouse gas emissions. "Trade policy and rules actually drive climate change in a very structural way in respect of food systems," Shrybman stressed in an interview.[31]

The habit of willfully erasing the climate crisis from trade agreements continues to this day: for instance, in early 2014, several negotiating documents for the proposed Trans-Pacific Partnership, a controversial new NAFTA-style trade deal spanning twelve countries, were released to the public via WikiLeaks and the Peruvian human rights group RedGE. A draft of the environment chapter had contained language stating that countries "acknowledge climate change as a global concern that requires collective action and recognize the importance of implementation of their respective commitments under the United Nations Framework Convention on Climate Change (UNFCCC)." The language was vague and nonbinding, but at least it was a tool that governments could use to defend themselves should their climate policies be challenged in a trade tribunal, as Ontario's plan was. But a later document showed that U.S. negotiators had proposed an edit: take out all the stuff about climate change and UNFCCC commitments. In other words, while trade has repeatedly been allowed to trump climate, under no circumstances would climate be permitted to trump trade.[32]

Nor was it only the trade negotiators who blocked out the climate crisis as they negotiated agreements that would send emissions soaring and make many solutions to this problem illegal. The climate negotiations exhibited their own special form of denial. In the early and mid-1990s, while the first climate protocol was being drafted, these negotiators, along with the Intergovernmental Panel on Climate Change, hashed out the details of precisely how countries should measure and monitor how much carbon they were emitting—a necessary process since governments were on the verge of pledging their first round of emission reductions, which would need to be reported and monitored.

The emissions accounting system on which they settled was an odd relic of the pre–free trade era that took absolutely no account of the revolutionary changes unfolding right under their noses regarding how (and where) the world's goods were being manufactured. For instance, emissions from the transportation of goods across borders—all those container ships, whose traffic has increased by nearly 400 percent over the last twenty years—are not formally attributed to any nation-state and therefore no one country is responsible for reducing their polluting impact. (And there remains little momentum at the U.N. for changing that, despite the reality that shipping emissions are set to double or even triple by 2050.)[33]

And fatefully, countries are responsible only for the pollution they create inside their own borders—not for the pollution produced in the manufacturing of goods that are shipped to their shores; those are attributed to the countries where the goods were produced.[34] This means that the emissions that went into producing, say, the television in my living room, appear nowhere on Canada's emissions ledger, but rather are attributed entirely to China's ledger, because that is where the set was made. And the international emissions from the container ship that carried my TV across the ocean (and then sailed back again) aren't entered into anyone's account book.

This deeply flawed system has created a vastly distorted picture of the drivers of global emissions. It has allowed rapidly de-industrializing wealthy states to claim that their emissions have stabilized or even gone down when, in fact, the emissions embedded in their consumption have soared during the free trade era. For instance, in 2011, the *Proceedings of the National Academy of Sciences* published a study of the emissions from industrialized countries that signed the Kyoto Protocol. It found that while their emissions had stopped growing, that was partly because international trade had allowed these countries to move their dirty production overseas. The researchers concluded that the rise in emissions from goods produced in developing countries but consumed in industrialized ones was *six times* greater than the emissions savings of industrialized countries.[35]

Cheap Labor, Dirty Energy: A Package Deal

As the free trade system was put in place and producing offshore became the rule, emissions did more than move—they multiplied. As mentioned earlier, before the neoliberal era, emissions growth had been slowing, from 4.5 percent annual increases in the 1960s to about 1 percent a year in the 1990s. But the new millennium was a watershed: between 2000 and 2008, the growth rate reached 3.4 percent a year, shooting past the highest IPCC projections of the day. In 2009, it dipped due to the financial crisis, but made up for lost time with the historic 5.9 percent increase in 2010 that left climate watchers reeling. (In mid-2014, two decades after the creation of the WTO, the IPCC finally acknowledged the reality of globalization and noted in its Fifth Assessment Report, "A growing share of total anthropogenic CO_2 emissions is released in the manufacture of products that are traded across international borders.")[36]

The reason for what Andreas Malm—a Swedish expert on the history of coal—describes as "the early 21st Century emissions explosion" is straightforward enough. When China became the "workshop of the world" it also became the coal-spewing "chimney of the world." By 2007, China was responsible for two thirds of the annual increase in global emissions. Some of that was the result of China's own internal development—bringing electricity to rural areas, and building roads. But a lot of it was directly tied to foreign trade: according to one study, between 2002 and 2008, 48 percent of China's total emissions was related to producing goods for export.[37]

"One of the reasons why we're in the climate crisis is because of this model of globalization," says Margrete Strand Rangnes, executive vice president at Public Citizen, a Washington-based policy institute that has been at the forefront of the fight against free trade. And that, she says, is a problem that requires "a pretty fundamental re-formation of our economy, if we're going to do this right."[38]

International trade deals were only one of the reasons that governments embraced this particular model of fast-and-dirty, export-led development, and every country had its own peculiarities. In many cases (though not China's), the conditions attached to loans from the International Monetary Fund and World Bank were a major factor, so was the economic orthodoxy imparted to elite students at schools like Harvard and the University of Chicago. All of these and other factors played a role in shaping what was (never ironically) referred to as the Washington Consensus. Underneath it all is the constant drive for endless economic growth, a drive that, as will be explored later on, goes much deeper than the trade history of the past few decades. But there is no question that the trade architecture and the economic ideology embedded within it played a central role in sending emissions into hyperdrive.

That's because one of the primary driving forces of the particular trade system designed in the 1980s and 1990s was always to allow multinationals the freedom to scour the globe in search of the cheapest and most exploitable labor force. It was a journey that passed through Mexico and Central America's sweatshop maquiladoras and had a long stopover in South Korea. But by the end of the 1990s, virtually all roads led to China, a country where wages were extraordinarily low, trade unions were brutally suppressed, and the state was willing to spend seemingly limitless funds on massive infrastructure projects—modern ports, sprawling highway systems, endless numbers of coal-fired power plants, massive dams—all to ensure that the lights stayed on in the factories and the goods made it from the assembly lines onto the container ships on time. A free trader's dream, in other words—and a climate nightmare.

A nightmare because there is a close correlation between low wages and high emissions, or as Malm puts it, "a causal link between the quest for cheap and disciplined labor power and rising CO_2 emissions." And why wouldn't there be? The same logic that is willing to work laborers to the bone for pennies a day will burn mountains of dirty coal while spending next to nothing on pollution controls because it's the cheapest way to produce. So when the factories moved to China, they also got markedly dirtier. As Malm points out, Chinese coal use was declining slightly between 1995 and 2000, only for the explosion in manufacturing to send it soaring once again. It's not that the companies moving their production to China wanted to drive up emissions: they were after the cheap labor, but exploited workers and an exploited planet are, it turns out, a package deal. A destabilized climate is the cost of deregulated, global capitalism, its unintended, yet unavoidable consequence.[39]

This connection between pollution and labor exploitation has been true since the earliest days of the Industrial Revolution. But in the past, when workers organized to demand better wages, and when city dwellers organized to demand cleaner air, the companies were pretty much forced to improve both working and environmental standards. That changed with the advent of free trade: thanks to the removal of virtually all barriers to capital flows, corporations could pick up and leave every time labor costs started rising. That's why many large manufacturers left South Korea for China in the late 1990s, and it's why many are now leaving China, where wages are climbing, for Bangladesh, where they are significantly lower. So while our clothes, electronics, and furniture may be made in China, the economic model was primarily made in the U.S.A.

And yet when the subject of climate change comes up in discussion in wealthy, industrialized countries, the instant response, very often, is that it's all China's fault (and India's fault and Brazil's fault and so on). Why bother cutting our own emissions when everyone knows that the fast developing economies are the real problem, opening more coal plants every month than we could ever close?[40] This argument is made as if we in the West are mere spectators to this reckless and dirty model of economic growth. As if it was not our governments and our multinationals that pushed a model of export-led development that made all of this possible. It is said as if it were not our own corporations who, with single-minded determination (and with full participation from China's autocratic rulers), turned the Pearl River Delta into their carbon-spewing special economic zone, with the goods going straight onto container ships headed to our superstores. All in the name of feeding the god of economic growth (via the altar of hyperconsumption) in every country in the world.

The victims in all this are regular people: the workers who lose their factory jobs in Juárez and Windsor; the workers who get the factory jobs in Shenzhen and Dhaka, jobs that are by this point so degraded that some employers install nets along the perimeters of roofs to catch employees when they jump, or where safety codes are so lax that workers are killed in the hundreds when buildings collapse. The victims are also the toddlers mouthing lead-laden toys; the Walmart employee expected to work over the Thanksgiving holiday only to be trampled by a stampede of frenzied customers, while still not earning a living wage. And the Chinese villagers whose water is contaminated by one of those coal plants we use as our excuse for inaction, as well as the middle class of Beijing and Shanghai whose kids are forced to play inside because the air is so foul.[41]

A Movement Digs Its Own Grave

The greatest tragedy of all is that so much of this was eminently avoidable. We knew about the climate crisis when the rules of the new trade system were being written. After all, NAFTA was signed just one year after governments, including the United States, signed the United Nations Framework Convention on Climate Change in Rio. And it was by no means inevitable that these deals would go through. A strong coalition of North American labor and environmental groups opposed NAFTA precisely because they knew it would drive down labor and environmental standards. For a time it even looked as if they would win.

Public opinion in all three countries was deeply divided, so much so that when Bill Clinton ran for president in 1992, he pledged that he would not sign NAFTA until it substantively reflected those concerns. In Canada, Jean Chrétien campaigned for prime minister against the deal in the election of 1993. Once both were in office, however, the deal was left intact and two toothless side agreements were tacked on, one for labor and one for environmental standards. The labor movement knew better than to fall for this ploy and continued to forcefully oppose the deal, as did many Democrats in the U.S. But for a complex set of reasons that will be explored later, having to do with a combination of reflexive political centrism and the growing influence of corporate "partners" and donors, the leadership of many large environmental organizations decided to play ball. "One by one, former NAFTA opponents and skeptics became enthusiastic supporters, and said so publicly," writes journalist Mark Dowie in his critical history of the U.S. environmental movement, *Losing Ground*. These Big Green groups even created their own pro-NAFTA organization, the Environmental Coalition for NAFTA—which included the National Wildlife Federation, the Environmental Defense Fund, Conservation International, the National Audubon Society, the Natural Resources Defense Council, and the World Wildlife Fund—which, according to Dowie provided its "unequivocal support to the agreement." Jay Hair, then head of the National Wildlife Federation, even flew to Mexico on an official U.S. trade mission to lobby his Mexican counterparts, while attacking his critics for "putting their protectionist polemics ahead of concern for the environment."[42]

Not everyone in the green movement hopped on the pro-trade bandwagon: Greenpeace, Friends of the Earth, and the Sierra Club, as well as many small organizations, continued to oppose NAFTA. But that didn't matter to the Clinton administration, which had what it wanted—the ability to tell a skeptical public that "groups representing 80 percent of national [environmental] group membership have endorsed NAFTA." And that was important, because Clinton faced an uphill battle getting NAFTA through Congress, with many in his own party pledging to vote against the deal. John Adams, then director of the Natural Resources Defense Council, succinctly described the extraordinarily helpful role played by groups like his: "We broke the back of the environmental opposition to NAFTA. After we established our position Clinton only had labor to fight. We did him a big favor."[43]

Indeed when the president signed NAFTA into law in 1993, he made a special point of thanking "the environmental people who came out and worked through this—many of them at great criticism, particularly in the environmental movement." Clinton also made it clear that this victory was about more than one agreement. "Today we have the chance to do what our parents did before us. We have the opportunity to remake the world." He explained that, "We are on the verge of a global economic expansion. . . . Already the confidence we've displayed by ratifying NAFTA has begun to bear fruit. We are now making real progress toward a worldwide trade agreement so significant that it could make the material gains of NAFTA for our country look small by comparison." He was referring to the World Trade Organization. And just in case anyone was still worried about the environmental consequences, Clinton offered his personal assurance. "We will seek new institutional arrangements to ensure that trade leaves the world cleaner than before."[44]

Standing by the president's side was his vice president, Al Gore, who had been largely responsible for getting so many Big Green groups on board. Given this history, it should hardly come as a surprise that the mainstream environmental movement has been in no rush to draw attention to the disastrous climate impacts of the free trade era. To do so would only highlight their own active role in helping the U.S. government to, in Clinton's words, "remake the world." Much better, as we will see later on, to talk about light bulbs and fuel efficiency.

The significance of the NAFTA signing was indeed historic, tragically so. Because if the environmental movement had not been so agreeable, NAFTA might have been blocked or renegotiated to set a different kind of precedent. A new trade architecture could have been built that did not actively sabotage the fragile global climate change consensus. Instead—as had been the promise and hope of the 1992 Rio Earth Summit—this new architecture could have been grounded in the need to fight poverty and reduce emissions at the same time. So for example, trade access to developing countries could have been tied to transfers of resources and green technology so that critical new electricity and transit infrastructure was low carbon from the outset. And the deals could have been written to ensure that any measures taken to support renewable energy would not be penalized and, in fact, could be rewarded. The global economy might not have grown as quickly as it did, but it also would not be headed rapidly off the climate cliff.

The errors of this period cannot be undone, but it is not too late for a new kind of climate movement to take up the fight against so-called free trade and build this needed architecture now. That doesn't—and never did—mean an end to economic exchange across borders. It does, however, mean a far more thoughtful and deliberate approach to why we trade and whom it serves. Encouraging the frenetic and indiscriminate consumption of essentially disposable products can no longer be the system's goal. Goods must once again be made to last, and the use of energy-intensive long-haul transport will need to be rationed—reserved for those cases where goods cannot be produced locally or where local production is more carbon-intensive. (For example, growing food in greenhouses in cold parts of the United States is often more energy intensive than growing it in warmer regions and shipping it by light rail.)[45]

According to Ilana Solomon, trade analyst for the Sierra Club, this is not a fight that the climate movement can avoid. "In order to combat climate change, there's a real need to start localizing our economies again, and thinking about how and what we're purchasing and how it's produced. And the most basic rule of trade law is you can't privilege domestic over foreign. So how do you tackle the idea of needing to incentivize local economies, tying together local green jobs policies with clean energy policies, when that is just a no-go in trade policy? . . . If we don't think about how the economy is structured, then we're actually never going to the real root of the problem."[46]

These kinds of economic reforms would be good news—for unemployed workers, for farmers unable to compete with cheap imports, for communities that have seen their manufacturers move offshore and their local businesses replaced with big box stores. And all of these constituencies would be needed to fight for these policies, since they represent the reversal of the thirty-year trend of removing every possible limit on corporate power.

From Frenetic Expansion to Steady States

Challenging free trade orthodoxy is a heavy lift in our political culture; anything that has been in place for that long takes on an air of inevitability. But, critical as these shifts are, they are not enough to lower emissions in time. To do that, we will need to confront a logic even more entrenched than free trade—the logic of indiscriminate economic growth. This idea has understandably inspired a good deal of resistance among more liberal climate watchers, who insist that the task is merely to paint our current growth-based economic model green, so it's worth examining the numbers behind the claim.

It is Kevin Anderson of the Tyndall Centre for Climate Change Research, and one of Britain's top climate experts, who has most forcefully built the case that our growth-based economic logic is now in fundamental conflict with atmospheric limits. Addressing everyone from the U.K. Department for International Development to the Manchester City Council, Anderson has spent more than a decade patiently translating the implications of the latest climate science to politicians, economists, and campaigners. In clear and understandable language, the spiky-haired former mechanical engineer (who used to work in the petrochemical sector) lays out a rigorous road map for cutting our emissions down to a level that provides a decent shot at keeping global temperature rise below 2 degrees Celsius.

But in recent years Anderson's papers and slide shows have become more alarming. Under titles such as "Climate Change: Going Beyond Dangerous . . . Brutal Numbers and Tenuous Hope," he points out that the chances of staying within anything like safe temperature levels are diminishing fast. With his colleague Alice Bows-Larkin, an atmospheric physicist and climate change mitigation expert at the Tyndall Centre, Anderson argues that we have lost so much time to political stalling and weak climate policies—all while emissions ballooned—that we are now facing cuts so drastic that they challenge the core expansionist logic at the heart of our economic system.[47]

They argue that, if the governments of developed countries want a fifty-fifty chance of hitting the agreed-upon international target of keeping warming below 2 degrees Celsius, and if reductions are to respect any kind of equity principle between rich and poor nations, then wealthy countries need to start cutting their greenhouse gas emissions by something like 8 to 10 percent a year—and they need to start right now. The idea that such deep cuts are required used to be controversial in the mainstream climate community, where the deadlines for steep reductions always seemed to be far off in the future (an 80 percent cut by 2050, for instance). But as emissions have soared and as tipping points loom, that is changing rapidly. Even Yvo de Boer, who held the U.N.'s top climate position until 2009, remarked recently that "the only way" negotiators "can achieve a 2-degree goal is to shut down the whole global economy."[48]

That is a severe overstatement, yet it underlines Anderson and Bows-Larkin's point that we cannot achieve 8 to 10 percent annual cuts with the array of modest carbon-pricing or green tech solutions usually advocated by Big Green. These measures will certainly help, but they are simply not enough. That's because an 8 to 10 percent drop in emissions, year after year, is virtually unprecedented since we started powering our economies with coal. In fact, cuts above 1 percent per year "have historically been associated only with economic recession or upheaval," as the economist Nicholas Stern put it in his 2006 report for the British government.[49]

Even after the Soviet Union collapsed, reductions of this duration and depth did not happen (the former Soviet countries experienced average annual reductions of roughly 5 percent over a period of ten years). Nor did this level of reduction happen beyond a single-year blip after Wall Street crashed in 2008. Only in the immediate aftermath of the great market crash of 1929 did the United States see emissions drop for several consecutive years by more than 10 percent annually, but that was the worst economic crisis of modern times.[50]

If we are to avoid that kind of carnage while meeting our science-based emissions targets, carbon reduction must be managed carefully through what Anderson and Bows-Larkin describe as "radical and immediate degrowth strategies in the US, EU and other wealthy nations."*[51]

Now, I realize that this can all sound apocalyptic—as if reducing emissions requires economic crises that result in mass suffering. But that seems so only because we have an economic system that fetishizes GDP growth above all else, regardless of the human or ecological consequences, while failing to place value on those things that most of us cherish above all—a decent standard of living, a measure of future security, and our relationships with one another. So what Anderson and Bows-Larkin are really saying is that there is still time to avoid catastrophic warming, but not within the rules of capitalism as they are currently constructed. Which is surely the best argument there has ever been for changing those rules.[52]

Rather than pretending that we can solve the climate crisis without rocking the economic boat, Anderson and Bows-Larkin argue, the time has come to tell the truth, to "liberate the science from the economics, finance and

---

* And they don't let developing countries like China and India off the hook. According to their projections, developing countries can have just one more decade to continue to increase their emissions to aid their efforts to pull themselves out of poverty while switching over to green energy sources. By 2025, they would need to be cutting emissions "at an unprecedented 7 per cent" a year as well.

astrology, stand by the conclusions however uncomfortable . . . we need to have the audacity to think differently and conceive of alternative futures."[53]

Interestingly, Anderson says that when he presents his radical findings in climate circles, the core facts are rarely disputed. What he hears most often are confessions from colleagues that they have simply given up hope of meeting the 2 degree temperature target, precisely because reaching it would require such a profound challenge to economic growth. "This position is shared by many senior scientists and economists advising government," Anderson reports.[54]

In other words, changing the earth's climate in ways that will be chaotic and disastrous is easier to accept than the prospect of changing the fundamental, growth-based, profit-seeking logic of capitalism. We probably shouldn't be surprised that some climate scientists are a little spooked by the radical implications of their own research. Most of them were quietly measuring ice cores, running global climate models, and studying ocean acidification, only to discover, as Australian climate expert and author Clive Hamilton puts it, that in breaking the news of the depth of our collective climate failure, they "were unwittingly destabilizing the political and social order."[55]

Nonetheless, that order has now been destabilized, which means that the rest of us are going to have to quickly figure out how to turn "managed degrowth" into something that looks a lot less like the Great Depression and a lot more like what some innovative economic thinkers have taken to calling "The Great Transition."[56]

———

Over the past decade, many boosters of green capitalism have tried to gloss over the clashes between market logic and ecological limits by touting the wonders of green tech, or the "decoupling" of environmental impacts from economic activity. They paint a picture of a world that can continue to function pretty much as it does now, but in which our power will come from renewable energy and all of our various gadgets and vehicles will become so much more energy-efficient that we can consume away without worrying about the impact.

If only humanity's relationship with natural resources was that simple.

While it is true that renewable technologies hold tremendous promise to lower emissions, the kinds of measures that would do so on the scale we need involve building vast new electricity grids and transportation systems, often from the ground up. Even if we started construction tomorrow, it would realistically take many years, perhaps decades, before the new systems were up and running. Moreover, since we don't yet have economies powered by clean energy, all that green construction would have to burn a lot of fossil fuels in the interim—a necessary process, but one that wouldn't lower our emissions fast enough. Deep emission cuts in the wealthy nations have to start immediately. That means that if we wait for what Bows-Larkin describes as the "whiz-bang technologies" to come online "it will be too little too late."[57]

So what to do in the meantime? Well, we do what we can. And what we can do—what doesn't require a technological and infrastructure revolution—is to consume less, right away. Policies based on encouraging people to consume less are far more difficult for our current political class to embrace than policies that are about encouraging people to consume green. Consuming green just means substituting one power source for another, or one model of consumer goods for a more efficient one. The reason we have placed all of our eggs in the green tech and green efficiency basket is precisely because these changes are safely within market logic—indeed, they encourage us to go out and buy more new, efficient, green cars and washing machines.

Consuming less, however, means changing how much energy we actually use: how often we drive, how often we fly, whether our food has to be flown to get to us, whether the goods we buy are built to last or to be replaced in two years, how large our homes are. And these are the sorts of policies that have been neglected so far. For instance, as researchers Rebecca Willis and Nick Eyre argue in a report for the U.K.'s Green Alliance, despite the fact that groceries represent roughly 12 percent of greenhouse gas emissions in Britain, "there is virtually no government policy which is aimed at changing the way we produce, incentivising farmers for low energy farming, or how we consume, incentivising consumption of local and seasonal food." Similarly, "there are incentives to drive more efficient cars, but very little is done to discourage car dependent settlement patterns."[58]

Plenty of people are attempting to change their daily lives in ways that do reduce their consumption. But if these sorts of demand-side emission reductions are to take place on anything like the scale required, they cannot be left to the lifestyle decisions of earnest urbanites who like going to farmers' markets on Saturday afternoons and wearing up-cycled clothing. We will need comprehensive policies and programs that make low-carbon choices easy and convenient for everyone. Most of all, these policies need to be fair, so that the people already struggling to cover the basics are not being asked to make additional sacrifice to offset the excess consumption of the rich. That means cheap public transit and clean light rail accessible to all; affordable, energy-efficient housing along those transit lines; cities planned for high-density living; bike lanes in which riders aren't asked to risk their lives to get to work; land management that discourages sprawl and encourages local, low-energy forms of agriculture; urban design that clusters essential services like schools and health care along transit routes and in pedestrian-friendly areas; programs that require manufacturers to be responsible for the electronic waste they produce, and to radically reduce built-in redundancies and obsolescences.*[59]

And as hundreds of millions gain access to modern energy for the first time, those who are consuming far more energy than they need would have to consume less. How much less? Climate change deniers like to claim that environmentalists want to return us to the Stone Age. The truth is that if we want to live within ecological limits, we would need to return to a lifestyle similar to the one we had in the 1970s, before consumption levels went crazy in the 1980s. Not exactly the various forms of hardship and deprivation evoked at Heartland conferences. As Kevin Anderson explains: "We need to give newly industrializing countries in the world the space to develop and improve the welfare and well-being of their people. This means more cuts in energy use by the developed world. It also means lifestyle changes which will have most impact on the wealthy. . . . We've done this in the past. In the 1960s and 1970s we enjoyed a healthy and moderate lifestyle and we need to return to this to keep emissions under control. It is a matter of the well-off 20 percent in a population taking the largest cuts. A

---

* A law passed by the European Parliament that would require that all cell phone manufacturers offer a common battery charger is a small step in the right direction. Similarly, requiring that electronics manufacturers use recycled metals like copper could save a great many communities from one of the most toxic mining processes in the world.

more even society might result and we would certainly benefit from a lower carbon and more sustainable way of life."[60]

There is no doubt that these types of policies have countless benefits besides lower emissions. They encourage civic space, physical activity, community building, as well as cleaner air and water. They also do a huge amount to reduce inequality, since it is low-income people, often people of color, who benefit most from improvements in public housing and public transit. And if strong living-wage and hire-local provisions were included in transition plans, they could also benefit most from the jobs building and running those expanded services, while becoming less dependent on jobs in dirty industries that have been disproportionately concentrated in low-income communities of color.

As Phaedra Ellis-Lamkins of the environmental justice organization Green for All puts it, "The tools we use to combat climate change are the same tools we can use to change the game for low-income Americans and people of color. . . . We need Congress to make the investments necessary to upgrade and repair our crumbling infrastructure—from building seawalls that protect shoreline communities to fixing our storm-water systems. Doing so will create family-sustaining, local jobs. Improving our storm-water infrastructure alone would put 2 million Americans to work. We need to make sure that people of color are a part of the business community and workforce building these new systems."[61]

Another way of thinking about this is that what is needed is a fundamental reordering of the component parts of Gross Domestic Product. GDP is traditionally understood to consist of *consumption* plus *investment* plus *government spending* plus *net exports*. The free market capitalism of the past three decades has put the emphasis particularly on consumption and trade. But as we remake our economies to stay within our global carbon budget, we need to see less consumption (except among the poor), less trade (as we relocalize our economies), and less private investment in producing for excessive consumption. These reductions would be offset by increased government spending, and increased public and private investment in the infrastructure and alternatives needed to reduce our emissions to zero. Implicit in all of this is a great deal more redistribution, so that more of us can live comfortably within the planet's capacity.

Which is precisely why, when climate change deniers claim that global warming is a plot to redistribute wealth, it's not (only) because they are paranoid. It's also because they are paying attention.

Growing the Caring Economy, Shrinking the Careless One

A great deal of thought in recent years has gone into how reducing our use of material resources could be managed in ways that actually improve quality of life overall—what the French call "selective degrowth."* Policies like luxury taxes could be put in place to discourage wasteful consumption.[62] The money raised could be used to support those parts of our economies that are already low-carbon and therefore do not need to contract. Obviously a huge number of jobs would be created in the sectors that are part of the green transition—in mass transit, renewable energy, weatherization, and ecosystem restoration. And those sectors that are not governed by the drive for increased yearly profit (the public sector, co-ops, local businesses, nonprofits) would expand their share of overall economic activity, as would those sectors with minimal ecological impact (such as the caregiving professions, which tend to be occupied by women and people of color and therefore underpaid). "Expanding our economies in these directions has all sorts of advantages," Tim Jackson, an economist at the University of Surrey and author of *Prosperity Without Growth,* has written. "In the first place, the time spent by these professions directly improves the quality of our lives. Making them more and more efficient is not, after a certain point, actually desirable. What sense does it make to ask our teachers to teach ever bigger classes? Our doctors to treat more and more patients per hour?"[63]

There could be other benefits too, like shorter work hours, in part to create more jobs, but also because overworked people have less time to engage in low-consumption activities like gardening and cooking (because they are just too busy). Indeed, a number of researchers have analyzed the

---

* In French, "decroissance" has the double meaning of challenging both growth, *croissance,* and *croire,* to believe—invoking the idea of choosing not to believe in the fiction of perpetual growth on a finite planet.

very concrete climate benefits of working less. John Stutz, a senior fellow at the Boston-based Tellus Institute, envisions that "hours of paid work and income could converge worldwide at substantially lower levels than is seen in the developed countries today." If countries aimed for somewhere around three to four days a week, introduced gradually over a period of decades, he argues, it could offset much of the emissions growth projected through 2030 while improving quality of life.[64]

Many degrowth and economic justice thinkers also call for the introduction of a basic annual income, a wage given to every person, regardless of income, as a recognition that the system cannot provide jobs for everyone and that it is counterproductive to force people to work in jobs that simply fuel consumption. As Alyssa Battistoni, an editor at the journal *Jacobin*, writes, "While making people work shitty jobs to 'earn' a living has always been spiteful, it's now starting to seem suicidal."[65]

A basic income that discourages shitty work (and wasteful consumption) would also have the benefit of providing much-needed economic security in the front-line communities that are being asked to sacrifice their health so that oil companies can refine tar sands oil or gas companies can drill another fracking well. Nobody wants to have their water contaminated or have their kids suffer from asthma. But desperate people can be counted on to do desperate things—which is why we all have a vested interest in taking care of one another so that many fewer communities are faced with those impossible choices. That means rescuing the idea of a safety net that ensures that everyone has the basics covered: health care, education, food, and clean water. Indeed, fighting inequality on every front and through multiple means must be understood as a central strategy in the battle against climate change.

This kind of carefully planned economy holds out the possibility of much more humane, fulfilling lifestyles than the vast majority of us are experiencing under our current system, which is what makes the idea of a massive social movement coalescing behind such demands a real possibility. But these policies are also the most politically challenging.

Unlike encouraging energy efficiency, the measures we must take to secure a just, equitable, and inspiring transition away from fossil fuels clash directly with our reigning economic orthodoxy at every level. As we will see, such a shift breaks all the ideological rules—it requires visionary long-term planning, tough regulation of business, higher levels of taxation for the affluent, big public sector expenditure, and in many cases reversals of core privatizations in order to give communities the power to make the changes they desire. In short, it means changing everything about how we think about the economy so that our pollution doesn't change everything about our physical world.

PUBLIC AND PAID FOR

Overcoming the Ideological Blocks to the Next Economy

"We have no option but to reinvent mobility . . . much of India still takes the bus, walks or cycles—in many cities as much as 20 percent of the population bikes. We do this because we are poor. Now the challenge is to reinvent city planning so that we can do this as we become rich."

—Sunita Narain, director general, Centre for Science and Environment, 2013 [1]

"The lady in the Rolls-Royce car is more damaging to morale than a fleet of Göring's bombing-planes."

—George Orwell, *The Lion and the Unicorn*, 1941 [2]

It was a tight vote but on September 22, 2013, residents of Germany's second largest city decided to take their power back. On that day, 50.9 percent of Hamburg's voters cast their ballots in favor of putting their electricity, gas, and heating grids under the control of the city, reversing a wave of corporate sell-offs that took place over a decade earlier. [3]

It's a process that has been given a few clunky names, including "re-municipalization" and "re-communalization." But the people involved tend to simply refer to their desire for "local power."

The Our Hamburg–Our Grid coalition made a series of persuasive arguments in favor of taking back the utilities. A locally controlled energy system would be concerned with public interests, not profits. Residents would have greater democratic say in their energy system, they argued, rather than having the decisions that affect them made in distant boardrooms. And money earned in the sale of energy would be returned to the city, rather than lost to the shareholders of multinationals that had control over the grids at the time—a definite plus during a time of relentless public austerity. "For people it's self-evident that goods on which everybody is dependent should belong to the public," campaign organizer Wiebke Hansen explained in an interview.[4]

There was something else driving the campaign as well. Many of Hamburg's residents wanted to be part of *Energiewende*: the fast-spreading transition to green, renewable energy that was sweeping the country, with nearly 25 percent of Germany's electricity in 2013 coming from renewables, dominated by wind and solar but also including some biogas and hydro—up from around 6 percent in 2000. In comparison, wind and solar made up just 4 percent of total U.S. electricity generation in 2013. The cities of Frankfurt and Munich, which had never sold off their energy grids, had already joined the transition and pledged to move to 100 percent renewable energy by 2050 and 2025, respectively. But Hamburg and Berlin, which had both gone the privatization route, were lagging behind. And this was a central argument for proponents of taking back Hamburg's grid: it would allow them to get off coal and nuclear and go green.[5]

Much has been written about Germany's renewable energy transition—particularly the speed at which it is being achieved, as well as the ambition of its future targets (the country is aiming for 55–60 percent renewables by 2035).[6] The weaknesses of the program have also been hotly debated, particularly the question of whether the decision to phase out nuclear energy has led to a resurgence of coal (more on that next chapter).

In all of this analysis, however, scarce attention has been paid to one key factor that has made possible what may be the world's most rapid shift to wind and solar power: the fact that in hundreds of cities and towns across the country, citizens have voted to take their energy grids back from the private corporations that purchased them. As Anna Leidreiter, a climate campaigner with the World Future Council, observed after the Hamburg vote, "This marks a clear reversal to the neoliberal policies of the 1990s, when large numbers of German municipalities sold their public services to large corporations as money was needed to prop up city budgets."[7]

Nor is this some small trend. According to a Bloomberg report, "More than 70 new municipal utilities have started up since 2007, and public operators have taken over more than 200 concessions to run energy grids from private companies in that time." And though there are no national statistics, the German Association of Local Utilities believes many more cities and towns than that have taken back control over their grids from outside corporations.[8]

Most surprising has been the force with which large parts of the German public have turned against energy privatization. In 2013 in Berlin 83 *percent* of participating voters cast their ballots in favor of switching to a publicly owned power utility based eventually on 100 percent renewable energy. Not enough people turned out to vote for the decision to be binding (though the campaign came very close), but the referendum made public opinion so clear that campaigners are still pushing for a nonprofit cooperative to take over the grid when the current contract ends.[9]

Energy privatization reversals—linked specifically to a desire for renewable energy—have started to spread beyond Germany in recent years, including to the United States. For instance, in the mid-2000s, residents and local officials in the liberal city of Boulder, Colorado, began lobbying their privatized power utility to move away from coal and toward renewable energy. The company, the Minneapolis-based Xcel Energy, wasn't particularly interested, so a coalition of environmentalists and an energetic youth group called New Era Colorado came to the same conclusion as the voters in Germany: they had to take their grid back. Steve Fenberg of New Era explains, "We have one of the most carbon-intensive energy supplies in the country, and [Boulder] is an environmentally minded community, and we wanted to change that. We realized that we had no control over that unless we controlled the energy supply."[10]

In 2011, despite being outspent by Xcel by ten to one, the pro-renewables coalition narrowly won two ballot measures that called on the city of Boulder to consider buying back its power system.[11] The vote did not immediately put the power utility under public control, but it gave the city the authority and financing to seriously consider the option (which it is currently doing). The coalition won another crucial vote in 2013 against an Xcel-supported initiative that would have blocked the formation of a new public utility, this time by a wide majority.

These were historic votes: other cities had reversed earlier privatizations because they were unhappy with the quality of the service or the pricing under the private operator. But this was the first time a U.S. city was taking these steps "for the sole purpose of reducing its impact on the planet," according to Tim Hillman, a Boulder-based environmental engineer. Indeed the pro–public forces had put fighting climate change front and center in their campaigns, accusing Xcel of being just another fossil fuel company standing in the way of much needed climate action. And according to Fenberg, their vision reaches beyond Boulder. "We want to show the world that you can actually power a city responsibly and not pay a lot for it," he now says. "We want this to be a model, not just do this one cool thing for ourselves in our community."[12]

What stands out about Boulder's experience is that, unlike some of the German campaigns, it did not begin with opposition to privatization. Boulder's local power movement began with the desire to switch to clean energy, regardless of who was providing it. Yet in the process of trying to achieve that goal, these residents discovered that they had no choice but to knock down one of the core ideological pillars of the free market era: that privately run services are always superior to public ones. It was an accidental discovery very similar to the one Ontario residents made when it became clear that their green energy transition was being undermined by free trade commitments signed long ago.

Though rarely mentioned in climate policy discussions, there is a clear and compelling relationship between public ownership and the ability of communities to get off dirty energy. Many of the countries with the highest commitments to renewable energy are ones that have managed to keep large parts of their electricity sectors in public (and often local) hands, including the Netherlands, Austria, and Norway. In the U.S., some of the cities that have set the most ambitious green energy targets also happen to have public utilities. Austin, Texas, for instance, is ahead of schedule for meeting its target of 35 percent renewable power by 2020, and Sacramento, California's, utility is gearing up to beat a similar target and has set a pioneering goal of reducing emissions by 90 percent by mid-century. On the other hand, according to John Farrell, senior researcher at the Minneapolis-based Institute for Local Self-Reliance, the attitude of most private players has been, "we're going to take the money that we make from selling fossil fuels, and use it to lobby as hard as we can against any change to the way that we do business."[13]

This does not mean that private power monopolies will not offer their customers the option of purchasing power from renewables as part of a mix that includes fossil fuels: many do offer that choice, usually at a premium price. And some offer renewable power exclusively, though this is invariably from large-scale hydropower. Nor is it the case that public power will always willingly go green—there are plenty of publicly owned power utilities that remain hooked on coal and are highly resistant to change.

However, many communities are discovering that while public utilities often need to be pressured hard to make emission reductions a priority (a process that may require fundamental reform to make them more democratic and accountable to their constituents) private energy monopolies offer no such option. Answerable chiefly to their shareholders and driven by the need for high quarterly profits, private companies will voluntarily embrace renewables only if it won't impact their earnings or if they are forced to by law. If renewables are seen as less profitable, at least in the short term, these bottom-line companies simply won't make the switch. Which is why, as German antinuclear activist Ralf Gauger puts it, more and more people are coming to the conclusion that, "Energy supply and environmental issues should not be left in the hands of private for-profit interests."[14]

This does not mean that the private sector should be excluded from a transition to renewables: solar and wind companies are already bringing clean energy to many millions of consumers around the world, including through innovative leasing models that allow customers to avoid the upfront costs of purchasing their own rooftop solar panels. But despite these recent successes, the market has proved extremely volatile and according to projections from the International Energy Agency, investment levels in clean energy need to quadruple by 2030 if we are to meet emission targets aimed at staying below 2 degrees Celsius of warming.[15]

It's easy to mistake a thriving private market in green energy for a credible climate action plan, but, though related, they are not the same thing. It's entirely possible to have a booming market in renewables, with a whole new generation of solar and wind entrepreneurs growing very wealthy—and for our countries to still fall far short of lowering emissions in line with science in the brief time we have left. To be sure of hitting those tough targets, we need systems that are more reliable than boom-and-bust private markets. And as a 2013 paper produced by a research team at the University of Greenwich explains, "Historically, the private sector has played little role in investing in renewable energy generation. Governments have been responsible for nearly all such investments. Current experience from around the world, including the markets of Europe, also shows that private companies and electricity markets cannot deliver investments in renewables on the scale required."[16]

Citing various instances of governments turning to the public sector to drive their transitions (including the German experience), as well as examples of large corporate-driven renewable projects that were abandoned by their investors midstream, the Greenwich research team concludes, "An active role for government and public sector utilities is thus a far more important condition for developing renewable energy than any expensive system of public subsidies for markets or private investors."[17]

Sorting out what mechanisms have the best chance of pulling off a dramatic and enormously high-stakes energy transition has become particularly pressing of late. That's because it is now clear that—at least from a technical perspective—it is entirely possible to rapidly switch our energy systems to 100 percent renewables. In 2009, Mark Z. Jacobson, a professor of civil and environmental engineering at Stanford University, and Mark A. Delucchi, a research scientist at the Institute of Transportation Studies at the University of California, Davis, authored a groundbreaking, detailed road map for "how 100 percent of the world's energy, for *all* purposes, could be supplied by wind, water and solar resources, by as early as 2030." The plan includes not only power generation but also transportation as well as heating and cooling. Later published in the journal *Energy Policy*, the road map is one of several credible studies that have come out in recent years that show how wealthy countries and regions can shift all, or almost all, of their energy infrastructure to renewables within a twenty-to-forty-year time frame.[18] Those studies demonstrating the potential for rapid progress include:

- In Australia, the University of Melbourne's Energy Institute and the nonprofit Beyond Zero Emissions have published a blueprint for achieving a 60 percent solar and 40 percent wind electricity system in an astonishing ten years.[19]
- By 2014, the U.S. National Oceanic and Atmospheric Administration (NOAA) had concluded from its own extensive research into weather patterns that cost-effective wind and solar could constitute nearly 60 percent of the U.S. electricity system by 2030.[20]
- Among more conservative projections, a major 2012 study by the U.S. Department of Energy's National Renewable Energy Laboratory argues that wind, solar, and other currently available green technologies could meet 80 percent of Americans' electricity needs by 2050.[21]

Most promising of all is new work by a team of researchers at Stanford, led by Mark Jacobson (who coauthored the 2009 global plan). In March 2013, they published a study in *Energy Policy* showing that New York state could meet all of its power needs with renewables by 2030. Jacobson and his colleagues are developing similar plans for every U.S. state, and have already published numbers for the country as a whole. "It's absolutely not true that we need natural gas, coal or oil—we think it's a myth," he told *The New York Times.*[22]

"This really involves a large scale transformation," he says. "It would require an effort comparable to the Apollo moon project or constructing the interstate highway system. But it is possible, without even having to go to new technologies. We really need to just decide collectively that this is the direction we want to head as a society." And he is clear on what stands in the way: "The biggest obstacles are social and political—what you need is the will to do it."[23]

In fact it takes more than will: it requires the profound ideological shift already discussed. Because our governments have changed dramatically since the days when ambitious national projects were conceived and implemented. And the imperatives created by the climate crisis are colliding with the dominant logic of our time on many other fronts.

Indeed every time a new, record-breaking natural disaster fills our screens with human horror, we have more reminders of how climate change demands that we invest in the publicly owned bones of our societies, made brittle by decades of neglect.

Rebuilding, and Reinventing, the Public Sphere

When I first spotted Nastaran Mohit, she was bundled in a long puffy black coat, a white toque pulled halfway over her eyes, barking orders to volunteers gathered in an unheated warehouse. "Take a sticky pad and write down what the needs are," the fast-talking thirty-year-old was telling a group newly designated as Team 1. "Okay, head on out. Who is Team 2?"[24]

It was ten days after Superstorm Sandy made landfall and we were in one of the hardest-hit neighborhoods in the Rockaways, a long, narrow strip of seaside communities in Queens, New York. The storm waters had receded but hundreds of basements were still flooded and power and cell phone service were still out. The National Guard patrolled the streets in trucks and Humvees, making sure curfew was observed, but when it came to offering help to those stranded in the cold and dark, the state and the big aid agencies were largely missing in action. (Or, more accurately, they were at the other, wealthier end of the Rockaway peninsula, where these organizations and agencies were a strong and helpful presence.)[25]

Seeing this abandonment, thousands of mostly young volunteers had organized themselves under the banner "Occupy Sandy" (many were veterans of Occupy Wall Street) and were distributing clothes, blankets, and hot food to residents of neglected areas. They set up recovery hubs in community centers and churches, and went door-to-door in the area's notorious, towering brick housing projects, some as high as twenty-three stories. "Muck" had become a ubiquitous verb, as in "Do you need us to come muck out your basement?" If the answer was yes, a team of eager twenty-somethings would show up on the doorstep with mops, gloves, shovels, and bleach, ready to get the job done.

Mohit had arrived in the Rockaways to help distribute basic supplies but quickly noticed a more pressing need: in some areas, absolutely no one was providing health care. And the need was so great, it scared her. Since the 1950s, the Rockaways—once a desirable resort destination—had become a dumping ground for New York's poor and unwanted: welfare recipients, the elderly, discharged mental patients. They were crammed into high-rises, many in a part of the peninsula known locally as the "Baghdad of Queens." [26]

As in so many places like it, public services in the Rockaways had been cut to the bone, and then cut some more. Just six months before the storm, Peninsula Hospital Center—one of only two hospitals in the area, which served a low-income and elderly population—had shut down after the state Department of Health refused to step in. Walk-in clinics had attempted to fill the gap but they had flooded during the storm and, along with the pharmacies, had not yet reopened. "This is just a dead-zone," Mohit sighed. [27]

So she and friends in Occupy Sandy called all the doctors and nurses they knew and asked them to bring in whatever supplies they could. Next, they convinced the owner of an old furrier, damaged in the storm, to let them convert his storefront on the neighborhood's main drag into a makeshift MASH unit. There, amidst the animal pelts hanging from the ceiling, volunteer doctors and nurses began to see patients, treat wounds, write prescriptions, and provide trauma counseling.

There was no shortage of patients; in its first two weeks, Mohit estimated that the clinic helped hundreds of people. But on the day I visited, worries were mounting about the people still stuck in the high-rises. As volunteers went door-to-door distributing supplies in the darkened projects, flashlights strapped to their foreheads, they were finding alarming numbers of sick people. Cancer and HIV/AIDS meds had run out, oxygen tanks were empty, diabetics were out of insulin, and addicts were in withdrawal. Some people were too sick to brave the dark stairwells and multiple flights of stairs to get help; some didn't leave because they had nowhere to go and no way to get off the peninsula (subways and buses were not operating); others feared that if they left their apartments, their homes would be burglarized. And without cell service or power for their TVs, many had no idea what was going on outside.

Most shockingly, residents reported that until Occupy Sandy showed up, no one had knocked on their doors since the storm. Not from the Health

Department, nor the city Housing Authority (responsible for running the projects), nor the big relief agencies like the Red Cross. "I was like 'Holy crap,'" Mohit told me. "There was just no medical attention at all."[*][28] Referring to the legendary abandonment of New Orleans's poor residents when the city flooded in 2005, she said: "This is Katrina 2.0."[29]

The most frustrating part was that even when a pressing health need was identified, and even when the volunteer doctors wrote the required prescriptions, "we bring it to the pharmacy and the pharmacy is sending it back to us because they need insurance information. And then we get as much information as we can and we bring it back and they say, 'Now we need their Social Security number.'"[30]

According to a 2009 Harvard Medical School study, as many as 45,000 people die annually in the United States because they lack health insurance. As one of the study's coauthors pointed out, this works out to about one death every twelve minutes. It's unclear how President Obama's stunted 2010 health care law will change those numbers, but watching the insurance companies continue to put money before human health in the midst of the worst storm in New York's history cast this preexisting injustice in a new, more urgent light. "We need universal health care," Mohit declared. "There is no other way around it. There is absolutely no other way around it." Anyone who disagreed should come to the disaster zone, she said, because this "is a perfect situation for people to really examine how nonsensical, inhumane, and barbaric this system is."[31]

The word "apocalypse" derives from the Greek *apokalypsis*, which means "something uncovered" or revealed. Besides the need for a dramatically better health care system, there was much else uncovered and revealed when the floodwaters retreated in New York that October. The disaster revealed how dangerous it is to be dependent on centralized forms of energy that can be knocked out in one blow. It revealed the life-and-death cost of social

---

* This was the situation not only in the Rockaways but seemingly wherever public housing was in the path of the storm. In Red Hook, Brooklyn, many residents were left without power for three weeks, during which time the Housing Authority never went systematically door-to-door. As sixty-year-old Wally Bazemore put it at an angry residents meeting: "We were literally in the dark and we were completely in the dark."

isolation, since it was the people who did not know their neighbors, or who were frightened of them, who were most at risk. Meanwhile, it was the tightest-knit communities, where neighbors took responsibility for one another's safety, that were best able to literally weather the storm.

The disaster also revealed the huge risks that come with deep inequality, since the people who were already the most vulnerable—undocumented workers, the formerly incarcerated, people in public housing—suffered most and longest. In low-income neighborhoods, homes filled not only with water but with heavy chemicals and detergents—the legacy of systemic environmental racism that allowed toxic industries to build in areas inhabited mostly by people of color. Public housing projects that had been left to decay—while the city bided its time before selling them off to developers—turned into death traps, their ancient plumbing and electrical systems giving way completely. As Aria Doe, executive director of the Action Center for Education and Community Development in the Rockaways, put it, the peninsula's poorest residents "were six feet under" before the storm even hit. "Right now, they're seven or eight feet under."[32]

All around the world, the hard realities of a warming world are crashing up against the brutal logic of austerity, revealing just how untenable it is to starve the public sphere at the very moment we need it most. The floods that hit the U.K. in the winter of 2013–2014, for instance, would have been trying for any government: thousands of homes and workplaces were inundated, hundreds of thousands of houses and other buildings lost power, farmland was submerged, several rail lines were down for weeks, all combining to create what one top official called an "almost unparalleled natural disaster." This as the country was still reeling from a previous devastating storm that had struck just two months before.[33]

But the floods were particularly awkward for the coalition government led by Conservative prime minister David Cameron because, in the three years prior, it had gutted the Environment Agency (EA), which was responsible for dealing with flooding. Since 2009, at least 1,150 jobs had been lost at the agency, with as many as 1,700 more on the chopping block, adding up to approximately a quarter of its total workforce. In 2012 *The Guardian* had revealed that "nearly 300 flood defence schemes across England [had] been left unbuilt due to government budget cuts." The head of the Environment Agency had stated plainly during the most recent round of cuts that "Flood risk maintenance will be impacted."[34]

Cameron is no climate change denier, which is what made it all the more incredible that he had hobbled the agency responsible for protecting the public from rising waters and more ferocious storms, two well-understood impacts of climate change. And his praise of the good works of the staff that had survived his axe provided cold comfort. "It is a disgrace that the Government is happy to put cost cutting before public safety and protecting family homes," announced the trade union representing EA workers in a scathing statement. "They can't have it both ways, praising the sterling work of members in the Agency in one breath, and in the next breath announcing further damaging cuts."[35]

During good times, it's easy to deride "big government" and talk about the inevitability of cutbacks. But during disasters, most everyone loses their free market religion and wants to know that their government has their backs. And if there is one thing we can be sure of, it's that extreme weather events like Superstorm Sandy, Typhoon Haiyan in the Philippines, and the British floods—disasters that, combined, pummeled coastlines beyond recognition, ravaged millions of homes, and killed many thousands—are going to keep coming.

Over the course of the 1970s, there were 660 reported disasters around the world, including droughts, floods, extreme temperature events, wildfires, and storms. In the 2000s, there were 3,322—a fivefold boost. That is a staggering increase in just over thirty years, and clearly global warming cannot be said to have "caused" all of it. But the climate signal is also clear. "There's no question that climate change has increased the frequency of certain types of extreme weather events," climate scientist Michael Mann told me in an interview, "including drought, intense hurricanes, and super typhoons, the frequency and intensity and duration of heat waves, and potentially other types of extreme weather though the details are still being debated within the scientific community."[36]

Yet these are the same three decades in which almost every government in the world has been steadily chipping away at the health and resilience of the public sphere. And it is this neglect that, over and over again, turns natural disasters into unnatural catastrophes. Storms burst through neglected levees. Heavy rain causes decrepit sewer systems to back up and overflow. Wildfires rage out of control for lack of workers and equipment to fight them (in Greece, fire departments can't afford spare tires for their trucks driving into forest blazes). Emergency responders are missing in action for days after a major hurricane. Bridges and tunnels, left in a state of disrepair, collapse under the added pressure.

The costs of coping with increasing weather extremes are astronomical. In the United States, each major disaster seems to cost taxpayers upward of a billion dollars. The cost of Superstorm Sandy is estimated at $65 billion. And that was just one year after Hurricane Irene caused around $10 billion in damage, just one episode in a year that saw fourteen billion-dollar disasters in the U.S. alone. Globally, 2011 holds the title as the costliest year ever for disasters, with total damages reaching at least $380 billion. And with policymakers still locked in the vise grip of austerity logic, these rising emergency expenditures are being offset with cuts to everyday public spending, which will make societies even more vulnerable during the next disaster—a classic vicious cycle.[37]

It was never a good idea to neglect the foundations of our societies in this way. In the context of climate change, however, that decision looks suicidal. There are many important debates to be had about the best way to respond to climate change—storm walls or ecosystem restoration? Decentralized renewables, industrial scale wind power combined with natural gas, or nuclear power? Small-scale organic farms or industrial food systems? There is, however, *no* scenario in which we can avoid wartime levels of spending in the public sector—not if we are serious about preventing catastrophic levels of warming, and minimizing the destructive potential of the coming storms.

It's no mystery where that public money needs to be spent. Much of it should go to the kinds of ambitious emission-reducing projects already discussed—the smart grids, the light rail, the citywide composting systems, the building retrofits, the visionary transit systems, the urban redesigns to keep us from spending half our lives in traffic jams. The private sector is ill suited to taking on most of these large infrastructure investments: if the services are to be accessible, which they must be in order to be effective, the profit margins that attract private players simply aren't there.

Transit is a good example. In March 2014, when air pollution in French cities reached dangerously high levels, officials in Paris made a snap decision to discourage car use by making public transit free for three days. Obviously private operators would strenuously resist such measures. And yet by all rights, our transit systems should be responding with the same kind of urgency to dangerously high levels of atmospheric carbon. Rather than allowing subway and bus fares to rise while service erodes, we need to be lowering prices and expanding services—regardless of the costs.

Public dollars also need to go to the equally important, though less glamorous projects and services that will help us prepare for the coming heavy weather. That includes things like hiring more firefighters and improving storm barriers. And it means coming up with new, nonprofit disaster insurance programs so that people who have lost everything to a hurricane or a forest fire are not left at the mercy of a private insurance industry that is already adapting to climate change by avoiding payouts and slapping victims with massive rate increases. According to Amy Bach, cofounder of the San Francisco–based advocacy group United Policyholders, disaster insurance is becoming "very much like health insurance. We're going to have to increasingly take the profit motive out of the system so that it operates efficiently and effectively, but without generating obscene executive salaries and bonuses and shareholder returns. Because it's not going to be a sustainable model. A publicly traded insurance company in the face of climate change is not a sustainable business model for the end user, the consumer."[38] It's that or a disaster capitalism free-for-all; those are the choices.

These types of improvements are of course in far greater demand in developing countries like the Philippines, Kenya, and Bangladesh that are already facing some of the most severe climate impacts. Hundreds of billions of dollars are urgently needed to build seawalls; storage and distribution networks for food, water, and medicine; early warning systems and shelters for hurricanes, cyclones, and tsunamis—as well as public health systems able to cope with increases in climate-related diseases like malaria.[39]

Though mechanisms to protect against government corruption are needed, these countries should not have to spend their health care and education budgets on costly disaster insurance plans purchased from transnational corporations, as is happening right now. Their people should be receiving direct compensation from the countries (and companies) most responsible for warming the planet.

The Polluter Pays

About now a sensible reader would be asking: how on earth are we going to pay for all this? It's the essential question. A 2011 survey by the U.N. Department of Economic and Social Affairs looked at how much it would cost for humanity to "overcome poverty, increase food production to eradicate hunger without degrading land and water resources, and avert the climate change catastrophe." The price tag was $1.9 trillion a year for the next forty years—and "at least one half of the required investments would have to be realized in developing countries."[40]

As we all know, public spending is going in the opposite direction almost everywhere except for a handful of fast-growing so-called emerging economies. In North America and Europe, the economic crisis that began in 2008 is still being used as a pretext to slash aid abroad and cut climate programs at home. All over Southern Europe, environmental policies and regulations have been clawed back, most tragically in Spain, which, facing fierce austerity pressure, drastically cut subsidies for renewables projects, sending solar projects and wind farms spiraling toward default and closure. The U.K. under David Cameron has also cut supports for renewable energy.

So if we accept that governments are broke, and they're not likely to introduce "quantitative easing" (aka printing money) for the climate system as they have for the banks, where is the money supposed to come from? Since we have only a few short years to dramatically lower our emissions, the only rational way forward is to fully embrace the principle already well established in Western law: the polluter pays.

The fossil fuel companies have known for decades that their core product was warming the planet, and yet they have not only failed to adapt to that reality, they have actively blocked progress at every turn. Meanwhile, oil and gas companies remain some of the most profitable corporations in history, with the top five oil companies pulling in $900 billion in profits from 2001 to 2010. ExxonMobil still holds the record for the highest corporate profits ever reported in the United States, earning $41 billion in 2011 and $45 billion in 2012. These companies are rich, quite simply, because they have dumped the cost of cleaning up their mess onto regular people around the world. It is this situation that, most fundamentally, needs to change.[41]

And it will not change without strong action. For well over a decade, several of the oil majors have claimed to be voluntarily using their profits to invest in a shift to renewable energy. In 2000, BP rebranded itself "Beyond Petroleum" and even changed its logo to a sunburst, called "the Helios mark after the sun god of ancient Greece." ("We are not an oil company," then–chief executive Sir John Browne said at the time, explaining that, "We are aware the world wants less carbon-intensive fuels. What we want to do is create options.") Chevron, for its part, ran a high-profile advertising campaign declaring, "It's time oil companies get behind renewables. . . . We agree." But according to a study by the Center for American Progress, just 4 percent of the Big Five's $100 billion in combined profits in 2008 went to "renewable and alternative energy ventures." Instead, they continue to pour their profits into shareholder pockets, outrageous executive pay (Exxon CEO Rex Tillerson makes more than $100,000 a day), and new technologies designed to extract even dirtier and more dangerous fossil fuels.[42]

And even as the demand for renewables increases, the percentage the fossil fuel companies spend on them keeps shrinking—by 2011, most of the majors were spending less than 1 percent of their overall expenditures on alternative energy, with Chevron and Shell spending a deeply unimpressive 2.5 percent. In 2014, Chevron pulled back even further. According to *Bloomberg Businessweek*, the staff of a renewables division that had almost doubled its target profits was told "that funding for the effort would dry up" and was urged "to find jobs elsewhere." Chevron also moved to sell off businesses that had developed green projects for governments and school districts. As oil industry watcher Antonia Juhasz has observed, "You wouldn't know it from their advertising, but the world's major oil companies have either entirely divested from alternative energy or significantly reduced their investments in favor of doubling down on ever-more risky and destructive sources of oil and natural gas."[43]

Given this track record, it's safe to assume that if fossil fuel companies are going to help pay for the shift to renewable energy, and for the broader costs of a climate destabilized by their pollution, it will be because they are forced to do so by law. Just as tobacco companies have been obliged to pay the costs of helping people to quit smoking, and BP has had to pay for much of the cleanup of its oil spill in the Gulf of Mexico, it is high time for the industry to at least split the bill for the climate crisis. And there is mounting evidence that the financial world understands that this is coming. In its 2013 annual report on "Global Risks," the World Economic Forum (host of the annual superelite gathering in Davos), stated plainly, "Although the Alaskan village of Kivalina—which faces being 'wiped out' by the changing climate—was unsuccessful in its attempts to file a US$ 400 million lawsuit against oil and coal companies, future plaintiffs may be more successful. Five decades ago, the U.S. tobacco industry would not have suspected that in 1997 it would agree to pay $368 billion in health-related damages." But it did.[44]

The question is: how do we stop fossil fuel profits from continuing to hemorrhage into executive paychecks and shareholder pockets—and how do we do it soon, before the companies are significantly less profitable or out of business because we have moved to a new energy system? As the Global Risks report suggests, communities severely impacted by climate change have made several attempts to use the courts to sue for damages, but so far they have been unsuccessful. A steep carbon tax would be a straightforward way to get a piece of the profits, as long as it contained a generous redistributive mechanism—a tax cut or income credit—that compensated poor and middle-class consumers for increased fuel and heating prices. As Canadian economist Marc Lee points out, designed properly, "It is possible to have a progressive carbon tax system that reduces inequality as it raises the price of emitting greenhouse gases."[45] An even more direct route to getting a piece of those pollution profits would be for governments to negotiate much higher royalty rates on oil, gas, and coal extraction, with the revenues going to "heritage trust funds" that would be dedicated to building the post–fossil fuel future, as well as to helping communities and workers adapt to these new realities.

Fossil fuel corporations can be counted on to resist any new rules that cut into their profits, so harsh penalties, including revoking corporate charters, would need to be on the table. Companies would threaten to pull out of certain operations, to be sure, but once a multinational like Shell has spent billions to build the mines and drilling platforms needed to extract fossil fuels, it is unlikely to abandon that infrastructure because royalties go up. (Though it will bitterly complain and may well seek damages at an investment tribunal.)

But the extractive industries shouldn't be the only targets of the "polluter pays" principle. The U.S. military is by some accounts the largest single consumer of petroleum in the world. In 2011, the Department of Defense released, at minimum, 56.6 million metric tons of CO_2 equivalent into the atmosphere, more than the U.S.-based operations of ExxonMobil and Shell combined.[46] So surely the arms companies should pay their share. The car companies have plenty to answer for too, as do the shipping industry and the airlines.

Moreover, there is a simple, direct correlation between wealth and emissions—more money generally means more flying, driving, boating, and powering of multiple homes. One case study of German consumers indicates that the travel habits of the most affluent class have an impact on climate 250 percent greater than that of their lowest-earning neighbors.[47]

That means any attempt to tax the extraordinary concentration of wealth at the very top of the economic pyramid, as documented so persuasively by Thomas Piketty among many others, would—if partially channeled into climate financing—effectively make the polluters pay. As journalist and climate and energy policy expert Gar Lipow puts it, "We should tax the rich more because it is the fair thing to do, and because it will provide a better life for most of us, and a more prosperous economy. However, providing money to save civilization and reduce the risk of human extinction is another good reason to bill the rich for their fair share of taxes." But it must be said that a "polluter pays" principle would have to reach beyond the super rich. According to Stephen Pacala, director of the Princeton Environmental Institute and codirector of Princeton's Carbon Mitigation Initiative, the roughly 500 million richest of us on the planet are responsible for about half of all global emissions. That would include the rich in every country in the world, notably in countries like China and India, as well significant parts of the middle classes in North America and Europe.*[48]

Taken together, there is no shortage of options for equitably coming up with the cash to prepare for the coming storms while radically lowering our emissions to prevent catastrophic warming.

Consider the following list, by no means complete:

- A "low-rate" financial transaction tax—which would hit trades of stocks, derivatives, and other financial instruments—could bring in nearly $650 billion at the global level each year, according to a 2011 resolution of the European Parliament (and it would have the added bonus of slowing down financial speculation).[49]
- Closing tax havens would yield another windfall. The U.K.-based Tax Justice Network estimates that in 2010, the private financial wealth of individuals stowed unreported in tax havens around the globe was somewhere between $21 trillion and $32 trillion. If that money were brought into the light and its earnings taxed at a 30 percent rate, it would yield at least $190 billion in income tax revenue each year.[50]
- A 1 percent "billionaire's tax," floated by the U.N., could raise $46 billion annually.[51]
- Slashing the military budgets of each of the top ten military spenders by 25 percent could free up another $325 billion, using 2012 numbers reported by the Stockholm International Peace Research Institute. (Granted, probably the toughest sell of all, particularly in the U.S.)[52]
- A $50 tax per metric ton of CO_2 emitted in developed countries would raise an estimated $450 billion annually, while a more modest

---

* This is why the persistent positing of population control as a solution to climate change is a distraction and moral dead end. As this research makes clear, the most significant cause of rising emissions is not the reproductive behavior of the poor but the consumer behaviors of the rich.

$25 carbon tax would still yield $250 billion per year, according to a 2011 report by the World Bank, the International Monetary Fund, and the Organisation for Economic Co-operation and Development (OECD), among others.[53]

- Phasing out fossil fuel subsidies globally would conservatively save governments a total $775 billion in a single year, according to a 2012 estimate by Oil Change International and the Natural Resources Defense Council.[54]

If these various measures were taken together, they would raise more than $2 trillion annually.[55] Certainly enough for a very healthy start to finance a Great Transition (and avoid a Great Depression). And that doesn't count any royalty increases on fossil fuel extraction. Of course, for any of these tax crackdowns to work, key governments would have to coordinate their responses so that corporations had nowhere to hide—a difficult task, though far from impossible, and one frequently bandied about at G20 summits.

In addition to the simple fact that the money is badly needed, there are practical political reasons why "polluter pays" should guide climate financing. As we have seen, responding to the climate crisis can offer real benefits to a majority of people, but real solutions will also, by definition, require short- and medium-term sacrifices and inconveniences. And what we know from past sacrifices made in the name of a crisis—most notably via rationing, conservation, and price controls during both world wars—is that success depends entirely on a perception of fairness.

In Britain and North America during World War II, for instance, every strata of society was required to make do with less, even the very rich. And in fact, though overall consumption in the U.K. dropped by 16 percent, caloric intake for the poor increased during the war, because the rations provided low-income people with more than they could otherwise afford.[56]

There was plenty of cheating and black market profiteering, of course, but these programs enjoyed broad-based support because they were, at least in theory, fair. The theme of equality pervaded government campaigns about these wartime programs: "Fair Shares for All" was a key slogan in the U.K, while the U.S. went with "Share and Share Alike" and "Produce, Conserve, Share and Play Square."[57] An Office of Price Administration pamphlet from 1942 argued that rationing was part of the American tradition. "What Is Rationing?" it asked.

> First, let's be sure what rationing is not. It is not starvation, long bread lines, shoddy goods. Rather, it is a community plan for dividing fairly the supplies we have among all who need them. Second, it is not "un-American." The earliest settlers of this country, facing scarcities of food and clothing, pooled their precious supplies and apportioned them out to everyone on an equal basis. It was an American idea then, and it is an American idea now, to share and share alike—to sacrifice, when necessary, but sacrifice together, when the country's welfare demands it.[58]

Governments also made sure that there were very public crackdowns on wealthy and well-connected individuals who broke the rules, sending the message that no one was exempt. In the U.K., movie stars, as well as corporations like Woolworth and Sainsbury, faced prosecution for rations violations. In the United States, cases were brought against some of the largest corporations in the country. It was no secret that many large U.S. manufacturers disliked the entire rationing system; they lobbied against it, because they believed it eroded their brand value. Yet they were forced to accept it all the same.[59]

This perception of fairness—that one set of rules applies to players big and small—has been entirely missing from our collective responses to climate change thus far. For decades, regular people have been asked to turn off their lights, put on sweaters, and pay premium prices for nontoxic cleaning products and renewable energy—and then watched as the biggest polluters have been allowed to expand their emissions without penalty. This has been the pattern ever since President Jimmy Carter addressed the American public in July 1979 about the fact that "too many of us now tend to worship self-indulgence and consumption. Human identity is no longer defined by what one does, but by what one owns." He urged Americans "for your good and for your nation's security to take no unnecessary trips, to use carpools or public transportation whenever you can, to park your car one extra day per week, to obey the speed limit, and to set your thermostats to save fuel. Every act of energy conservation like this is more than just common sense—I tell you it is an act of patriotism."[60]

The address was initially well received but came to be derided as the "malaise" speech and is frequently cited as one of the reasons Carter lost his reelection bid to Ronald Reagan. And though he was not talking about climate change but rather a broad "crisis of confidence" against a backdrop of energy scarcity, the speech is still invoked as proof that any politician who asks voters to sacrifice to solve an environmental crisis is on a suicide mission. Indeed this assessment has shaped the win-win messaging of environmentalists ever since.

So it's interesting to note that the late intellectual Christopher Lasch, who was one of Carter's key advisors on the infamous speech, was also one of its most pointed critics. The author of *The Culture of Narcissism* had strongly urged the president to temper his message of personal austerity with assurances of fundamental fairness and social justice. As Lasch revealed to an interviewer years later, he had told Carter to "put a more populist construction in his indictment of American consumerism. . . . What was needed was a program that called for sacrifices all right, but made it clear that the sacrifices would be distributed in an equitable fashion." And that, Lasch said, "would mean that those most able to make sacrifices would be the ones on whom the sacrifices fell. That's what I mean by populism."[61]

We cannot know if the reaction might have differed had Carter listened to that advice and presented a plan for conservation that began with those pushing and profiting most from wasteful consumption. We do know that responses to climate change that continue to put the entire burden on individual consumers are doomed to fail. For instance, the annual "British Social Attitudes" survey, conducted by the independent NatCen Social Research, asked a set of questions about climate policies in the year 2000, and then again in 2010. It found that, "Whereas, 43 per cent a decade ago said they would be willing to pay higher prices to protect the environment, this is nowadays only true of 26 per cent. There has been a similar fall in the proportion prepared to pay higher taxes (31 to 22 per cent), but a smaller decline in relation to cuts in the standard of living (26 per cent to 20 per cent)."[62]

These results, and others like them, have been cited as proof that during times of economic hardship, people's environmental concerns go out the window. But that is not what these polls prove. Yes, there has been a drop in the willingness of individuals to bear the financial burden of responding to climate change, but not simply because economic times are hard. Western governments have responded to these hard times—which have been created by rampant greed and corruption among their wealthiest citizens—by asking those least responsible for the current conditions to bear the burden. After paying for the crisis of the bankers with cuts to education, health care, and social safety nets, is it any wonder that a beleaguered public is in no mood to bail out the fossil fuel companies from the crisis that they not only created but continue to actively worsen?

Most of these surveys, notably, don't ask respondents how they feel about raising taxes on the rich and removing fossil fuel subsidies, yet these are some of the most reliably popular policies around. And it's worth noting that a U.S. poll conducted in 2010—with the country still reeling from economic crisis—asked voters whether they would support a plan that "would make oil and coal companies pay for the pollution they cause. It would encourage the creation of new jobs and new technologies in cleaner energy like wind, solar, and nuclear power. The proposal also aims to protect working families, so it refunds almost all of the money it collects directly to the American people, like a tax refund, and most families end up better off." The poll found that three quarters of voters, including the vast majority of Republicans, supported the ideas as outlined, and only 11 percent strongly opposed it. The plan was similar to a proposal, known as "cap and dividend," being floated by a pair of senators at the time, but it was never seriously considered by the U.S. Senate.[63]

And when, in June 2014, Obama finally introduced plans to use the Environmental Protection Agency to limit greenhouse gas emissions from existing power plants, the coal lobby howled with indignation but public opinion was solidly supportive. According to one poll, 64 percent of Americans, including a great many Republicans, backed such a policy even though it would likely mean paying more for energy every month.[64]

The lesson from all this is not that people won't sacrifice in the face of the climate crisis. It's that they have had it with our culture of *lopsided* sacrifice, in which individuals are asked to pay higher prices for supposedly green choices while large corporations dodge regulation and not only refuse to change their behavior, but charge ahead with ever more polluting activities. Witnessing this, it is perfectly sensible for people to shed much of the keener enthusiasm that marked the early days of the climate movement, and to make it clear that no more sacrifice will be made until the policy solutions on the table are perceived as just. This does not mean the middle class is off the hook. To fund the kind of social programs that will make a just transition possible, taxes will have to rise for everyone but the poor. But if the funds raised go toward social programs and services that reduce inequality and make lives far less insecure and precarious, then public attitudes toward taxation would very likely shift as well.

———

To state the obvious: it would be incredibly difficult to persuade governments in almost every country in the world to implement the kinds of redistributive climate mechanisms I have outlined. But we should be clear about the nature of the challenge: it is not that "we" are broke or that we lack options. It is that our political class is utterly unwilling to go where the money is (unless it's for a campaign contribution), and the corporate class is dead set against paying its fair share.

Seen in this light, it's hardly surprising that our leaders have so far failed to act to avert climate chaos. Indeed even if aggressive "polluter pays" measures were introduced, it isn't at all clear that the current political class would know what to do with the money. After all, changing the building blocks of our societies—the energy that powers our economies, how we move around, the designs of our major cities—is not about writing a few checks. It requires bold long-term planning at every level of government, and a willingness to stand up to polluters whose actions put us all in danger. And that won't happen until the corporate liberation project that has shaped our political culture for three and a half decades is buried for good.

Just as the climate change deniers I met at the Heartland Institute fear, there is a direct relationship between breaking fossilized free market rules and making swift progress on climate change. Which is why, if we are to collectively meet the enormous challenges of this crisis, a robust social movement will need to demand (and create) political leadership that is not only committed to making polluters pay for a climate-ready public sphere, but willing to revive two lost arts: long-term public planning, and saying no to powerful corporations.

4

PLANNING AND BANNING

Slapping the Invisible Hand, Building a Movement

"Post-modernism has cut off the present from all futures. The daily media adds to this by cutting off the past. Which means that critical opinion is often orphaned in the present."

—John Berger, *Keeping a Rendezvous*, 1991[1]

"A reliably green company is one that is required to be green by law."

—Gus Speth, former dean of the Yale School of Forestry and Environmental Studies, 2008[2]

To understand how free market ideology continues to suffocate the potential for climate action, it's useful to look back on the most recent moment when transformative change of the scope required actually seemed like a real possibility, even in the United States. That time was 2009, the peak of the world financial crisis and the first year of the Obama presidency.

Hindsight is easy, granted, but bear with me: imagining what might have been can help clarify what the future might still create.

This was a moment when history was unfolding in fast-forward, when almost anything seemed possible, for better and worse. A large part of what made better scenarios seem possible was the decisive democratic mandate that Obama had just earned. He had been elected on a platform promising to rebuild the "Main Street" economy and to treat climate change as, in his words, "an opportunity, because if we create a new energy economy, we can create five million new jobs. . . . It can be an engine that drives us into the future the same way the computer was the engine for economic growth over the last couple of decades."[3] Both the fossil fuel companies and the environmental movement took it as a given that the new president would introduce a bold piece of climate legislation early in his presidency.

The financial crisis, meanwhile, had just shattered public faith in laissez-faire economics around the world—so much so that there was tremendous support even in the U.S. for breaking long-standing ideological taboos against intervening directly in the market to create good jobs. That gave Obama the leverage to design a stimulus program worth about $800 billion (and he probably could have asked for more) to get the economy moving again.

The other extraordinary factor in this moment was the weak state of the banks: in 2009, they were still on their knees, dependent on trillions in bailout funds and loan guarantees. And there was a live debate unfolding about how those banks should be restructured in exchange for all that taxpayer generosity (there was even serious discussion of nationalization). The other factor worth remembering is that starting in 2008, two of the Big Three automakers—companies at the very heart of the fossil fuel economy—had so badly mismanaged their affairs that they too had landed in the hands of the government, which had been tasked with securing their viability.

All told, three huge economic engines—the banks, the auto companies, and the stimulus bill—were in a state of play, placing more economic power in the hands of Obama and his party than any U.S. government since the administration of Franklin Delano Roosevelt. Imagine, for a moment, if his administration had been willing to invoke its newly minted democratic mandate to build the new economy promised on the campaign trail—to treat the stimulus bill, the broken banks, and the shattered car companies as the building blocks of that green future. Imagine if there had been a powerful social movement—a robust coalition of trade unions, immigrants, students, environmentalists, and everyone else whose dreams were getting crushed by the crashing economic model—demanding that Obama do no less.

The stimulus package could have been used to build the best public transit systems and smart grids in the world. The auto industry could have been dramatically reengineered so that its factories built the machinery to power that transition—not just a few token electric cars (though those too) but also vast streetcar and high-speed rail systems across an underserved nation. Just as a shuttered auto parts factory in Ontario had reopened as the Silfab solar plant, similar transitions could have been made in closed and closing factories across the continent. This transformation was proposed at the time by one of the most important intellectuals of the North American labor movement, Sam Gindin, who served for many years as research director for the Canadian Auto Workers Union:

> If we are serious about incorporating environmental needs into the economy, this means changing everything about how we produce and consume and how we travel and live. The potential work to be done in this regard—in the tool and die shops that are closing, the component plants that have the capacity to make more than a specific component, and by a workforce anxious to do useful work—is limitless.
>
> The equipment and skills can be used to not only build different cars, and different car components, but to expand public transit and develop new transportation systems. They can participate in altering, in line with environmental demands, the machinery in every workplace and the motors that run the machinery. They can be applied to new systems of production that recycle used materials and final products (such as cars). Homes will have to be retrofitted and appliances modified. The use of solar panels and wind turbines will spread, new electricity grids will have to be developed, and urban infrastructure will have to be reinvented to accommodate the changes in transportation and energy use.
>
> What better time to launch such a project than now, in the face of having to overcome both the immediate economic crisis and the looming environmental crisis? And what greater opportunity to insist that we cannot lose valuable facilities and equipment, nor squander the creativity, knowledge and abilities of engineers, skilled trades and production workers?[4]

Retrofitting factories on that scale is expensive, to be sure, and that's where the bailed-out banks could have come in. A government unafraid to use its newfound power could have used the leverage it had over the banks (having just pulled them from the precipice) to enlist them—kicking and screaming if necessary—in this great transformation. As every banker knows, when you loan someone money, you acquire a fair bit of power over them. Does a factory need some capital to make the transition from dirty to clean? If it has a credible business plan, especially one that supports the stimulus vision, then the bailed-out banks could have been mandated by the state as part of the bailout to give that factory a loan. If one refused, it could have been nationalized, as several major banks were around the world in the period.

Many of the previous factory owners would not have been interested in sticking around for this kind of transition, since the profit margins, at least at first, would have been small. But that is no reason to allow useful machines to be sold off as scrap. The workers at these plants, as Gindin suggested, could have been given the chance to run their old factories as cooperatives, as happened in several hundred abandoned factories in Argentina after that country's economic crisis in 2001. I lived in Buenos Aires for two years while making a documentary film about those factories, called *The Take*. One of the stories we told was about a group of workers who took over their shuttered auto-parts plant and turned it into a thriving co-op. It was a highly emotional journey, as workers took big risks and discovered new skills they had not known they possessed. And over a decade later, we still receive reports about how well things are going at the factory. Most of Argentina's "recovered factories"—as the hundreds of worker-run co-ops are called—are still in production, churning out everything from kitchen tiles to men's suits.* This decentralized ownership model has the added benefit of pushing against the trend toward utterly unsustainable wealth inequality; rather than simply propping up the current global system in which eighty-five people control as much wealth as half the world's population, the ability to create wealth is gradually dispersed to the workers themselves, and the communities sustained by the presence of well-paying jobs.[5]

---

* Workers in the U.S. and Europe have attempted to emulate this model in recent years during several plant closures, most notably the high-profile Republic Windows and Doors factory in Chicago, which was shut down during the economic crisis and then occupied by its workers. Today many of those former employees are now worker-owners at the reborn New Era Windows Cooperative.

If that kind of coherent and sweeping vision had emerged in the United States in that moment of flux as the Obama presidency began, right-wing attempts to paint climate action as an economy killer would have fallen flat. It would have been clear to all that climate action is, in fact, a massive job creator, as well as a community rebuilder, and a source of hope in moments when hope is a scarce commodity indeed. But all of this would have required a government that was unafraid of bold long-term economic planning, as well as social movements that were able to move masses of people to demand the realization of that kind of vision. (The mainstream climate organizations in the U.S., in this crucial period, were instead narrowly focused on a failed attempt to get a piece of carbon-trading energy legislation through Congress, not on helping to build a broad movement.)

In the absence of those factors, that rarest of historical moments—so pregnant with potential—slipped away. Obama let the failed banks do what they liked, despite the fact that their gross mismanagement had put the entire economy at risk. The fundamentals of the car industry were also left intact, with little more than a fresh wave of downsizing to show for the crisis. The industry lost nearly 115,000 manufacturing jobs between 2008 and 2014.[6]

To be fair, there was significant support for wind and solar and for green initiatives like energy efficient building upgrades in the stimulus bill; without question, as journalist Michael Grunwald shows in *The New New Deal*, the funding amounted to "the biggest and most transformative energy bill in U.S. history." But public transit was still inexplicably shortchanged and the biggest infrastructure winner was the national highway system, precisely the wrong direction from a climate perspective. This failure was not only Obama's; as University of Leeds ecological economist Julia Steinberger observes, it was global. The financial crisis that began in 2008 "should have been an opportunity to invest in low-carbon infrastructure for the 21st century. Instead, we fostered a lose-lose situation: carbon emissions rocketing to unprecedented levels, alongside increases in joblessness, energy costs, and income disparities."[7]

What stopped Obama from seizing his historical moment to stabilize the economy and the climate at the same time was not lack of resources, or a lack of power. He had plenty of both. What stopped him was the invisible confinement of a powerful ideology that had convinced him—as it has convinced virtually all of his political counterparts—that there is something wrong with telling large corporations how to run their businesses even when they are running them into the ground, and that there is something sinister, indeed vaguely communist, about having a plan to build the kind of economy we need, even in the face of an existential crisis.

This is, of course, yet another legacy bequeathed to us by the free market counterrevolution. As recently as the early 1970s, a Republican president—Richard Nixon—was willing to impose wage and price controls to rescue the U.S. economy from crisis, popularizing the notion that "We are all Keynesians now."[8] But by the 1980s, the battle of ideas waged out of the same Washington think tanks that now deny climate change had successfully managed to equate the very idea of industrial planning with Stalin's five-year plans. Real capitalists don't plan, these ideological warriors insisted—they unleash the power of the profit motive and let the market, in its infinite wisdom, create the best possible society for all.

Obama, obviously, does not share this extreme vision: as his health care and other social policies suggest, he believes government should nudge business in the right direction. And yet he is still sufficiently a product of his anti-planning era that when he had the banks, the auto companies, and the stimulus in his hands, he saw them as burdens to be rid of as soon as possible, rather than as a rare chance to build an exciting new future.

If there is a lesson in this tremendous missed opportunity, it is this: if we are going to see climate action of the scale and speed required, the left is going to have to quickly learn from the right. Conservatives have managed to stall and roll back climate action amidst economic crisis by making climate about economics—about the pressing need to protect growth and jobs during difficult times (and they are always difficult). Progressives can easily do the same: by showing that the real solutions to the climate crisis are also our best hope of building a much more stable and equitable economic system, one that strengthens and transforms the public sphere, generates plentiful, dignified work, and radically reins in corporate greed.

But before that can happen, it's clear that a core battle of ideas must be fought about the right of citizens to democratically determine what kind of economy they need. Policies that simply try to harness the power of the market—by minimally taxing or capping carbon and then getting out of the way—won't be enough. If we are to rise to a challenge that involves altering the very foundation of our economy, we will need every policy tool in the democratic arsenal.

Planning for Jobs

Some policymakers already understand this, which is why so many of the climate disputes being dragged in front of WTO tribunals hinge on attempts by governments, whether in Ontario or India, to reintroduce some measure of industrial planning to their economies. These governments are saying to industry: we will support you, but only if you support the communities from which you profit, by providing well-paying local jobs, and sourcing your products locally.

The reason governments turn to buy-local or hire-local policies such as these is because they make political sense. Any response to the climate crisis that has a chance of success will create not just winners but also a significant number of losers—industries that can no longer exist in their current form and workers whose jobs will disappear. There is little hope of bringing the fossil fuel companies onside to a green transition; the profits they stand to lose are simply too great. That is not the case, however, for the workers whose salaries are currently tied to fossil fuel extraction and combustion.

What we know is this: trade unions can be counted on to fiercely protect jobs, however dirty, if these are the only jobs on offer. On the other hand, when workers in dirty sectors are offered good jobs in clean sectors (like the former autoworkers at the Silfab factory in Toronto), and are enlisted as active participants in a green transition, then progress can happen at lightning speed.

The potential job creation is huge. For instance, a plan put forward by the U.S. BlueGreen Alliance, a body that brings together unions and environmentalists, estimated that a $40 billion annual investment in public transit and high-speed rail for six years would produce more than 3.7 million jobs during that period. And we know that investments in public transit pay off: a 2011 study by research and policy organization Smart Growth America found they create 31 percent more jobs per dollar than investment in new road and bridge construction. Investing in the maintenance and repair of roads and bridges creates 16 percent more jobs per dollar than investment in new road and bridge construction.[9] All of which means that making existing transportation infrastructure work better for more people is a smarter investment from both a climate and an economic perspective than covering more land with asphalt.

Renewable energy is equally promising, in part because it creates more jobs per unit of energy delivered than fossil fuels. In 2012, the International Labour Organization estimated that about five million jobs had already been created in the sector worldwide—and that is with only the most scattershot and inadequate levels of government commitment to emission reduction.[10] If industrial policy were brought in line with climate science, the supply of energy through wind, solar, and other forms of renewable energy (geothermal and tidal power, for example) would generate huge numbers of jobs in every country—in manufacturing, construction, installation, maintenance, and operation.

Similar research in Canada has found that an investment of $1.3 billion (the amount the Canadian government spends on subsidies to oil and gas companies) could create seventeen to twenty thousand jobs in renewable energy, public transit, or energy efficiency—six to eight times as many jobs as that money generates in the oil and gas sector. And according to a 2011 report for the European Transport Workers Federation, comprehensive policies to reduce emissions in the transport sector by 80 percent would create seven million new jobs across the continent, while another five million clean energy jobs in Europe could slash electricity emissions by 90 percent. A bold coalition in South Africa, meanwhile, going under the banner of One Million Climate Jobs, is calling for mass job creation programs in areas ranging from renewable energy to public transit to ecosystem restoration to small-scale sustainable farming. "By placing the interests of workers and the poor at the forefront of strategies to combat climate change we can simultaneously halt climate change and address our jobs bloodbath," the campaign states.[11]

These are not, however, the kinds of jobs that the market will create on its own. They will be created on this scale only by thoughtful policy and planning. And in some cases, having the tools to make those plans will require citizens doing what the residents of so many German cities and towns have done: taking back control over electricity generation so that the switch to renewables can be made without delay, while any profits generated go not to shareholders but back into supporting hungry public services.

And it's not only power generation that should receive this treatment. If the private companies that took over the national railways are cutting back and eroding services at a time when the climate crisis demands expanded low-carbon transportation alternatives to keep more of us out of planes, then these services too must be reclaimed. And after more than two decades of hard experience with privatizations—which has too often involved diminished services combined with higher prices—a great many people are ready to consider that option. For instance, a British poll released in November 2013 found "voters of all politics united in their support for nationalisation of energy and rail. 68 per cent of the public say the energy companies should be run in the public sector, while only 21 per cent say they should remain in private hands. 66 per cent support nationalising the railway companies while 23 per cent think they should be run privately." One of the most surprising aspects of the poll was the amount of support for nationalization among self-described Conservative voters: 52 percent favored taking back both the energy companies and the rails.[12]

Planning for Power

The climate case for rethinking private ownership is particularly strong when it comes to natural gas, which is currently being touted by many governments as a "bridge fuel." The theory is that, in the time it takes for us to make a full switch to zero carbon sources of energy, gas can serve as an alternative to dirtier fossil fuels like coal and oil. It is far from clear that this bridge is necessary, given the speed of the shift to renewables in countries like Germany. And there are many problems, as we will see, with the whole idea of natural gas being clean. But from a planning perspective, the most immediate problem is that for the bridge concept to work, ways would have to be found to ensure that natural gas was being used *only* as a replacement for coal and oil—and not to undercut renewable energy. And this is a very real concern: in the U.S., the deluge of cheap natural gas thanks to fracking has already hurt the country's wind market, with wind power's share of the new electricity coming online plummeting from at least 42 percent in 2009 to 25 percent in 2010 and 32 percent in 2011—the key years that fracking skyrocketed.[13] Moreover, once the "bridge" to a renewable future has been built, there would have to be a way to phase out gas extraction completely, since it is a major emitter of greenhouse gases.

There are various ways to design a system that would meet these specific goals. Governments could mandate "combined-cycle" plants that are better at ramping up and down to support wind and solar when available, for example, and they could firmly link any new gas plants to coal plants taken off the grid. Also crucial, says the Canadian Centre for Policy Alternatives' Ben Parfitt, an expert on fracking impacts, would be "regulations in place at the state and the national levels that made the link between where the gas is being produced and how it is being produced, and the ultimate production of the power," meaning that power plants could only source gas that was proven to have lower life-cycle emissions than coal.[14] And that could well rule out fracked gas completely. Barriers would also need to be placed on the ability of companies to export their gas, in order to prevent it from being burned in countries that place no such restrictions. These measures would limit many, though by no means all, of the risks associated with natural gas, but they would also seriously eat into the profitability of the sector.

Which raises the question: why would notoriously ruthless for-profit companies accept a business model that relies on them not competing with large parts of the energy sector (wind and solar), requires that they submit to a huge range of costly regulation, all with the eventual goal of putting themselves out of business? The answer is that they would not. Treating natural gas as a truly temporary transition fuel is anathema to the profit-seeking imperative that drives these corporations. After all, who is doing the fracking? It's companies like BP and Chevron, with their long track records of safety violations and fending off tough regulation. These are companies whose business model requires that they replace the oil and gas they have in production with new reserves of fossil fuels or face a shareholder rebellion. That same growth-above-all model demands that they occupy as much of the energy market as possible—which means competing not just with coal but with every player in the energy market, including vulnerable renewables. To quote John Browne when he was chief executive of BP (he now heads the gas giant Cuadrilla): "Corporations have to be responsive to price signals. We are not public service."[15] True enough—but that was neither always the case with our energy companies, nor must it remain so.

The bottom line is simple. No private company in the world wants to put itself out of business; its goal is to expand its market. Which is why, if natural gas is to serve as a short-term transition fuel, that transition must be tightly managed by—and for—the public, so that the profits from current sales are reinvested in renewable technologies for the future, and the sector is constrained from indulging in the kind of exponential growth it is currently enjoying amidst the shale gas boom.[16]

The solution is most emphatically not energy nationalization on existing models. The big publicly owned oil companies—from Brazil's Petrobras to Norway's Statoil to PetroChina—are just as voracious in pursuing high-risk pools of carbon as their private sector counterparts.[17] And in the absence of a credible transition plan to harness the profits for a switch to renewable energy, having the state as the major shareholder in these companies has profoundly corrupting effects, creating an addiction to easy petrodollars that makes it even less likely that policymakers will introduce measures that hurt fossil fuel profits in any way. In short, these centralized monsters are fossils in every sense of the word, and need to be broken up and phased out whether they are held in public or private hands.

A better model would be a new kind of utility—run democratically, by the communities that use them, as co-ops or as a "commons," as author and activist David Bollier and others have outlined.[18] This kind of structure would enable citizens to demand far more from their energy companies than they are able to now—for example, that they direct their profits away from new fossil fuel exploration and obscene executive compensation and shareholder returns and into building the network of complementary renewables that we now know has the potential to power our economies in our lifetimes.

The rapid rise of renewables in Germany makes a powerful case for this model. The transition has occurred, first of all, within the context of a sweeping, national feed-in tariff program that includes a mix of incentives designed to ensure that anyone who wants to get into renewable power generation can do so in a way that is simple, stable, and profitable. Providers are guaranteed priority access to the grid, and offered a guaranteed price so the risk of losing money is low.

This has encouraged small, noncorporate players to become renewable energy providers—farms, municipalities, and hundreds of newly formed co-ops. That has decentralized not just electrical power, but also political power and wealth: roughly half of Germany's renewable energy facilities are in the hands of farmers, citizen groups, and almost nine hundred energy cooperatives. Not only are they generating power but they also have the chance to generate revenue for their communities by selling back to the grid. Over all, there are now 1.4 million photovoltaic installations and about 25,000 windmills. Nearly 400,000 jobs have been created.[19]

Each one of these measures represents a departure from neoliberal orthodoxy: the government is engaging in long-term national planning; it is deliberately picking winners in the market (renewables over nuclear power, which it is simultaneously closing down); it is fixing prices (a clear market interference); and creating a fair playing field for any potential renewable energy producer—big or small—to enter the market. And yet despite—or rather because of—these ideological heresies, Germany's transition is among the fastest in the world. According to Hans Thie, the advisor on economic policy for the Left Party in the German parliament, who has been intensely involved in the transition, "Virtually all expansion estimates have been surpassed. The speed of expansion is considerably higher than had been expected."[20]

Nor can this success be dismissed as a one-off. Germany's program mirrors one implemented in Denmark in the 1970s and 1980s, which helped switch more than 40 percent of the country's electricity consumption to renewables, mostly wind. Up to around 2000, roughly 85 percent of Danish wind turbines were owned by small players like farmers and co-ops. Though large offshore wind operators have entered the market in recent years, this remains a striking commonality between Denmark and Germany: it's neither big nationally owned monopolies nor large corporate-owned wind and solar operators that have the best track record for spurring renewable energy turnarounds—it's communities, co-ops, and farmers, working within the context of an ambitious, well-designed national framework.[21] Though often derided as the impractical fantasy of small-is-beautiful dreamers, decentralization delivers, and not on a small scale but on the largest scale of any model attempted thus far, and in highly developed postindustrial nations.

It is also surely no coincidence that Denmark, a deeply social democratic country, introduced these policies well before it began its halfhearted embrace of neoliberalism, or that Germany—while prescribing brutal austerity to debtor countries like Greece and Spain—has never fully followed these prescriptions at home. These examples make clear that when governments are willing to introduce bold programs and put goals other than profit making at the forefront of their policymaking, change can happen with astonishing speed.

Decentralized control over energy is also important for very practical reasons. There are plenty of examples of large-scale, privately owned renewable energy projects that fell apart because they were imposed from the outside without local input or profit sharing. Indeed, when communities are excluded in this way, there is a very good chance that they will rebel against the noise and "unsightliness" of wind turbines, or the threats—some real, some imagined—to wildlife and ecosystems posed by solar arrays. These objections are often dismissed as NIMBY-ism (Not in My Backyard) and are used as more evidence of humanity's tendency toward selfishness and shortsightedness.

But in several regions, these objections have been entirely neutralized with thoughtful planning. As Preben Maegaard, former president of the World Wind Energy Association, once put it, "When local people own the wind farms, and share in the benefits, they will support them. It won't be NIMBY (Not In My Back Yard), it will be POOL (Please On Our Land)."[22]

This is particularly true in times of unending public austerity. "The future is something that is not relevant at the moment for some people because they're surviving for the present," Dimitra Spatharidou, a Greek climate change activist engaged in that country's broader anti-austerity movement, told me. "It's difficult to understand the concept of sustainability when people are fighting for food and to have energy to heat their homes." Because of these pressing concerns, her work is "not about preaching about what happens when climate change hits Greece, it's about what's happening now and how we can change our economies and our societies into something better, to something more equitable and to something fair."[23] For Spatharidou, that has meant showing how community-controlled renewable energy can be cheaper than dirtier alternatives, and can even be a source of income when energy is fed back into the grid. It has also meant resisting a government push to privatize municipal water supplies, pushing instead for community ownership, an idea with broad support in Greece. The key, she says, is to offer people something the current system doesn't: the tools and the power to build a better life for themselves.

This relationship between power decentralization and successful climate action points to how the planning required by this moment differs markedly from the more centralized versions of the past. There is a reason, after all, why it was so easy for the right to vilify state enterprises and national planning: many state-owned companies were bureaucratic, cumbersome, and unresponsive; the five-year plans cooked up under state socialist governments were indeed top-down and remote, utterly disconnected from local needs and experiences, just as the plans issued by the Communist Party of China's Central Committee are today.

The climate planning we need is of a different sort entirely. There is a clear and essential role for national plans and policies—to set overall emission targets that keep each country safely within its carbon budget, and to introduce policies like the feed-in tariffs employed in Germany, Ontario, and elsewhere, that make renewable energy affordable. Some programs, like national energy grids and effective rail services, must be planned, at least in part, at the national level. But if these transitions are to happen as quickly as required, then the best way to win widespread buy-in is for the actual implementation of a great many of the plans to be as decentralized as possible. Communities should be given new tools and powers to design the methods that work best for them—much as worker-run co-ops have the capacity to play a huge role in an industrial transformation. And what is true for energy and manufacturing can be true for many other sectors: transit systems accountable to their riders, water systems overseen by their users, neighborhoods planned democratically by their residents, and so on.

Most critically, farming—a major source of greenhouse gas emissions—can also become an expanded sector of decentralized self-sufficiency and poverty reduction, as well as a key tool for emission reduction. Currently, much of the debate about agriculture and climate change focuses on contrasting the pros and cons of industrial agriculture versus local and organic farming, with one side emphasizing higher yields and the other emphasizing lower chemical inputs and often (though not always) shorter supply lines. Coming up through the middle is "agroecology," a less understood practice in which small-scale farmers use sustainable methods based on a combination of modern science and local knowledge.

Based on the principle that farming should maximize species diversity and enhance natural systems of soil protection and pest control, agroecology looks different wherever its holistic techniques are practiced. But a report in *National Geographic* provides a helpful overview of how these principles translate in a few different contexts: the integration of "trees and shrubs into crop and livestock fields; solar-powered drip irrigation, which delivers water directly to plant roots; intercropping, which involves planting two or more crops near each other to maximize the use of light, water, and nutrients; and the use of green manures, which are quick-growing plants that help prevent erosion and replace nutrients in the soil."[24]

These methods and many others maintain healthy soil while producing nutritious food—more than industrial agriculture does, per unit area—and limit the need for farmers to buy expensive products like chemical pesticides, fertilizers, and patented seeds. But many farmers who have long used these methods have realized that they also have a triple climate benefit: they sequester carbon in the soil, avoid fossil fuel–based fertilizers, and often use less carbon for transportation to market, in addition to better withstanding extreme weather and other climate impacts. And communities that can feed themselves are far less vulnerable to price shocks within the broader globalized food system. Which is why La Via Campesina, a global network of small farmers with 200 million members, often declares, "Agroecology is the solution to solve the climate crisis." Or "small farmers cool the planet."[25]

In recent years, a phalanx of high-level food experts has come to similar conclusions. "A large segment of the scientific community now acknowledges the positive impacts of agroecology on food production, poverty alleviation and climate change mitigation—and this is what is needed in a world of limited resources," says Olivier De Schutter, who served as the UN Special Rapporteur on the Right to Food from 2008 to 2014.[26]

Just as they dismiss decentralized energy as too small, defenders of Big Agribusiness maintain that local organic agriculture simply cannot feed a world of 7 billion and growing—but those claims are generally based on comparisons between yields from industrial, often genetically engineered monocrops, and organic monocrops. Agroecology is left out of the picture. That's a problem because as De Schutter notes, "Today's scientific evidence demonstrates that agroecological methods outperform the use of chemical fertilizers in boosting food production where the hungry live—especially in unfavorable environments." He cites the example of Malawi, where a recent turn to agroecology has led to a doubling or tripling of maize yields in some areas, and adds that "to date, agroecological projects have shown an average crop yield increase of 80% in 57 developing countries, with an average increase of 116% for all African projects. Recent projects conducted in 20 African countries demonstrated a doubling of crop yields over a period of 3–10 years."[27]

All this amounts to a compelling case against the claim, frequently voiced by powerful philanthropists like Bill Gates, that the developing world, particularly Africa, needs a "New Green Revolution"—a reference to philanthropic and government efforts in the mid-twentieth century to introduce industrial agriculture in Asia and Latin America. "It's often claimed, particularly by those who'd like to see it rebooted, that the Green Revolution saved the world from hunger," sociologist Raj Patel, author of *Stuffed and Starved*, told me. "The problem is that even with the Green Revolution, starvation continues—particularly in India, where the revolution was most intense. Hunger isn't about the amount of food around—it's about being able to afford and control that food. After all, the U.S. has more food than it knows what to do with, and still 50 million people are food insecure."[28]

And he adds, "The tragedy here is that there are thousands of successful experiments, worldwide, showing how climate-smart agriculture can work. They're characterized not by expensive fertilizer from Yara and proprietary seeds from Monsanto, but knowledge developed and shared by peasants freely and equitably." And, Patel says, "In its finest moments, agroecology gets combined with 'food sovereignty,' with democratic control of the food system, so that not only is more food produced, but it's distributed so that *everyone* gets to eat it too."[29]

About That German Miracle . . .

We now have a few models to point to that demonstrate how to get far-reaching decentralized climate solutions off the ground with remarkable speed, while fighting poverty, hunger, and joblessness at the same time. But it's also clear that, however robust, these tools and incentives are not enough to lower emissions in time. And this brings us to what has most definitely *not* worked about the German energy transition.

In 2012—with its renewable sector soaring to new heights—German emissions actually went up from the previous year. Preliminary data suggest that the same thing happened in 2013. The country's emissions are still 24 percent below what they were in 1990, so these two years may turn out to have been a short-term blip, but the fact that the dramatic rise of renewables is not corresponding to an equally dramatic drop in greenhouse gas emissions is cause for great concern.[30] It also tells us something critical about the limits of economic plans based on incentives and market mechanisms alone.

Many have attributed the emissions rise to Germany's decision to phase out nuclear power, but the facts are not nearly so simple. It's true that in 2011, in the wake of the Fukushima disaster, the government of Chancellor Angela Merkel—under intense pressure from the country's powerful antinuclear movement—announced that it would phase out nuclear power by 2022, and took aggressive action to begin the process. But at the same time, the government took no similar action to phase out coal and even allowed coal companies to export power to other countries. So even though Germans have indeed been moving in ever greater numbers to renewable energy, coal power continued to grow, with some of it displacing nuclear power, some of it displacing gas, and some of it being exported. And much of the coal in Germany is lignite, often referred to as brown coal, a low-grade variety with particularly high emissions.[31]

As we have already seen, the latest research on renewable energy, most notably by Mark Jacobson's team at Stanford, shows that a global transition to 100 percent renewable energy—"wind, water and solar"—is both technically and economically feasible "by as early as 2030." That means lowering greenhouse emissions in line with science-based targets does not have to involve building a global network of new nuclear plants. In fact that could well slow down the transition, since renewable energy is faster and cheaper to roll out than nuclear, critical factors given the tightness of the timeframe. Moreover, says Jacobson, in the near-term nuclear is "not carbon-free, no matter what the advocates tell you. Vast amounts of fossil fuels must be burned to mine, transport and enrich uranium and to build the nuclear plant. And all that dirty power will be released during the 10 to 19 years that it takes to plan and build a nuclear plant. (A wind farm typically takes two to five years.)" He concludes that "if we invest in nuclear versus true renewables, you can bet that the glaciers and polar ice caps will keep melting while we wait, and wait, for the nuclear age to arrive. We will also guarantee a riskier future for us all." Indeed, renewable installations present dramatically lower risks than either fossil fuels or nuclear energy to those who live and work next to them. As comedian Bill Maher once observed, "You know what happens when windmills collapse into the sea? A splash."[*][32]

That said, about 12 percent of the world's power is currently supplied by nuclear energy, much of it coming from reactors that are old and obsolete.[33] From a climate perspective, it would certainly be preferable if governments staggered their transitions away from high-risk energy sources like nuclear,

---

[*] Much of the support for nuclear power as a solution to global warming is based on the promise of "next generation" nuclear technologies, which range from more efficient reactors cooled with gas instead of water, to "fast reactor" designs that can run on spent fuel or "breed" more fuel in addition to consuming it—or even nuclear fusion, in which atomic nuclei are forced together (as occurs in the sun) rather than split. Boosters of these groundbreaking technologies assure us that they eliminate many of the risks currently associated with nuclear energy, from meltdowns to longterm waste storage to weaponization of enriched uranium. And perhaps they do have the potential to eliminate some of those risks. But since these technologies are untested, and some may carry even greater risks, the onus is on the boosters, not on the rest of us, to demonstrate their safety. All the more so because we have *proven* clean, renewable technologies available, and democratic, participatory models for their implementation, that demand no such risks.

prioritizing fossil fuels for cuts because the next decade is so critical for getting us off our current trajectory toward 4–6 degrees Celsius of warming. That would be compatible with a moratorium on new nuclear facilities, a decommissioning of the oldest plants and then a full nuclear phase-out once renewables had decisively displaced fossil fuels.

And yet it must also be acknowledged that it was the power of Germany's antinuclear movement that created the conditions for the renewables revolution in the first place (as was the case in Denmark in the 1980s), so there might have been no energy transition to debate without that widespread desire to get off nuclear due to its many hazards. Moreover, many German energy experts are convinced that the speed of the transition so far proves that it is possible to phase out both nuclear and fossil fuels simultaneously. A 2012 report by the German National Center for Aerospace, Energy and Transport Research (DLR), for instance, demonstrated that 67 percent of the electicity in all of the EU could come from renewables by 2030, with that number reaching 96 percent by 2050.[34] But, clearly, this will become a reality only if the right policies are in place.

For that to happen, the German government would have to be willing to do to the coal industry what it has been willing to do to the nuclear power industry: introduce specific, top-down regulations to phase it out. Instead, because of the vast political power of the German coal lobby, the Merkel government has relied on the weak market mechanism of carbon trading, through the European emissions trading system, to try to put negative pressure on coal.[35] When the European carbon market fell apart, and the price of carbon plummeted, this strategy proved disastrous. Coal was cheap, there was no real penalty to burning it, and there were no blocks on exporting coal power, and so key years that should have been triumphs over pollution became setbacks.

Tadzio Mueller, a Berlin-based researcher and climate expert, put the problem to me like this: "German emissions are not up because nuclear power is down. They're up because nobody told the German power companies not to burn coal, and as long as they can profitably sell the electricity somewhere, they'll burn the coal—even if most electricity consumed in Germany was renewable. What we need are strict rules against the extraction and burning of coal. Period."[36]

It is critical for governments to put creative incentives in place so that communities around the world have tools to say yes to renewable energy. But what the German experience shows is that all that progress will be put at risk unless policymakers are willing simultaneously to say no to the ever rapacious fossil fuel industry.

Remembering How to Say No

Even before I saw the giant mines, when the landscape out the window was still bright green boggy marshes and lush boreal forest, I could feel them—a catch in the back of my throat. Then, up and over a small elevation, there they were: the notorious Alberta tar sands, a parched, gray desert stretching to the horizon. Mountains of waste so large workers joke that they have their own weather systems. Tailing ponds so vast they are visible from space. The second largest dam in the world, built to contain that toxic water. The earth, skinned alive.

Science fiction is rife with fantasies of terraforming—humans traveling to lifeless planets and engineering them into earthlike habitats. The Canadian tar sands are the opposite: terra-deforming. Taking a habitable ecosystem, filled with life, and engineering it into a moonscape where almost nothing can live. And if this goes on, it could impact an area roughly the size of England. All to access a semisolid form of "unconventional" oil known as bitumen that is so difficult and energy-intensive to extract that the process is roughly three to four times as greenhouse gas intensive as extracting conventional oil.[37]

In June 2011, I cosigned a letter drafted by author and climate activist Bill McKibben that called on people to come to Washington, D.C., "in the hottest and stickiest weeks of the summer" to get arrested protesting the proposed Keystone XL pipeline. Amazingly, more than 1,200 people did just that, making it the largest act of civil disobedience in the history of the North American climate movement.[38]

For over a year, a coalition of ranchers and Indigenous people who lived along the proposed route of the pipeline had been campaigning hard against the project. But the action in Washington took the campaign national, and turned it into a flashpoint for a resurgent U.S. climate movement.

The science for singling out Keystone XL was clear enough. The pipeline would be carrying oil from the Alberta tar sands, and James Hansen, then still working at NASA, had recently declared that if the bitumen trapped in the tar sands was all dug up and burned, it would be "game over for the climate."[39] But there was also some political strategy at work: unlike so many other key climate policies, which either required approval from Congress or were made at the state level, the decision about whether to approve the Keystone XL pipeline was up to the State Department and, ultimately, the president himself, based on whether he determined the project to be in the "national interest." On this one, Obama would have to give his personal yes or no, and it seemed to us that there was value in extracting either answer.

If he said no, that would be a much needed victory on which to build at a time when the U.S. climate movement, bruised from the failure to get energy legislation through Congress, badly needed some good news. If he said yes, well, that too would be clarifying. Climate activists, almost all of whom had worked to get Obama elected, would have to finally abandon the hopes they had pinned on the young senator who had proclaimed that his election would be remembered as "the moment when the rise of the oceans began to slow and our planet began to heal."[40] Letting go of that faith would be disillusioning for many, but at least tactics could be adjusted accordingly. And it seemed we would not have to wait long for a verdict: the president would be in a position to make his decision by early September, which is why the civil disobedience was called for the end of August.

It never occurred to us in those early strategy sessions at 350.org, the climate organization that McKibben cofounded and where I am a board member, that three years later we would still be waiting for the president's yes or no. Three years during which Obama waffled and procrastinated, while his administration ordered more environmental reviews, then reviews of those reviews, then reviews of those too.

A great deal of intellectual energy has been expended trying to interpret the president's mixed signals on Keystone XL—at times he seemed to be sending a clear message that he was going to give his approval, as when he arranged for a photo op in front of a raft of metal pipeline waiting to be laid down; other times he seemed to be suggesting that he was leaning toward rejection, as when he declared, in one of his more impassioned speeches about climate change, that Keystone would be approved "only if this project does not significantly exacerbate the problem of carbon pollution."[41]

But whichever way the decision eventually goes (and one can hope that we will know the answer by the time you read this), the drawn-out saga made at least one thing absolutely clear. Like Angela Merkel, Obama has a hell of a hard time saying no to the fossil fuel industry. And that's a very big problem because to lower emissions as rapidly and deeply as required, we need to keep large, extremely profitable pools of carbon in the ground— resources that the fossil fuel companies are fully intending to extract.

That means our governments are going to have to start putting strict limits on the industry—limits ranging from saying no to pipelines linked to expanded extraction, to caps on the amount of carbon corporations can emit, to banning new coal-fired power plants, to winding down dirty-energy extraction projects like the Alberta tar sands, to saying no to demands to open up new carbon frontiers (like the oil trapped under melting Arctic ice).

———

In the 1960s and 1970s, when a flurry of environmental legislation was passed in the U.S. and in other major industrial countries, saying no to dirty industry was, though never easy, an accepted part of the balancing act of government. That is simply no longer the case, as is evident from the howls of outrage from Republicans and many Democrats over the mere suggestion that Obama might reject Keystone XL, a moderate-sized infrastructure project that, by the president's own admission, would create so few lasting jobs that they represent "a blip relative to the need."[42] Given how wrenchingly difficult that yes-or-no regulatory decision proved to be, it should not be at all surprising that broader, more forceful controls on how much carbon should be extracted and emitted have thus far been entirely elusive.

Obama's much-heralded move in June 2014 mandating emission reductions from power plants was certainly the right direction, but the measures were still much too timid to bring the U.S. in line with a safe temperature trajectory. As author and long-time climate watcher Mark Hertsgaard observed at the time, "President Obama clearly grasps the urgency of the climate crisis and has taken important steps to address it. But it is his historical fate to be in power at a time when good intentions and important steps are no longer enough. . . . Perhaps all this places an unfair burden on President Obama. But science does not care about fair, and leaders inherit the history they inherit." And yet as Hertsgaard acknowledges, the kind of policies that would be enough "seem preposterous to the political and economic status quo."[43]

This state of affairs is, of course, yet another legacy of the free market counterrevolution. In virtually every country, the political class accepts the premise that it is not the place of government to tell large corporations what they can and cannot do, even when public health and welfare—indeed the habitability of our shared home—are clearly at stake. The guiding ethos of light-touch regulation, and more often of active deregulation, has taken an enormous toll in every sector, most notably the financial one. It has also blocked commonsense responses to the climate crisis at every turn— sometimes explicitly, when regulations that would keep carbon in the ground are rejected outright, but mostly implicitly, when those kinds of regulations are not even proposed in the first place, and so-called market solutions are favored for tasks to which they are wholly unequipped.

It's true that the market is great at generating technological innovation and, left to its own devices, R&D departments will continue to come up with impressive new ways to make solar modules and electrical appliances more efficient. But at the same time, market forces will also drive new and innovative ways to get hard-to-reach fossil fuels out of the deep ocean and hard shale—and those dirty innovations will make the green ones essentially irrelevant from a climate change perspective.

At the Heartland conference, Cato's Patrick Michaels inadvertently made that point when he argued that, though he believes climate change is happening, the real solution is to do nothing and wait for a technological miracle to rain down from the heavens. "Doing nothing *is* actually doing something," he proclaimed, assuring the audience that "technologies of the future" would save the day. His proof? "Two words: Shale gas. . . . That's what happens if you allow people to use their intellect, and their inquisitiveness, and their drive, in order to produce new energy sources." And of course the Heartland audience cheered earnestly for the intellectual breakthrough that is hydraulic fracturing (aka fracking) combined with horizontal drilling, the technology that has finally allowed the fossil fuel industry to screw us sideways.[44]

And it's these "unconventional" methods of extracting fossil fuels that are the strongest argument for forceful regulation. Because one of the greatest misconceptions in the climate debate is that our society is refusing to change, protecting a status quo called "business-as-usual." The truth is that there is no business-as-usual. The energy sector is changing dramatically all the time—but the vast majority of those changes are taking us in precisely the wrong direction, toward energy sources with even higher planet-warming emissions than their conventional versions.

Take fracking. Natural gas's reputation as a clean alternative to coal and oil is based on emissions measurements from gas extracted through conventional drilling practices. But in April 2011, a new study by leading scientists at Cornell University showed that when gas is extracted through fracking, the emissions picture changes dramatically.[45]

The study found that methane emissions linked to fracked natural gas are at least 30 percent higher than the emissions linked to conventional gas. That's because the fracking process is leaky—methane leaks at every stage of production, processing, storage, and distribution. And methane is an extraordinarily dangerous greenhouse gas, thirty-four times more effective at trapping heat than carbon dioxide, based on the latest Intergovernmental Panel on Climate Change estimates. According to the Cornell study, this means that fracked gas has a greater greenhouse gas impact than oil and may well have as much of a warming impact as coal when the two energy sources are examined over an extended life cycle.[46]

Furthermore, Cornell biogeochemist Robert Howarth, the lead author of the study, points out that methane is an even more efficient trapper of heat in the first ten to fifteen years after it is released—indeed it carries a warming potential that is *eighty-six times* greater than that of carbon dioxide. And given that we have reached "decade zero," that matters a great deal. "It is in this shorter time frame that we risk locking ourselves into very rapid warming," Howarth explains, especially because huge liquid natural gas export terminals currently planned or being built in Australia, Canada, and the United States are not being constructed to function for only the next decade but for closer to the next half century. So, to put it bluntly, in the key period when we need to be looking for ways to cut our emissions rapidly, the global gas boom is in the process of constructing a network of ultra-powerful atmospheric ovens.[47]

The Cornell study was the first peer-reviewed research on the greenhouse gas footprint of shale production, including from methane emissions, and its lead author was quick to volunteer that his data were inadequate (largely due to the industry's lack of transparency). Still, the study was a bombshell, and though it remains controversial, a steady stream of newer work has bolstered the case for a high rate of methane leakage in the fracking process.* [48]

The gas industry isn't the only one turning to dirtier, higher-risk methods. Like Germany, the Czech Republic and Poland are increasingly relying on and expanding production of extra-dirty lignite coal.[49] And the major oil companies are rushing into various tar sands deposits, most notably in Alberta, all with significantly higher carbon footprints than conventional oil. They are also moving into ever deeper and icier waters for offshore drilling, carrying the risk of not just more catastrophic spills, as we saw with BP's Deepwater Horizon disaster, but spills that are simply impossible to clean up. Increasingly, these extreme extraction methods—blasting oil and gas out of rock, steaming oil out of tarlike dirt —are being used together, as when fracked natural gas is piped in to superheat the water that melts the bitumen in the tar sands, to cite just one example from the energy death spiral. What industry calls innovation, in other words, looks more like the final suicidal throes of addiction. We are blasting the bedrock of

---

* There is a great deal of confusion about the climate benefits of natural gas because the fuel is often given credit for a 12 percent drop in U.S. carbon dioxide emissions since 2007. But this good news does not address the fact that methane emissions have been rising over the past decade, or the fact that U.S. methane emissions are very likely underestimated, since leakage has been extremely poorly accounted for. Moreover, many experts and modelers warn that any climate gains from the shale boom will continue to be undercut not only by potent methane emissions, but also by the tendency of cheap natural gas to displace wind and solar. Similarly, as coal generation is displaced by natural gas in the U.S., coal companies are simply exporting their dirty product overseas, which according to one analysis by the CO_2 Scorecard Group has "more than offset" the emissions savings from natural gas since 2007.

our continents, pumping our water with toxins, lopping off mountaintops, scraping off boreal forests, endangering the deep ocean, and scrambling to exploit the melting Arctic—all to get at the last drops and the final rocks. Yes, some very advanced technology is making this possible, but it's not innovation, it's madness.

The fact that fossil fuel companies have been permitted to charge into unconventional fossil fuel extraction over the past decade was not inevitable, but rather the result of very deliberate regulatory decisions—decisions to grant these companies permits for massive new tar sands and coal mines; to open vast swaths of the United States to natural gas fracking, virtually free from regulation and oversight; to open up new stretches of territorial waters and lift existing moratoriums on offshore drilling. These various decisions are a huge part of what is locking us into disastrous levels of planetary warming. These decisions, in turn, are the product of intense lobbying by the fossil fuel industry, motivated by the most powerful driver of them all: the will to survive.

As a rule, extracting and refining unconventional energy is a far more expensive and involved industrial process than doing the same for conventional fuels. So, for instance, Imperial Oil (of which Exxon owns a majority share) sank about $13 billion to open the sprawling Kearl open-pit mine in the Alberta tar sands. At two hundred square kilometers, it will be one of the largest open-pit mines in Canada, more than three times the size of Manhattan. And it is only a fraction of the new construction planned for the tar sands: the Conference Board of Canada projects that a total of $364 billion will be invested through 2035.[50]

In Brazil, meanwhile, Britain's BG Group is expected to make a $30 billion investment over the next decade, much of it going into ultra-deepwater "subsalt" projects in which oil is extracted from depths of approximately three thousand meters (ten thousand feet). But the prize for fossil fuel lock-in surely goes to Chevron, which is spending a projected $54 billion on a gas development on Barrow Island, a "Class A Nature Reserve" off the northwest coast of Australia. The project will release so much natural gas from the earth that it is appropriately named Gorgon, after the terrifying, snake-haired female monster of Greek mythology. One of Chevron's partners in the project is Shell, which is reportedly spending an additional

$10–12 billion to build the largest floating offshore facility ever constructed (longer than four soccer fields) in order to extract natural gas from a different location off the northwest coast of Australia.[51]

These investments won't be recouped unless the companies that made them are able to keep extracting for decades, since the up-front costs are amortized over the life of the projects. Chevron's Australia project is expected to keep producing natural gas for at least thirty years, while Shell's floating gas monstrosity is built to function on that site for up to twenty-five years. Exxon's Alberta mine is projected to operate for forty years, as is BP/Husky Energy's enormous Sunrise project, also in the tar sands. This is only a small sampling of mega-investments taking place around the world in the frantic scramble for hard-to-extract oil, gas, and coal. The long time frames attached to all these projects tell us something critical about the assumptions under which the fossil fuel industry is working: it is betting that governments are not going to get serious about emissions cuts for the next twenty-five to forty years. And yet climate experts tell us that if we want to have a shot at keeping warming below 2 degrees Celsius, then developed country economies need to have begun their energy turnaround by the end of this decade and to be almost completely weaned from fossil fuels before 2050.[52]

If the companies have miscalculated and we do get serious about leaving carbon in the ground, these huge projects will become what is known as "stranded assets"—investments that lose their projected value as a result of, for example, dramatic changes in environmental policy. When a company has a great deal of expensive stranded assets on its books, the stock market takes notice, and responds by bidding down the share price of the company that made these bad bets.

This problem goes well beyond a few specific projects and is integrated into the way that the market assigns value to companies that are in the business of extracting finite resources from the earth. In order for the value of these companies to remain stable or grow, oil and gas companies must always be able to prove to their shareholders that they have fresh carbon reserves to exploit after they exhaust those currently in production. This process is as crucial for extractive companies as it is for a company that sells cars or clothing to show their shareholders that they have preorders for their future products. At minimum, an energy company is expected to have as much oil and gas in its proven reserves as it does in current production, which would give it a "reserve-replacement ratio" of 100 percent. As the popular site Investopedia explains, "A company's reserve replacement ratio must be at least 100% for the company to stay in business long-term; otherwise, it will eventually run out of oil."[53]

Which is why investors tend to get quite alarmed when the ratio drops below that level. For instance, in 2009, on the same day that Shell announced that its reserve-replacement ratio for the previous year had ominously dipped to 95 percent, the company scrambled to reassure the market that it was not in trouble. It did this, tellingly, by declaring that it would cease new investments in wind and solar energy. At the same time, it doubled down on a strategy of adding new reserves from shale gas (accessible only through fracking), deepwater oil, and tar sands. All in all, Shell managed that year to add a record 3.4 billion barrels of oil equivalent in new proven reserves—nearly three times its production in 2009, or a reserve-replacement ratio of 288 percent. Its stock price went up accordingly.[54]

For a fossil fuel major, keeping up its reserve-replacement ratio is an economic imperative; without it, the company has no future. It has to keep moving just to stand still. And it is this structural imperative that is pushing the industry into the most extreme forms of dirty energy; there are simply not enough conventional deposits left to keep up the replacement ratios. According to the International Energy Agency's annual World Energy Outlook report, global conventional oil production from "existing fields" will drop from 68 million barrels per day in 2012 to an expected 27 million in 2035.[55]

That means that an oil company looking to reassure shareholders that it has a plan for what to do, say, when the oil in Alaska's Prudhoe Bay runs out, will be forced to go into higher-risk, dirtier territories. It is telling, for instance, that *more than half* of the reserves Exxon added in 2011 come from a single oil project: the massive Kearl mine being developed in the Alberta tar sands.[56] This imperative also means that, so long as this business model is in place, no coastline or aquifer will be safe. Every victory against the fossil fuel companies, no matter how hard won, will be temporary, just waiting to be overtaken with howls of "Drill, Baby, Drill." It won't be enough even when we can walk across the Gulf of Mexico on the oil rigs, or when Australia's Great Barrier Reef is a parking lot for coal tankers, or when Greenland's melting ice sheet is stained black from a spill we have no idea how to clean up. Because these companies will always need more reserves to top up their replacement ratios, year after year after year.

From the perspective of a fossil fuel company, going after these high-risk carbon deposits is not a matter of choice—it is its fiduciary responsibility to shareholders, who insist on earning the same kinds of mega-profits next year as they did this year and last year. And yet fulfilling that fiduciary responsibility virtually guarantees that the planet will cook.

This is not hyperbole. In 2011, a think tank in London called the Carbon Tracker Initiative conducted a breakthrough study that added together the reserves claimed by all the fossil fuel companies, private and state-owned. It found that the oil, gas, and coal to which these players had already laid claim—deposits they have on their books and which were already making money for shareholders—represented 2,795 gigatons of carbon (a gigaton is 1 billion metric tons). That's a very big problem because we know roughly how much carbon can be burned between now and 2050 and still leave us a solid chance (roughly 80 percent) of keeping warming below 2 degrees Celsius. According to one highly credible study, that amount of carbon is 565 gigatons between 2011 and 2049. And as Bill McKibben points out, "The thing to notice is, 2,795 is five times 565. It's not even close." He adds: "What those numbers mean is quite simple. This industry has announced, in filings to the SEC and in promises to shareholders, that they're determined to burn five times more fossil fuel than the planet's atmosphere can begin to absorb."[57]

Those numbers also tell us that the very thing we must do to avert catastrophe—stop digging—is the very thing these companies cannot contemplate without initiating their own demise. They tell us that getting serious about climate change, which means cutting our emissions radically, is simply not compatible with the continued existence of one of the most profitable industries in the world.

And the amounts of money at stake are huge. The total amount of carbon in reserve represents roughly $27 trillion—more than ten times the annual GDP of the United Kingdom. If we were serious about keeping warming below 2 degrees, approximately 80 percent of that would be useless, stranded assets. Given these stakes, it is no mystery why the fossil fuel companies fight furiously to block every piece of legislation that would point us in the right emissions direction, and why some directly fund the climate change denier movement.[58]

It also helps that these companies are so profitable that they have money not just to burn, but to bribe—especially when that bribery is legal. In 2013 in the United States alone, the oil and gas industry spent just under $400,000 *a day* lobbying Congress and government officials, and the industry doled out a record $73 million in federal campaign and political donations during the 2012 election cycle, an 87 percent jump from the 2008 elections.[59]

In Canada, corporations are not required to disclose how much money they spend on lobbying, but the number of times they communicate with public officials is a matter of public record. A 2012 report found that a single industry organization—the Canadian Association of Petroleum Producers—spoke with federal government officials 536 times between 2008 and 2012, while TransCanada, the company behind the Keystone XL pipeline, had 279 communications. The Climate Action Network, on the other hand, the country's broadest coalition devoted to emission reductions, only logged six communications in the same period. In the U.K., the energy industry met with the Department of Energy and Climate Change roughly eleven times more frequently than green groups did during David Cameron's first year in office. In fact, it has become increasingly difficult to discern where the oil and gas industry ends and the British government begins. As *The Guardian* reported in 2011, "At least 50 employees of companies including EDF Energy, npower and Centrica have been placed within government to work on energy issues in the past four years. . . . The staff are provided free of charge and work within the departments for secondments of up to two years."[60]

What all this money and access means is that every time the climate crisis rightfully triggers our collective self-preservation instinct, the incredible monetary power of the fossil fuel industry—driven by its own, more immediate self-preservation instinct—gets in the way. Environmentalists often speak about contemporary humanity as the proverbial frog in a pot of boiling water, too accustomed to the gradual increases in heat to jump to safety. But the truth is that humanity has tried to jump quite a few times. In Rio in 1992. In Kyoto in 1997. In 2006 and 2007, when global concern rose yet again after the release of *An Inconvenient Truth* and with the awarding of the Nobel Peace Prize to Al Gore and the Intergovernmental Panel on Climate Change. In 2009, in the lead up to the United Nations climate summit in Copenhagen. The problem is that the money that perverts the political process acts as a kind of lid, intercepting that survival instinct and keeping us all in the pot.

The influence wielded by the fossil fuel lobby goes a long way toward explaining why the sector is so very unconcerned about the nonbinding commitments made by politicians at U.N. climate summits to keep temperatures below 2 degrees Celsius. Indeed the day the Copenhagen summit concluded—when the target was made official—the share prices of some of the largest fossil fuel companies hardly reacted at all.[61]

Clearly, intelligent investors had determined that the promises governments made in that forum were nothing to worry about—that they were not nearly as important as the actions of their powerful energy departments back home that grant mining and drilling permits. Indeed in March 2014, ExxonMobil confirmed as much when the company came under pressure from activist shareholders to respond to reports that much of its reserves would become stranded assets if governments kept promises to keep warming below 2 degrees by passing aggressive climate legislation. The company explained that it had determined that restrictive climate policies were "highly unlikely" and, "based on this analysis, we are confident that none of our hydrocarbon reserves are now or will become 'stranded.'"[62]

Those working inside government understand these dynamics all too well. John Ashton, who served as special representative for climate change to three successive U.K. governments between 2006 and 2012, told me that he would often point out to his colleagues making energy policy that their approach to the development of fossil fuels contradicted the government's claim to be "running a 2 degree climate policy." But when he did, they "simply ignored my efforts and carried on as before—I might as well have been speaking in Attic Greek." From this Ashton concluded, "In government it is usually easy to rectify a slight misalignment between two policies but near impossible to resolve a complete contradiction. Where there is a contradiction, the forces of incumbency start with a massive advantage."[63]

This dynamic will shift only when the power (and wealth) of the fossil fuel industry is seriously eroded. Which is very tough to do: the handy thing about selling natural resources upon which entire economies have been built—and about having so far succeeded in blocking policies that would offer real alternatives—is that most people keep having to buy your products whether they like you or not. So since these companies are going to continue being rich for the foreseeable future, the best hope of breaking the political deadlock is to radically restrict their ability to spend their profits buying, and bullying, politicians.

The good news for the climate movement is that there are a whole lot of other sectors that also have an active interest in curtailing the influence of money over politics, particularly in the U.S., the country that has been the most significant barrier to climate progress. After all, climate action has failed on Capitol Hill for the same reasons that serious financial sector reform didn't pass after the 2008 meltdown and the same reasons gun reform didn't pass after the horrific 2012 school shooting in Newtown, Connecticut. Which in turn are the same reasons why Obama's health reform failed to take on the perverting influence of the medical insurance and pharmaceutical companies. All these attempts to fix glaring and fundamental flaws in the system have failed because large corporations wield far too much political power—a power exerted through corporate campaign contributions, many of them secret; through almost unfettered access to regulators via their lobbyists; through the notorious revolving door between business and government; as well as through the "free speech" rights these corporations have been granted by the U.S. Supreme Court. And though U.S. politics are particularly far gone in this regard, no Western democracy has a level playing field when it comes to political access and power.

Because these distortions have been in place for so long—and harm so many diverse constituencies—a great many smart people have done a huge amount of thinking about what it would take to clean up the system. As with responses to climate change, the problem is not an absence of "solutions"—the solutions are clear. Politicians must be prohibited from receiving donations from the industries they regulate, or from accepting jobs in lieu of bribes; political donations need to be both fully disclosed and tightly capped; campaigns must be given the right to access the public airwaves; and, ideally, elections should be publicly funded as a basic cost of having a democracy.

Yet among large sections of the public, a sense of fatalism pervades: how can you convince politicians to vote for reforms designed to free them from the binds of corporate influence when those binds are still tightly in place? It's tough, to be sure, but the only thing politicians fear more than losing donations is losing elections. And this is where the power of climate change—and its potential for building the largest possible political tent—comes into play. As we have seen, the scientific warnings that we are running out of time to avert climate disaster are coming from a galaxy of credible scientific organizations and establishment international agencies— from the American Association for the Advancement of Science to NASA to Britain's Royal Society to the Intergovernmental Panel on Climate Change to the U.S. National Academy of Sciences to the World Bank to the International Energy Agency. A resurgent climate movement could use those warnings to light a fire under the call to kick corporate money out of politics—not just fossil fuel money, but money from all the deep-pocketed barriers to progress from the National Rifle Association to the fast food industry to the private-prison complex. Such a rallying cry could bring together all of the various constituencies that would benefit from reducing corporate power over politics—from health care workers to parents worried about their children's safety at school. There are no guarantees that this coalition could succeed where other attempts at similar reforms have failed. But it certainly seems worth expending at least as much energy and money as the U.S. climate movement did trying, unsuccessfully, to push through climate legislation that it knew was wholly inadequate, precisely because it was written to try to neutralize opposition from fossil fuel companies (more on that later).

Not an "Issue," a Frame

The link between challenging corruption and lowering emissions is just one example of how the climate emergency could—by virtue of its urgency and the fact that it impacts, well, everyone on earth—breathe new life into a political goal for which there is already a great deal of public support. The same holds true for many of the other issues discussed so far—from raising taxes on the rich to blocking harmful new trade deals to reinvesting in the public sphere. But before those kinds of alliances can be built, some very bad habits will need to be abandoned.

Environmentalists have a long history of behaving as if no issue is more important than the Big One—why, some wonder (too often out loud), is everyone wasting their time worrying about women's rights and poverty and wars when it's blindingly obvious that none of this matters if the planet decides to start ejecting us for poor behavior? When the first Earth Day was declared in 1970, one of the movement's leaders, Democratic senator Gaylord Nelson, declared that the environmental crisis made "Vietnam, nuclear war, hunger, decaying cities, and all other major problems one could name . . . relatively insignificant by comparison." Which helps explains why the great radical journalist I. F. Stone described Earth Day as "a gigantic snowjob" that was using "rock and roll, idealism and noninflammatory social issues to turn the youth off from more urgent concerns which might really threaten our power structure."[64]

They were both wrong. The environmental crisis—if conceived sufficiently broadly—neither trumps nor distracts from our most pressing political and economic causes: it supercharges each one of them with existential urgency. As Yotam Marom, an organizer with Occupy Wall Street in New York, wrote in July 2013, "The fight for the climate isn't a separate movement, it's both a challenge and an opportunity for *all* of our movements. We don't need to become climate activists, we *are* climate activists. We don't need a separate climate movement; we need to seize the climate *moment*."[65]

The nature of the moment is familiar but bears repeating: whether or not industrialized countries begin deeply cutting our emissions this decade will determine whether we can expect the same from rapidly developing nations like China and India next decade. That, in turn, will determine whether or not humanity can stay within a collective carbon budget that will give us a decent chance of keeping warming below levels that our own governments have agreed are unacceptably dangerous. In other words, we don't have another couple of decades to talk about the changes we want while being satisfied with the occasional incremental victory. This set of hard facts calls for strategy, clear deadlines, dogged focus—all of which are sorely missing from most progressive movements at the moment.

Even more importantly, the climate moment offers an overarching narrative in which everything from the fight for good jobs to justice for migrants to reparations for historical wrongs like slavery and colonialism can all become part of the grand project of building a nontoxic, shockproof economy before it's too late.

And it is also worth remembering because it's so very easy to forget: the alternative to such a project is not the status quo extended indefinitely. It is climate-change-fueled disaster capitalism—profiteering disguised as emission reduction, privatized hyper-militarized borders, and, quite possibly, high-risk geoengineering when things spiral out of control.

So how realistic is it to imagine that the climate crisis could be a political game changer, a unifier for all these disparate issues and movements? Well, there is a reason hard-right conservatives are putting so much effort into denying its existence. Their political project is not, after all, as sturdy as it was in 1988, when climate change first pierced public consciousness. Free market ideology may still bind the imaginations of our elites, but for most of the general public, it has been drained of its powers to persuade. The disastrous track record of the past three decades of neoliberal policy is simply too apparent. Each new blast of statistics about how a tiny band of global oligarchs controls half the world's wealth exposes the policies of privatization and deregulation for the thinly veiled license to steal that they always were. Each new report of factory fires in Bangladesh, soaring pollution in China, and water cut-offs in Detroit reminds us that free trade was exactly the race to the bottom that so many warned it would be. And each news story about an Italian or Greek pensioner who took his or her own life rather than try to survive under another round of austerity is a reminder of how many lives continue to be sacrificed for the few.

The failure of deregulated capitalism to deliver on its promises is why, since 2009, public squares around the world have turned into rotating semipermanent encampments of the angry and dispossessed. It's also why there are now more calls for fundamental change than at any point since the 1960s. It's why a challenging book like Thomas Piketty's *Capital in the Twenty-First Century*, exposing the built-in structures of ever-increasing wealth concentration, can sit atop bestseller lists for months, and why when comedian and social commentator Russell Brand went on the BBC and called for "revolution," his appearance attracted more than ten million YouTube views.[66]

Climate change pits what the planet needs to maintain stability against what our economic model needs to sustain itself. But since that economic model is failing the vast majority of the people on the planet on multiple fronts that might not be such a bad thing. Put another way, if there has ever been a moment to advance a plan to heal the planet that also heals our broken economies and our shattered communities, this is it.

Al Gore called climate change "an inconvenient truth," which he defined as an inescapable fact that we would prefer to ignore. Yet the truth about climate change is inconvenient only if we are satisfied with the status quo except for the small matter of warming temperatures. If, however, we see the need for transformation quite apart from those warming temperatures, then the fact that our current road is headed toward a cliff is, in an odd way, convenient—because it tells us that we had better start making that sweeping turn, and fast.

Not surprisingly, the people who understand this best are those whom our economic model has always been willing to sacrifice. The environmental justice movement, the loose network of groups working with communities on the toxic front lines of extractive industries—next to refineries, for instance, or downstream from mines—has always argued that a robust response to emission reduction could form the basis of a transformative economic project. In fact the slogan long embraced by this movement has been "System Change, Not Climate Change"—a recognition that these are the two choices we face.[67]

"The climate justice fight here in the U.S. and around the world is not just a fight against the [biggest] ecological crisis of all time," Miya Yoshitani, executive director of the Oakland-based Asian Pacific Environmental Network (APEN), explains. "It is the fight for a new economy, a new energy system, a new democracy, a new relationship to the planet and to each other, for land, water, and food sovereignty, for Indigenous rights, for human rights and dignity for all people. When climate justice wins we win the world that we want. We can't sit this one out, not because we have too much to lose but because we have too much to gain. . . . We are bound to-gether in this battle, not just for a reduction in the parts per million of CO_2, but to transform our economies and rebuild a world that we want today."[68]

This is what many liberal commentators get wrong when they assume that climate action is futile because it asks us to sacrifice in the name of far-off benefits. "How can you persuade the human race to put the future ahead of the present?" asked *Observer* columnist Nick Cohen despondently.[69] The answer is that you don't. You point out, as Yoshitani does, that for a great many people, climate action is their best hope for a better present, and a future far more exciting than anything else currently on offer.

Yoshitani is part of a vibrant activist scene in the San Francisco Bay Area that is ground zero of the green jobs movement most prominently championed by former Obama advisor Van Jones. When I first met Yoshi-tani, the Asian Pacific Environmental Network was working closely with Asian immigrants in Oakland to demand affordable housing close to a mass transit station to make sure that gentrification didn't displace the people who actually use subways and buses. And APEN has also been part of an initiative to help create worker co-ops in the solar energy sector in nearby Richmond, so that there are jobs on offer other than the ones at the local Chevron oil refinery.

More such connections between climate action and economic justice are being made all the time. As we will see, communities trying to stop danger-ous oil pipelines or natural gas fracking are building powerful new alliances with Indigenous peoples whose territories are also at risk from these activi-ties. And several large environmental organizations in the U.S.—including Greenpeace, the Sierra Club, the BlueGreen Alliance, and 350.org—took stands in support of demands for comprehensive reform of the U.S. immi-gration system, in part because migration is increasingly linked to climate and also because members of immigrant communities are often prevented from defending themselves against heightened environmental risks since doing so could lead to incarceration or deportation.[70]

These are encouraging signs, and there are plenty of others. Yet the kind of counter-power that has a chance of changing society on anything close to the scale required is still missing. It is a painful irony that while the right is forever casting climate change as a left-wing plot, most leftists and liberals are still averting their eyes, having yet to grasp that climate science has handed them the most powerful argument against unfettered capitalism since William Blake's "dark Satanic Mills" blackened England's skies (which, incidentally, was the beginning of climate change). By all rights, this reality should be filling progressive sails with conviction, lending new confidence to the demands for a more just economic model. And yet when demonstrators are protesting the various failures of this system in Athens, Madrid, Istanbul, and New York, climate change is too often little more than a footnote when it could be the coup de grâce.[71]

The mainstream environmental movement, meanwhile, generally stands apart from these expressions of mass frustration, choosing to define climate activism narrowly—demanding a carbon tax, say, or even trying to stop a pipeline. And those campaigns are important. But building a mass movement that has a chance of taking on the corporate forces arrayed against science-based emission reduction will require the broadest possible spectrum of allies. That would include the public sector workers—firefighters, nurses, teachers, garbage collectors—fighting to protect the services and infrastructure that will be our best protection against climate change. It would include antipoverty activists trying to protect affordable housing in downtown cores, rather than allowing low-income people to be pushed by gentrification into sprawling peripheries that require more driving. As Colin Miller of Oakland-based Bay Localize told me, "Housing is a climate issue." And it would include transit riders fighting against fare increases at a time when we should be doing everything possible to make subways and buses more comfortable and affordable for all. Indeed when masses of people take to the streets to stop such fare hikes and demand free public transit—as they did in Brazil in June and July of 2013—these actions should be welcomed as part of a global effort to fight climate chaos, even if those populist movements never once use the words "climate change."[72]

Perhaps it should be no surprise that a sustained and populist climate movement has not yet emerged—a movement like that has yet to be sustained to counter any of the other failures of this economic model. Yes, there have been periods when mass outrage in the face of austerity, corruption, and inequality has spilled into the streets and the squares for weeks and months on end. Yet if the recent years of rapid-fire rebellions have demonstrated anything, it is that these movements are snuffed out far too quickly, whether by repression or political cooptation, while the structures they opposed reconstitute themselves in more terrifying and dangerous forms. Witness Egypt. Or the inequalities that have grown even more obscene since the 2008 economic crisis, despite the many movements that rose up to resist the bailouts and austerity measures.

I have, in the past, strongly defended the right of young movements to their amorphous structures—whether that means rejecting identifiable leadership or eschewing programmatic demands. And there is no question that old political habits and structures must be reinvented to reflect new realities, as well as past failures. But I confess that the last five years immersed in climate science has left me impatient. As many are coming to realize, the fetish for structurelessness, the rebellion against any kind of institutionalization, is not a luxury today's transformative movements can afford.

The core of the problem comes back to the same inescapable fact that has both blocked climate action and accelerated emissions: all of us are living in the world that neoliberalism built, even if we happen to be critics of neoliberalism.

In practice that means that, despite endless griping, tweeting, flash mobbing, and occupying, we collectively lack many of the tools that built and sustained the transformative movements of the past. Our public institutions are disintegrating, while the institutions of the traditional left—progressive political parties, strong unions, membership-based community service organizations—are fighting for their lives.

And the challenge goes deeper than a lack of institutional tools and reaches into our very selves. Contemporary capitalism has not just accelerated the behaviors that are changing the climate. This economic model has changed a great many of us as individuals, accelerated and uprooted and dematerialized us as surely as it has finance capital, leaving us at once everywhere and nowhere. These are the hand-wringing clichés of our time— What is Twitter doing to my attention span? What are screens doing to our relationships?—but the preoccupations have particular relevance to the way we relate to the climate challenge.

Because this is a crisis that is, by its nature, slow moving and intensely place based. In its early stages, and in between the wrenching disasters, climate is about an early blooming of a particular flower, an unusually thin layer of ice on a lake, the late arrival of a migratory bird—noticing these small changes requires the kind of communion that comes from knowing a place deeply, not just as scenery but also as sustenance, and when local knowledge is passed on with a sense of sacred trust from one generation to the next. How many of us still live like that? Similarly, climate change is also about the inescapable impact of the actions of past generations not just on the present, but on generations in the future. These time frames are a language that has become foreign to a great many of us. Indeed Western culture has worked very hard to erase Indigenous cosmologies that call on the past and the future to interrogate present-day actions, with long-dead ancestors always present, alongside the generations yet to come.

In short: more bad timing. Just when we needed to slow down and notice the subtle changes in the natural world that are telling us that something is seriously amiss, we have sped up; just when we needed longer time horizons to see how the actions of our past impact the prospects for our future, we entered into the never-ending feed of the perpetual now, slicing and dicing our attention spans as never before.

To understand how we got to this place of profound disconnection from our surroundings and one another, and to think about how we might build a politics based on reconnection, we will need to go back a good deal further than 1988. Because the truth is that, while contemporary, hyper-globalized capitalism has exacerbated the climate crisis, it did not create it. We started treating the atmosphere as our waste dump when we began using coal on a commercial scale in the late 1700s and engaged in similarly reckless ecological practices well before that.

Moreover, humans have behaved in this shortsighted way not only under capitalist systems, but under systems that called themselves socialist as well (whether they were or not remains a subject of debate). Indeed the roots of the climate crisis date back to core civilizational myths on which post-Enlightenment Western culture is founded—myths about humanity's duty to dominate a natural world that is believed to be at once limitless and entirely controllable. This is not a problem that can be blamed on the political right or on the United States; these are powerful cultural narratives that transcend geography and ideological divides.

I have, so far, emphasized the familiarity of many of the deep solutions to the climate crisis and there is real comfort to take from that. It means that in many of our key responses, we would not be embarking on this tremendous project from scratch but rather drawing on more than a century of progressive work. But truly rising to the climate challenge—particularly its challenge to economic growth—will require that we dig even deeper into our past, and move into some distinctly uncharted political territory.

BEYOND EXTRACTIVISM

Confronting the Climate Denier Within

"The best thing about the Earth is if you poke holes in it oil and gas comes out."

—Republican U.S. Congressman Steve Stockman, 2013[1]

"The open veins of Latin America are still bleeding."

—Bolivian Indigenous leader Nilda Rojas Huanca, 2014[2]

"It is our predicament that we live in a finite world, and yet we behave as if it were infinite. Steady exponential material growth with no limits on resource consumption and population is the dominant conceptual model used by today's decision makers. This is an approximation of reality that is no longer accurate and [has] started to break down."

—Global systems analyst Rodrigo Castro and colleagues, paper presented at a scientific modeling conference, 2014[3]

For the past few years, the island of Nauru has been on a health kick. The concrete walls of public buildings are covered in murals urging regular exercise and healthy eating, and warning against the danger of diabetes. Young people are asking their grandparents how to fish, a lost skill. But there is a problem. As Nerida-Ann Steshia Hubert, who works at a diabetes center on the island, explains, life spans on Nauru are short, in part because of an epidemic of the disease. "The older folks are passing away early and we're losing a lot of the knowledge with them. It's like a race against time—trying to get the knowledge from them before they die."[4]

For decades, this tiny, isolated South Pacific island, just twenty-one square kilometers and home to ten thousand people, was held up as a model for the world—a developing country that was doing everything right. In the early 1960s, the Australian government, whose troops seized control of Nauru from the Germans in 1914, was so proud of its protectorate that it made promotional videos showing the Micronesians in starched white Bermuda shorts, obediently following lessons in English-speaking schools, settling their disputes in British-style courts, and shopping for modern conveniences in well-stocked grocery stores.[5]

During the 1970s and 1980s, after Nauru had earned independence, the island was periodically featured in press reports as a place of almost obscene riches, much as Dubai is invoked today. An Associated Press article from 1985 reported that Nauruans had "the world's highest per capita gross national product . . . higher even than Persian Gulf oil Sheikdoms." Everyone had free health care, housing, and education; homes were kept cool with air-conditioning; and residents zoomed around their tiny island—it took twenty minutes to make the entire loop—in brand-new cars and motorcycles. A police chief famously bought himself a yellow Lamborghini. "When I was young," recalls Steshia Hubert, "we would go to parties where people would throw thousands of dollars on the babies. Extravagant parties—first, sixteenth, eighteenth, twenty-first, and fiftieth birthdays. . . . They would come with gifts like cars, pillows stuffed with hundred-dollar bills—for one-year-old babies!"[6]

All of Nauru's monetary wealth derived from an odd geological fact. For hundreds of thousands of years, when the island was nothing but a cluster of coral reefs protruding from the waves, Nauru was a popular pit stop for migrating birds, who dropped by to feast on the shellfish and mollusks. Gradually, the bird poop built up between the coral towers and spires, eventually hardening to form a rocky landmass. The rock was then covered over in topsoil and dense forest, creating a tropical oasis of coconut palms, tranquil beaches, and thatched huts so beatific that the first European visitors dubbed the island Pleasant Isle.[7]

For thousands of years, Nauruans lived on the surface of their island, sustaining themselves on fish and black noddy birds. That began to change when a colonial officer picked up a rock that was later discovered to be made of almost pure phosphate of lime, a valuable agricultural fertilizer. A German-British firm began mining, later replaced by a British–Australian–New Zealand venture.[8] Nauru started developing at record speed—the catch was that it was, simultaneously, commiting suicide.

By the 1960s, Nauru still looked pleasant enough when approached from the sea, but it was a mirage. Behind the narrow fringe of coconut palms circling the coast lay a ravaged interior. Seen from above, the forest and topsoil of the oval island were being voraciously stripped away; the phosphate mined down to the island's sharply protruding bones, leaving behind a forest of ghostly coral totems. With the center now uninhabitable and largely infertile except for some minor scrubby vegetation, life on Nauru unfolded along the thin coastal strip, where the homes and civic structures were located.[9]

Nauru's successive waves of colonizers—whose economic emissaries ground up the phosphate rock into fine dust, then shipped it on ocean liners to fertilize soil in Australia and New Zealand—had a simple plan for the country: they would keep mining phosphate until the island was an empty shell. "When the phosphate supply is exhausted in thirty to forty years' time, the experts predict that the estimated population will not be able to live on this pleasant little island," a Nauruan council member said, rather stiffly, in a sixties-era black-and-white video produced by the Australian government. But not to worry, the film's narrator explained: "Preparations are being made now for the future of the Nauruan people. Australia has offered them a permanent home within her own shores. . . . Their prospects are bright; their future is secure."[10]

Nauru, in other words, was developed to disappear, designed by the Australian government and the extractive companies that controlled its fate as a disposable country. It's not that they had anything against the place, no genocidal intent per se. It's just that one dead island that few even knew existed seemed like an acceptable sacrifice to make in the name of the progress represented by industrial agriculture.

When the Nauruans themselves took control of their country in 1968, they had hopes of reversing these plans. Toward that end, they put a large chunk of their mining revenues into a trust fund that they invested in what seemed like stable real estate ventures in Australia and Hawaii. The goal was to live off the fund's proceeds while winding down phosphate mining and beginning to rehabilitate their island's ecology—a costly task, but perhaps not impossible.[11]

The plan failed. Nauru's government received catastrophically bad investment advice, and the country's mining wealth was squandered. Meanwhile, Nauru continued to disappear, its white powdery innards loaded onto boats as the mining continued unabated. Meanwhile, decades of easy money had taken a predictable toll on Nauruans' life and culture. Politics was rife with corruption, drunk driving was a leading cause of death, average life expectancy was dismally low, and Nauru earned the dubious honor of being featured on a U.S. news show as "the fattest place on Earth" (half the adult population suffers from type 2 diabetes, the result of a diet comprised almost exclusively of imported processed food). "During the golden era when the royalties were rolling in, we didn't cook, we ate in restaurants," recalls Steshia Hubert, a health care worker. And even if the Nauruans had wanted to eat differently, it would have been hard: with so much of the island a latticework of deep dark holes, growing enough fresh produce to feed the population was pretty much impossible. A bitterly ironic infertility for an island whose main export was agricultural fertilizer.[12]

By the 1990s, Nauru was so desperate for foreign currency that it pursued some distinctly shady get-rich-quick schemes. Aided greatly by the wave of financial deregulation unleashed in this period, the island became a prime money-laundering haven. For a time in the late 1990s, Nauru was the titular "home" to roughly four hundred phantom banks that were utterly unencumbered by monitoring, oversight, taxes, and regulation. Nauru-registered shell banks were particularly popular among Russian gangsters, who reportedly laundered a staggering $70 billion of dirty money through the island nation (to put that in perspective, Nauru's entire GDP is $72 million, according to most recent figures). Giving the country partial credit for the collapse of the Russian economy, a *New York Times Magazine* piece in 2000 pronounced that "amid the recent proliferation of money-laundering centers that experts estimate has ballooned into a $5 trillion shadow economy, Nauru is Public Enemy #1."[13]

These schemes have since caught up with Nauru too, and now the country faces a double bankruptcy: with 90 percent of the island depleted from mining, it faces ecological bankruptcy; with a debt of at least $800 million, Nauru faces financial bankruptcy as well. But these are not Nauru's only problems. It now turns out that the island nation is highly vulnerable to a crisis it had virtually no hand in creating: climate change and the drought, ocean acidification, and rising waters it brings. Sea levels around Nauru have been steadily climbing by about 5 millimeters per year since 1993, and much more could be on the way if current trends continue. Intensified droughts are already causing severe freshwater shortages.[14]

A decade ago, Australian philosopher and professor of sustainability Glenn Albrecht set out to coin a term to capture the particular form of psychological distress that sets in when the homelands that we love and from which we take comfort are radically altered by extraction and industrialization, rendering them alienating and unfamiliar. He settled on "solastalgia," with its evocations of solace, destruction, and pain, and defined the new word to mean, "the homesickness you have when you are still at home." He explained that although this particular form of unease was once principally familiar to people who lived in sacrifice zones—lands decimated by open-pit mining, for instance, or clear-cut logging—it was fast becoming a universal human experience, with climate change creating a "new abnormal" wherever we happen to live. "As bad as local and regional negative transformation is, it is the big picture, the Whole Earth, which is now a home under assault. A feeling of global dread asserts itself as the planet heats and our climate gets more hostile and unpredictable," he writes.[15]

Some places are unlucky enough to experience both local and global solastalgia simultaneously. Speaking to the 1997 U.N. climate conference that adopted the Kyoto Protocol, Nauru's then-president Kinza Clodumar described the collective claustrophobia that had gripped his country: "We are trapped, a wasteland at our back, and to our front a terrifying, rising flood of biblical proportions."[16] Few places on earth embody the suicidal results of building our economies on polluting extraction more graphically than Nauru. Thanks to its mining of phosphate, Nauru has spent the last century disappearing from the inside out; now, thanks to our collective mining of fossil fuels, it is disappearing from the outside in.

In a 2007 cable about Nauru, made public by WikiLeaks, an unnamed U.S. official summed up his government's analysis of what went wrong on the island: "Nauru simply spent extravagantly, never worrying about to-morrow."[17] Fair enough, but that diagnosis is hardly unique to Nauru; our entire culture is extravagantly drawing down finite resources, never worrying about tomorrow. For a couple of hundred years we have been telling ourselves that we can dig the midnight black remains of other life forms out of the bowels of the earth, burn them in massive quantities, and that the airborne particles and gases released into the atmosphere—because we can't see them—will have no effect whatsoever. Or if they do, we humans, brilliant as we are, will just invent our way out of whatever mess we have made.

And we tell ourselves all kinds of similarly implausible no-consequences stories all the time, about how we can ravage the world and suffer no adverse effects. Indeed we are always surprised when it works out otherwise. We extract and do not replenish and wonder why the fish have disappeared and the soil requires ever more "inputs" (like phosphate) to stay fertile. We occupy countries and arm their militias and then wonder why they hate us. We drive down wages, ship jobs overseas, destroy worker protections, hollow out local economies, then wonder why people can't afford to shop as much as they used to. We offer those failed shoppers subprime mortgages instead of steady jobs and then wonder why no one foresaw that a system built on bad debts would collapse.

At every stage our actions are marked by a lack of respect for the powers we are unleashing—a certainty, or at least a hope, that the nature we have turned to garbage, and the people we have treated like garbage, will not come back to haunt us. And Nauru knows all about this too, because in the past decade it has become a dumping ground of another sort. In an effort to raise much needed revenue, it agreed to house an offshore refugee detention center for the government of Australia. In what has become known as "the Pacific Solution," Australian navy and customs ships intercept boats of migrants and immediately fly them three thousand kilometers to Nauru (as well as to several other Pacific islands). Once on Nauru, the migrants—most from Afghanistan, Sri Lanka, Iraq, Iran, and Pakistan—are crammed into a rat-infested guarded camp made up of rows of crowded, stiflingly hot tents. The island imprisonment can last up to five years, with the migrants in a state of constant limbo about their status, something the Australian government hopes will serve as a deterrent to future refugees.[18]

The Australian and Nauruan governments have gone to great lengths to limit information on camp conditions and have prevented journalists who make the long journey to the island from seeing where migrants are being housed. But the truth is leaking out nonetheless: grainy video of prisoners chanting "We are not animals"; reports of mass hunger strikes and suicide attempts; horrifying photographs of refugees who had sewn their own mouths shut, using paper clips as needles; an image of a man who had badly mutilated his neck in a failed hanging attempt. There are also images of toddlers playing in the dirt and huddling with their parents under tent flaps for shade (originally the camp had housed only adult males, but now hundreds of women and children have been sent there too). In June 2013, the Australian government finally allowed a BBC crew into the camp in order to show off its brand-new barracks—but that PR attempt was completely upstaged one month later by the news that a prisoner riot had almost completely destroyed the new facility, leaving several prisoners injured.[19]

Amnesty International has called the camp on Nauru "cruel" and "degrading," and a 2013 report by the United Nations High Commissioner for Refugees concluded that those conditions, "coupled with the protracted period spent there by some asylum-seekers, raise serious issues about their compatibility with international human rights law, including the prohibition against torture and cruel, inhuman or degrading treatment." Then, in March 2014, a former Salvation Army employee named Mark Isaacs, who had been stationed at the camp, published a tell-all memoir titled *The Undesirables*. He wrote about men who had survived wars and treacherous voyages losing all will to live on Nauru, with one man resorting to swallowing cleaning fluids, another driven mad and barking like a dog. Isaacs likened the camp to "death factories," and said in an interview that it is about "taking resilient men and grinding them into the dust." On an island that itself was systematically ground to dust, it's a harrowing image. As harrowing as enlisting the people who could very well be the climate refugees of tomorrow to play warden to the political and economic refugees of today.[20]

Reviewing the island's painful history, it strikes me that so much of what has gone wrong on Nauru—and goes on still—has to do with its location, frequently described as "the middle of nowhere" or, in the words of a 1921 *National Geographic* dispatch, "perhaps the most remote territory in the world," a tiny dot "in lonely seas." The nation's remoteness made it a convenient trash can—a place to turn the land into trash, to launder dirty money, to disappear unwanted people, and now a place that may be allowed to disappear altogether.[21]

This is our relationship to much that we cannot easily see and it is a big part of what makes carbon pollution such a stubborn problem: we can't see it, so we don't really believe it exists. Ours is a culture of disavowal, of simultaneously knowing and not knowing—the illusion of proximity coupled with the reality of distance is the trick perfected by the fossil-fueled global market. So we both know and don't know who makes our goods, who cleans up after us, where our waste disappears to—whether it's our sewage or electronics or our carbon emissions.

But what Nauru's fate tells us is that there is no middle of nowhere, nowhere that doesn't "count"—and that nothing ever truly disappears. On some level we all know this, that we are part of a swirling web of connections. Yet we are trapped in linear narratives that tell us the opposite: that we can expand infinitely, that there will always be more space to absorb our waste, more resources to fuel our wants, more people to abuse.

These days, Nauru is in a near constant state of political crisis, with fresh corruption scandals perpetually threatening to bring down the government, and sometimes succeeding. Given the wrong visited upon the nation, the island's leaders would be well within their rights to point fingers outward—at their former colonial masters who flayed them, at the investors who fleeced them, and at the rich countries whose emissions now threaten to drown them. And some do. But several of Nauru's leaders have also chosen to do something else: to hold up their country as a kind of warning to a warming world.

In *The New York Times* in 2011, for instance, then-president Marcus Stephen wrote that Nauru provides "an indispensable cautionary tale about life in a place with hard ecological limits." It shows, he claimed, "what can happen when a country runs out of options. The world is headed down a similar path with the relentless burning of coal and oil, which is altering the planet's climate, melting ice caps, making oceans more acidic and edging us ever closer to a day when no one will be able to take clean water, fertile soil or abundant food for granted." In other words, Nauru isn't the only one digging itself to death; we all are.[22]

But the lesson Nauru has to teach is not only about the dangers of fossil fuel emissions. It is about the mentality that allowed so many of us, and our ancestors, to believe that we could relate to the earth with such violence in the first place—to dig and drill out the substances we desired while thinking little of the trash left behind, whether in the land and water where the extraction takes place, or in the atmosphere, once the extracted material is burned. This carelessness is at the core of an economic model some political scientists call "extractivism," a term originally used to describe economies based on removing ever more raw materials from the earth, usually for export to traditional colonial powers, where "value" was added. And it's a habit of thought that goes a long way toward explaining why an economic model based on endless growth ever seemed viable in the first place. Though developed under capitalism, governments across the ideological spectrum now embrace this resource-depleting model as a road to development, and it is this logic that climate change calls profoundly into question.

Extractivism is a nonreciprocal, dominance-based relationship with the earth, one purely of taking. It is the opposite of stewardship, which involves taking but also taking care that regeneration and future life continue. Extractivism is the mentality of the mountaintop remover and the old-growth clear-cutter. It is the reduction of life into objects for the use of others, giving them no integrity or value of their own—turning living complex ecosystems into "natural resources," mountains into "overburden" (as the mining industry terms the forests, rocks, and streams that get in the way of its bulldozers). It is also the reduction of human beings either into labor to be brutally extracted, pushed beyond limits, or, alternatively, into social burden, problems to be locked out at borders and locked away in prisons or reservations. In an extractivist economy, the interconnections among these various objectified components of life are ignored; the consequences of severing them are of no concern.

Extractivism is also directly connected to the notion of sacrifice zones—places that, to their extractors, somehow don't count and therefore can be poisoned, drained, or otherwise destroyed, for the supposed greater good of economic progress. This toxic idea has always been intimately tied to imperialism, with disposable peripheries being harnessed to feed a glittering center, and it is bound up too with notions of racial superiority, because in order to have sacrifice zones, you need to have people and cultures who count so little that they are considered deserving of sacrifice. Extractivism ran rampant under colonialism because relating to the world as a frontier of conquest—rather than as home—fosters this particular brand of irresponsibility. The colonial mind nurtures the belief that there is always somewhere else to go to and exploit once the current site of extraction has been exhausted.

These ideas predate industrial-scale extraction of fossil fuels. And yet the ability to harness the power of coal to power factories and ships is what, more than any single other factor, enabled these dangerous ideas to conquer the world. It's a history worth exploring in more depth, because it goes a long way toward explaining how the climate crisis challenges not only capitalism but the underlying civilizational narratives about endless growth and progress within which we are all, in one way or another, still trapped.

The Ultimate Extractivist Relationship

If the modern-day extractive economy has a patron saint, the honor should probably go to Francis Bacon. The English philosopher, scientist, and statesman is credited with convincing Britain's elites to abandon, once and for all, pagan notions of the earth as a life-giving mother figure to whom we owe respect and reverence (and more than a little fear) and accept the role as her dungeon master. "For you have but to follow and as it were hound nature in her wanderings," Bacon wrote in *De Augmentis Scientiarum* in 1623, "and you will be able, when you like, to lead and drive her afterwards to the same place again. . . . Neither ought a man to make scruple of entering and penetrating into these holes and corners, when the inquisition of truth is his sole object."[23] (Not surprisingly, feminist scholars have filled volumes analyzing the ex–Lord Chancellor's metaphor choices.)

These ideas of a completely knowable and controllable earth animated not only the Scientific Revolution but, critically, the colonial project as well, which sent ships crisscrossing the globe to poke and prod and bring the secrets, and wealth, back to their respective crowns. The mood of human invincibility that governed this epoch was neatly encapsulated in the words of clergyman and philosopher William Derham in his 1713 book *Physico-Theology*: "We can, if need be, ransack the whole globe, penetrate into the bowels of the earth, descend to the bottom of the deep, travel to the farthest regions of this world, to acquire wealth."[24]

And yet despite this bravado, throughout the 1700s, the twin projects of colonialism and industrialization were still constrained by nature on several key fronts. Ships carrying both slaves and the raw materials they harvested could sail only when winds were favorable, which could lead to long delays in the supply chain. The factories that turned those raw materials into finished products were powered by huge water wheels. They needed to be located next to waterfalls or rapids which made them dependent on the flow and levels of rivers. As with high or low winds at sea, an especially dry or wet spell meant that working hours in the textile, flour, and sugar mills had to be adjusted accordingly—a mounting annoyance as markets expanded and became more global.

Many water-powered factories were, by necessity, spread out around the countryside, near bodies of fast-moving water. As the Industrial Revolution matured and workers in the mills started to strike and even riot for better wages and conditions, this decentralization made factory owners highly vulnerable, since quickly finding replacement workers in rural areas was difficult.

Beginning in 1776, a Scottish engineer named James Watt perfected and manufactured a power source that offered solutions to all these vulnerabilities. Lawyer and historian Barbara Freese describes Watt's steam engine as "perhaps the most important invention in the creation of the modern world"—and with good reason.[25] By adding a separate condenser, air pump, and later a rotary mechanism to an older model, Watt was able to make the coal-fired steam engine vastly more powerful and adaptable than its predecessors. In contrast, the new machines could power a broad range of industrial operations, including, eventually, boats.

For the first couple of decades, the new engine was a tough sell. Water power, after all, had a lot going for it compared with coal. For one thing, it was free, while coal needed to be continually re-purchased. And contrary to the widespread belief that the steam engine provided more energy than water wheels, the two were actually comparable, with the larger wheels packing several times more horsepower than their coal-powered rivals. Water wheels also operated more smoothly, with fewer technical breakdowns, so long as the water was flowing. "The transition from water to steam in the British cotton industry did not occur because water was scarce, less powerful, or more expensive than steam," writes Swedish coal expert Andreas Malm. "To the contrary, steam gained supremacy *in spite of water being abundant, at least as powerful, and decidedly cheaper.*"[26]

As Britain's urban population ballooned, two factors tipped the balance in favor of the steam engine. The first was the new machine's insulation from nature's fluctuations: unlike water wheels, steam engines worked at the same rate all the time, so long as there was coal to feed them and the machinery wasn't broken. The flow rates of rivers were of no concern. Steam engines also worked anywhere, regardless of the geography, which meant that factory owners could shift production from more remote areas to cities like London, Manchester, and Lancaster, where there were gluts of willing industrial workers, making it far easier to fire troublemakers and put down strikes. As an 1832 article written by a British economist explained, "The invention of the steam-engine has relieved us from the necessity of building factories in inconvenient situations merely for the sake of a waterfall." Or as one of Watt's early biographers put it, the generation of power "will no longer depend, as heretofore, on the most inconstant of natural causes—on atmospheric influences."[27]

Similarly, when Watt's engine was installed in a boat, ship crews were liberated from having to adapt their journeys to the winds, a development that rapidly accelerated the colonial project and the ability of European powers to easily annex countries in distant lands. As the Earl of Liverpool put it in a public meeting to memorialize James Watt in 1824, "Be the winds friendly or be they contrary, the power of the Steam Engine overcomes all difficulties. . . . Let the wind blow from whatever quarter it may, let the destination of our force be to whatever part of the world it may, you have the power and the means, by the Steam Engine, of applying that force at the proper time and in the proper manner."[28] Not until the advent of electronic trading would commerce feel itself so liberated from the constraints of living on a planet bound by geography and governed by the elements.

Unlike the energy it replaced, power from fossil fuel always required sacrifice zones—whether in the black lungs of the coal miners or the poisoned waterways surrounding the mines. But these prices were seen as worth paying in exchange for coal's intoxicating promise of freedom from the physical world—a freedom that unleashed industrial capitalism's full force to dominate both workers and other cultures. With their portable energy creator, the industrialists and colonists of the 1800s could now go wherever labor was cheapest and most exploitable, and wherever resources were most plentiful and valuable. As the author of a steam engine manual wrote in the mid-1830s, "Its mighty services are always at our command, whether in winter or in summer, by day or by night—it knows of no intermission but what our wishes dictate."[29] Coal represented, in short, total domination, of both nature and other people, the full realization of Bacon's dream at last. "Nature can be conquered," Watt reportedly said, "if we can but find her weak side."[30]

Little wonder then that the introduction of Watt's steam engine coincided with explosive levels of growth in British manufacturing, such that in the eighty years between 1760 and 1840, the country went from importing 2.5 million pounds of raw cotton to importing 366 million pounds of raw cotton, a genuine revolution made possible by the potent and brutal combination of coal at home and slave labor abroad.[31]

This recipe produced more than just new consumer products. In *Ecological Economics*, Herman Daly and Joshua Farley point out that Adam Smith published *The Wealth of Nations* in 1776—the same year that Watt produced his first commercial steam engine. "It is no coincidence," they write, "that the market economy and fossil fuel economy emerged at essentially the exact same time. . . . New technologies and vast amounts of fossil energy allowed unprecedented production of consumer goods. The need for new markets for these mass-produced consumer goods and new sources of raw material played a role in colonialism and the pursuit of empire. The market economy evolved as an efficient way of allocating such goods, and stimulating the production of even more."[32] Just as colonialism needed coal to fulfill its dream of total domination, the deluge of products made possible by both coal and colonialism needed modern capitalism.

The promise of liberation from nature that Watt was selling in those early days continues to be the great power of fossil fuels. That power is what allows today's multinationals to scour the globe for the cheapest, most exploitable workforce, with natural features and events that once appeared as obstacles—vast oceans, treacherous landscapes, seasonal fluctuations—no longer even registering as minor annoyances. Or so it seemed for a time.

———

It is often said that Mother Nature bats last, and this has been poignantly the case for some of the men who were most possessed by the ambition of conquering her. A perhaps apocryphal story surrounds the death of Francis Bacon: in an attempt to test his hypothesis that frozen meat could be prevented from rotting, he traipsed around in chilly weather stuffing a chicken full of snow. As a result, it is said, the philosopher caught pneumonia, which eventually led to his demise.[33] Despite some controversy, the anecdote survives for its seeming poetic justice: a man who thought nature could be bent to his will died from simple exposure to the cold.

A similar story of comeuppance appears to be unfolding for the human race as a whole. Ralph Waldo Emerson called coal "a portable climate"— and it has been a smash success, carrying countless advantages, from longer life spans to hundreds of millions freed from hard labor.[34] And yet precisely because our bodies are so effectively separated from our geographies, we who have access to this privilege have proven ourselves far too capable of ignoring the fact that we aren't just changing our personal climate but the entire planet's climate as well, warming not just the indoors but the outdoors too. And yet the warming is no less real for our failure to pay attention.

The harnessing of fossil fuel power seemed, for a couple of centuries at least, to have freed large parts of humanity from the need to be in constant dialogue with nature, having to adjust its plans, ambitions, and schedules to natural fluctuations and topographies. Coal and oil, precisely because they were fossilized, seemed entirely possessable forms of energy. They did not behave independently—not like wind, or water, or, for that matter, workers. Just as Watt's engine promised, once purchased, they produced power wherever and whenever their owners wished—the ultimate nonreciprocal relationship.

But what we have learned from atmospheric science is that the give-and-take, call-and-response that is the essence of all relationships in nature was not eliminated with fossil fuels, it was merely delayed, all the while gaining force and velocity. Now the cumulative effect of those centuries of burned carbon is in the process of unleashing the most ferocious natural tempers of all.

As a result, the illusion of total power and control Watt and his cohorts once peddled has given way to the reality of near total powerlessness and loss of control in the face of such spectacular forces as Hurricane Sandy and Typhoon Haiyan. Which is just one of the reasons climate change is so deeply frightening. Because to confront this crisis truthfully is to confront ourselves—to reckon, as our ancestors did, with our vulnerability to the elements that make up both the planet and our bodies. It is to accept (even embrace) being but one porous part of the world, rather than its master or machinist, as Bacon long ago promised. There can be great well-being in that realization of interconnection, pleasure too. But we should not underestimate the depth of the civilizational challenge that this relationship represents. As Australian political scientist Clive Hamilton puts it, facing these truths about climate change "means recognizing that the power relation between humans and the earth is the reverse of the one we have assumed for three centuries."[35]

For one of those centuries, a huge white marble statue of James Watt dominated St. Paul's chapel in Westminster Abbey, commemorating a man who "enlarged the resources of his Country" and "increased the power of Man." And Watt certainly did that: his engine massively accelerated the Industrial Revolution and the steamships his engine made possible subsequently opened sub-Saharan Africa and India to colonial pillage. So while making Europe richer, he also helped make many other parts of the world poorer, carbon-fueled inequalities that persist to this day. Indeed, coal was the black ink in which the story of modern capitalism was written.

But all the facts were not yet in when Watt was being memorialized in marble in 1825. Because it is the cumulative impact of the carbon emissions that began in those early mills and mines that has already engraved itself in the geologic record—in the levels of the oceans, in their chemical composition, in the slow erasure of islands like Nauru; in the retreat of glaciers, the collapse of ice shelves, the thawing of permafrost; in the disturbed soil cycles and in the charred forests.

Indeed, it turns out that coal's earliest casualties—the miners who died from black lung, the workers in the Satanic Mills—were not merely the price of progress. They were also an early warning that we were unleashing a poisonous substance onto the world. "It has become clear over the last century," writes Ecuadorian ecologist Esperanza Martínez, "that fossil fuels, the energy sources of capitalism, destroy life—from the territories where they are extracted to the oceans and the atmosphere that absorb the waste."[36]

Jean-Paul Sartre called fossil fuels "capital bequeathed to mankind by other living beings"; they are quite literally the decayed remnants of long-dead life-forms. It's not that these substances are evil; it's just that they belong where they are: in the ground, where they are performing valuable ecological functions. Coal, when left alone, helpfully sequesters not just the carbon long ago pulled out of the air by plants, but all kinds of other toxins. It acts, as world-renowned Australian climate scientist Tim Flannery puts it, like "a natural sponge that absorbs many substances dissolved in groundwater, from uranium to cadmium and mercury."[37]

When coal is dug up and burned, however, those toxins are released in the ecosystem, eventually making their way into the oceans, where they are absorbed by krill and plankton, then by fish, and then by us. The released carbon, meanwhile, enters the atmosphere, causing global warming (not to mention coal's contribution to the smog and particulate pollution that have plagued urban society since the Industrial Revolution, afflicting untold numbers of people with respiratory, heart, and other diseases).

Given this legacy, our task is not small, but it is simple: rather than a society of grave robbers, we need to become a society of life amplifiers, deriving our energy directly from the elements that sustain life. It's time to let the dead rest.

The Extractivist Left

The braided historical threads of colonialism, coal, and capitalism shed significant light on why so many of us who are willing to challenge the injustices of the market system remain paralyzed in the face of the climate threat. Fossil fuels, and the deeper extractivist mind-set that they represent, built the modern world. If we are part of industrial or postindustrial societies, we are still living inside the story written in coal.

Ever since the French Revolution, there have been pitched ideological battles within the confines of this story: communists, socialists, and trade unions have fought for more equal distribution of the spoils of extraction, winning major victories for the poor and working classes. And the human rights and emancipation movements of this period have also fought valiantly against industrial capitalism's treatment of whole categories of our species as human sacrifice zones, no more deserving of rights than raw commodities. These struggles have also won major victories against the dominance-based paradigm—against slavery, for universal suffrage, for equality under the law. And there have been voices in all of these movements, moreover, that identified the parallels between the economic model's abuse of the natural world and its abuse of human beings deemed worthy of being sacrificed, or at least uncounted. Karl Marx, for instance, recognized capitalism's "irreparable rift" with "the natural laws of life itself," while feminist scholars have long recognized that patriarchy's dual war against women's bodies and against the body of the earth were connected to that essential, corrosive separation between mind and body—and between body and earth—from which both the Scientific Revolution and Industrial Revolution sprang.[38]

These challenges, however, were mainly in the intellectual realm; Bacon's original, biblically inspired framework remained largely intact—the right of humans to place ourselves above the ecosystems that support us and to abuse the earth as if it were an inanimate machine. The strongest challenges to this worldview have always come from outside its logic, in those historical junctures when the extractive project clashes directly with a different, older way of relating to the earth—and that older way fights back. This has been true from the earliest days of industrialization, when English and Irish peasants, for instance, revolted against the first attempts to enclose communal lands, and it has continued in clashes between colonizers and Indigenous peoples through the centuries, right up to—as we will see—the Indigenous-led resistance to extreme fossil fuel extraction gaining power today.

But for those of us born and raised inside this system, though we may well see the dead-end flaw of its central logic, it can remain intensely difficult to see a way out. And how could it be otherwise? Post-Enlightenment Western culture does not offer a road map for how to live that is not based on an extractivist, nonreciprocal relationship with nature.

This is where the right-wing climate deniers have overstated their conspiracy theories about what a cosmic gift global warming is to the left. It is true, as I have outlined, that many climate responses reinforce progressive support for government intervention in the market, for greater equality, and for a more robust public sphere. But the deeper message carried by the ecological crisis—that humanity has to go a whole lot easier on the living systems that sustain us, acting regeneratively rather than extractively—is a profound challenge to large parts of the left as well as the right. It's a challenge to some trade unions, those trying to freeze in place the dirtiest jobs, instead of fighting for the good clean jobs their members deserve. And it's a challenge to the overwhelming majority of center-left Keynesians, who still define economic success in terms of traditional measures of GDP growth, regardless of whether that growth comes from rampant resource extraction. (This is all the more baffling because Keynes himself, like John Stuart Mill, advocated a transition to a post-growth economy.)

It's a challenge, too, to those parts of the left that equated socialism with the authoritarian rule of the Soviet Union and its satellites (though there was always a rich tradition, particularly among anarchists, that considered Stalin's project an abomination of core social justice principles). Because the fact is that those self-described socialist states devoured resources with as much enthusiasm as their capitalist counterparts, and spewed waste just as recklessly. Before the fall of the Berlin Wall, for instance, Czechs and Russians had even higher carbon footprints per capita than Canadians and Australians. Which is why one of the only times the developed world has seen a precipitous emissions drop was after the economic collapse of the former Soviet Union in the early 1990s. Mao Zedong, for his part, openly declared that "man must conquer nature," setting loose a devastating onslaught on the natural world that transitioned seamlessly from clear-cuts under communism to mega-dams under capitalism. Russia's oil and gas companies, meanwhile, were as reckless and accident-prone under state socialist control as they are today in the hands of the oligarchs and Russia's corporatist state.[39]

And why wouldn't they be? Authoritarian socialism and capitalism share strong tendencies toward centralizing (one in the hands of the state, the other in the hands of corporations). They also both keep their respective systems going through ruthless expansion—whether through production for production's sake, in the case of Soviet-era socialism, or consumption for consumption's sake, in the case of consumer capitalism.

One possible bright spot is Scandinavian-style Social Democracy, which has undoubtedly produced some of the most significant green breakthroughs in the world, from the visionary urban design of Stockholm, where roughly 74 percent of residents walk, bike, or take public transit to work, to Denmark's community-controlled wind power revolution. And yet Norway's late-life emergence as a major oil producer—with majority state-owned Statoil tearing up the Alberta tar sands and gearing up to tap massive reserves in the Arctic—calls into question whether these countries are indeed charting a path away from extractivism.[40]

In Latin America and Africa, moving away from overdependence on raw resource extraction and export, and toward more diversified economies, has always been a central piece of the postcolonial project. And yet some countries where left and center-left governments have come to power over the last decade are moving in the opposite direction. The fact that this tendency is little discussed outside the continent should not be surprising. Progressives around the world have rightfully cheered Latin America's electoral "pink tide," with government after government coming to power promising to reduce inequality, tackle extreme poverty, and take back control over the extractive industries of their respective countries. And purely from the perspective of poverty reduction, the results have often been stunning.

Since the election of Luiz Inácio Lula da Silva, and now under the leadership of his former chief of staff, Dilma Rousseff, Brazil has reduced its extreme poverty rate by 65 percent in a single decade, according to the government. More than thirty million people have been lifted out of poverty. After the election of Hugo Chávez, Venezuela slashed the percentage of the population living in extreme poverty by more than half—from 16.6 percent in 1999 to 7 percent in 2011, according to government statistics.

College enrollment has doubled since 2004. Ecuador under Rafael Correa has dropped its poverty rates by 32 per cent, according to the World Bank. In Argentina, urban poverty plummeted from 54.7 percent in 2003 to 6.5 percent in 2011, according to government data collected by the U.N.[41]

Bolivia's record, under the presidency of Evo Morales, is also impressive. It has reduced the proportion of its population living in extreme poverty from 38 percent in 2005 to 21.6 percent in 2012, according to government figures.[42] And unemployment rates have been cut in half. Most importantly, while other developing countries have used growth to create societies of big winners and big losers, Bolivia is actually succeeding in building a more equal society. Alicia Bárcena Ibarra, executive secretary of the U.N. Economic Commission for Latin America and the Caribbean, observes that in Bolivia "the gap between rich and poor has been hugely narrowed."[43]

All of this is a marked improvement over what came before, when the wealth extracted from each of these countries was overwhelmingly concentrated among a tiny elite, with far too much of it fleeing the continent entirely. And yet these left and center-left governments have so far been unable to come up with economic models that do not require extremely high levels of extraction of finite resources, often at tremendous ecological and human cost. This is true for Ecuador, with its growing oil dependence, including oil from the Amazon; Bolivia, with its huge dependence on natural gas; Argentina, with its continued support for open-pit mining and its "green deserts" of genetically modified soy and other crops; Brazil, with its highly contentious mega-dams and forays into high-risk offshore oil drilling; and of course it has always been the case for petro-dependent Venezuela. Moreover, most of these governments have made very little progress on the old dream of diversifying their economies away from raw resource exports—in fact, between 2004 and 2011, raw resources as a percentage of overall exports increased in all of these countries except Argentina, though some of this increase was no doubt due to rising commodity prices. It hasn't helped that China has been throwing easy credit around the continent, in some cases demanding to be paid back in oil.[44]

This reliance on high risk and ecologically damaging forms of extraction is particularly disappointing in the governments of Evo Morales in Bolivia and Rafael Correa in Ecuador. In their first terms, both had signaled that a new, nonextractive chapter was beginning in their countries. Part of this involved granting real respect to the Indigenous cultures that had survived centuries of marginalization and oppression and that form powerful political constituencies in both countries. Under Morales and Correa, the Indigenous concepts of *sumak kawsay* and *buen vivir*, which strive to build societies in harmony with nature (in which everyone has enough, rather than more and more), became the discourse of government, even recognized in law. But in both cases, escalating industrial-scale development and extraction has overtaken this promising rhetoric. According to Ecuador's Esperanza Martínez, "Since 2007, Correa's has been the most extractivist government in the history of the country, in terms of oil and now also mining." Indeed Latin American intellectuals have invented a new term to describe what they are experiencing: "progressive extractivism."[45]

The governments claim they have no choice—that they need to pursue extractive policies in order to pay for programs that alleviate poverty. And in many ways this explanation comes back to the question of climate debt: Bolivia and Ecuador have been at the forefront of the coalition of governments asking that the countries responsible for the bulk of historical greenhouse gas emissions help to pay for the Global South's transition away from dirty energy and toward low-carbon development. These calls have been alternatively ignored and dismissed. Forced to choose between poverty and pollution, these governments are choosing pollution, but those should not be their only options.

The default overreliance on dirty extraction is not only a problem for progressives in the developing world. In Greece in May 2013, for instance, I was surprised to discover that the left-wing Syriza party—then the country's official opposition and held up by many progressive Europeans as the great hope for a real political alternative on the continent—did not oppose the governing coalition's embrace of new oil and gas exploration. Instead, it argued that any funds raised by the effort should be spent on pensions, not used to pay back creditors. In other words: they were not providing an alternative to extractivism but simply had better plans for distributing the spoils.

Far from seeing climate change as an opportunity to argue for their socialist utopia, as conservative climate change deniers fear, Syriza had simply stopped talking about global warming altogether.

This is something that the party's leader, Alexis Tsipras, admitted to me quite openly in an interview: "We were a party that had the environment and climate change in the center of our interest," he said. "But after these years of depression in Greece, we forgot climate change."[46] At least he was honest.

The good news, and it is significant, is that large and growing social movements in all of these countries are pushing back against the idea that extraction-and-redistribution is the only route out of poverty and economic crisis. There are massive movements against gold mining in Greece, so large that Syriza has become a significant opponent of the mines. In Latin America, meanwhile, progressive governments are increasingly finding themselves in direct conflict with many of the people who elected them, facing accusations that their new model of what Hugo Chávez called "Twenty-first-Century Socialism" simply isn't new enough. Huge hydro dams in Brazil, highways through sensitive areas in Bolivia, and oil drilling in the Ecuadorian Amazon have all become internal flashpoints. Yes, the wealth is better distributed, particularly among the urban poor, but outside the cities, the ways of life of Indigenous peoples and peasants are still being endangered without their consent, and they are still being made landless by ecosystem destruction. What is needed, writes Bolivian environmentalist Patricia Molina, is a new definition of development, "so that the goal is the elimination of poverty, and not of the poor."[47]

This critique represents more than just the push and pull of politics; it is a fundamental shift in the way an increasingly large and vocal political constituency views the goal of economic activity and the meaning of development. Space is opening up for a growing influence of Indigenous thought on new generations of activists, beginning, most significantly, with Mexico's Zapatista uprising in 1994, and continuing, as we will see, with the important leadership role that Indigenous land-rights movements are playing in pivotal anti-extraction struggles in North America, Latin America, Australia, and New Zealand. In part through these struggles, non-Indigenous progressive movements are being exposed to worldviews based on relationships of reciprocity and interconnection with the natural world that are the antithesis of extractivism. These movements have truly heard the message of climate change and are winning battles to keep significant amounts of carbon in the ground.

Some Warnings, Unheeded

There is one other group that might have provided a challenge to Western culture's disastrous view of nature as a bottomless vending machine. That group, of course, is the environmental movement, the network of organizations that exists to protect the natural world from being devoured by human activity. And yet the movement has not played this role, at least not in a sustained and coherent manner.

In part, that has to do with the movement's unusually elite history, particularly in North America. When conservationism emerged as a powerful force in the late nineteenth and early twentieth centuries, it was primarily about men of privilege who enjoyed fishing, hunting, camping, and hiking and who recognized that many of their favorite wilderness spots were under threat from the rapid expansion of industrialization. For the most part, these men did not call into question the frenetic economic project that was devouring natural landscapes all over the continent—they simply wanted to make sure that some particularly spectacular pockets were set aside for their recreation and aesthetic appreciation. Like the Christian missionaries who traveled with traders and soldiers, most early preservationists saw their work as a civilizing addendum to the colonial and industrial projects—not as a challenge to them. Writing in 1914, Bronx Zoo director William Temple Hornaday summed up this ethos, urging American educators to "take up their share of the white man's burden" and help to "preserve the wild life of our country."[48]

This task was accomplished not with disruptive protests, which would have been unseemly for a movement so entrenched in the upper stratum of society. Instead, it was achieved through quiet lobbying, with well-bred men appealing to the noblesse oblige of other men of their class to save a cherished area by turning it into a national or state park, or a private family preserve—often at the direct expense of Indigenous people who lost access to these lands as hunting and fishing grounds.

There were those in the movement, however, who saw in the threats to their country's most beautiful places signs of a deeper cultural crisis. For instance, John Muir, the great naturalist writer who helped found the Sierra Club in 1892, excoriated the industrialists who dammed wild rivers and drowned beautiful valleys. To him they were heathens—"devotees of ravaging commercialism" who "instead of lifting their eyes to the God of the mountains, lift them to the Almighty Dollar."[49]

He was not the only heretic. A strain of radicalism drove some of the early Western ecological thinkers to argue for doing more than protecting isolated landscapes. Though frequently unacknowledged, these thinkers often drew heavily on Eastern beliefs about the interconnectedness of all life, as well as on Native American cosmologies that see all living creatures as our "relations."

In the mid-1800s, Henry David Thoreau wrote that, "The earth I tread on is not a dead, inert mass. It is a body, has a spirit, is organic, and fluid to the influence of its spirit, and to whatever particle of that spirit is in me."[*] This was a straight repudiation of Francis Bacon's casting of the earth as an inert machine whose mysteries could be mastered by the human mind. And almost a century after Thoreau, Aldo Leopold, whose book *A Sand County Almanac* was the touchstone for a second wave of environmentalists, similarly called for an ethic that "enlarges the boundaries of the community to include soils, waters, plants, and animals" and that recognizes "the individual is a member of a community of interdependent parts." A "land ethic," as he called it, "changes the role of *Homo sapiens* from conqueror of the land-community to plain member and citizen of it. It implies respect for his fellow-members, and also respect for the community as such."[50]

These ideas were hugely influential in the evolution of ecological thought, but unattached to populist movements, they posed little threat to galloping industrialization. The dominant worldview continued to see humans as a conquering army, subduing and mechanizing the natural world. Even so, by the 1930s, with socialism on the rise around the world, the more conservative elements of the growing environmental movement sought to

---

[*] "In the morning I bathe my intellect in the stupendous and cosmogonal philosophy of the Bhagvat Geeta," wrote Thoreau in *Walden* of the famous Indian scripture. He continued, "I lay down the book and go to my well for water and lo! there I meet the servant of the Brahmin, priest of Brahma and Vishnu and Indra, who still sits in his temple on the Ganges reading the Vedas, or dwells at the root of a tree with his crust and water jug. . . . The pure Walden water is mingled with the sacred water of the Ganges."

distance themselves from Leopold's "radical" suggestion that nature had an inherent value beyond its utility to man. If watersheds and old-growth forests had a "right to continued existence," as Leopold argued (a preview of the "rights of nature" debates that would emerge several decades later), then an owner's right to do what he wished with his land could be called into question. In 1935, Jay Norwood "Ding" Darling, who would later help found the National Wildlife Federation, wrote to Leopold warning him, "I can't get away from the idea that you are getting us out into water over our depth by your new philosophy of wildlife environment. The end of that road leads to socialization of property."[51]

By the time Rachel Carson published *Silent Spring* in 1962, the attempts to turn nature into a mere cog in the American industrial machine had grown so aggressive, so overtly militaristic, that it was no longer possible to pretend that combining capitalism with conservation was simply a matter of protecting a few pockets of green. Carson's book boiled over with righteous condemnations of a chemical industry that used aerial bombardment to wipe out insects, thoughtlessly endangering human and animal life in the process. The marine biologist-turned-social-critic painted a vivid picture of the arrogant "control men" who, enthralled with "a bright new toy," hurled poisons "against the fabric of life."[52]

Carson's focus was DDT, but for her the problem was not a particular chemical; it was a logic. "The 'control of nature,'" Carson wrote, "is a phrase conceived in arrogance, born of the Neanderthal age of biology and philosophy, when it was supposed that nature exists for the convenience of man. . . . It is our alarming misfortune that so primitive a science has armed itself with the most modern and terrible weapons, and that in turning them against the insects it has also turned them against the earth."[53]

Carson's writing inspired a new, much more radical generation of environmentalists to see themselves as part of a fragile planetary ecosystem rather than as its engineers or mechanics, giving birth to the field of Ecological Economics. It was in this context that the underlying logic of extractivism—that there would always be more earth for us to consume—began to be forcefully challenged within the mainstream. The pinnacle of this debate came in 1972 when the Club of Rome published *The Limits to Growth*, a runaway best-seller that used early computer models to predict that if natural systems continued to be depleted at their current rate, humanity would overshoot the planet's carrying capacity by the middle of the twenty-first century. Saving a few beautiful mountain ranges wouldn't be enough to get us out of this fix; the logic of growth itself needed to be confronted.

As author Christian Parenti observed recently of the book's lasting influence, "*Limits* combined the glamour of Big Science—powerful MIT computers and support from the Smithsonian Institution—with a focus on the interconnectedness of things, which fit perfectly with the new countercultural zeitgeist." And though some of the book's projections have not held up over time—the authors underestimated, for instance, the capacity of profit incentives and innovative technologies to unlock new reserves of finite resources—*Limits* was right about the most important limit of all. On "the limits of natural 'sinks,' or the Earth's ability to absorb pollution," Parenti writes, "the catastrophically bleak vision of *Limits* is playing out as totally correct. We may find new inputs—more oil or chromium—or invent substitutes, but we have not produced or discovered more natural sinks. The Earth's capacity to absorb the filthy byproducts of global capitalism's voracious metabolism is maxing out. That warning has always been the most powerful part of *The Limits to Growth*."[54]

And yet in the most powerful parts of the environmental movement, in the key decades during which we have been confronting the climate threat, these voices of warning have gone unheeded. The movement did not reckon with limits of growth in an economic system built on maximizing profits, it instead tried to prove that saving the planet could be a great new business opportunity.

The reasons for this political timidity have plenty to do with the themes already discussed: the power and allure of free market logic that usurped so much intellectual life in the late 1980s and 1990s, including large parts of the conservation movement. But this persistent unwillingness to follow science to its conclusions also speaks to the power of the cultural narrative that tells us that humans are ultimately in control of the earth, and not the other way around. This is the same narrative that assures us that, however bad things get, we are going to be saved at the last minute—whether by the market, by philanthropic billionaires, or by technological wizards—or best of all, by all three at the same time. And while we wait, we keep digging in deeper.

Only when we dispense with these various forms of magical thinking will we be ready to leave extractivism behind and build the societies we need within the boundaries we have—a world with no sacrifice zones, no new Naurus.

PART TWO

MAGICAL THINKING

"Vast economic incentives exist to invent pills that would cure alcoholism or drug addiction, and much snake oil gets peddled claiming to provide such benefits. Yet substance abuse has not disappeared from society. Given the addiction of modern civilization to cheap energy, the parallel ought to be unnerving to anyone who believes that technology alone will allow us to pull the climate rabbit out of the fossil-fuel hat. . . . The hopes that many Greens place in a technological fix are an expression of high-modernist faith in the unlimited power of science and technology as profound—and as rational—as Augustine's faith in Christ."

—Political scientist William Barnes and intellectual historian Nils Gilman, 2011[1]

"The leaders of the largest environmental groups in the country have become all too comfortable jet-setting with their handpicked corporate board members, a lifestyle they owe to those same corporate moguls. So it is little wonder that instead of prodding their benefactors to do better, these leaders—always hungry for the next donation—heap praise on every corporate half measure and at every photo opportunity."

—Christine MacDonald, former employee of Conservation International, 2008[2]

FRUITS, NOT ROOTS

The Disastrous Merger of Big Business and Big Green

"Our arguments must translate into profits, earnings, productivity, and economic incentives for industry."

—Former National Wildlife Federation President Jay Hair, 1987[1]

"I know this seems antithetical, but the bottom line here is not whether new coal-fired plants are built. . . . If the new coal plants are coming online under a cap that is bringing total emissions down, then it is not the worst thing in the world. Coal isn't the enemy. Carbon emissions are."

—Environmental Defense Fund President Fred Krupp, 2009[2]

Before the twentieth century, as many as a million Attwater's prairie chickens made their homes in the tall grasses along the coasts of Texas and Louisiana.[3] During mating season, they were quite a spectacle. To attract females, the males stomped their feet in little staccato motions, made loud, spooky cooing noises (known as "booming"), and inflated bright yellow air sacs on the sides of their necks, giving them the appearance of having swallowed two golden eggs.

But as the native prairie was turned into subdivisions and sliced up by oil and gas development, the Attwater's prairie chicken population began to crash. Local birders mourned the loss and in 1965, The Nature Conservancy—renowned for buying up ecologically important tracts of land and turning them into preserves—opened a Texas chapter. Early on, one of its major priorities was saving the Attwater's prairie chicken from extinction.[4]

It wasn't going to be easy, even for what would become the richest environmental organization in the world. One of the last remaining breeding grounds was located on 2,303 acres in southeast Texas on the shore of Galveston Bay, a property that happened to be owned by Mobil (now ExxonMobil). The fossil fuel giant hadn't yet covered the land in oil and gas infrastructure, but there were active wells on its southern edge, closing in on the breeding grounds of the endangered bird. Then in 1995, came some surprisingly good news. Mobil was donating its Galveston Bay property to The Nature Conservancy —"the last best hope of saving one of the world's most endangered species," as the company put it. The conservancy, which named the land the Texas City Prairie Preserve, would make "the recovery of the Attwater's prairie chicken" its "highest priority." To all appearances, it was a shining conservation success story—proof that a non-confrontational, partnership-based approach to environmentalism could yield tangible results.[5]

But four years later, something very strange happened. The Nature Conservancy began to do the very thing that its supporters thought it was there to prevent: it began extracting fossil fuels on the preserve. In 1999, the conservancy commissioned an oil and gas operator to sink a new gas well inside the preserve, which would send millions in revenue flowing directly into the environmental organization's coffers. And while the older oil and gas wells—those drilled before the land was designated a bird preserve—were mostly clustered far from the habitat of the Attwater's prairie chickens, that was decidedly not the case for the new well. According to Aaron Tjelmeland, the current manager of the preserve, the spot where the conservancy allowed drilling was relatively near the areas where the endangered birds nested, as well as performed their distinctive mating rituals. Of all the wells, this drilling pad was "the closest to where the prairie chickens normally hung out, or normally boomed," he said in an interview.[6]

For about three years, The Nature Conservancy's foray into the fossil fuel business attracted relatively little public controversy. That changed in 2002, when a piece in the *Los Angeles Times* exposed the drilling. For traditional conservationists, it was a little like finding out that Amnesty International had opened its own prison wing at Guantánamo. "They're exploiting the Attwater's prairie chicken to make money," fumed Clait E. Braun, then president of the Wildlife Society, and a leading expert on prairie chickens.

Then, in May 2003, *The Washington Post* followed up with a scathing investigation into the organization's questionable land deals, delving deeper into the surprising fact that on the Texas City Prairie Preserve, one of the most respected environmental organizations in the United States was now moonlighting as a gas driller.[7]

The Nature Conservancy, sounding like pretty much everyone in the oil and gas business, insisted that, "We can do this drilling without harming the prairie chickens and their habitat."[8] But the track record on the preserve makes that far from clear. In addition to the increased traffic, light, and noise that are part of any drilling operation, there were several points when drilling and wildlife preservation seemed to come into direct conflict.

For instance, because Attwater's prairie chickens are so endangered, there is a public-private program that breeds them in captivity and then releases them into the wild, an initiative in which The Nature Conservancy was participating on the Texas City Prairie Preserve. But at one point early on in its drilling foray, a delay in the construction of a gas pipeline led the conservancy to postpone the release of the captive-bred chicks by three months—a dicey call because migrating raptors and other predators appear to have been waiting for them.[9]

The bird release that year was a disaster. According to an internal Nature Conservancy report, all seventeen of the chicks "died shortly after their delayed release." The science director of the Texas chapter wrote that the months of waiting had subjected the birds "to higher probability of death from raptor predation." According to *The Washington Post* report, by 2003 there were just sixteen Attwater's prairie chickens that The Nature Conservancy knew about on the preserve, down from thirty-six before the drilling began. Though top conservancy officials insisted that the birds had not been adversely affected by its industrial activities, it was a dismal record.[10]

When I first came across the decade-old story, I assumed that The Nature Conservancy's extraction activities had stopped when they were exposed, since the revelation had ignited a firestorm of controversy and forced the organization to pledge not to repeat this particular fundraising technique. After the story broke, the organization's then president stated clearly, "We won't initiate any new oil, gas drilling, or mining of hard rock minerals on preserves that we own. We've only done that twice in 52 years but we thought, nonetheless, we should, for appearances' sake, not do that again."[11]

Turns out I was wrong. In fact, as of this book's writing, the conservancy was *still* extracting hydrocarbons on the Texas preserve that it rescued from Mobil back in 1995. In a series of communications, conservancy spokespeople insisted that the organization was required to continue fossil fuel extracting under the terms of the original drilling lease. And it's true that the 2003 pledge had been carefully worded, promising not to initiate "any new" drilling activities, and containing a proviso that it would honor "existing contracts."[12]

But The Nature Conservancy has not simply continued extracting for gas in that same well. A 2010 paper presented at a Society of Petroleum Engineers conference, and coauthored by two conservancy officials, reveals that the original well "died in March 2003, and was unable to flow due to excessive water production," leading to the drilling of a replacement well in the same area in late 2007. It also turns out that while the original well was for gas, the new one is now producing only oil.[13]

Given that close to five years elapsed between the death of the Nature Conservancy's first well and the drilling of the replacement, it seems possible that the organization had the legal grounds to extricate itself from the original lease if it had been sufficiently motivated to do so. The lease I have seen states clearly that in the event that oil or gas production ever stops in a given "well tract," the operator has a 180-day window to begin "reworking" the well or to start drilling a new one. If it fails to do so, the lease for that area is automatically terminated. If The Nature Conservancy causes a delay in the operator's work—which the organization claims has regularly occurred, since it restricts drilling to a few months per year—then the 180-day window is extended by the equivalent amount of time. So, the organization insists that though it was "concerned" about initial plans for the new 2007 well, due to the proposed well's proximity to the Attwater's habitat, it believed it was "bound by the existing lease and required to permit the drilling of the replacement well," albeit in a different location. James Petterson, director of marketing strategies at the conservancy, told me that the organization had sought "an outside legal opinion from an oil and gas expert" that confirmed this view. Yet in an internal explanatory document on drilling entitled "Attwater's Prairie Chicken Background," the organization emphasizes that it maintains the power to control what can and cannot occur on the preserve. "Given the birds' endangered status," the document states, "no activity can take place that is deemed likely to harm the species." Petterson insists that "bird experts were consulted" and "nobody [here] would want to do anything to harm an endangered species, particularly one as endangered as the Attwater's Prairie Chicken . . . nobody is going to choose oil and gas development over the last remaining handful of birds on the planet."[14]

Regardless of whether the conservancy resumed drilling for oil in Texas because it had no choice or because it wanted to get the petro dollars flowing again after the initial controversy had died down, the issue has taken on new urgency of late. That's because, in November 2012, and with little fanfare, the last of the Attwater's prairie chickens disappeared from the Preserve. Aaron Tjelmeland, the preserve manager, said of the birds that there are "none that we know about." It is worth underlining this detail: under the stewardship of what *The New Yorker* describes as "the biggest environmental nongovernmental organization in the world"—boasting over one million members and assets of roughly $6 billion and operating in thirty-five countries—an endangered species has been completely wiped out from one of its last remaining breeding grounds, on which the organization earned millions drilling for and pumping oil and gas. Amazingly, the website for the Texas City Prairie Preserve continues to boast that the "land management techniques the conservancy utilizes at the preserve are best practices that we export to other preserves." And though it mentions in passing that there are no more Attwater's prairie chickens on the land, it says nothing about its side business in oil and gas.[15]

The disappearance of the prairie chickens is no doubt the result of a combination of factors—invasive species, low numbers of captive-bred birds, drought (possibly linked to climate change), and the relatively small size of the reserve (the conservancy's preferred explanation). It's possible that the oil and gas drilling played no role at all.

So let's set the birds aside for a moment. Even if a few had survived, and even if a few return in the future, the fact remains that The Nature Conservancy has been in the oil and gas business for a decade and half. That this could happen in the age of climate change points to a painful reality behind the environmental movement's catastrophic failure to effectively battle the economic interests behind our soaring emissions: large parts of the movement aren't actually fighting those interests—they have merged with them.

The Nature Conservancy, I should stress, is the only green group (that I know of, at least) to actually sink its own oil and gas wells. But it is far from the only group to have strong ties with the fossil fuel sector and other major polluters. For instance, Conservation International, The Nature Conservancy, and the Conservation Fund have all received money from Shell and BP, while American Electric Power, a traditional dirty-coal utility, has donated to the Conservation Fund and The Nature Conservancy. WWF (originally the World Wildlife Fund) has had a long relationship with Shell, and the World Resources Institute has what it describes as "a long-term, close strategic relationship with the Shell Foundation." Conservation International has partnerships with Walmart, Monsanto, Australian-based mining and petroleum giant BHP Billiton (a major extractor of coal), as well as Shell, Chevron, ExxonMobil, Toyota, McDonald's, and BP (according to *The Washington Post* BP has channeled $2 million to Conservation International over the years).* And that is the barest of samplings.[16]

The relationships are also more structural than mere donations and partnerships. The Nature Conservancy counts BP America, Chevron, and Shell among the members of its Business Council and Jim Rogers, chairman of the board and former CEO of Duke Energy, one of the largest U.S. coal-burning utilities, sits on the organization's board of directors (past board members include former CEOs of General Motors and American Electric Power).[17]

There is yet another way in which some green groups have entangled their fates with the corporations at the heart of the climate crisis: by investing their own money with them. For instance, while investigating The Nature Conservancy's foray into oil and gas drilling, I was struck by a line item in its 2012 financial statements: $22.8 million of the organization's endowment—one of the largest in the U.S.—was invested in "energy" companies (that figure has since gone up to $26.5 million). Energy, of course,

---

* By 2011, the situation had become so surreal that Conservation International (CI) was the target of an embarrassing prank. A couple of activist/journalists posed as executives of the weapons giant Lockheed Martin and told the director of corporate relations for CI that they were looking for help greening their company's image. Rather than cutting their emissions, they said they were thinking of sponsoring an endangered species. Without missing a beat, the CI representative was recorded helpfully suggesting a bird of prey, to make the "link with aviation." ("We do not help companies with their image," CI later maintained, stressing that Lockheed would have needed to undergo a "due diligence process.")

means oil, gas, coal, and the like.* Curious, I soon discovered that most big conservation groups did not have policies prohibiting them from investing their endowments in fossil fuel companies. The hypocrisy is staggering: these organizations raise mountains of cash every year on the promise that the funds will be spent on work that is preserving wildlife and attempting to prevent catastrophic global warming. And yet some have turned around and invested that money with companies that have made it abundantly clear, through their reserves, that they intend to extract several times more carbon than the atmosphere can absorb with any degree of safety. It must be stated that these choices, made unilaterally by the top tier of leadership at the big green groups, do not represent the wishes or values of the millions of members who support them through donations or join genuinely community supported campaigns to clean up polluted rivers, protect beloved pieces of wilderness, or support renewables legislation. Indeed, many have been deeply alarmed to discover that groups they believed to be confronting polluters were in fact in business with them.[18]

There are, moreover, large parts of the green movement that have never engaged in these types of arrangements—they don't have endowments to invest or they have clear policies prohibiting fossil fuel holdings, and some have equally clear policies against taking donations from polluters. These groups, not coincidentally, tend also to be the ones with track records of going head-to-head with big oil and coal: Friends of the Earth and Greenpeace have been battling Shell's and Chevron's alleged complicity with horrific human rights abuses in the Niger Delta since the early 1990s (though Shell has agreed to pay out $15.5 million to settle a case involving these claims, it continues to deny wrongdoing, as does Chevron); Rainforest Action Network has been at the forefront of the international campaign against Chevron for the disaster left behind in the Ecuadorian Amazon; Food & Water Watch has helped secure big victories against fracking; 350. org helped launch the fossil fuel divestment movement and has been at the forefront of the national mobilization against the Keystone XL pipeline.

---

* After my article on the subject appeared in *The Nation*, The Nature Conservancy adopted a policy to "divest from companies that derive a significant percentage of their revenue from fossil fuels with the highest carbon content and will support a shift to carbon-free energy in the longer term."

The Sierra Club is a more complex case: it has also been a part of these campaigns and is the bane of the U.S. coal industry—but between 2007 and 2010, the group secretly took millions from a natural gas company. But under new leadership—and facing pressure from the grass roots—it has cut ties with the fossil fuel sector.[19]

Even so, almost no one's hands are clean. That's because many of the top foundations that underwrite much of the environmental movement—including groups and projects with which I have been involved—come from fortunes, like the Rockefeller family's, that are linked with fossil fuels. And though these foundations do fund campaigns that confront big polluters most do not prohibit their own endowments from being invested with coal and oil. So, for example, the Ford Foundation, which has supported the Environmental Defense Fund and Natural Resources Defense Council (and helped support a film that is accompanying this book), reported in 2013 that it had nearly $14 million in Shell and BP stocks alone (another multimillion-dollar stock holding is Norway's Statoil).[20] In North America and Europe, it's virtually impossible to do public interest work of any scale—in academia or journalism or activism—without taking money of questionable origin, whether the origin is the state, corporations, or private philanthropy. And though more accountable grassroots movement financing models are desperately needed (and crowdfunding is a promising start), the fact of these financial ties is not what is particularly noteworthy, nor proof of some nefarious corruption.

Where following the financial ties between funders and public interest work becomes relevant is when there is a compelling reason to believe that funding is having undue influence—shaping the kinds of research undertaken, the kinds of policies advanced, as well as the kinds of questions that get asked in the first place. And since it is generally accepted that fossil fuel money and conservative foundations have shaped the climate change denial movement, it seems fair to ask whether fossil fuel money and the values of centrist foundations have shaped parts of the movement that are in the business of proposing solutions. And there is a good deal of evidence that these ties have indeed had a decisive influence.

The big, corporate-affiliated green groups don't deny the reality of climate change, of course—many work hard to raise the alarm. And yet several of these groups have consistently, and aggressively, pushed responses

to climate change that are the least burdensome, and often directly beneficial, to the largest greenhouse gas emitters on the planet—even when the
policies come at the direct expense of communities fighting to keep fossil
fuels in the ground. Rather than advancing policies that treat greenhouse
gases as dangerous pollutants demanding clear, enforceable regulations that
would restrict emissions and create the conditions for a full transition to renewables, these groups have pushed convoluted market-based schemes that
have treated greenhouse gases as late-capitalist abstractions to be traded,
bundled, speculated upon, and moved around the globe like currency or
subprime debt.

And many of these same groups have championed one of the main fossil fuels—natural gas—as a supposed solution to climate change, despite
mounting evidence that in the coming decades, the methane it releases,
particularly through the fracking process, has the potential to help lock
us into catastrophic levels of warming (as explained in chapter four). In
some cases, large foundations have collaborated to explicitly direct the
U.S. green movement toward these policies. Most infamously within the
movement, a 2007 road map titled "Design to Win: Philanthropy's Role
in the Fight Against Global Warming"—which was sponsored by six large
foundations—advocated carbon trading as a response to climate change
and supported both natural gas and expanded nuclear power. And as these
policies were being turned into political campaigns, the message sent to
green groups was essentially "step in line, or else you're not going to get
your share of the money," recalls Jigar Shah, a renowned solar entrepreneur, former Greenpeace USA board member, and one-time director of the
industry-focused Carbon War Room.[21]

The "market-based" climate solutions favored by so many large foundations and adopted by many greens have provided an invaluable service to
the fossil fuel sector as a whole. For one, they succeeded in taking what
began as a straightforward debate about shifting away from fossil fuels and
put it through a jargon generator so convoluted that the entire climate issue
came to seem too complex and arcane for nonexperts to understand, seriously undercutting the potential to build a mass movement capable of taking on powerful polluters. As Drexel University sociologist Robert Brulle
has observed, "The movement to technical and market-based analyses as
the core of reform environmentalism gutted whatever progressive vision"

the movement had previously held. "Rather than engaging the broader public, reform environmentalism focuses debate among experts in the scientific, legal, and economic communities. It may provide technical solutions to specific problems but it neglects the larger social dynamics that underlie environmental degradation."[22]

These policies have also fed the false perception that a full transition to renewable energy is technically impossible—since if it were possible, why would all these well-meaning green groups be spending so much of their time pushing trading schemes and singing the praises of natural gas, even when extracted through the ecologically destructive method of fracking?

Often these compromises are rationalized according to the theory of "low-hanging fruit." This strategy holds, in essence, that it's hard and expensive to try to convince politicians to regulate and discipline the most powerful corporations in the world. So rather than pick that very tough fight, it's wiser and more effective to begin with something easier. Asking consumers to buy a more expensive, less toxic laundry detergent, for instance. Making cars more fuel-efficient. Switching to a supposedly cleaner fossil fuel. Paying an Indigenous tribe to stop logging a forest in Papua New Guinea to offset the emissions of a coal plant that gets to stay open in Ohio.

With emissions up by about 57 percent since the U.N. climate convention was signed in 1992, the failure of this polite strategy is beyond debate. And yet still, at the upper echelons of the climate movement, our soaring emissions are never blamed on anything as concrete as the fossil fuel corporations that work furiously to block all serious attempts to regulate emissions, and certainly not on the economic model that demands that these companies put profit before the health of the natural systems upon which all life depends. Rather the villains are always vague and unthreatening—a lack of "political will," a deficit of "ambition"—while fossil fuel executives are welcomed at U.N. climate summits as key "partners" in the quest for "climate solutions."[23]

This upside-down world reached new levels of absurdity in November 2013 at the annual U.N. climate summit held in Warsaw, Poland. The gathering was sponsored by a panoply of fossil fuel companies, including a major miner of lignite coal, while the Polish government hosted a parallel "Coal & Climate Summit," which held up the dirtiest of all the fossil fuels as part of the battle against global warming. The official U.N. climate negotiation process gave its tacit endorsement of the coal event when its highest official—Christiana Figueres, executive secretary of the United Nations Framework Convention on Climate Change—agreed to deliver a keynote address to the gathering, defying calls from activists to boycott. "The summit's focus on continued reliance on coal is directly counter to the goal of these climate negotiations," said Alden Meyer of the Union of Concerned Scientists, "which is to dramatically reduce emissions of heat-trapping gases in order to avoid the worst impacts of climate change."[24]

A great many progressives have opted out of the climate change debate in part because they thought that the Big Green groups, flush with philanthropic dollars, had this issue covered. That, it turns out, was a grave mistake. To understand why, it's necessary to return, once again, to the epic case of bad historical timing that has plagued this crisis since the late eighties.

The Golden Age of Environmental Law

I. F. Stone may have thought that environmentalism was distracting the youth of the 1960s and early 1970s from more urgent battles, but by today's standards, the environmentalists of that era look like fire-breathing radicals. Galvanized by the 1962 publication of *Silent Spring* and the 1969 Santa Barbara oil spill (the Deepwater Horizon disaster of its day), they launched a new kind of North American environmentalism, one far more confrontational than the gentlemen's conservationism of the past.

In addition to the newly formed Friends of the Earth (created in 1969) and Greenpeace (launched in 1971), the movement also included groups like the Environmental Defense Fund, then an idealistic gang of scrappy scientists and lawyers determined to heed Rachel Carson's warnings. The group's unofficial slogan was, "Sue the bastards," and so they did. The EDF fought for and filed the original lawsuit that led to the U.S. ban on DDT as an insecticide, resulting in the revival of many species of birds, including the bald eagle.[25]

This was a time when intervening directly in the market to prevent harm was still regarded as a sensible policy option. Confronted with unassailable evidence of a grave collective problem, politicians across the political spectrum still asked themselves: "What can we do to stop it?" (Not:

"How can we develop complex financial mechanisms to help the market fix it for us?")

What followed was a wave of environmental victories unimaginable by today's antigovernment standards. In the United States, the legislative legacy is particularly striking: the Clean Air Act (1963), the Wilderness Act (1964), the Water Quality Act (1965), the Air Quality Act (1967), the Wild and Scenic Rivers Act (1968), the National Environmental Policy Act (1970), the revised Clean Air Act (1970), the Occupational Safety and Health Act (1970), the Clean Water Act (1972), the Marine Mammal Protection Act (1972), the Endangered Species Act (1973), the Safe Drinking Water Act (1974), the Toxic Substances Control Act (1976), the Resource Conservation and Recovery Act (1976). In all, twenty-three federal environmental acts became law over the course of the 1970s alone, culminating in the Superfund Act in 1980, which required industry, through a small levy, to pay the cost of cleaning up areas that had become toxic.

These victories spilled over into Canada, which was also experiencing a flurry of environmental activism. The federal government passed its own Water Act (1970) and Clean Air Act (1971), and gave teeth to the nineteenth-century Fisheries Act a few years later, turning it into a powerful force for combating marine pollution and protecting habitats. Meanwhile, the European Community declared environmental protection a top priority as early as 1972, laying the groundwork for its leadership in environmental law in the decades to follow. And in the wake of the U.N. Conference on the Human Environment in Stockholm that same year, the 1970s became a foundational decade for international environmental law, producing such landmarks as the Convention on the Prevention of Marine Pollution by Dumping of Wastes and Other Matter (1972), the Convention on International Trade in Endangered Species of Wild Fauna and Flora (1973), and the Convention on Long-Range Transboundary Air Pollution (1979).

Although robust environmental law would not begin to take hold in much of the developing world for another decade or so, direct environmental defense also intensified in the 1970s among peasant, fishing, and Indigenous communities across the Global South —the origins of what economist Joan Martínez Alier and others have described as the "environmentalism of the poor." This stretched from creative, women-led campaigns against deforestation in India and Kenya, to widespread resistance to nuclear power plants, dams, and other forms of industrial development in Brazil, Colombia, and Mexico.[26]

Simple principles governed this golden age of environmental legislation: ban or severely limit the offending activity or substance and where possible, get the polluter to pay for the cleanup. As journalist Mark Dowie outlines in his history of the U.S. environmental movement, *Losing Ground*, the real-world results of this approach were concrete and measurable. "Tens of millions of acres have been added to the federal wilderness system, environmental impact assessments are now required for all major developments, some lakes that were declared dead are living again. . . . Lead particulates have been impressively reduced in the atmosphere; DDT is no longer found in American body fat, which also contains considerably fewer polychlorinated biphenyls (PCBs) than it once did. Mercury has virtually disappeared from Great Lakes sediment; and Strontium 90 is no longer found in either cows' milk or mothers' milk." And Dowie stressed: "What all these facts have in common is that they are the result of outright bans against the use or production of the substances in question."[*][27]

These are the tough tools with which the environmental movement won its greatest string of victories. But with that success came some rather significant changes. For a great many groups, the work of environmentalism stopped being about organizing protests and teach-ins and became about drafting laws, then suing corporations for violating them, as well as challenging governments for failing to enforce them. In rapid fashion, what had been a rabble of hippies became a movement of lawyers, lobbyists, and U.N. summit hoppers. As a result, many of these newly professional environmentalists came to pride themselves on being the ultimate insiders, able to wheel and deal across the political spectrum. And so long as the victories kept coming, their insider strategy seemed to be working.

Then came the 1980s. "A tree is a tree," Ronald Reagan famously said in the midst of a pitched battle over logging rights. "How many more do you need to look at?" With Reagan's arrival in the White House, and the ascendency of many think-tank ideologues to powerful positions in his adminis-

---

* It's worth keeping this history in mind when free market ideologues treat a cleaner environment as a natural stage in capitalist development. In fact it is the result of specific sets of regulations, ones that run directly counter to hard-right ideology.

tration, the goalposts were yanked to the right. Reagan filled his inner circle with pro-industry scientists who denied the reality of every environmental ill from acid rain to climate change. And seemingly overnight, banning and tightly regulating harmful industrial practices went from being bipartisan political practice to a symptom of "command and control environmentalism." Using messaging that would have fit right in at a Heartland conference three decades later, James Watt, Reagan's much despised interior secretary, accused greens of using environmental fears "as a tool to achieve a greater objective," which he claimed was "centralized planning and control of the society." Watt also warned darkly about where that could lead: "Look what happened to Germany in the 1930s. The dignity of man was subordinated to the powers of Nazism. The dignity of man was subordinated in Russia. Those are the forces that this thing can evolve into."[28]

For the Big Green groups, all this came as a rude surprise. Suddenly they were on the outside looking in, being red-baited by the kinds of people with whom they used to have drinks. Worse, the movement's core beliefs about the need to respond to environmental threats by firmly regulating corporations were being casually cast into the dustbin of history. What was an insider environmentalist to do?

Extreme 1980s Makeover

There were options, as there always are. The greens could have joined coalitions of unions, civil rights groups, and pensioners who were also facing attacks on hard-won gains, forming a united front against the public sector cutbacks and deregulation that was hurting them all. And they could have kept aggressively using the courts to sue the bastards. There was, throughout the 1980s, mounting public concern even among Republicans about Reagan's environmental rollbacks (which is how Planet Earth ended up on the cover of *Time* in early 1989).*[29]

---

* By the end of the 1980s, the majority of self-identified Republicans were telling pollsters that they thought there was "too little" spent protecting the environment. By 1990, the percentage of Republicans who agreed with that statement topped 70 percent.

And some did take up that fight. As Reagan launched a series of attacks on environmental regulations, there was resistance, especially at the local level, where African American communities in particular were facing an aggressive new wave of toxic dumping. These urgent, health-based struggles eventually coalesced into the environmental justice movement, which held the First National People of Color Environmental Leadership Summit in October 1991, a historic gathering that adopted a set of principles that remains a movement touchstone to this day.[30] At the national and international levels, groups like Greenpeace continued to engage in direct action throughout the 1980s, though much of their energy was understandably focused on the perils of both nuclear energy and weapons.

But many green groups chose a very different strategy. In the 1980s, extreme free market ideology became the discourse of power, the language that elites were speaking to one another, even if large parts of the general public remained un-persuaded. That meant that for the mainstream green movement, confronting the antigovernment logic of market triumphalism head-on would have meant exiling themselves to the margins. And many of the big-budget green groups—having grown comfortable with their access to power and generous support from large, elite foundations—were unwilling to do that. Gus Speth, who co-founded the Natural Resources Defense Council and served as a top environmental advisor to Jimmy Carter during his presidency, described the problem like this: "We didn't adjust with Reagan. We kept working within a system but we should have tried to change the system and root causes."[31] (After years in high-level jobs inside the U.N. system and as a dean of Yale's School of Forestry and Environmental Studies, Speth has today thrown his lot in with the radicals, getting arrested to protest the Keystone XL pipeline and co-founding an organization questioning the logic of economic growth.)

Part of what increased the pressure for ideological conformity in the 1980s was the arrival of several new groups on the environmental scene, competing for limited philanthropic dollars. These groups pitched themselves as modern environmentalists for the Reagan era: pro-business, non-confrontational, and ready to help polish even the most tarnished corporate logos. "Our approach is one of collaboration, rather than confrontation. We are creative, entrepreneurial, and partnership-driven. We don't litigate," explains the Conservation Fund, founded in 1985. Two years later came Conservation International, which claims to have "single-handedly redefined conservation" thanks largely to a philosophy of working "with companies large and small to make conservation part of their business model."[32]

This open-for-business approach was so adept at attracting big donors and elite access that many older, more established green groups raced to get with the agreeable program, taking an "if you can't beat 'em, join 'em" attitude to brazen extremes. It was in this period that the Nature Conservancy started loosening its definition of "preservation" so that conservation lands would eventually accommodate such dissonant activities as mansion building and oil drilling (laying the foundation for the group to get in on the drilling action itself). "I used to say that the only things not allowed on Nature Conservancy reserves were mining and slavery, and I wasn't sure about the latter," said Kierán Suckling of the Center for Biological Diversity. "Now I may have to withdraw the former as well."[33]

Indeed the pro-corporate conversion of large parts of the green movement in the 1980s led to deep schisms inside the environmental movement. Some activists grew so disillusioned with the willingness of the big groups to partner with polluters that they broke away from the mainstream movement completely. Some formed more militant, confrontation-oriented groups like Earth First!, whose members attempted to stop loggers with sabotage and direct action.

The debates, for the most part, took place behind the scenes but on April 23, 1990, they spilled into the headlines. It was the day after Earth Day—at that time an annual ritual of mass corporate greenwashing—and around one thousand demonstrators stormed the New York and the Pacific Stock Exchanges to draw attention to the "institutions responsible for much of the ecological devastation which is destroying the planet." Members of grassroots groups like the Love Canal Homeowners Association, the Bhopal Action Resource Group, and the National Toxics Campaign handed out pamphlets that read in part, "Who is destroying the earth—are we all equally to blame? No! We say go to the source. We say take it to Wall Street!" The pamphlets went on: "The polluters would have us believe that we are all just common travelers on Spaceship Earth, when in fact a few of them are at the controls, and the rest of us are choking on their exhaust."[34]

This confrontational rhetoric—a foreshadowing of Occupy Wall Street two decades later, as well as the fossil fuel divestment movement—was an explicit critique of the corporate infiltration of the green movement. Daniel Finkenthal, a spokesperson for the anticorporate protests, declared, "Real environmental groups are disgusted with the corporate buyout of Earth Day," telling one journalist that sponsors are "spending more money on Earth Day promotion than they are on actual corporate reform and the environment."[35]

Climate Policy and the Price of Surrender

Of all the big green groups that underwent pro-business makeovers in the 1980s, none attracted more acrimony or disappointment than the Environmental Defense Fund, the once combative organization that had spent its early years translating Rachel Carson's ideas into action. In the mid-1980s, a young lawyer named Fred Krupp took the reins of the organization and he was convinced that the group's "sue the bastards" motto was so out of step with the times that it belonged at a garage sale next to dog-eared copies of The Limits to Growth. Under Krupp's leadership, which continues to this day, the EDF's new goal became: "creating markets for the bastards," as his colleague Eric Pooley would later characterize it.[36] And it was this transformation, more than any other, that produced a mainstream climate movement that ultimately found it entirely appropriate to have coal and oil companies sponsor their most important summits, while investing their own wealth with these same players.

The new era was officially inaugurated on November 20, 1986, when Krupp published a cocky op-ed in The Wall Street Journal. In it he announced that a new generation of pro-business environmentalists had arrived and with it "a new strategy in the movement." Krupp explained that his generation rejected the old-fashioned idea that "either the industrial economy wins or the environment wins, with one side's gain being the other's loss. The new environmentalism does not accept 'either-or' as inevitable and has shown that in many critical instances it is a fallacy." Rather than attempting to ban harmful activities, as Krupp's own organization had helped to do with DDT, the EDF would now form partnerships with polluters—or "coalitions of former enemies"—and persuade them that there are cost savings as well as new markets in going green. In time, Walmart, McDonald's, FedEx, and AT&T would all enjoy high-profile partnerships with this storied environmental pioneer.[37]

The group prided itself on putting "results" above ideology, but in truth Krupp's EDF was highly ideological—it's just that its ideology was the pro-corporate groupthink of the day, one that holds that private, market-based solutions are inherently superior to simple regulatory ones. A turning point came in 1988 when George H. W. Bush came to power promising action on acid rain. The old way of addressing the problem would have been straightforward: since sulfur dioxide emissions were the primary cause of acid rain, the solution would have been to require their reduction by a fixed amount across the board. Instead, the EDF pushed for the first full-fledged cap-and-trade system. These rules did not tell polluters that they had to cut their sulfur emissions but, instead, set a nationwide cap on sulfur dioxide, beneath which big emitters like coal-fired power plants could do as they pleased—pay other companies to make reductions for them, purchase allowances permitting them to pollute as much as they had before, or make a profit by selling whatever permits they didn't use.[38]

The new approach worked and it was popular among foundations and private donors, particularly on Wall Street, where financiers were understandably attracted to the idea of harnessing the profit motive to solve environmental ills. Under Krupp's leadership, the EDF's annual budget expanded from $3 million to roughly $120 million. Julian Robertson, founder of the hedge fund Tiger Management, underwrote the EDF's work to the tune of $40 million, a staggering sum for a single benefactor.*[39]

The Environmental Defense Fund has always insisted that it does not take donations from the companies with which it forms partnerships—that, writes EDF senior vice president for strategy and communications Eric Pooley, "would undermine our independence and integrity." But the

---

* Indeed the worlds of finance and Big Green would become so entangled in the years to come—between donations, board members, and partnerships—that when The Nature Conservancy needed a new CEO in 2008, it recruited not from within the nonprofit world but from Goldman Sachs. Its current director, Mark Tercek, had been working at the notorious investment bank for some twenty-five years before moving over to the NGO, where he has consistently advanced a model of conservation based on bringing ever more parts of the natural world into the market.

policy doesn't bear much scrutiny. For instance, one of the EDF's flagship partnerships is with Walmart, with whom it collaborates to "make the company more sustainable." And it's true that Walmart doesn't donate to the EDF directly. However, the Walton Family Foundation, which is entirely controlled by members of the family that founded Walmart, gave the EDF $65 million between 2009 and 2013. In 2011, the foundation provided the group with nearly 15 percent of its funding. Meanwhile, Sam Rawlings Walton, grandson of Walmart founder Sam Walton, sits on the EDF's board of trustees (identified merely as "Boatman, Philanthropist, Entrepreneur" on the organization's website).[40]

The EDF claims that it "holds Walmart to the same standards we would any other company." Which, judging by Walmart's rather dismal environmental record since this partnership began—from its central role in fueling urban sprawl to its steadily increasing emissions—is not a very high standard at all.[41]

Nor is the Environmental Defense Fund the only environmental organization to have benefited from the Walton family's largesse. Their foundation is one of the top green funders, handing out more than $71 million in grants for environmental causes in 2011, with about half of the money going to the EDF, Conservation International, and the Marine Stewardship Council. All have partnerships with Walmart, whether to lower emissions, stamp an eco label on some of the seafood the company sells, or to co-launch a line of "mine to market" jewelry. Stacy Mitchell, a researcher with the Institute for Local Self-Reliance, observes that having large parts of the green movement so dependent on the scions of a company that almost singlehandedly supersized the retail sector and exported the model around the world has had profound political implications. "Walmart's money is exerting significant influence in setting the agenda, defining the problems, and elevating certain kinds of approaches—notably those that reinforce, rather than challenge, the power of large corporations in our economy and society," she writes.[42]

And this is the heart of the issue—not simply that a group that gets a large portion of its budget from the Walton family fortune is unlikely to be highly critical of Walmart. The 1990s was the key decade when the contours of the climate battle were being drawn—when a collective strategy for rising to the challenge was developed and when the first wave of supposed solutions was presented to the public. It was also the period when Big Green became most enthusiastically pro-corporate, most committed to a low-friction model of social change in which everything had to be "win-win." And in the same period many of the corporate partners of groups like the EDF and the Nature Conservancy—Walmart, FedEx, GM—were pushing hard for the global deregulatory framework that has done so much to send emissions soaring.

This alignment of economic interests—combined with the ever powerful desire to be seen as "serious" in circles where seriousness is equated with toeing the pro-market line—fundamentally shaped how these green groups conceived of the climate challenge from the start. Global warming was not defined as a crisis being fueled by overconsumption, or by high emissions industrial agriculture, or by car culture, or by a trade system that insists that vast geographical distances do not matter—root causes that would have demanded changes in how we live, work, eat, and shop. Instead, climate change was presented as a narrow technical problem with no end of profitable solutions within the market system, many of which were available for sale at Walmart.[*]

The effect of this "bounding of the debate," as the Scottish author and environmentalist Alastair McIntosh describes it, reaches far beyond a few U.S. groups. "In my experience," writes McIntosh, "most international climate change agency personnel take the view that 'we just can't go there' in terms of the politics of cutting consumerism." This is usually framed as an optimistic faith in markets, but in fact it "actually conceals pessimism because it keeps us in the displacement activity of barking up the wrong tree. It is an evasion of reality, and with it, the need to fundamentally appraise the human condition in order to seek the roots of hope."[43] Put another way, the refusal of so many environmentalists to consider responses to the cli-

---

[*] This is one of the many ironies of the Heartlanders' claim that greens are closet socialists. If so, then they are deep in the closet. In reality, many mainstream environmentalists bristle at the suggestion that they are part of the left at all, fearing (correctly) that such an identification would hurt their chances with foundation funders and corporate donors. Far from using climate change as a tool to alter the American way of life, many of the large environmental organizations spend their days doing everything in their power to furiously protect that way of life, at the direct expense of demanding the levels of change required by science.

mate crisis that would upend the economic status quo forces them to place their hopes in solutions—whether miracle products, or carbon markets, or "bridge fuels"—that are either so weak or so high-risk that entrusting them with our collective safety constitutes what can only be described as magical thinking.

I do not question the desire on the part of these self-styled pragmatists to protect the earth from catastrophic warming. But between the Heartlanders who recognize that climate change is a profound threat to our economic and social systems and therefore deny its scientific reality, and those who claim climate change requires only minor tweaks to business-as-usual and therefore allow themselves to believe in its reality, it's not clear who is more deluded.

Shopping Our Way Out of It

For a few years around the 2006 release of Al Gore's *An Inconvenient Truth*, it seemed as if climate change was finally going to inspire the transformative movement of our era. Public belief in the problem was high, and the issue seemed to be everywhere. Yet on looking back on that period, what is strange is that all the energy seemed to be coming from the very top tier of society. In the first decade of the new millennium, climate talk was a strikingly elite affair, the stuff of Davos panels and gee-whiz TED Talks, of special green issues of *Vanity Fair* and celebrities arriving at the Academy Awards in hybrid cars. And yet behind the spectacle, there was virtually no discernible movement, at least not of the sort that anyone involved in the civil rights, antiwar, or women's movements would recognize. There were few mass marches, almost no direct action beyond the occasional media-friendly stunt, and no angry leaders (other than a former vice president of the United States).

In a sense, the period represented a full-circle return to the gentlemen's clubhouse in which the conservation movement began, with Sierra Club cofounder John Muir persuading President Theodore Roosevelt to save large parts of Yosemite while the two men talked around the bonfire on a camping trip. And though the head of Conservation International did not go camping on the melting glaciers with George W. Bush in order to impress upon him the reality of climate change, there were plenty of post-modern equivalents, including celebrity-studded eco-cruises that allowed Fortune 500 CEOs to get a closer look at endangered coral reefs.

It wasn't that there was no role for the public. We were called upon periodically to write letters, sign petitions, turn off our lights for an hour, make a giant human hourglass that could be photographed from the sky. And of course we were always asked to send money to the Big Green groups that were supposedly just on the cusp of negotiating a solution to climate change on our behalf. But most of all, regular, noncelebrity people were called upon to exercise their consumer power—not by shopping less but by discovering new and exciting ways to consume more.* And if guilt set in, well, we could click on the handy carbon calculators on any one of dozens of green sites and purchase an offset, and our sins would instantly be erased.[44]

In addition to not doing much to actually lower emissions, these various approaches also served to reinforce the very "extrinsic" values that we now know are the greatest psychological barriers to climate action—from the worship of wealth and fame for their own sakes to the idea that change is something that is handed down from above by our betters, rather than something we demand for ourselves. They may even have played a role in weakening public belief in the reality of human-caused climate change. Indeed a growing number of communications specialists now argue that because the "solutions" to climate change proposed by many green groups in this period were so borderline frivolous, many people concluded that the groups must have been exaggerating the scale of the problem. After all, if climate change really was as dire as Al Gore argued it was in *An Inconvenient Truth*, wouldn't the environmental movement be asking the public to

---

* The Nature Conservancy, ever the envelope pusher, has been particularly enthusiastic in this regard, hiring its chief marketing officer straight from World Wrestling Entertainment and participating in the marketing frenzy that accompanied the release of Universal Pictures' film version of *The Lorax* (which used Dr. Seuss's anti-consumerism classic to hawk IHOP pancakes and Mazda SUVs). In 2012, the conservancy managed to outrage many of its female staffers by partnering with the online luxury goods retailer Gilt to promote the *Sports Illustrated* Swimsuit Edition (the magazine explained that "whether you decide to buy a bikini, surfboards or tickets to celebrate at our parties, any money you spend . . . will help The Nature Conservancy ensure we have beaches to shoot Swimsuit on for another half-century").

do more than switch brands of cleaning liquid, occasionally walk to work, and send money? Wouldn't they be trying to shut down the fossil fuel companies?

"Imagine that someone came up with a brilliant new campaign against smoking. It would show graphic images of people dying of lung cancer followed by the punch line: 'It's easy to be healthy—smoke one less cigarette a month.' We know without a moment's reflection that this campaign would fail," wrote British climate activist and author George Marshall. "The target is so ludicrous, and the disconnection between the images and the message is so great, that most smokers would just laugh it off."[45]

It would be one thing if, while individuals were being asked to voluntarily "green" the minutiae of their lives, the Big Green NGOs had simultaneously gone after the big polluters, demanding that they match our individual small cuts in carbon emissions with large-scale, industry-wide reductions. And some did. But many of the most influential green groups did precisely the opposite. Not only did they help develop complex financial mechanisms to allow these corporations to keep emitting, they also actively campaigned to expand the market for one of the three main fossil fuels.

Fracking and the Burning Bridge

The gas industry itself came up with the pitch that it could be a "bridge" to a clean energy future back in the early 1980s. Then in 1988, with climate change awareness breaking into the mainstream, the American Gas Association began to explicitly frame its product as a response to the "greenhouse effect."[46]

In 1992, a coalition of progressive groups—including the Natural Resources Defense Council, Friends of the Earth, Environmental Action, and Public Citizen—officially embraced the idea, presenting a "Sustainable Energy Blueprint" to the incoming administration of Bill Clinton that included a significant role for natural gas. The NRDC was a particularly strong advocate, going on to call natural gas "the bridge to greater reliance on cleaner and renewable forms of energy."[47]

And at the time, it seemed to make a good deal of sense: renewable technology was less mature than it is now, and the gas in question was being extracted through conventional drilling methods. Today, the landscape has shifted dramatically on both counts. Renewable technologies have become radically more efficient and affordable, making a full transition to the power they provide both technologically and economically possible within the next few decades. The other key change is that the vast majority of new gas projects in North America rely on hydraulic fracturing—not conventional drilling—and fracking-based exploration and production are on the rise around the world.[48]

These developments have significantly weakened the climate case for natural gas—especially fracked natural gas. We now know that fracked natural gas may leak enough methane to make its warming impact, especially in the near term, comparable to that of coal. Anthony Ingraffea, who coauthored the breakthrough Cornell study on methane leakage and describes himself as "a longtime oil and gas engineer who helped develop shale fracking techniques for the Energy Department," wrote in *The New York Times*, "The gas extracted from shale deposits is not a 'bridge' to a renewable energy future—it's a gangplank to more warming and away from clean energy investments."[49]

We also know, from experience in the U.S., that cheap and abundant natural gas doesn't replace only coal but also potential power from renewables. This has led the Tyndall Centre's Kevin Anderson to conclude, "If we are serious about avoiding dangerous climate change, the only safe place for shale gas remains in the ground." Biologist Sandra Steingraber of New Yorkers Against Fracking puts the stark choice like this: we are "standing at an energy crossroads. One signpost points to a future powered by digging fossils from the ground and lighting them on fire. The other points to renewable energy. You cannot go in both directions at once. Subsidizing the infrastructure for one creates disincentives for the other."[50]

Even more critically, many experts are convinced that we do not need unconventional fuels like fracked gas to make a full transition to renewables. Mark Z. Jacobson, the Stanford engineering professor who coauthored the road map for reaching 100 percent renewable energy by 2030, says that conventional fossil fuels can power the transition and keep the lights on in the meantime. "We don't need unconventional fuels to produce the infrastructure to convert to entirely clean and renewable wind, water, and solar power for all purposes. We can rely on the existing infrastructure plus the new infrastructure [of renewable generation] to provide the energy for producing the rest of the clean infrastructure that we'll need," he said in an interview, adding, "Conventional oil and gas is much more than enough."[51]

How have the Big Green groups responded to this new information? Some, like the NRDC, have cooled off from their earlier support, acknowledging the risks and pushing for tougher regulations while still advocating natural gas as a replacement for coal and other dirty fuels. But others have chosen to dig in even deeper. The Environmental Defense Fund and the Nature Conservancy, for example, have responded to revelations about the huge risks associated with natural gas by undertaking a series of initiatives that give the distinct impression that fracking is on the cusp of becoming clean and safe. And as usual, much of the funding for this work has strong links to the fossil fuel sector.

The Nature Conservancy, for its part, has received hundreds of thousands of dollars from JP Morgan to come up with voluntary rules for fracking. JP Morgan, unsurprisingly, is a leading financier of the industry, with at least a hundred major clients who frack, according to the bank's top environmental executive, Matthew Arnold. ("We are number one or number two in any given year in the oil and gas industry worldwide," Arnold told *The Guardian* in February 2013.) The conservancy also has a high-profile partnership with BP in Wyoming's Jonah Field, a huge fracking-for-gas operation in an area rich with vulnerable wildlife. The Nature Conservancy's job has been to identify habitat preservation and conservation projects to "offset the impacts of oil and gas drilling pads and infrastructure." From a climate change perspective, this is an absurd proposition, since these projects have no hope of offsetting the most damaging impact of all: the release of heat-trapping gases into the atmosphere. Which is why the most important preservation work that any environmental group can do is preserving the carbon in the ground, wherever it is. (Then again, this is The Nature Conservancy, which has its very own gas well in the middle of a nature preserve in Texas).[52]

Similarly, the EDF has teamed up with several large energy companies to open the Center for Sustainable Shale Development (CSSD)—and as many have pointed out, the very name of the center makes it clear that it will not be questioning whether "sustainable" extraction of fossil fuels from shale is possible in the age of climate change. The center has advanced a set of voluntary industry standards that its members claim will gradually make fracking safer. But as then–Demos senior policy analyst J. Mijin Cha pointed out, "The Center's new standards . . . are not enforceable. If anything, they provide cover for oil and gas interests that want to derail the transition to a clean economy powered by renewable energy."[53]

One of the center's key funders is the Heinz Endowments, which as it turns out, was no disinterested party. A June 2013 investigation by the Public Accountability Initiative reported that, "The Heinz Endowments, has significant, undisclosed ties to the natural gas industry. . . . Heinz Endowments president Robert F. Vagt is currently a director at Kinder Morgan, a natural gas pipeline company, and owns more than $1.2 million in company stock. This is not disclosed on the Heinz Endowments website or the website of CSSD, where Vagt serves as a director. Kinder Morgan has cited increased regulation of fracking as a key business risk in recent corporate filings." (After the controversy broke, Heinz Endowments appeared to move away from some of its earlier pro-gas positions and went through a significant staffing shakeup, including the resignation of Vagt as foundation president in early 2014.)[54]

The EDF has also received a $6 million grant from the foundation of New York's billionaire ex-mayor Michael Bloomberg (who is strongly pro-fracking), specifically to develop and secure regulations intended to make fracking safe—once again, not to impartially assess whether such an outcome is even possible. And Bloomerg is no impartial observer in all this. The former mayor's personal and philanthropic fortune—worth over $30 billion—is managed by investment firm Willett Advisors, which was established by Bloomberg and his associates. According to *Bloomberg Businessweek*, and confirmed by Bloomberg Philanthropies (which shares a building with the firm), Willett "invests in real assets focusing on oil and natural gas areas." Michael Bloomberg did not respond to repeated requests for comment.[55]

The EDF has done more than help the fracking industry appear to be taking environmental concerns seriously. It also led research that has been used to counter claims that high methane leakage disqualifies fracked natural gas as a climate solution. The EDF has partnered with Shell, Chevron, and other top energy companies on one in a series of studies on methane leaks with the clear goal, as one EDF official put it, of helping "natural gas to be an accepted part of a strategy for improving energy security and moving to a clean energy future." When the first study arrived in September 2013, published in *Proceedings of the National Academy of Sciences*, it made news by identifying fugitive methane leakage rates from gas extraction that were ten to twenty times lower than those in most other studies to date.[56]

But the study's design contained serious limitations, the most glaring of which was allowing the gas companies to choose the wells they wanted inspected. Robert Howarth, the lead author of the breakthrough 2011 Cornell study on the same subject, pointed out that the EDF's findings were "based only on evaluation of sites and times chosen by industry," and that the paper "must be viewed as a best-case scenario," rather than a reflection of how the industry functions as a whole. He added, "The gas industry can produce gas with relatively low emissions, but they very often do not do so. They do better when they know they are being carefully watched." These concerns, however, were entirely upstaged by the priceless headlines inspired by the Environmental Defense Fund study: "Study: Leaks at Natural Gas Wells Less Than Previously Thought" (*Time*); "Study: Methane Leaks from Gas Drilling Not Huge" (Associated Press); "Fracking Methane Fears Overdone" (*The Australian*); and so on.[57]

The result of all this has been a great deal of public uncertainty. Is fracking safe after all? Is it about to become safe? Is it clean or dirty? Like the well-understood strategy of sowing doubts about the science of climate change, this confusion effectively undermines the momentum away from fossil fuels and toward renewable energy. As Josh Fox, the director of the Academy Award–nominated documentary on fracking, *Gasland*, puts it: "I think that what's happening here is a squandering of the greatest political will that we've ever had towards getting off of fossil fuels."[58]

Because while green groups battle over the research and voluntary codes, the gas companies are continuing to drill, leak, and pour billions of dollars into new infrastructure designed to last for many decades.

Trading in Pollution

When governments began negotiating the international climate treaty that would become the Kyoto Protocol, there was broad consensus about what the agreement needed to accomplish. The wealthy, industrialized countries responsible for the lion's share of historical emissions would have to lead by capping their emissions at a fixed level and then systematically reducing them. The European Union and developing countries assumed that governments would do this by putting in place strong domestic measures to reduce emissions at home, for example by taxing carbon, and beginning a shift to renewable energy.

But when the Clinton administration came to the negotiations, it proposed an alternate route: create a system of international carbon trading modeled on the cap-and-trade system used to address acid rain (in the run-up to Kyoto, the EDF worked closely on the plan with Al Gore's office).[59] Rather than straightforwardly requiring all industrialized countries to lower their greenhouse gas emissions by a fixed amount, the scheme would issue pollution permits, which they could use, sell if they didn't need them, or purchase so that they could pollute more. National programs would be set up so that companies could similarly trade these permits, with the country staying within an overall emissions cap. Meanwhile, projects that were employing practices that claimed to be keeping carbon out of the atmosphere— whether by planting trees that sequester carbon, or by producing low carbon energy, or by upgrading a dirty factory to lower its emissions—could qualify for carbon credits. These credits could be purchased by polluters and used to offset their own emissions.

The U.S. government was so enthusiastic about this approach that it made the inclusion of carbon trading a deal breaker in the Kyoto negotiations. This led to what France's former environment minister Dominique Voynet described as "radically antagonistic" conflicts between the United States and Europe, which saw the creation of a global carbon market as tantamount to abandoning the climate crisis to "the law of the jungle." Angela Merkel, then Germany's environment minister, insisted, "The aim cannot be for industrialized countries to satisfy their obligations solely through emissions trading and profit."[60]

It is one of the great ironies of environmental history that the United States—after winning this pitched battle at the negotiating table—would fail to ratify the Kyoto Protocol, and that the most important emissions market would become a reality in Europe, where it was opposed from the outset. The European Union's Emissions Trading System (ETS) was launched in 2005 and would go on to become closely integrated with the United Nations' Clean Development Mechanism (CDM), which was written into the Kyoto Protocol. At least initially, the markets seemed to take off. From 2005 through 2010, the World Bank estimates that the various carbon markets around the globe saw over $500 billion in trades (though some experts believe those estimates are inflated). Huge numbers of projects around the world, meanwhile, are generating carbon credits—the CDM alone had an estimated seven-thousand-plus registered projects in early 2014.[61]

But it didn't take long for the flaws in the plan to show. Under the U.N. system, all kinds of dodgy industrial projects can generate lucrative credits. For instance, oil companies operating in the Niger Delta that practice "flaring"—setting fire to the natural gas released in the oil drilling process because capturing and using the potent greenhouse gas is more expensive than burning it—have argued that they should be paid if they stop engaging in this enormously destructive practice. And indeed some are already registered to receive carbon credits under the U.N. system for no longer flaring—despite the fact that gas flaring has been illegal in Nigeria since 1984 (it's a law filled with holes and is largely ignored).[62] Even a highly polluting factory that installs a piece of equipment that keeps a greenhouse gas out of the atmosphere can qualify as "green development" under U.N. rules. And this, in turn, is used to justify more dirty emissions somewhere else.

The most embarrassing controversy for defenders of this model involves coolant factories in India and China that emit the highly potent greenhouse gas HFC-23 as a by-product. By installing relatively inexpensive equipment to destroy the gas (with a plasma torch, for example) rather than venting it into the air, these factories—most of which produce gases used for air-conditioning and refrigeration—have generated tens of millions of dollars in emission credits every year. The scheme is so lucrative, in fact, that it has triggered a series of perverse incentives: in some cases, companies can earn twice as much by destroying an unintentional by-product as they can from making their primary product, which is itself emissions intensive. In the most egregious instance of this, selling carbon credits constituted a jaw-dropping 93.4 percent of one Indian firm's total revenues in 2012.[63]

According to one group that petitioned the U.N. to change its policies on HFC-23 projects, there is "overwhelming evidence that manufacturers are gaming" the system "by producing more potent greenhouse gases just so they can get paid to destroy them."[64] But it gets worse: the primary product made by these factories is a type of coolant that is so damaging to the ozone that it is being phased out under the Montreal Protocol on ozone depletion.

And this is not some marginal piece of the world emissions market—as of 2012, the U.N. system awarded these coolant manufacturers its largest share of emission credits, more than any genuinely clean energy projects.[65] Since then, the U.N. has enacted some partial reforms, and the European Union has banned credits from these factories in its carbon market.

It should hardly be surprising that so many questionable offset projects have come to dominate the emissions market. The prospect of getting paid real money based on projections of how much of an invisible substance is kept out of the air tends to be something of a scam magnet. And the carbon market has attracted a truly impressive array of grifters and hustlers who scour biologically rich but economically poor nations like Papua New Guinea, Ecuador, and Congo, often preying on the isolation of Indigenous people whose forests can be classified as offsets. These carbon cowboys, as they have come to be called, arrive bearing aggressive contracts (often written in English, with no translation) in which large swaths of territory are handed over to conservation groups on the promise of money for nothing. In the bush of Papua New Guinea, carbon deals are known as "sky money"; in Madagascar, where the promised wealth has proved as ephemeral as the product being traded, the Betsimisaraka people talk of strangers who are "selling the wind."[66]

A notorious carbon cowboy is Australian David Nilsson, who runs a particularly fly-by-night operation; in one recent incarnation, his carbon credit enterprise reportedly consisted only of an answering service and a web domain. After Nilsson tried to convince the Matsés people in Peru to sign away their land rights in exchange for promises of billions in revenues from carbon credits, a coalition of Indigenous people in the Amazon Basin called for Nilsson to be expelled from the country. And they alleged that Nilsson's pitch was "similar to 100 other carbon projects" which were "dividing our people with non-existent illusions of being millionaires."* Some Indigenous leaders even say that it is easier to deal with big oil and mining companies, because at least people understand who these companies are and what they want; less so when the organization after your land is a virtuous-seeming NGO and the product it is trying to purchase is something that cannot be seen or touched.[67]

This points to a broader problem with offsets, one that reaches beyond the official trading systems and into a web of voluntary arrangements administered by large conservation groups in order to unofficially "offset" the emissions of big polluters. Particularly in the early days of offsetting, after forest conservation projects began appearing in the late 1980s and early 1990s, by far the most persistent controversy was that—in the effort to quantify and control how much carbon was being stored so as to assign a monetary value to the standing trees—the people who live in or near those forests were sometimes pushed onto reservation-like parcels, locked out of their previous ways of life.[68] This locking out could be literal, complete with fences and armed men patrolling the territory looking for trespassers. The NGOs claim that they were merely attempting to protect the resources and the carbon they represented, but all this was seen, quite understandably, as a form of land grabbing.

For instance, in Paraná, Brazil, at a project providing offsets for Chevron, GM, and American Electric Power and administered by The Nature Conservancy and a Brazilian NGO, Indigenous Guarani were not allowed to forage for wood or hunt in the places they'd always occupied, or even fish in nearby waterways. As one local put it, "They want to take our home from us." Cressant Rakotomanga, president of a community organization in Madagascar where the Wildlife Conservation Society is running an offset program, expressed a similar sentiment. "People are frustrated because

---

* Interestingly, before Nilsson got into the carbon game, he was investigated by a member of Queensland's parliament for selling what appeared to be entirely fictional Australian real estate to unlucky marks in none other than Nauru.

before the project, they were completely free to hunt, fish and cut down the forests."[69]

Indeed the offset market has created a new class of "green" human rights abuses, wherein peasants and Indigenous people who venture into their traditional territories (reclassified as carbon sinks) in order to harvest plants, wood, or fish are harassed or worse. There is no comprehensive data available about these abuses, but the reported incidents are piling up. Near Guaraqueçaba, Brazil, locals have reported being shot at by park rangers while they searched the forest for food and plants inside the Paraná offset project hosted by The Nature Conservancy. "They don't want human beings in the forest," one farmer told the investigative journalist Mark Schapiro. And in a carbon-offset tree-planting project in Uganda's Mount Elgon National Park and Kibale National Park, run by a Dutch organization, villagers described a similar pattern of being fired upon and having their crops uprooted.[70]

In the wake of such reports, some of the green groups involved in offsetting now stress their dedication to Indigenous rights. However, dissatisfaction remains and controversies continue to crop up. For example, in the Bajo Aguán region of Honduras, some owners of palm oil plantations have been able to register a carbon offset project that claims to capture methane. Spurred by the promise of cash for captured gas, sprawling tree farms have displaced local agriculture, leading to a violent cycle of land occupations and evictions that has left as many as a hundred local farmers and their advocates dead as of 2013. "The way we see it, it has become a crime to be a farmer here," says Heriberto Rodríguez of the Unified Campesino Movement of Aguán, which places part of the blame for the deaths on the carbon market itself. "Whoever gives the finance to these companies also becomes complicit in all these deaths. If they cut these funds, the landholders will feel somewhat pressured to change their methods."[71]

Though touted as a classic "win-win" climate solution, there are very few winners in these farms and forests. In order for multinational corporations to protect their freedom to pollute the atmosphere, peasants, farmers, and Indigenous people are losing their freedom to live and sustain themselves in peace. When the Big Green groups refer to offsets as the "low-hanging fruit" of climate action, they are in fact making a crude cost-benefit analysis that concludes that it's easier to cordon off a forest inhabited by politically weak people in a poor country than to stop politically powerful corporate emitters in rich countries—that it's easier to pick the fruit, in other words, than dig up the roots.

The added irony is that many of the people being sacrificed for the carbon market are living some of the most sustainable, low-carbon lifestyles on the planet. They have strong reciprocal relationships with nature, drawing on local ecosystems on a small scale while caring for and regenerating the land so it continues to provide for them and their descendants. An environmental movement committed to real climate solutions would be looking for ways to support these ways of life—not severing deep traditions of stewardship and pushing more people to become rootless urban consumers.

Chris Lang, a British environmentalist based in Jakarta who runs an offset watchdog website called REDD-Monitor, told me that he never thought his job would involve exposing the failings of the green movement. "I hate the idea of the environmental movement fighting among itself instead of fighting the oil companies," he said. "It's just that these groups don't seem to have any desire to take on the oil companies, and with some of them, I'm not sure they really are environmentalists at all." [72]

―――――

This is not to say that every project being awarded carbon credits is somehow fraudulent or actively destructive to local ways of life. Wind farms and solar arrays are being built, and some forests classified as offsets are being preserved. The problem is that by adopting this model of financing, even the very best green projects are being made ineffective as climate responses because for every ton of carbon dioxide the developers keep out of the atmosphere, a corporation in the industrialized world is able to pump a ton into the air, using offsets to claim the pollution has been neutralized. One step forward, one step back. At best, we are running in place. And as we will see, there are other, far more effective ways to fund green development than the international carbon market.

Geographer Bram Büscher coined the term "liquid nature" to refer to what these market mechanisms are doing to the natural world. As he describes it, the trees, meadows, and mountains lose their intrinsic, place-based meaning and become deracinated, virtual commodities in a global trading system. The carbon-sequestering potential of biotic life is virtually poured into polluting industries like gas into a car's tank, allowing them to keep on emitting. Once absorbed into this system, a pristine forest may look as lush and alive as ever, but it has actually become an extension of a dirty power plant on the other side of the planet, attached by invisible financial transactions. Polluting smoke may not be billowing from the tops of its trees but it may as well be, since the trees that have been designated as carbon offsets are now allowing that pollution to take place elsewhere.[73]

The mantra of the early ecologists was "everything is connected"—every tree a part of an intricate web of life. The mantra of the corporate-partnered conservationists, in sharp contrast, may as well be "everything is disconnected," since they have successfully constructed a new economy in which the tree is not a tree but rather a carbon sink used by people thousands of miles away to appease our consciences and maintain our levels of economic growth.

But the biggest problem with this approach is that carbon markets have failed even on their own terms, as markets. In Europe, the problems began with the decision to entice companies and countries to join the market by handing out a huge number of cheap carbon permits. When the economic crisis hit a few years later, it caused production and consumption to contract and emissions to drop on their own. That meant the new emissions market was drowning in excess permits, which in turn caused the price of carbon to drop dramatically (in 2013, a ton of carbon was trading for less than €4, compared to the target price of €20). That left little incentive to shift away from dirty energy or to buy carbon credits. Which helps explain why, in 2012, coal's share of the U.K.'s electricity production rose by more than 30 percent, while in Germany, as we have already seen, emissions from coal went up despite the country's rapid embrace of renewable power. Meanwhile, the United Nations Clean Development Mechanism has fared even worse: indeed it has "essentially collapsed," in the words of a report commissioned by the U.N. itself. "Weak emissions targets and the economic downturn in wealthy nations resulted in a 99 percent decline in carbon credit prices between 2008 and 2013," explains Oscar Reyes, an expert on climate finance at the Institute for Policy Studies.[74]

This is a particularly extreme example of the boom-and-bust cycle of markets, which are volatile and high-risk by nature. And that's the central flaw with this so-called solution: it is simply too risky, and time is too short, for us to put our collective fate in such an inconstant and unreliable force. John Kerry has likened the threat of climate change to a "weapon of mass destruction," and it's a fair analogy.[75] But if climate change poses risks on par with nuclear war, then why are we not responding with the seriousness that that comparison implies? Why aren't we ordering companies to stop putting our future at risk, instead of bribing and cajoling them? Why are we gambling?

Tired of this time wasting, in February 2013, more than 130 environmental and economic justice groups called for the abolition of the largest carbon-trading system in the world, the EU's Emissions Trading System (ETS), in order "to make room for climate measures that work." The declaration stated that, seven years into this experiment, "The ETS has not reduced greenhouse gas emissions . . . the worst polluters have had little to no obligation to cut emissions at source. Indeed, offset projects have resulted in an *increase* of emissions worldwide: even conservative sources estimate that between ⅓ and ⅔ of carbon credits bought into the ETS 'do not represent real carbon reductions.'"[76]

The system has also allowed power companies and others to pass on the cost of compliance to their consumers, especially in the early years of the market, leading to a 2008 estimate by Point Carbon of windfall profits between $32 and $99 billion for electric utilities in the U.K., Germany, Spain, Italy, and Poland over a span of just five years. One report found airline companies raked in a windfall of up to $1.8 billion in their first year on the market in 2012. In short, rather than getting the polluters to pay for the mess they have created—a basic principle of environmental justice—taxpayers and ratepayers have heaped cash on them and for a scheme that hasn't even worked.[77]

———

In the context of the European debacle, the fact that the U.S. Senate failed to pass climate legislation in 2009 should not be seen, as it often is, as the climate movement's greatest defeat, but rather as a narrowly dodged bullet.

The cap-and-trade bills under consideration in the U.S. House and Senate in Obama's first term would have repeated all the errors of the European and U.N. emission trading systems, and then added some new ones of their own.

Both laws were based on proposals crafted by a coalition put together by the Environmental Defense Fund's Fred Krupp, which had brought large polluters (General Electric, Dow Chemical, Alcoa, ConocoPhillips, BP, Shell, the coal giant Duke Energy, DuPont, and many more) together with a handful of Big Green groups (The Nature Conservancy, the National Wildlife Federation, the Natural Resources Defense Council, the World Resources Institute, and what was then called the Pew Center on Global Climate Change). Known as the United States Climate Action Partnership (USCAP), the coalition had been guided by the familiar defeatist logic that there is no point trying to take on the big emitters directly so it's better to try to get them onside with a plan laden with corporate handouts and loopholes.[78]

The deal that ultimately emerged out of USCAP—touted as a historic compromise between greens and industry—handed out enough free allowances to cover 90 percent of emissions from energy utilities, including coal plants, meaning they could keep on emitting that amount and pay no price at all. "We're not going to get a better deal," Duke Energy's then CEO Jim Rogers boasted. "Ninety percent is terrific." Congressman Rick Boucher, a Democrat representing coal-rich southwestern Virginia, gushed that the bill had so many giveaways that it "ushered in a new golden age of coal."[79]

These "free allowances" to burn or trade carbon were, in essence, bribes. As solar entrepreneur Jigar Shah put it: "When you look at these companies that were in USCAP, they were not interested in regulating carbon. They were interested in a huge amount of wealth being transferred to their companies in exchange for their vote on climate change."[80] Needless to say, a deal that made fossil fuel interests this happy would have brought us nowhere near the deep cuts to our greenhouse gas emissions that scientists tell us are required to have a good chance of keeping warming below 2 degrees Celsius. And yet the green groups in USCAP didn't merely stand back and let the corporations in a direct conflict of interest write U.S. climate policy—they actively recruited them to do so.

And the saddest irony in all this pandering is that it still wasn't enough for the polluters. Working with USCAP to help draft climate legislation was, for many of the big corporate players who joined the coalition, a hedge. In 2007, when the coalition was formed, climate legislation looked extremely likely, and these companies wanted to be sure that whatever bill passed Congress was riddled with enough loopholes to be essentially meaningless—a classic Beltway strategy. They also knew that getting be-hind cap-and-trade was the best way of blocking the worrying prospect of a newly elected president using the Environmental Protection Agency (EPA) to put firm limits on the amount of carbon companies could emit. In fact, Waxman-Markey, the primary piece of climate legislation based on the coalition's blueprint, specifically barred the EPA from regulating car-bon from many major pollution sources, including coal-fired power plants. Michael Parr, senior manager of government affairs at DuPont, summa-rized the corporate strategy succinctly: "You're either at the table or on the menu."[81]

The problem for Fred Krupp and his colleagues was that these compa-nies were sitting at plenty of other tables at the same time. Many continued to be members of the American Petroleum Institute, the National Associa-tion of Manufacturers, and the U.S. Chamber of Commerce—all of which actively opposed climate legislation. When Barack Obama took office in January 2009, it looked like the corporate hard-liners were going to lose. But then, in the summer of 2009, with USCAP still trying to push cap-and-trade through the U.S. Senate, the political climate abruptly shifted. The economy was still deeply troubled, Obama's popularity was tanking, and a new political force came to centerstage. Flush with oil money from the Koch brothers and pumped up by Fox News, the Tea Party stormed town-hall meetings across the country, shouting about how Obama's health-care reform was part of a sinister plan to turn the United States into an Islamic/Nazi/socialist utopia. In short order, the president started sending signals that he was reluctant to pick another major legislative fight.[82]

That's when many of the key corporate members of USCAP began to realize that they now had a solid chance of scuttling climate legislation altogether. Caterpillar and BP dropped out of the coalition, as did Cono-coPhillips, after having complained of "unrecoverable costs . . . on what is historically a low-margin business." (ConocoPhillips revenues the year after it left USCAP totaled $66 billion, with a tidy net income of $12.4 billion.) And some of these companies didn't just leave Krupp's coalition of "former enemies": by directing their formidable firepower squarely at the legislation that they had helped craft, they made it abundantly clear that they had never stopped being its enemies. ConocoPhillips, for instance, set up a dedicated webpage to encourage visitors (including its roughly thirty thousand employees) to tell legislators how much they opposed the climate bill. "Climate change legislation will result in higher direct energy costs for the typical American family," the site warned, further claiming (outlandishly) that it "could result in a net loss of more than two million U.S. jobs each year." As for fellow defector BP, company spokesman Ronnie Chappell explained, "The lowest-cost option for reducing emissions is the increased use of natural gas."[83]

In other words, thinking they were playing a savvy inside game, Big Green was outmaneuvered on a grand scale. The environmentalists who participated in USCAP disastrously misread the political landscape. They chose a stunningly convoluted approach to tackling climate change, one that would have blocked far more effective strategies, specifically because it was more appealing to big emitters—only to discover that the most appealing climate policy to polluters remained none at all. Worse, once their corporate partners fled the coalition, they had no shortage of ammo to fire at their former friends. The climate bill was a boondoggle, they claimed (it was), filled with handouts and subsidies (absolutely), and it would pass on higher energy costs to cash-strapped consumers (likely).* To top it all off, as pro-oil Republican congressman Joe Barton put it, "The environmental benefit is nonexistent" (as the left flank of the green movement had been saying all along).[84]

It was a classic double-cross, and it worked. In January 2010, the climate legislation modeled on USCAP's proposals died in the Senate, as it deserved to—but not before it discredited the very idea of climate action in the minds of many.[85]

---

* Heartland regular Chris Horner called the bill "crony capitalism" on the Enron model—and Horner should know, because he used to work there.

Plenty of postmortems have been written about what the greens did wrong in the cap-and-trade fight but the hardest hitting came in a scathing report by Harvard University sociologist Theda Skocpol. She concluded that a major barrier to success was the absence of a mass movement applying pressure from below. "To counter fierce political opposition, reformers will have to build organizational networks across the country, and they will need to orchestrate sustained political efforts that stretch far beyond friendly Congressional offices, comfy boardrooms, and posh retreats."[86] As we will see, a resurgent grassroots climate movement has now arrived that is doing precisely that—and it is winning a series of startling victories against the fossil fuel sector as a result.

But old habits die hard. When the cap-and-trade fight in the U.S. Congress was finally over, with around half a billion dollars spent pushing the policy (ultimately down the drain), the man who led the pro-business revolution in the green movement offered his version of what went wrong. Fred Krupp—in a sharp gray suit, his well-styled hair now white after two and a half decades leading the Environmental Defense Fund—explained that climate legislation had failed because greens had been too hard-line, too "shrill," and needed to be more "humble" and more bipartisan.[87] In other words, compromise some more, tone it down even further, assert ideas with less confidence, and try to be even more palatable to their opponents. Never mind that that is precisely what groups like EDF have been doing since Reagan.

Fittingly enough, Krupp chose to share these pearls of wisdom during the annual Brainstorm Green session hosted by *Fortune*, a magazine devoted to the celebration of wealth, and sponsored by, among others, Shell Oil.[88]

7

NO MESSIAHS

The Green Billionaires Won't Save Us

"I had always got away with breaking rules and thought this was no different. I would have got away with it as well if I hadn't been greedy."

—Richard Branson, on getting caught dodging taxes in the early 1970s[1]

"You gotta lead from the front. Nobody is going to start it from the grassroots."

—Former New York mayor Michael Bloomberg, 2013[2]

In his autobiography/New Age business manifesto *Screw It, Let's Do It*, Richard Branson, the flamboyant founder of Virgin Group, shared the inside story of what he describes as his "Road to Damascus" conversion to the fight against climate change. It was 2006 and Al Gore, on tour with *An Inconvenient Truth*, came to the billionaire's home to impress upon him the dangers of global warming, and to try to convince Branson to use Virgin Airlines as a catalyst for change.[3]

"It was quite an experience having a brilliant communicator like Al Gore give me a personal PowerPoint presentation," Branson writes of the meeting. "Not only was it one of the best presentations I have ever seen in my life, but it was profoundly disturbing to become aware that we are potentially facing the end of the world as we know it. . . . As I sat there and listened to Gore, I saw that we were looking at Armageddon."[4]

As he tells it, Branson's first move following his terrifying epiphany was to summon Will Whitehorn, then Virgin Group's corporate and brand development director. Together "we carefully discussed these issues and took the decision to change the way Virgin operates on a corporate and global level. We called this new Virgin approach to business Gaia Capitalism in honor of James Lovelock and his revolutionary scientific view" (a reference to Lovelock's theory that the earth is "one single enormous living organism and every single part of the ecosystem reacted with every other part"). Not only would Gaia Capitalism "help Virgin to make a real difference in the next decade and not be ashamed to make money at the same time," but Branson believed it held the potential to become "a new way of doing business on a global level."[5]

Before the year was out, he was ready to make his grand entrance onto the green scene (and Branson knows how to make an entrance—by parachute, by hot air balloon, by Jet Ski, by kite-sail with a naked model clinging to his back . . .). At the 2006 Clinton Global Initiative annual meeting in New York City—the highest power event on the philanthropic calendar—Branson pledged to spend roughly $3 billion over the next decade to develop biofuels as an alternative to oil and gas, and on other technologies to battle climate change. The sum alone was staggering but the most elegant part was where the money would be coming from: Branson would divert the funds from profits generated by Virgin's fossil fuel–burning transportation lines. As Branson explained in an interview, "Any dividends or share sales or any money that we make from our airlines or trains will be ploughed back into tackling global warming, into investing in finding new, clean fuels and investing in trying to find fuels for jet engines so that we can hopefully reverse the inevitability of, you know, of destroying the world if we let it carry on the way it's going."[6]

In short, Branson was volunteering to do precisely what our governments have been unwilling to legislate—require that the profits being earned from warming the planet be channeled into the costly transition away from these dangerous energy sources. The director of the Natural Resources Defense Council's Move America Beyond Oil campaign said of Virgin's renewable energy initiatives, "This is exactly what the whole industry should be looking at." Furthermore, Branson pledged that if its transportation divisions weren't profitable enough to meet the $3 billion target, "the money will come out of our existing businesses." He would do "whatever it takes" to fulfill the commitment because "what is the point of holding back when there will be no businesses" if we fail to act?[7]

Bill Clinton was dazzled, calling the $3 billion pledge "groundbreaking, not only because of the price tag—which is phenomenal—but also because of the statement that he is making." *The New Yorker* described it as "by far the biggest such commitment that has yet been made to fight global warming."[8]

But Branson wasn't finished. A year later he was back in the news with the "Virgin Earth Challenge"—a $25 million prize that would go to the first inventor to figure out how to sequester one billion tons of carbon a year from the air "without countervailing harmful effects." He described it as "the largest ever science and technology prize to be offered in history." This, Branson pronounced, was "the best way to find a solution to the problem of climate change," elaborating in an official statement that, "If the greatest minds in the world today compete, as I'm sure they will, for The Virgin Earth Challenge, I believe that a solution to the CO_2 problem could hopefully be found—a solution that could save our planet—not only for our children but for all the children yet to come."[9]

And the best part, he said, is that if these competing geniuses crack the carbon code, the "'doom and gloom' scenario vanishes. We can carry on living our lives in a pretty normal way—we can drive our cars, we can fly our planes, life can carry on as normal."[10] Indeed the idea that we can solve the climate crisis without having to change our lifestyles in any way—certainly not by taking fewer Virgin flights—seemed to be the underlying assumption of all of Branson's various climate initiatives.

With the $3 billion pledge, he would try to invent a low carbon fuel that could keep his airlines running at full capacity. If that failed, and carbon still needed to be burned to keep the planes in the air, then the prize would surely help invent a way to suck the heat-trapping gas out of the sky before it's too late. To cover one more base, in 2009 Branson launched the Carbon War Room, an industry group looking for ways that different sectors could lower their emissions voluntarily, and save money in the process. "Carbon is the enemy," Branson declared. "Let's attack it in any possible way we can, or many people will die just like in any war."[11]

Billionaires and Broken Dreams

For many mainstream greens, Branson seemed to be a dream come true: a flashy, media-darling billionaire out to show the world that fossil-fuel-intensive companies can lead the way to a green future using profit as the most potent tool—and proving just how serious he was by putting striking amounts of his own cash on the line. As Branson explained to *Time*, "If the government can't deliver, it's up to industries to [do it] themselves. We have to make it a win-win for all concerned."[12] This is what groups like the Environmental Defense Fund had been saying since the 1980s in explaining why they partnered with big polluters, and what they had attempted to prove with the carbon market. But never before had there been a single figure willing to use his own multibillion-dollar empire as a test case. Branson's personal account of the impact of Gore's PowerPoint also seemed to confirm the notion, cherished in many green circles, that transforming the economy away from fossil fuels is not about confronting the rich and powerful but simply about reaching them with sufficiently persuasive facts and figures and appealing to their sense of humanity.

There had been major green philanthropists before. Men like financier Jeremy Grantham, who underwrites a large portion of the U.S. and British green movement, as well as a lot of related academic research, with wealth from Grantham, Mayo, Van Otterloo & Co., the investment management firm he cofounded.* But these funders tended to stay behind the scenes. And unlike Branson, Grantham has not attempted to turn his own financial firm into proof that the quest for short-term profits can be reconciled with his personal concerns about ecological collapse. On the contrary, Grantham is known for his bleak quarterly letters, in which he has mused on our economic model's collision course with the planet. "Capitalism, by ignoring the finite nature of resources and by neglecting the long-term well-being of the planet and its potentially crucial biodiversity, threatens our existence," Grantham wrote in 2012—but that doesn't mean savvy investors can't get very rich on the way down, both from the final scramble for fossil fuels, and by setting themselves up as disaster capitalists.[13]

Take Warren Buffett, for instance. For a brief time, he too seemed to be auditioning for the role of Great Green Hope, stating, in 2007, that, "the odds are good that global warming is serious" and that even if there

---

* The Grantham Foundation for the Protection of the Environment has funded a broad range of large environmental groups, ranging from The Nature Conservancy to Greenpeace, the Environmental Defense Fund to 350.org.

is a chance it won't be "you have to build the ark before the rains come. If you have to make a mistake, err on the side of the planet. Build a margin of safety to take care of the only planet we have."[14] But it soon became clear that Buffett was not interested in applying this logic to his own corporate assets. On the contrary, Berkshire Hathaway has done its best in subsequent years to make sure those rains come with ferocity.

Buffett owns several huge coal-burning utilities and holds large stakes in ExxonMobil and the tar sands giant Suncor. Most significantly, in 2009 Buffett announced that his company would spend $26 billion to buy what it didn't already own of the Burlington Northern Santa Fe railroad. Buffett called the deal—the largest in Berkshire Hathaway history—"a bet on the country."[15] It was also a bet on coal: BNSF is one of the biggest coal haulers in the United States and one of the primary engines behind the drive to greatly expand coal exports to China.

Investments like these push us further down the road toward catastrophic warming, of course—and Buffett is poised to be one of the biggest winners there too. That's because he is a major player in the reinsurance business, the part of the insurance sector that stands to profit most from climate disruption. As Eli Lehrer, the insurance industry advocate who defected from the Heartland Institute after its controversial billboard campaign, explains, "A large reinsurer like Warren Buffett's Berkshire Hathaway might simultaneously underwrite the risk of an industrial accident in Japan, a flood in the U.K., a hurricane in Florida, and a cyclone in Australia. Since there's almost no chance that all of these events will happen at the same time, the reinsurer can profit from the premiums it earns on one type of coverage even when it pays out mammoth claims on another." Perhaps it's worth remembering that Noah's Ark was not built to hold everyone, but just the lucky few.[16]

The newest billionaire raising big hopes in the climate scene is Tom Steyer, a major donor to various climate and anti–tar sands campaigns, as well as to the Democratic Party. Steyer, who made his fortune with the fossil-fuel-heavy hedge fund Farallon Capital Management, has made some serious attempts to bring his business dealings in line with his climate concerns. But unlike Branson, Steyer has done this by leaving the business that he founded, precisely because, as *The Globe and Mail* reported, "it valued a company's bottom line, not its carbon footprint." He further explained, "I have a passion to push for what I believe is the right thing. And I couldn't do it in good conscience and hold down a job—and get paid very well for doing a job—where I wasn't directly doing the right thing."* This stance is very different from Branson's, who is actively trying to prove that it is possible for a fossil-fuel–based company not just to do the right thing but to lead the transition to a clean economy.[17]

Nor is Branson in precisely the same category as Michael Bloomberg and Bill Gates, who have both used their philanthropy to aggressively shape the kinds of climate solutions on offer. Bloomberg, for instance, has been held up as a hero for his large donations to green groups like the Sierra Club and EDF, and for the supposedly enlightened climate policies he introduced while mayor of New York City.†[18]

But while talking a good game about carbon bubbles and stranded assets (his company is set to introduce the "Bloomberg Carbon Risk Valuation Tool" to provide data and analysis to its clients about how fossil fuel stocks would be impacted by a range of climate actions), Bloomberg has made no discernible attempt to manage his own vast wealth in a manner that reflects these concerns. On the contrary, as previously mentioned, he helped set up Willett Advisors, a firm specializing in oil and gas assets, for both his personal and philanthropic holdings. Brad Briner, director of real assets for Willett, stated plainly in May 2013 that, "We are natural gas bulls. We think oil is well priced," citing new drilling investments on the horizon.[19]

It's not simply that Bloomberg is actively snapping up fossil fuel assets even as he funds reports warning that climate change makes for "risky busi-

---

* It should be noted that though Steyer has separated out his personal funds from Farallon, he remains a limited partner and has also promoted the use of natural gas, helping to fund the EDF research that went into its pro-fracking study and enthusiastically endorsing natural gas in *The Wall Street Journal*.

† Those policies have been criticized for favoring big developers over vulnerable communities and for using a green veneer to push through mega real estate development projects with dubious environmental benefits, as Hunter College urban affairs professor Tom Angotti and others have written. Communities heavily impacted by Hurricane Sandy, meanwhile, have claimed that Bloomberg's post-disaster rebuilding plans were made with only token input from them.

ness." It's that those gas assets may well have increased in value as a result of Bloomberg's environmental giving, what with EDF championing natural gas as a replacement for coal and the Sierra Club spending tens of millions of Bloomberg's dollars shutting down coal plants. Was funding the war on coal at least partially about boosting the share price of gas? Or was that just a bonus? Perhaps there is no connection between his philanthropic priorities and his decision to entrust so much of his fortune to the oil and gas sector. But these investment choices do raise uncomfortable questions about Bloomberg's status as a climate hero, as well as his 2014 appointment as a United Nations special envoy for cities and climate change (questions Bloomberg has not answered despite repeated requests). At the very least, they demonstrate that seeing the risks climate change poses to financial markets in the long term may not be enough to curtail the temptation to profit from planet destabilization in the short-term.[20]

Bill Gates has a similar firewall between mouth and money. Though he professes great concern about climate change, the Gates Foundation had at least $1.2 billion invested in just two oil giants, BP and ExxonMobil, as of December 2013, and those are only the beginning of his fossil fuel holdings.[21]

Gates's approach to the climate crisis, meanwhile, shares a fair amount with Branson's. When Gates had his climate change epiphany, he too immediately raced to the prospect of a silver-bullet techno-fix in the future, without pausing to consider viable—if economically challenging—responses in the here and now. In TED Talks, op-eds, interviews, and in his much-discussed annual letters, Gates repeats his call for governments to massively increase spending on research and development with the goal of uncovering "energy miracles." By miracles, Gates means nuclear reactors that have yet to be invented (he is a major investor and chairman of nuclear start-up TerraPower); he means machines to suck carbon out of the atmosphere (he is also a primary investor in at least one such prototype); and he means direct climate manipulation (Gates has spent millions of his own money funding research into various schemes to block the sun, and his name is listed on several hurricane-suppression patents). At the same time, he has been dismissive of the potential of existing renewable technologies. "We focus too much on deployment of stuff that we have today," Gates claims, writing off energy solutions like rooftop solar as "cute" and "noneco-nomic" (despite the fact that these cute technologies are already providing 25 percent of Germany's electricity).[22]

The real difference between Gates and Branson is that Branson still has a hands-on leadership role at Virgin and Gates left the top job at Microsoft years ago. Which is why when Branson entered the climate fray, he was re-ally in a category of his own—promising to turn a major multinational, one with fossil fuels at its center, into an engine for building the next economy. The only other figure who had raised similar hopes was the brash Texas oil-man T. Boone Pickens. In 2008, he launched "The Pickens Plan," which, backed by a huge advertising budget in print and television, promised to end U.S. dependence on foreign oil by massively boosting wind and solar, and converting vehicles to natural gas. "I've been an oil man my whole life," Pickens said in his commercials, with his heavy Texas twang. "But this is one emergency we can't drill our way out of."[23]

The kind of policies and subsidies Pickens was advocating were ones from which the billionaire's energy hedge fund, BP Capital, was poised to profit—but for the greens cheering him on, that wasn't the point. Carl Pope, then heading the Sierra Club, joined the billionaire on his private Gulfstream jet to help him sell the plan to reporters. "To put it plainly, T. Boone Pickens is out to save America," he pronounced.[24]

Or not. Shortly after Pickens's announcement, the fracking frenzy took off, and suddenly powering the grid with unconventional natural gas looked a lot more appealing to BP Capital than relying on wind. Within a couple of years the Pickens Plan had radically changed. It now had almost noth-ing to do with renewable energy and everything to do with pushing for more gas extraction no matter the cost. "You're stuck with hydrocarbons—come on, get real," Pickens told a group of reporters in April 2011, while questioning the seriousness of human-caused global warming to boot. By 2012, he was extolling the virtues of the tar sands and the Keystone XL pipeline. As David Friedman, then research director for the Clean Vehicles Program at the Union of Concerned Scientists, put it: Pickens "kept say-ing that this wasn't about private interests, it was about the nation and the world. But to dump the part that actually had the greatest potential to cut global warming and pollution and help create new jobs in the U.S., in favor of the piece that really does most benefit his bottom line, was a disappointment."[25]

———————

Which leaves us with Branson—his pledge, his prize, and his broader vision of voluntarily changing capitalism so that it is in keeping with the laws of "Gaia." Almost a decade after Branson's PowerPoint epiphany, it seems like a good time to check in on the "win-win" crusade. It's too much to expect Branson to have changed the way business is done in less than a decade, of course. But given the hype, it does seem fair to examine how his attempts to prove that industry can lead us away from climate catastrophe without heavy government intervention have progressed. Because given the dismal track record of his fellow green billionaires on that front, it may be fair to conclude that if Branson can't do it, no one can.

The Pledge That Turned into a "Gesture"

Let's start with Branson's "firm commitment" to spend $3 billion over a decade developing a miracle fuel. Despite press reports portraying the pledge as a gift, the original concept was more like straight-up vertical integration. And integration is Branson's hallmark: the first Virgin business was record sales, but Branson built his global brand by making sure that he not only owned the music stores, but the studio where the bands recorded, and the record label that represented them. Now he was applying the same logic to his airlines. Why pay Shell and Exxon to power Virgin planes and trains when Virgin could invent its own transport fuel? If it worked, the gambit would not only turn Branson into an environmental hero but also make him a whole lot richer.

So the first tranche of money Branson diverted from his transportation divisions went to launch a new Virgin business, originally called Virgin Fuels and since replaced by a private equity firm called the Virgin Green Fund. In keeping with his pledge, Branson started off by investing in various agrofuel businesses, including making a very large bet of roughly $130 million on corn ethanol.* And Virgin has attached its name to several biofuel pilot projects—one to derive jet fuel from eucalyptus trees, another from fermented gas waste—though it has not gone in as an investor. (Instead it mainly offers PR support, and a pledge to purchase the fuel if it becomes viable.) But by Branson's own admission, the miracle fuel he was looking for "hasn't been invented yet" and the biofuels sector has stalled, thanks in part to the influx of fracked oil and gas. In response to written interview questions, Branson conceded, "It's increasingly clear that this is a question of creating the market conditions that would allow a diverse portfolio of different renewable fuel producers, suppliers and customers to all work in the same way that conventional fuel supply chains work today. It's one of the issues that the Carbon War Room's renewable jet fuels operation is looking to solve." [26]

Perhaps this is why Branson's green investment initiative appears to have lost much of its early interest in alternative fuels. Today, the Virgin Green Fund continues to invest in one biofuel company, but the rest of its investments are a grab bag of vaguely green-hued projects, from water desalination to energy-efficient lighting, to an in-car monitoring system to help drivers conserve gas. Evan Lovell, a partner in the Virgin Green Fund, acknowledged in an interview that the search for a breakthrough fuel has given way to a "much more incremental" approach, one with fewer risks and more short-term return. [27]

Diversifying his holdings to get a piece of the green market is Branson's prerogative, of course. But hundreds of venture capitalists have made the same hedge, as have all the major investment banks. It hardly would seem to merit the fanfare inspired by Branson's original announcement. Especially because the investments themselves have been so unremarkable. Jigar Shah, a Branson supporter who ran the Carbon War Room, is frank about this: "I don't think that he's made a lot of great investments in the climate change space. But the fact that he's passionate about it is a good thing." [28]

---

* Aided by investments like these, the ethanol boom was responsible for 20–40 percent of the spike in agricultural commodity prices in 2007–2009, according to a survey by the National Academy of Sciences.

Then there is the small matter of dollar amounts. When Branson made his pledge, he said that he would "invest 100 percent of all future proceeds of the Virgin Group from our transportation businesses into tackling global warming for an estimated value of $3 billion over the next 10 years."[29] That was 2006. If Branson is to make it to $3 billion by 2016, by this point, at least $2 billion should have been spent. He's not even close.

In 2010—four years into the pledge—Branson told *The Economist* that he had so far only invested "two or three hundred million dollars in clean energy," blaming poor profits in the airline sector. In February 2014, he told *The Observer* that "we have invested hundreds of millions in clean technology projects." Not much progress, in other words. And it may be even less: according to Virgin Green Fund partner Lovell, Virgin has still only contributed around $100 million (outsider investors had matched that), on top of the original ethanol investment, which as of 2013, brings the total Branson investment to something around $230 million. (Lovell confirmed that "we are the primary vehicle" for Branson's promise.) Add to that an undisclosed, but likely modest personal investment in an algae company called Solazyme, and we are still looking at well under $300 million dollars, seven years into the ten-year pledge that is supposed to reach $3 billion. As of this writing, no major new investments had been announced.[30]

Branson refused to answer direct questions about how much he had spent, writing that "it's very hard to quantify the total amount we've invested in relation to climate change across the Group," and his labyrinthine holdings make it hard to come up with independent estimates. "I'm not very good with figures," the billionaire has said about another murky corner of Virgin's empire, adding "I failed my elementary maths." Part of the confusion stems from the fact that it's unclear what should be counted toward the original $3 billion pledge. It began as a targeted quest for a miracle green fuel, then expanded to become a search for clean technologies generally, then, apparently, eco-anything. Branson now says that he is counting "investments made by individual Virgin companies in sustainability measures, such as more efficient fleets" of planes. More recently, Branson's fight against global warming has centered around various attempts to "green" his two private islands in the Caribbean, one of which serves as his deluxe family compound and the other a $60,000 a night hotel. Branson claims the model he is setting will help nearby Caribbean nations to switch to renewable power themselves. Perhaps it will, but it's all a long way from the pledge to transform capitalism back in 2006.[31]

The Virgin boss now plays down his original commitment, no longer referring to it as a "pledge" but rather a "gesture." In 2009, he told *Wired* magazine, "In a sense, whether it's $2 billion, $3 billion or $4 billion is not particularly relevant." Branson told me that when the deadline rolls around, "I suspect it will be less than $1 billion right now." That too may prove an exaggeration: if the publicly available information is correct, he would have to more than triple the green energy investments he has made so far. When asked, Branson blamed the shortfall on everything from high oil prices to the global financial crisis: "The world was quite different back in 2006. . . . In the last eight years our airlines have lost hundreds of millions of dollars."[32]

Given these various explanations for falling short, it is worth taking a look at some of the things for which Richard Branson and Virgin *did* manage to find money in this key period. Like, for instance, a massive global push to put more carbon-spewing planes in the skies adorned with stylized "V"s on their tails.

———

When Branson met with Al Gore he warned the former vice president that however alarmed he became about what he learned about climate change, he was about to launch a new air route to Dubai and wasn't about to change that. That wasn't the half of it. In 2007, just one year after seeing the climate light with Gore and deciding, as he put it, that "my new goal in life is to work at reducing carbon emissions," Branson launched his most ambitious venture in years: Virgin America, a brand-new airline competing in the U.S. domestic market. Even by the standards of a new venture, Virgin America's growth rates in its first five years have been startling: from forty flights a day to five destinations in its first year, it reached 177 flights a day to twenty-three destinations in 2013. And the airline has plans to add forty more planes to its fleet by the middle of the next decade. In 2010, *The Globe and Mail* reported that Virgin America was heading for "the most aggressive expansion of any North American airline in an era when most domestic carriers have been retrenching."[33]

Branson's capacity to expand so quickly has been boosted by rock-bottom seat sales, including offering tickets for just $60.[34] With prices like that, Branson was not just poaching passengers from United and American, he was getting more people up in the air. The new airline, however, has been a hugely costly endeavor, generating hundreds of millions in losses. Bad news for the Green Fund, which needed Virgin's transportation businesses to do well in order to get topped up.

Branson hasn't only been expanding his transportation businesses in the Americas. The number of people flying on various Virgin-branded Australian airlines increased by 27 percent in the five years following his climate pledge, from fifteen million passengers in 2007 to nineteen million in 2012. And in 2009, he launched a whole new long-haul airline, V Australia. Then, in April 2013, Branson unveiled yet another ambitious new venture: Little Red, a domestic airline for the U.K., starting with twenty-six flights a day. In true Branson style, when he launched the new airline in Edinburgh, he dressed in a kilt and flashed his underwear to reporters, which he had emblazoned with the words "stiff competition."* But as with Virgin America, this was not just about competing with rivals for existing fliers: Virgin was so keen on expanding the number of people who could use the most carbon-intensive form of transportation that it offered a celebratory seat sale for some flights that charged customers no fares at all, only taxes—which came to about half the price of a cab from central London to Heathrow during rush hour.[35]

So this is what Branson has done on his climate change Road to Damascus: he went on an airplane-procurement spree. When Virgin's various expansions are tallied up, around 160 hardworking planes have been added to its global fleet since Branson's epiphany with Al Gore—quite possibly more than that. And the atmospheric consequences are entirely predict-

---

* Puerile humor of this sort is a recurring theme in Branson's PR machine (the company once painted "Mine's Bigger than Yours" on the side of a new Airbus A340-600; boasted of its business-class seats that "size does matter"; and even flew a blimp over London emblazoned with the slogan "BA [British Airways] Can't Get It Up!!").

able. In the years after his climate pledge, Virgin airlines' greenhouse gas emissions soared by approximately 40 percent. Virgin Australia's emissions jumped by 81 percent between 2006–2007 and 2012–2013, while Virgin America's emissions shot up by 177 percent between 2008 and 2012. (The only bright spot in Branson's emissions record was a dip at Virgin Atlantic between 2007 and 2010—but that likely was less the result of visionary climate policy than the global economic downturn and the massive volcanic eruption in Iceland, which hit airlines indiscriminately.)[36]

Much of the sharp overall rise in Virgin's emissions was due to the airlines' rapid growth rate—but that wasn't the only factor. A study by the International Council on Clean Transportation on the relative fuel efficiency of fifteen U.S. domestic airlines in 2010 found that Virgin America clocked in at ninth place.[37] This is quite a feat considering that, unlike its much older competitors, the brand-new airline could have built the best fuel efficiency practices into its operations from day one. Clearly Virgin chose not to.

And it's not just airplanes. While he has been publicly waging his carbon war, Branson unveiled Virgin Racing to compete in Formula One (he claimed he had entered the sport only because he saw opportunities to make it greener but quickly lost interest). He also invested heavily in Virgin Galactic, his own personal dream of launching the first commercial flights into space, for a mere $250,000 per passenger. Not only is leisure space travel a pointless waste of (planet-warming) energy, it is also yet another money pit: according to *Fortune*, by early 2013 Branson had spent "more than $200 million" on the vanity project, with much more in the works. That would be more than he appears to have spent on the search for a green fuel to power his planes.* [38]

When asked about the status of his $3 billion climate pledge, Branson tends to plead poor, pointing to the losses posted by his transportation busi-

---

* To quote one scathing assessment of the project by sociologist Salvatore Babones, "If two words can capture the extraordinary redistribution of wealth from workers to the wealthy over the past forty years, the flagrant shamelessness of contemporary conspicuous consumption, the privatization of what used to be public privileges and the wanton destruction of our atmosphere that is rapidly leading toward the extinction of nearly all non-human life on earth, all covered in a hypocritical pretense of pious environmental virtue . . . those two words are Virgin Galactic."

nesses.[39] But given the manic level of growth in these sectors, it's an excuse that rings hollow. Not only have his trains been doing quite well, but given the flurry of new routes and new airlines, there was clearly no shortage of surplus money to spend. It's just that the Virgin Group decided to follow the basic imperative of capital: grow or die.

It's also worth remembering that Branson was very clear when he announced the pledge that if his transportation divisions were not profitable enough to meet the target, he would divert funds from other parts of the Virgin empire. And here we run into another kind of problem: Branson's corporate modus operandi is somewhat nontraditional. He tends to pull in relatively modest profits (or even losses) while spending a great deal of money (his own, his partners', and taxpayers') building up flashy extensions of the Virgin brand. Then, when a new company is established, he sells all or part of his stake for a hefty sum, and a lucrative brand licensing deal. This money isn't posted as profits from those businesses, but it helps explain how Branson's net worth rose from an estimated $2.8 billion in 2006, the year he met with Gore, to an estimated $5.1 billion in 2014. Musing on his passion for environmentalism to John Vidal in *The Observer*, Branson said, "I find it interests me a lot more than making a few more bucks; it's much more satisfactory." And yet a few more bucks he has certainly made.[40]

Meanwhile, with the ten-year deadline fast approaching, it seems we are no closer to a miracle fuel to power Branson's planes, which are burning significantly more carbon than when the pledge period began. But fear not, because Branson has what he describes as his "fallback insurance policy." So how is that going?[41]

The Incredible Disappearing Earth Challenge

After the original hoopla over Branson's $25 million Virgin Earth Challenge (or Earth Prize, as it is more frequently called), the initiative seemed to go dormant for a while. When journalists remembered to ask the Virgin chief about the search for a miracle technology to suck large amounts of carbon from the air, he seemed to subtly lower expectations, much as he has done with green fuels. And he had always cautioned that there was a chance that no one would win the prize. In November 2010, Branson revealed that Virgin had received something on the order of 2,500 entries. Nick Fox, Branson's spokesperson, explained that many ideas had to be ruled out because they were too risky and seemingly safer ones were not "developed enough to be commercialized right now." In Branson's words, there was no "slam dunk winner yet."[42]

Fox also mentioned that far more than $25 million was needed to determine some ideas' large-scale viability, something more on the order of $2.5 billion.[43]

Branson claims he hasn't completely given up on awarding the prize at some future point, saying, "We hope it's only a matter of time before there's a winner." He has, however, changed his role from straight-up patron to something more akin to a celebrity judge on a reality TV show, giving his blessing to the most promising ideas and helping them land high-level advice, investment, and other opportunities flowing from their association with the Virgin brand.[44]

This new incarnation of the Earth Challenge was unveiled (to significantly less fanfare than the first time around) in November 2011, at an energy conference in Calgary, Alberta. Appearing by video link, Branson announced the eleven most promising entries. Four were machines that directly sucked carbon out of the air (though none at anywhere near the scale needed); three were companies using the biochar process, which turns carbon-sequestering plant matter or manure into charcoal and then buries it in the soil and is controversial on a mass scale; and among the miscellaneous ideas was a surprisingly low-tech one involving revamping livestock grazing to boost the carbon-sequestering potential of soil.[45]

According to Branson, none of these finalists was ready yet to win the $25 million prize but they were being showcased like beauty queens at the energy conference so that "the best engineers, investors, opinion formers and policy makers [would] work together on this challenge. Only then will the potential be realised. I see Calgary as a great city to start in."[46]

It was certainly a revealing choice. Calgary is the economic heart of Canada's tar sands boom. Oil from those dirty deposits has made the city one of the richest metropolises in the world, but its ongoing prosperity is entirely contingent on continuing to find customers for its product. And that depends very much on getting controversial pipelines like Keystone XL constructed through increasingly hostile territory, as well as on dissuading foreign governments from passing laws that would penalize Alberta's particularly high-carbon fuel.

Enter Alan Knight, Richard Branson's sustainability advisor and the man he put in charge of the Earth Challenge. Knight took great pride in being Branson's go-to green guy, but their relationship was far from exclusive. Shell and Statoil (two of the biggest players in the tar sands) were among the other clients in Knight's consultancy. So too was, in his words, "Calgary City and the Alberta oil sands industry," specifically the Oil Sands Leadership Initiative (OSLI), an industry trade group comprised of ConocoPhillips, Nexen, Shell, Statoil, Suncor Energy, and Total. Knight boasted of being given "private access to their meetings," and explained that he advised his clients in the Alberta oil patch on how to allay mounting concerns about the enormous ecological costs of an extraction process that is three to four times more greenhouse gas intensive than conventional crude.[47]

His suggestion? Adopt a "narrative" about how their "awesome" technology can be used not just to extract dirty oil but to solve the environmental problems of tomorrow. And, he says, the choice of Calgary to host the next phase of Branson's Earth Challenge was "no coincidence"; indeed it appeared to be a way for him to serve the interests of some of his biggest clients all at once—the tar sands giants as well as Richard Branson. In an interview, Knight explained that "you've got a lot of very good engineers and you've got a lot of very highly financed companies who should be looking at this technology."[48]

But what exactly would they be looking at this technology to do? Not simply to suck out the carbon that they are putting into the air, but also, it turns out, to add even more carbon. Because in Calgary, the Virgin Earth Challenge was "reeingineered," to use Knight's word. While previously the goal had been to find technology capable of removing large amounts of carbon and safely storing it, Knight started referring to the prize as "an initiative to develop technology to recycle CO_2 direct from the air into commercially viable products."[49]

It made a certain amount of sense: removing carbon out of the air has long been technically possible. The problems have always been finding a means of removal that was not prohibitively costly, as well as storage and scale. In a market economy that means finding customers interested in buying a whole lot of captured carbon. Which is where the decision to pitch the eleven most promising entries in Calgary started to gel. Since the mid-2000s, the oil industry has been increasing its use of a method known as Enhanced Oil Recovery (EOR)—a set of techniques that mostly use high-pressure gas or steam injections to squeeze more oil out of existing fields. Most commonly, wells are injected with CO_2 and research shows that this use of CO_2 could cause the U.S. proven oil reserves to double or even, with "next-generation" technologies, quadruple. But there is a problem (other than the obvious planet-cooking one): according to Tracy Evans, former president of the Texas oil and gas company Denbury Resources, "The single largest deterrent to expanding production from EOR today is the lack of large volumes of reliable and affordable CO_2."[50]

With this in mind, several of Branson's group of eleven finalists have pitched themselves as the start-ups best positioned to supply the oil industry with the steady stream of carbon dioxide it needs to keep the oil flowing. Ned David, president of Kilimanjaro Energy, one of Branson's finalists, claimed that machines like his have the potential to release huge volumes of oil once assumed untappable, similar to what fracking did for natural gas. It could be, he said, "a money gusher." He told *Fortune*, "The prize is nearly 100 billion barrels of U.S. oil if you can economically capture CO_2 from air. That's \$10 trillion of oil."[51]

David Keith, who has been studying geoengineering for twenty-five years, and who is the inventor of another one of the carbon-capture machines to make Branson's list, was slightly more circumspect. He explained that if carbon that was removed from the air were used to extract oil, "you're making hydrocarbon fuel with a very low life-cycle [of] carbon emissions." Maybe not so low, because according to a study from the U.S. Department of Energy's National Energy Technology Laboratory, EOR techniques are estimated to be almost three times as greenhouse-gas intensive as conventional extraction. And the oil is still going to be burned, thereby contributing to climate change. While more research is needed on the overall carbon footprint of EOR, one striking modeling study examined a similar proposal that would use CO_2 captured not from the air but directly from coal plants. It found that the emissions benefit of sequestering CO_2 would be more than canceled out by all that extra oil: on a system-wide basis, the process could still end up releasing about four times as much CO_2 as it would save.[52]

Moreover, much of this is oil that is currently considered unrecoverable—i.e., not even counted in current proven reserves, which as we know already represents five times more than we can safely burn. Any technology that can quadruple proven reserves in the U.S. alone is a climate menace, not a climate solution. As David Hawkins of the Natural Resources Defense Council puts it, air capture has "morphed very rapidly from a technology whose purpose is to remove CO_2 to a technology whose purpose is to produce CO_2."[53] And Richard Branson has gone from promising to help get us off oil to championing technologies aimed at extracting and burning much more of it. Some prize.

A Regulatory Avoidance Strategy?

There was something else worth noting about Branson's decision to allow his Earth Challenge to cobrand with the Alberta oil sector. The Calgary event took place at a moment when the San Francisco–based Forest Ethics had been upping the pressure on large corporations to boycott oil derived from the Alberta tar sands because of its high-carbon footprint. A fierce debate was also unfolding over whether new European fuel standards would effectively ban the sale of tar sands oil in Europe. And as early as 2008, the NRDC had sent open letters to fifteen U.S. and Canadian airlines asking them "to adopt their own corporate 'Low Carbon Fuel Standard' and to publicly oppose the expansion" of fuel from the tar sands and other unconventional sources, and to avoid these fuels in their own fleets. The group made a special appeal to Branson, citing his leadership in "combating global warming and developing alternative fuels."[54]

It seemed like a fair enough demand: the Virgin chief had enjoyed enormous publicity for his very public climate-change promises. None of them had yielded much, but surely, while he was waiting for an algae-based jet fuel to materialize or for someone to win the Earth Prize, Branson could make the relatively minor concession of refusing to power his rapidly expanding fleet of airplanes with one of the most carbon-intensive fuels on the market.

Branson did not make that commitment. Alan Knight publicly stated, "I do not believe supporting a boycott is fair" and claimed that it was "impossible for an airline to boycott fuel made from oil sands"—a position contradicted by many experts.*[55] But Branson went further than refusing to participate in the boycott. By bringing the Earth Challenge to Calgary, Branson in effect did for the tar sands what his grand (but largely ephemeral) climate gestures have been doing for Virgin all these years: dangled the prospect of a miracle technological fix for carbon pollution just over the horizon in order to buy time to continue escalating emissions, free of meddlesome regulation. Indeed it can be argued—and some do—that Branson's planet-savior persona is an elaborate attempt to avoid the kind of tough regulatory action that was on the horizon in the U.K. and Europe precisely when he had his high-profile green conversion.

After all, 2006 was a pivotal year for the climate change debate. Public concern was rising dramatically, particularly in the U.K. where the movement's radical, grassroots flank was dominated by young activists who were determined to stop the expansion of the fossil fuel economy in its tracks. Much as they oppose fracking today, these activists used daring direct action to oppose new airports, as well as the highly controversial proposed new runway at Heathrow, which the airport claims would increase its number of flights by more than 50 percent.[56]

At the same time, the U.K. government was considering a broad climate change bill that would have impacted the airline sector, and Gordon Brown, Britain's chancellor at the time, had attempted to discourage flying with a marginal increase of the air passenger duty. In addition, the EU was entertaining a proposal to lift the airline industry's exemption from paying

---

* Including sustainability consultant Brendan May, founder of the Robertsbridge Group. "Of course you can segregate fuel according to its source," May writes. "If there's a will, there's a way. . . . At present, there's just no will."

a Value Added Tax and introduce an additional tax on aviation fuel. All these measures taken together posed a significant threat to the profit margins of Branson's chosen industry.[57]

Branson often talks a good game about supporting government regulation (saying he favors a global carbon tax, for instance), but he consistently opposes serious climate regulations when they are actually on the table. He has, for instance, been a hyperbolic, even bullying advocate of British airport expansion, including that new runway at Heathrow. He is so hungry for the expansion, in fact, that he has claimed, at various points, that its absence "will turn us into a third world country," that "global corporations will turn their back on London in favour of better connected cities," and that "Heathrow will become a symbol of British decline.")*[58]

This wasn't the only time Branson's claims to be committed to waging war on carbon came into conflict with his hard-nosed business instincts. He came out against the proposed climate tax in Australia and blasted a plan for a global tax on airlines, claiming it "would tax the industry out of existence."†[59]

It's this pattern that convinced Mike Childs of Friends of the Earth U.K. that Branson's reinvention as a guilt-ridden planet wrecker volunteering to use his carbon profits to solve the climate crisis was little more than a cynical ploy. "It comes across as a charitable act," Childs warned of the $3 billion pledge, "but I think he is also trying to take some of the heat out of aviation as a political issue. If you are running a transport company, you must have realized by now that climate change is going to be a massive issue for you."[60]

Was he right? Well, one immediate impact of Branson's pledge was that,

---

* In 2012, he went so far as to offer to invest roughly $8 billion in an expansion of Virgin Atlantic's operations at Heathrow if the government would approve the new runway—a prospect that once again raises questions about Branson's claims to be too broke to keep up with his $3 billion climate pledge.

† Branson is apparently no great fan of paying taxes generally, as his byzantine network of offshore holding companies in the Channel Islands and British Virgin Islands attests. Indeed he spent a night in jail and received a hefty fine after getting caught in an illegal cross-border tax avoidance scam when running his first company in 1971. "I was a criminal," Branson writes of his jailhouse revelation in his autobiography.

all of a sudden, you could feel good about flying again—after all, the profits from that ticket to Barbados were going into Branson's grand plan to discover a miracle green fuel. It was an even more effective conscience cleaner than carbon offsets (though Virgin sold those, too). As for punitive regulations and taxes, who would want to get in the way of an airline whose proceeds were going to such a good cause? And this was always Branson's argument: let him grow, unencumbered by regulation, and he will use that growth to finance our collective green transition. "If you hold industry back, we will not, as a nation, have the resources to come up with the new clean-energy solutions we need," Branson argued. "Business is the key to solving the financial and environmental crisis."[61]

So the skeptics might be right: Branson's various climate adventures may indeed prove to have all been a spectacle, a Virgin production, with everyone's favorite bearded billionaire playing the part of planetary savior to build his brand, land on late night TV, fend off regulators, and feel good about doing bad. It is certainly noteworthy that the show has been significantly less voluble ever since the Conservative-led government of David Cameron came to power in the U.K. and made it clear that Branson and his ilk faced no serious threat of top-down climate regulations.

But even if the constantly moving goalposts in Branson's climate initiatives merit that kind of cynicism, there is also a more charitable interpretation of what has gone wrong. This interpretation would grant Branson his obvious love of nature (whether it's watching the tropical birds on his private island or ballooning over the Himalayas). And it would credit him with genuinely trying to figure out ways to reconcile running carbon-intensive businesses with a profound personal desire to help slow species extinction and avert climate chaos. It would acknowledge, too, that between the pledge, the prize, and the Carbon War Room, Branson has thought up some rather creative mechanisms to try to channel profits generated from warming the planet into projects that could help keep it cool.

But if we do grant Branson these good intentions, then the fact that all of these projects have failed to yield results is all the more relevant. Branson set out to harness the profit motive to solve the climate crisis—but the temptation to profit from practices worsening the crisis proved too great to resist. Again and again, the demands of building a successful empire trumped the climate imperative—whether that meant lobbying against needed regulation, or putting more planes in the air, or pitching oil companies on using his pet miracle technologies to extract more oil.

The idea that capitalism and only capitalism can save the world from a crisis created by capitalism is no longer an abstract theory; it's a hypothesis that has been tested and retested in the real world. We are now able to set theory aside and take a hard look at the results: at the celebrities and media conglomerates that were supposed to model chic green lifestyles who have long since moved on to the next fad; at the green products that were shunted to the back of the supermarket shelves at the first signs of recession; at the venture capitalists who were supposed to bankroll a parade of innovation but have come up far short; at the fraud-infested, boom-and-bust carbon market that has failed miserably to lower emissions; at the natural gas sector that was supposed to be our bridge to renewables but ended up devouring much of their market instead. And most of all, at the parade of billionaires who were going to invent a new form of enlightened capitalism but decided that, on second thought, the old one was just too profitable to surrender.

We've tried it Branson's way. (And Buffett's, Bloomberg's, Gates's, and Pickens's way.) The soaring emissions speak for themselves. There will, no doubt, be more billionaire saviors who make splashy entrances, with more schemes to rebrand capitalism. The trouble is, we simply don't have another decade to lose pinning our hopes on these sideshows. There is plenty of room to make a profit in a zero-carbon economy; but the profit motive is not going to be the midwife for that great transformation.

This is important because Branson was onto something with his pledge. It makes perfect sense to make the profits and proceeds from the businesses that are most responsible for exacerbating the climate crisis help pay for the transition to a safer, greener future. Branson's original idea—to spend 100 percent of the proceeds from his trains and airlines on figuring out a way to get off fossil fuels—was, at least in theory, exactly the kind of thing that needs to take place on a grand scale. The problem is that under current business models, once the shareholders have taken a slice, once the executives have given themselves yet another raise, once Richard Branson has launched yet another world-domination project and purchased another private island, there doesn't seem to be much left over to fulfill the promise.

Similarly, Alan Knight was onto something when he told his tar sands clients that they should use their technological prowess to invent the low-carbon and renewable energy sources of the future. As he said, "The potential narrative is perfect."[62] The hitch, of course, is that, so long as this vision is left to the enlightened self-interest of oil and airline executives, it will remain just that: a narrative—or rather, a fairy tale. Meanwhile, the industry will use its technology and resources to develop ever more ingenious and profitable new ways to extract fossil fuels from the deepest recesses of the earth, even as it fiercely defends its public subsidies and resists the kind of minor increases to its tax and royalty rates that would allow governments to fund green transitions without its help.

In this regard, Virgin is particularly brazen. The National Union of Rail, Maritime and Transport Workers estimates that Virgin Trains has received more than £3 billion (the equivalent of $5 billion) in subsidies since British railways were privatized in the late 1990s—significantly more than Branson pledged to the green fund. As recently as 2010, Branson and the Virgin Group received £18 million in dividends from Virgin Trains. Branson insists that the characterization of him as a freeloader is "garbage," pointing to sharp increases in the number of passengers on Virgin trains and writing, "far from receiving subsidies, we now pay more than £100m a year to the taxpayer." But paying taxes is part of doing business. So when Branson pays into the Green Fund, whose money is it really—his own or the taxpayers'? And if a substantial portion of it originally belonged to taxpayers, wouldn't it have been a better arrangement never to have sold off the rails in the first place?[63]

If that were the case, the British people—with the climate crisis in mind—might have long ago decided to reinvest rail profits back into improving their public transit system, rather than allowing trains to become outdated and fares to skyrocket while shareholders of private rail companies like Branson's pocketed hundreds of millions in returns from their taxpayer-subsidized operations. And rather than gambling on the invention of a miracle fuel, they might have decided to make it a top political priority to shift the entire system over to electric trains, with that power coming from renewable energy, rather than have the system partially powered by diesel, as it is now. No wonder 66 percent of British residents tell pollsters they support renationalizing the railway companies.[64]

Richard Branson got at least one thing right. He showed us the kind of bold model that has a chance of working in the tight time frame left: the profits from our dirtiest industries must be diverted into the grand and hopeful project of cleaning up their mess. But if there is one thing Branson has demonstrated, it is that it won't happen on a voluntary basis or on the honor system. It will have to be legislated—using the kinds of tough regulations, higher taxes, and steeper royalty rates these sectors have resisted all along.

———

Of course there is still a chance that one of Branson's techno-schemes could pan out. He might yet stumble across a zero-carbon jet fuel, or a magic machine for safely and cheaply removing carbon from the skies. Time, however, is not on our side. David Keith, inventor of one of those carbon-sucking machines, in which Bill Gates is a key investor, estimates that the technology is still decades away from being taken to scale. "There's no way you can do a useful amount of carbon dioxide removal in less than a third of a century or maybe half a century," he says.[65] As always with matters related to climate change, we have to keep one eye on the ticking clock—and what that clock tells us is that if we are to have a solid chance of avoiding catastrophic warming, we will need to be burning strictly minimal amounts of fossil fuels a half century from now. If we spend the precious years between now and then dramatically expanding our emissions (as Branson is doing with his airlines, and as Knight's clients are doing in the tar sands), then we are literally betting the habitability of the planet on the faint hope of a miracle cure.

And yet Branson (a notorious risk addict with a penchant for crash-landing hot air balloons) is far from the only one willing to stake our collective future on this kind of high-stakes gamble. Indeed the reason his various far-fetched schemes have been taken as seriously as they have over the years is that he, alongside Bill Gates with his near mystical quest for energy "miracles," taps into what may be our culture's most intoxicating narrative: the belief that technology is going to save us from the effects of our actions. Post–market crash and amidst ever more sinister levels of inequality, most of us have come to realize that the oligarchs who were minted by the era of deregulation and mass privatization are not, in fact, going to use their vast wealth to save the world on our behalf. Yet our faith in techno wizardry persists, embedded inside the superhero narrative that at the very last minute our best and brightest are going to save us from disaster.

This is the great promise of geoengineering and it remains our culture's most powerful form of magical thinking.

DIMMING THE SUN

The Solution to Pollution Is . . . Pollution?

"Geoengineering holds forth the promise of addressing global warming concerns for just a few billion dollars a year."

—Newt Gingrich, former speaker of the U.S. House of Representatives, 2008 [1]

"Our science is a drop, our ignorance a sea."

—William James, 1895 [2]

It's March 2011 and I have just arrived at a three-day retreat about geoengineering in the Buckinghamshire countryside, about an hour and a half northwest of London. The meeting has been convened by the Royal Society, Britain's legendary academy of science, which has counted among its fellows Isaac Newton, Charles Darwin, and Stephen Hawking.

In recent years, the society has become the most prominent scientific organization to argue that, given the lack of progress on emission reduction, the time has come for governments to prepare a technological Plan B. In a report published in 2009, it called upon the British government to devote significant resources to researching which geoengineering methods might prove most effective. Two years later it declared that planetary-scale engineering interventions that would block a portion of the sun's rays "may be the only option for reducing global temperatures quickly in the event of a climate emergency." [3]

The retreat in Buckinghamshire has a relatively narrow focus: How should research into geoengineering, as well as eventual deployment, be governed? What rules should researchers follow? What bodies, if any, will regulate these experiments? National governments? The United Nations? What constitutes "good governance" of geoengineering? To answer these questions and others, the society has teamed up with two cosponsors for the retreat: the World Academy of Sciences based in Italy, which focuses on promoting scientific opportunities in the developing world, and the Environmental Defense Fund, which has described geoengineering as a "bridging tool" (much as it has described natural gas).[4] That makes this conference both the most international gathering about geoengineering to date, and the first time a major green group has publicly offered its blessing to the exploration of radical interventions into the earth's climate system as a response to global warming.

The venue for this futuristic discussion is an immaculately restored sixty-two-room redbrick Georgian mansion called Chicheley Hall, once a set in a BBC production of *Pride and Prejudice*, and the Royal Society's newly acquired retreat center. The effect is wildly anachronistic: the estate's sprawling bright green lawns, framed by elaborately sculpted hedges, seem to cry out for women in corseted silk gowns and parasols discussing their suitors—not disheveled scientists discussing a parasol for the planet. And yet geoengineering has always had a distinctly retro quality, not quite steampunk, but it definitely harkens back to more confident times, when taking control over the weather seemed like the next exciting frontier of scientific innovation—not a last-ditch attempt to save ourselves from incineration.

After dinner, consumed under towering oil paintings of plump-faced men in silver wigs, the delegates are invited to the wood-paneled library. There, about thirty scientists, lawyers, environmentalists, and policy wonks gather for the opening "technical briefing" on the different geoengineering schemes under consideration. A Royal Society scientist takes us through a slide show that includes "fertilizing" oceans with iron to pull carbon out of the atmosphere; covering deserts with vast white sheets in order to reflect sunlight back to space; and building fleets of machines like the ones competing for Richard Branson's Earth Challenge that would suck carbon out of the air.

The scientist explains that there are too many such schemes to evaluate in depth, and each presents its own particular governing challenge. So for the next three days, we will zero in on the geoengineering methods the scientists here consider most plausible and promising. These involve various means of injecting particles into the atmosphere in order to reflect more sunlight back to space, thereby reducing the amount of heat that reaches the earth. In geoengineering lingo, this is known as Solar Radiation Management (SRM)—since these methods would be attempting to literally "manage" the amount of sunlight that reaches earth.

There are various possible sun-dimming approaches. The most gleefully sci-fi is space mirrors, which is quickly dismissed out of hand. Another is "cloud brightening": spraying seawater into the sky (whether from fleets of boats or from towers on shore) to create more cloud cover or to make clouds more reflective and longer lasting. The most frequently discussed option involves spraying sulfate aerosols into the stratosphere, whether via specially retrofitted airplanes or a very long hose suspended by helium balloons (some have even suggested using cannons).

The choice to focus exclusively on SRM is somewhat arbitrary given that ocean fertilization experiments have been conducted on several occasions, including a heavily reported "rogue" test off the coast of British Columbia in 2012. But SRM is attracting the lion's share of serious scientific interest: sun blocking has been the subject of over one hundred peer-reviewed papers, and several high-level research teams are poised to run open-air field trials, which would test the mechanics of these schemes using ships, planes, and very long hoses. If rules and guidelines aren't developed soon (including, as some are suggesting, banning field tests outright), we could end up with a research Wild West.[5]

Spraying sulfate into the stratosphere is often referred to as "the Pinatubo Option," after the 1991 eruption of Mount Pinatubo in the Philippines. Most volcanic eruptions send ash and gases into the lower atmosphere, where sulfuric acid droplets are formed that simply fall down to earth. (That was the case, for instance, with the 2010 Icelandic volcano that grounded many European flights.) But certain, much rarer eruptions—Mount Pinatubo among them—send high volumes of sulfur dioxide all the way up to the stratosphere.

When that happens, the sulfuric acid droplets don't fall back down: they remain in the stratosphere, and within weeks can circulate to surround the entire planet. The droplets act like tiny, light-scattering mirrors, preventing the full heat of the sun from reaching the planet's surface. When these larger volcanic eruptions occur in the tropics, the aerosols stay suspended in the stratosphere for roughly one to two years, and the global cooling effects can last even longer.

That's what happened after Pinatubo. The year after the eruption, global temperatures dropped by half a degree Celsius, and as Oliver Morton noted in *Nature*, "Had there not been a simultaneous El Niño, 1992 would have been 0.7 degrees cooler, worldwide, than 1991."[6] That figure is notable because we have warmed the earth by roughly the same amount thus far with our greenhouse gas emissions. Which is why some scientists have become convinced that if they could just find a way to do artificially what those large eruptions do naturally, then they could force down the temperature of the earth to counteract global warming.

The scientist leading the briefing starts with the pros of this approach. He observes that the technology to pull this off already exists, though it needs to be tested; it's relatively cheap; and, if it worked, the cooling effects would kick in pretty quickly. The cons are that, depending on which sun-blocking method is used and how intensively, a permanent haze could appear over the earth, potentially making clear blue skies a thing of the past.[7] The haze could prevent astronomers from seeing the stars and planets clearly and weaker sunlight could reduce the capacity of solar power generators to produce energy (irony alert).

But the biggest problem with the Pinatubo Option is that it does nothing to change the underlying cause of climate change, the buildup of heat-trapping gases, and instead treats only the most obvious symptom—warmer temperatures. That might help control something like glacial melt, but would do nothing about the increased atmospheric carbon that the ocean continues to soak up, causing rapid acidification that is already taking a heavy toll on hard-shelled marine life from coral to oysters, and may have cascading impacts through the entire aquatic food chain. On the other hand, we hear, there could be some advantages to allowing atmospheric carbon dioxide levels to increase while keeping temperatures artificially cool, since plants like carbon dioxide (so long as it's not accompanied by scorching heat and drought) and they might well do better in what would essentially become an artificial global greenhouse.

Oh, and another con: once you start spraying material into the stratosphere to block the sun, it would basically be impossible to stop because if you did, all the warming that you had artificially suppressed by putting up that virtual sunshade would hit the planet's surface in one single tidal wave of heat, with no time for gradual adaptation. Think of the wicked witches of fairy tales, staying young by drinking ill-gotten magical elixirs, only to decay and wither all at once when the supply is abruptly cut off.

One solution to this "termination problem," as our British guide politely describes it, would be to suck a whole lot of carbon out of the atmosphere while the shade was still up so that when the particles dissipate and the sun beams down full bore, there is less heat-trapping gas in the atmosphere to augment the warming. Which would be fine except for the fact that we don't actually know how to do that on anything close to the required scale (as Richard Branson has discovered).

Listening to all this, a grim picture emerges. Nothing on earth would be outside the reach of humanity's fallible machines, or even fully outside at all. We would have a roof, not a sky—a milky, geoengineered ceiling gazing down on a dying, acidified sea.

And it gets worse, because our guide has saved the biggest con for last. A slide comes up showing a map of the world, with regions color-coded based on projections showing how severely their rainfall will be affected by injecting sulfur dioxide into the stratosphere. Precipitation in Europe and North America appears minimally changed, but Africa's equatorial region is lit up red, an indication of serious drought. And though the borders are hazy, parts of Asia appear to be in trouble as well because the drop in land temperature caused by a weaker sun could also weaken the summer monsoons, the main source of rainfall in these regions.

Up to this point, the audience has been quietly listening, but this news seems to wake up the room. One participant interrupts the presentation: "Let's put aside the science and talk about the ethics," he says, clearly upset. "I come from Africa and I don't like what I'm seeing with precipitation."*

---

* The retreat took place under the Chatham House Rule, which allows those attending to report on what was said in sessions, but not on who said what. (Any interviews conducted outside of the official sessions are exempt from these rules.)

Indeed, one of the society's own reports on geoengineering acknowledges that Solar Radiation Management "could conceivably lead to climate changes that are worse than the 'no SRM' option."[8]

The African delegate shakes his head. "I don't know how many of us will sleep well tonight."

Warming Up to "Horrifying"

Schemes for deliberately intervening in the climate system to counteract the effects of global warming have been around for half a century at least. In fact, when the President's Science Advisory Committee issued a report warning Lyndon B. Johnson about climate change in 1965, the authors made no mention of cutting emissions. The only potential solutions considered were technological schemes like modifying clouds and littering oceans with reflective particles.[9]

And well before it was seen as a potential weapon against global warming, weather modification was simply seen as a weapon. During the Cold War, U.S. physicists imagined weakening the nation's enemies by stealthily manipulating rainfall patterns, whether by causing droughts or by generating targeted storms that would turn a critical supply route into a flooded mess, as was attempted during the Vietnam War.[10]

So it's little wonder that mainstream climate scientists have, until quite recently, shied away from even discussing geoengineering. In addition to the Dr. Strangelove baggage, there was a widespread fear of creating a climate moral hazard. Just as bankers take greater risks when they know governments will bail them out, the fear was that the mere suggestion of an emergency techno-fix—however dubious and distant—would feed the dangerous but prevalent belief that we can keep ramping up our emissions for another couple of decades.

More out of despair than conviction, the geoengineering taboo has been gradually eroding over the past decade. A significant turning point came in 2006 when Paul Crutzen, who won the Nobel Prize in chemistry for his breakthrough research on the deterioration of the ozone layer, wrote an essay arguing that the time had come to consider injecting sulfur into the stratosphere as an emergency escape route from severe global warming. "If sizeable reductions in greenhouse gas emissions will not happen and temperatures rise rapidly, then climatic engineering . . . is the only option available to rapidly reduce temperature rises and counteract other climatic effects," he wrote.[11]

Crutzen created some space for preliminary research to take place, but geoengineering's real breakthrough came after the Copenhagen summit flopped in 2009, the same year that climate legislation tanked in the U.S. Senate. Soaring levels of hope had been pinned on both processes and when neither panned out, would-be planet hackers came out of their labs, positioning even the most seemingly outlandish ideas as the only realistic options left—especially with a world economic crisis making costly energy transformations seem politically untenable.

The Pinatubo Option has become a media favorite thanks in large part to the work of Nathan Myhrvold, the excitable former Microsoft chief technology officer who now runs Intellectual Ventures, a company that specializes in eclectic high-tech inventions and is often described as a vehicle for patent trolling.[12] Myhrvold is a made-for-TV character—a child prodigy turned physicist turned tech star, as well as an avid dinosaur hunter and wildlife photographer. Not to mention a formally trained amateur cook who spent millions researching and co-writing a six-volume bible on molecular gastronomy.

In 2009, Myhrvold and his team unveiled details for a contraption they called the "StratoShield," which would use helium balloons to suspend a sulfur dioxide–spraying tube thirty kilometers into the sky. And he wasted no time pitching it as a substitute for government action: just two days after the Copenhagen summit concluded, Myhrvold was on CNN boasting that his device—which he said could deliver a "Mount Pinatubo on demand"— had the power to "negate global warming as we have it today."[13]

Two months earlier, Steven D. Levitt and Stephen J. Dubner's global bestseller *SuperFreakonomics* had come out, devoting an entire awestruck chapter to Myhrvold's hose to the sky. And whereas most scientists engaged in this research are careful to present sun blocking as a worst-case scenario—a Plan B to be employed only if Plan A (emission cuts) proves insufficient—Levitt and Dubner declared that the Pinatubo Option was straight-up preferable to getting off fossil fuels. "For anyone who loves cheap and simple solutions, things don't get much better."[14]

Most of those calling for more geoengineering research do so with significantly less glee. In September 2010, the New America Foundation and *Slate* magazine held a one-day forum in Washington, D.C., titled "Geoengineering: The Horrifying Idea Whose Time Has Come?"[15] That one sentence pretty much sums up the tone of grim resignation that has characterized the steady stream of conferences and government reports that have inched geoengineering into the political mainstream.

This gathering at Chicheley Hall is another milestone in this gradual process of normalization. Rather than debate whether or not to engage in geoengineering research—as most previous gatherings have done—this conference seems to take some kind of geoengineering activity as a given (or else why would it need to be "governed"). Adding to the sense not just of inevitability but general banality, the organizers have even given this process a clunky acronym: SRMGI, the Solar Radiation Management Governance Initiative.

Geoengineering debate generally takes place within a remarkably small and incestuous world, with the same group of scientists, inventors, and funders promoting each other's work and making the rounds to virtually every relevant discussion of the topic. (Science journalist Eli Kintisch, who wrote one of the first books on geoengineering, calls them the "Geoclique.") And many of the members of that clique are in attendance here. There is David Keith, the wiry, frenetic physicist, then at the University of Calgary (now at Harvard), whose academic work has a major focus on SRM, and whose carbon-sucking machine—blessed by both Richard Branson and Bill Gates—stands to make him rather rich should the idea of a techno fix for global warming take off. This kind of vested interest is a recurring theme: many of the most aggressive advocates of geoengineering research are associated with planet-hacking start-ups, or hold patents on various methods. This, says Colby College science historian James Fleming, gives them "skin in the game" since these scientists stand "to make an incredible amount of money if their technique goes forward."[16]

Here too is Ken Caldeira, a prominent atmospheric scientist from the Carnegie Institution for Science, and one of the first serious climate scientists to run computer models examining the impact of deliberately dimming the sun. In addition to his academic work, Caldeira has an ongoing relationship with Nathan Myhrvold's Intellectual Ventures as a "Senior Inventor."[17] Another player present is Phil Rasch, a climate scientist at the Pacific Northwest National Laboratory in Washington state, who has been preparing to launch perhaps the first cloud-brightening field experiment.

Bill Gates isn't here, but he provided much of the cash for the gathering, allocated through a fund administered by Keith and Caldeira. Gates has given the scientists at least $4.6 million specifically for climate-related research that wasn't getting funding elsewhere. Most of it has gone to geoengineering themes, with Keith, Caldeira, and Rasch all receiving large shares. Gates is also an investor in Keith's carbon capture company, as well as in Intellectual Ventures, where his name appears on several geoengineering patents (alongside Caldeira's), while Nathan Myhrvold serves as vice chairman at TerraPower, Gates's nuclear energy start-up. Branson's Carbon War Room has sent a delegate and is supporting this work in various ways.[18] If that all sounds confusing and uncomfortably clubby, especially for so global and high stakes a venture, well, that's the Geoclique for you.

Because governing geoengineering, as opposed to just testing it, is the focus of this retreat, the usual club has been temporarily expanded to include several climate scientists from Africa and Asia, as well as legal ethicists, experts in international treaties and conventions, and staffers from several green NGOs, including Greenpeace and WWF-UK (Greenpeace does not support geoengineering, but WWF-UK has come out in cautious support of "research into geo-engineering approaches in order to find out what is possible").[19]

The organizers have also invited a couple of outspoken critics. Alan Robock, a famously gruff white-bearded climatologist from Rutgers University, is here. When I last saw him in action, he was presenting a slide show titled "20 Reasons Why Geoengineering May Be a Bad Idea," ranging from "Whitening of the sky" (#7) to "Rapid warming if deployment stops" (#10). Most provocative is Australian climate expert Clive Hamilton, who has wondered aloud whether "the geoengineers [are] modern-day Phaetons, who dare to regulate the sun, and who must be struck down by Zeus before they destroy the earth?"[20]

In the end, the conference manages to agree on nothing of substance—not even the need for small-scale field trials to take place. But throwing this group of people together in a country mansion for three days does make for some interesting intellectual fireworks.

What Could Possibly Go Wrong?

After a night's sleep, the guests at Chicheley Hall are ready to dive into the debates. In a sleek slate-and-glass lecture hall located in the old coach house, the organizers separate the group into breakout sessions. Everyone receives a sheet of paper with a triangle on it, and on each point is a different word: "Promote," "Prohibit," "Regulate." The instructions say "Mark where you feel your current perspective best fits on the triangle." Do you want further research into sun-shielding banned? Aggressively promoted? Promoted with some measure of regulation?

I spend the morning eavesdropping on the different breakouts and before long a pattern emerges. The scientists already engaged in geoengineering research tend to categorize their positions somewhere between "regulate" and "promote," while most everyone else leans toward "prohibit" and "regulate." Several of the participants express a desire to promote more research, but only to establish that geoengineering *isn't* a viable option that we can bank on to save the day. "We particularly need to know if it's not going to work," one environmentalist pleads to the scientists in his session. "Right now we're struggling in the dark."

But in one breakout group, things have gone off the rails. A participant flatly refuses to place his views on the triangle and instead, helps himself to a large piece of poster paper. On it he writes three questions in blue marker:

- Is the human that gave us the climate crisis capable of properly/ safely regulating SRM?
- In considering SRM regulation, are we not in danger of perpetuating the view that the earth can be manipulated in our interests?
- Don't we have to engage with these questions before we place ourselves in the triangle?

When the groups come back together to discuss their triangular mind maps, these questions are never acknowledged, let alone answered. They just hang on the wall of the lecture hall as a sort of silent rebuke. It's too bad, because the Royal Society, with its long and storied history of helping to both launch the Scientific Revolution and the age of fossil fuels, offers a unique vantage point from which to ponder these matters.

The Royal Society was founded in 1660 as an homage to Francis Bacon. Not only is the organization's motto—*Nullius in verba*—"take nothing on authority"—inspired by Bacon but, somewhat bizarrely, much of the society's basic structure was modeled on the fictional scientific society portrayed in Bacon's proto-sci-fi/utopian novel *New Atlantis*, published in 1627. The institution was at the forefront of Britain's colonial project, sponsoring voyages by Captain James Cook (including the one in which he laid claim to New Zealand), and for over forty years the Royal Society was led by one of Cook's fellow explorers, the wealthy botanist Joseph Banks, described by a British colonial official as "the staunchest imperialist of the day."[21] During his tenure, the society counted among its fellows James Watt, the steam engine pioneer, and his business partner, Matthew Boulton—the two men most responsible for launching the age of coal.

As the questions hanging on the wall imply, these are the tools and the logic that created the crisis geoengineering is attempting to solve—not just the coal-burning factories and colonial steam ships, but Bacon's twisted vision of the Earth as a prone woman and Watt's triumphalism at having found her "weak side." Given this, does it really make sense to behave as if, with big enough brains and powerful enough computers, humans can master and control the climate crisis just as humans have been imagining they could master the natural world since the dawn of industrialization—digging, damming, drilling, dyking. Is it really as simple as adding a new tool to our nature-taming arsenal: dimming?

This is the strange paradox of geoengineering. Yes, it is exponentially more ambitious and more dangerous than any engineering project humans have ever attempted before. But it is also very familiar, nearly a cliché, as if the past five hundred years of human history have been leading us, ineluctably, to precisely this place. Unlike cutting our emissions in line with the scientific consensus, succumbing to the logic of geoengineering does not require any change from us; it just requires that we keep doing what we have done for centuries, only much more so.

Wandering the perfectly manicured gardens at Chicheley Hall—through the trees sculpted into lollipops, through the hedges chiseled into daggers—I realize that what scares me most is not the prospect of living on a "designer planet," to use a phrase I heard at an earlier geoengineering conference. My fear is that the real-world results will be nothing like this garden, or even like anything we saw in that technical briefing, but rather something far, far worse. If we respond to a global crisis caused by our pollution with more pollution—by trying to fix the crud in our lower atmosphere by pumping a different kind of crud into the stratosphere—then geoengineering might do something far more dangerous than tame the last vestiges of "wild" nature. It may cause the earth to go wild in ways we cannot imagine, making geoengineering not the final engineering frontier, another triumph to commemorate on the walls of the Royal Society, but the last tragic act in this centuries-long fairy tale of control.

A great many of our most brilliant scientists have taken the lessons of past engineering failures to heart, including the failure of foresight represented by climate change itself, which is one of the primary reasons there is still so much resistance to geoengineering among biologists and climate scientists. To quote Sallie Chisholm, a world-renowned expert on marine microbes at MIT, "Proponents of research on geoengineering simply keep ignoring the fact that the biosphere is a player (not just a responder) in whatever we do, and its trajectory cannot be predicted. It is a living breathing collection of organisms (mostly microorganisms) that are evolving every second—a 'self-organizing, complex, adaptive system' (the strict term). These types of systems have emergent properties that simply cannot be predicted. We all know this! Yet proponents of geoengineering research leave that out of the discussion."[22]

Indeed in my time spent among the would-be geoengineers, I have been repeatedly struck by how the hard-won lessons about humility before nature that have reshaped modern science, particularly the fields of chaos and complexity theory, do not appear to have penetrated this particular bubble. On the contrary, the Geoclique is crammed with overconfident men prone to complimenting each other on their fearsome brainpower. At one end you have Bill Gates, the movement's sugar daddy, who once remarked that it was difficult for him to decide which was more important, his work on computer software or inoculations, because they both rank "right up there with the printing press and fire." At the other end is Russ George, the U.S. entrepreneur who has been labeled a "rogue geoengineer" for dumping some one hundred tons of iron sulphate off the coast of British Columbia in 2012. "I am the champion of this on the planet," he declared after the experiment was exposed, the only one with the guts to "step forward to save the oceans." In the middle are scientists like David Keith, who often comes off as deeply conflicted about "opening up Pandora's Box"—but once said of the threat of weakened monsoons from Solar Radiation Management that "hydrological stresses" can be managed "a little bit by irrigation."[23]

The ancients called this hubris; the great American philosopher, farmer and poet Wendell Berry calls it "arrogant ignorance," adding, "We identify arrogant ignorance by its willingness to work on too big a scale, and thus to put too much at risk."[*][24]

It doesn't provide much reassurance that just two weeks before we all gathered at Chicheley Hall, three nuclear reactors at Fukushima melted down in the wake of a powerful tsunami. The story was still leading the news the entire time we met. And yet the extent to which the would-be geoengineers acknowledged the disaster was only to worry that opponents of nuclear energy would seize upon the crisis to block new reactors. They never entertained the idea that Fukushima might serve as a cautionary tale for their own high-risk engineering ambitions.

Which brings us back to that slide showing parts of Africa lit up red that caused such a stir on opening night: is it possible that geoengineering, far from a quick emergency fix, could make the impacts of climate change even worse for a great many people? And if so, who is most at risk and who gets to decide to take those risks?

---

* It's particularly troubling that within the small group of scientists, engineers, and inventors who dominate the geoengineering debate, there have been a disproportionate share of big public errors in the past. Take, for instance, Lowell Wood, co-creator of Myhrvold's StratoShield. Before becoming a prominent proponent of the "Pinatubo Option," Wood was best known for coming up with some of the more fantastical elements of Ronald Reagan's "Star Wars" missile defense program, widely discredited as expensive and reckless.

Like Climate Change, Volcanoes *Do* Discriminate

Boosters of Solar Radiation Management tend to speak obliquely about the "distributional consequences" of injecting sulfur dioxide into the stratosphere, and of the "spatial heterogeneity" of the impacts. Petra Tschakert, a geographer at Penn State University, calls this jargon "a beautiful way of saying that some countries are going to get screwed."[25] But which countries? And screwed precisely how?

Having reliable answers to those key questions would seem like a prerequisite for considering deployment of such a world-altering technology. But it's not at all clear that obtaining those answers is even possible. Keith and Myhrvold can test whether a hose or an airplane is a better way to get sulfur dioxide into the stratosphere. Others can spray saltwater from boats or towers and see if it brightens clouds. But you'd have to deploy these methods on a scale large enough to impact the *global* climate system to be certain about how, for instance, spraying sulfur in the Arctic or the tropics will impact rainfall in the Sahara or southern India. But that wouldn't be a test of geoengineering; it would actually be conducting geoengineering.[26]

Nor could the necessary answers be found from a brief geoengineering stint—pumping sulfur for, say, one year. Because of the huge variations in global weather patterns from one year to the next (some monsoon seasons are naturally weaker than others, for instance), as well as the havoc already being wreaked by global warming, it would be impossible to connect a particular storm or drought to an act of geoengineering. Sulfur injections would need to be maintained long enough for a clear pattern to be isolated from both natural fluctuations and the growing impacts of greenhouse gases. That likely means keeping the project running for a decade or more.[*][27]

As Martin Bunzl, a Rutgers philosopher and climate change expert, points out, these facts alone present an enormous, perhaps insurmountable ethical problem for geoengineering. In medicine, he writes, "You can test a vaccine on one person, putting that person at risk, without putting every-

---

[*] That said, we would be wise to anticipate even small amounts of geoengineering unleashing a new age of weather-related geopolitical recrimination, paranoia, and possibly retaliation, with every future natural disaster being blamed—rightly or wrongly—on the people in faraway labs playing god.

one else at risk." But with geoengineering, "You can't build a scale model of the atmosphere or tent off part of the atmosphere. As such you are stuck going directly from a model to full scale planetary-wide implementation." In short, you could not conduct meaningful tests of these technologies without enlisting billions of people as guinea pigs—for years. Which is why science historian James Fleming calls geoengineering schemes "untested and untestable, and dangerous beyond belief."[28]

Computer models can help, to be sure. That's how we get our best estimates of how earth systems will be impacted by the emission of greenhouse gases. And it's straightforward enough to add a different kind of emission—sulfur in the stratosphere—to those models and see how the results change. Several research teams have done just that, with some very disturbing results. Alan Robock, for instance, has run different SRM scenarios through supercomputers. The findings of a 2008 paper he coauthored in the *Journal of Geophysical Research* were blunt: sulfur dioxide injections "would disrupt the Asian and African summer monsoons, reducing precipitation to the food supply for billions of people." Those monsoons provide precious freshwater to an enormous share of the world's population. India alone receives between 70 and 90 percent of its total annual rainfall during its June through September monsoon season.[29]

Robock and his colleagues aren't the only ones coming up with these alarming projections. Several research teams have produced models that show significant losses of rainfall as a result of SRM and other sunlight-reflecting geoengineering methods. One 2012 study shows a 20 percent reduction in rainfall in some areas of the Amazon after a particularly extreme use of SRM. When another team modeled spraying sulfur from points in the Northern Hemisphere for a 2013 study, the results projected a staggering 60–100 percent drop in a key measure of plant productivity in the African countries of the Sahel (Burkina Faso, Chad, Mali, Niger, Senegal, and Sudan)—that means, potentially, a complete crop collapse in some areas.[30]

This is not some minor side effect or "unintended consequence." If only some of these projections were to come true, that would transform a process being billed as an emergency escape from catastrophic climate change into a mass killer in its own right.

One might think all of this alarming research would be enough to put a serious damper on the upbeat chatter surrounding the Pinatubo Option. The problem is that—though computer models have proven remarkably accurate at predicting the broad patterns of climate change—they are not infallible. As we have seen from the failure to anticipate the severity of summer sea ice loss in the Arctic as well as the rate of global sea level rise in recent decades, computer models have tended to underestimate certain risks, and overstate others.[31] Most significantly, climate models are at their weakest when predicting specific regional impacts—how much more southern Somalia will warm than the central United States, say, or the precise extent to which drought will impact crop production in India or Australia. This uncertainty has allowed some would-be geoengineers to scoff at findings that make SRM look like a potential humanitarian disaster, insisting that regional climate models are inherently unreliable, while simultaneously pointing to other models that show more reassuring results. And if the controversy were just a matter of dueling computer models, perhaps we could call it a draw. But that is not the case.

History as Teacher—and Warning

Without being able to rely on either models or field tests, only one tool remains to help forecast the risks of sun blocking, and it is distinctly low-tech. That tool is history, specifically the historical record of weather patterns following major volcanic eruptions. The relevance of history is something all sides of the debate appear to agree on. Ken Caldeira has described the 1991 eruption of Mount Pinatubo "as a natural test of some of the concepts underlying solar radiation management" since it sent so much sulfur dioxide into the stratosphere. And David Keith assured me, "It's pretty clear that just putting a lot of sulfur in the stratosphere isn't terrible. After all, volcanoes do it." Likewise, Lowell Wood, Myhrvold's partner in the invention of the StratoShield, has argued that because his hose-to-the-sky would attempt to imitate a natural volcano, there is "a proof of harmlessness."[32]

Levitt and Dubner have stressed the relevance of historical precedent most forcefully, writing in *SuperFreakonomics* that not only did the earth cool after Pinatubo, but "forests around the world grew more vigorously because trees prefer their sunlight a bit diffused. And all that sulfur dioxide in the stratosphere created some of the prettiest sunsets that people had ever seen." They do not, however, appear to believe that history offers any cautionary lessons: aside from a reference to the "relatively small" number of deaths in the immediate aftermath of the eruption due to storms and mud slides, they make no mention in the book of any negative impact from Pinatubo.[33]

Critics of sun shielding also draw on history to bolster their arguments, and when they look back, they see much more than pretty sunsets and "proof of harmlessness." In fact, a great deal of compelling research shows a connection between large volcanic eruptions and precisely the kinds of droughts some computer models are projecting for SRM. Take the 1991 eruption of Mount Pinatubo itself. When it erupted, large swaths of Africa were already suffering from drought due to natural fluctuations. But after the eruption, the situation grew much worse. In the following year, there was a 20 percent reduction in precipitation in southern Africa and a 10–15 percent reduction in precipitation in South Asia. The United Nations Environment Programme (UNEP) described the drought as "the most severe in the last century"; an estimated 120 million people were affected. The *Los Angeles Times* reported crop losses of 50–90 percent, and half the population of Zimbabwe required food aid.[34]

At the time, few linked these disastrous events to the Pinatubo eruption since isolating such climate signals takes time. But more recent research looking at rainfall and streamflow patterns from 1950 to 2004 has concluded that only the sulfur dioxide that Pinatubo sent into the stratosphere can account for the severity of the drop in rainfall that followed the eruption. Aiguo Dai, an expert in global drought at the State University of New York, Albany, stresses that though the drought had additional causes, "Pinatubo contributed significantly to the drying." A 2007 paper cowritten by Dai and Kevin Trenberth, head of the Climate Analysis Section at the Colorado-based National Center for Atmospheric Research, concluded "that the Pinatubo eruption played an important role in the record decline in land precipitation and discharge, and the associated drought conditions in 1992."[35]

If Pinatubo was the only large eruption to have been followed by severe and life-endangering drought, that might not be enough to draw clear conclusions. But it fits neatly into a larger pattern. Alan Robock, a leading expert on the effect of volcanoes on climate, points in particular to two other eruptions—Iceland's Laki in 1783 and Alaska's Mount Katmai in 1912. Both were sufficiently powerful to send a high volume of sulfur dioxide into the stratosphere and, like Pinatubo, it turns out that both were followed by a series of terrible, or badly worsening regional droughts.

Reliable records of rainfall go back only roughly one hundred years, but as Robock informed me, "There's one thing that's been measured for 1,500 years, and that's the flow of the Nile River. And if you look back at the flow of the Nile River in 1784 or 1785"—the two years following Laki's eruption in Iceland—"it was much weaker than normal." The usual floods that could be counted on to carry water and precious fertilizing nutrients into farmers' fields barely took place, the devastating consequences of which were recounted in the eighteenth-century travel memoirs of French historian Constantin-François Volney. "Soon after the end of November, the famine carried off, at Cairo, nearly as many as the plague; the streets, which before were full of beggars, now afforded not a single one: all had perished or deserted the city." Volney estimated that in two years, one sixth of the population in Egypt either died or fled the country.[36]

Scholars have noted that in the years immediately following the eruption, drought and famine gripped Japan and India, claiming millions of lives, although there is much debate and uncertainty surrounding Laki's contribution. In Western and Central Europe, meanwhile, a brutally cold winter led to flooding and high mortality rates. Expert estimates of the global death toll from the eruption and the resulting extreme weather range widely, from over one-and-a-half million to as many as six million people. At a time when world population was less than one billion, those are stunningly high numbers, making Laki quite possibly the deadliest volcano in recorded history.[37]

Robock found something similar when he delved into the aftermath of the 1912 Katmai eruption in Alaska. Once again, his team looked at the historical record of the flow of the Nile and discovered that the year after Katmai saw "the lowest flow for the twentieth century." Robock and his colleagues also "had found a significant weakening of the Indian monsoon in response to the 1912 Katmai volcanic eruption in Alaska, which resulted from the decreased temperature gradient between Asia and the Indian Ocean." But it was in Africa where the impact of the great eruption took the heaviest human toll. In Nigeria, sorghum, millet, and rice crops withered in the fields while speculators hoarded what grains survived. The result was a massive famine in 1913–1914 that took the lives of at least 125,000 in western Africa alone.[38]

These are not the only examples of deadly droughts seemingly triggered by large volcanic eruptions. Robock has looked at how such eruptions have impacted "the water supply for Sahel and northern Africa" over the past two thousand years. "You get the same story from every [eruption] you look at," he said, adding, "there haven't been that many big eruptions but they all tell you the same stories. . . . The global average precipitation went down. In fact, if you look at global average precipitation for the last fifty years, the three years with the lowest global precipitation were after the three largest volcanic eruptions. Agung in 1963, El Chichón in 1982, and Pinatubo in 1991." The connections are so clear, Robock and two coauthors argued in one paper, that the next time there is a large "high-latitude volcanic erup-tion," policymakers should start preparing food aid immediately, "allowing society time to plan for and remediate the consequences."[39]

So how, given all this readily available evidence, could geoengineering boosters invoke the historical record for "proof of harmlessness"? The truth is the mirror opposite: of all the extreme events the planet periodically lobs our way—from earthquakes and tsunamis to hurricanes and floods—powerful volcanic eruptions may well be the most threatening to human life. Because the people in the immediate path of an eruption are not the only ones at risk; the lives of billions of others scattered throughout the globe can be destroyed by lack of food and water in the drier years to come. No naturally occurring disaster short of an asteroid has such global reach.

This grim track record makes the cheerful talk of a Pinatubo Option distinctly bizarre, if not outright sinister—especially because what is being contemplated is simulating the cooling effects of an eruption like Pinatubo not once but *year after year for decades*, which could obviously magnify the significant risks that have been documented in the aftermath of one-off eruptions.

The risks can be debated and contested, of course—and they are. The most common response is that, yes, there could be negative impacts, but not as negative as the impacts of climate change itself. David Keith goes further, arguing that we have the power to effectively minimize the risks with appropriate design; he proposes an SRM program that would slowly ramp up and then down again, "in combination with cutting emissions and with a goal to reduce—but not eliminate—the rate of temperature rise." As he explains in his 2013 book, *A Case for Climate Engineering*, "Crop losses, heat stress and flooding are the impacts of climate change that are likely to fall most harshly on the world [sic] poorest. The moderate amounts of geo-engineering contemplated in this slow ramp scenario are likely to reduce each of these impacts over the next half century, and so it will benefit the poor and politically disadvantaged who are most vulnerable to rapid environmental change. This potential for reducing climate risk is the reason I take geoengineering seriously."[40]

But when climate models and the historical record tell such a similar story about what could go wrong (and of course it wouldn't be scientists but politicians deciding how to use these technologies), there is ample cause for focusing on the very real risks. Trenberth and Dai, authors of the study on Pinatubo's harrowing legacy, are blunt. "The central concern with geo-engineering fixes to global warming is that the cure could be worse than the disease." And they stress, "Creating a risk of widespread drought and reduced freshwater resources for the world to cut down on global warming does not seem like an appropriate fix."[41]

It's hard not to conclude that the willingness of many geoegineering boosters to gloss over the extent of these risks, and in some cases, to ignore them entirely, has something to do with who appears to be most vulnerable. After all, if the historical record, backed by multiple models, indicated that injecting sulfur into the stratosphere would cause widespread drought and famine in North America and Germany, as opposed to the Sahel and India, is it likely that this Plan B would be receiving such serious consideration?

It's true that it might be technically possible to conduct geoengineering in a way that distributed the risks more equitably. For instance, the same 2013 study that found that the African Sahel could be devastated by SRM done in the Northern Hemisphere—a common assumption about where the sulfur injections would take place—found that the Sahel could actually see an increase in rainfall if the injections happened in the Southern Hemisphere instead. However, in this scenario, the United States and the Caribbean could see a 20 percent increase in hurricane frequency, and northeastern Brazil could see its rainfall plummet. In other words, it might be possible to tailor some of these technologies to help the most vulnerable people on the planet, and those who contributed least to the creation of the climate crisis—but not without endangering some of the wealthiest and most powerful regions. So we are left with a question less about technology than about politics: does anyone actually believe that geoengineering will be used to help Africa if that help could come only by putting North America at greater risk of extreme weather?[42]

In contrast, it is all too easy to imagine scenarios wherein geoengineering could be used in a desperate bid to, say, save corn crops in South Dakota, even if it very likely meant sacrificing rainfall in South Sudan. And we can imagine it because wealthy-country governments are already doing this, albeit more passively, by allowing temperatures to increase to levels that are a danger to hundreds of millions of people, mostly in the poorest parts of the world, rather than introducing policies that interfere with short-term profits. This is why African delegates at U.N. climate summits have begun using words like "genocide" to describe the collective failure to lower emissions. And why Mary Ann Lucille Sering, climate change secretary for the Philippines, told the 2013 summit in Warsaw, Poland, "I am beginning to feel like we are negotiating on who is to live and who is to die." Rob Nixon, an author and University of Wisconsin English professor, has evocatively described the brutality of climate change as a form of "slow violence"; geoengineering could well prove to be a tool to significantly speed that up.[43]

Geoengineering as Shock Doctrine

All of this may still seem somewhat abstract but it's critical to reckon with these harrowing risks now. That's because if geoengineering were ever deployed, it would almost surely be in an atmosphere of collective panic with scarce time for calm deliberation. Its defenders readily concede as much.

Bill Gates describes geoengineering as "just an insurance policy," something to have "in the back pocket in case things happen faster." Nathan Myhrvold likens SRM to "having fire sprinklers in a building"—you hope you won't need it, "but you also need something to fall back on in case the fire occurs anyway."[44]

In a true emergency, who would be immune to this logic? Certainly not me. Sure, the idea of spraying sulfur dioxide into the stratosphere like some kind of cosmic umbrella seems crazy to me now. But if my city were so hot that people were dropping dead in the thousands, and someone was peddling a quick and dirty way to cool it off, wouldn't I beg for that relief in the same way that I reach for the air conditioner on a sweltering day, knowing full well that by turning it on I am contributing to the very problem I am trying to escape?

This is how the shock doctrine works: in the desperation of a true crisis all kinds of sensible opposition melts away and all manner of high-risk behaviors seem temporarily acceptable. It is only outside of a crisis atmosphere that we can rationally evaluate the future ethics and risks of deploying geoengineering technologies should we find ourselves in a period of rapid change. And what those risks tell us is that dimming the sun is nothing like installing a sprinkler system—unless we are willing to accept that some of those sprinklers could very well spray gasoline instead of water. Oh—and that, once turned on, we might not be able to turn off the system without triggering an inferno that could burn down the entire building. If someone sold you a sprinkler like that, you'd definitely want a refund.

Perhaps we do need to find out all we possibly can about these technologies, knowing that we will never know close to enough to deploy them responsibly. But if we accept that logic, we also have to accept that small field tests often turn into bigger ones. It may start with just checking the deployment hardware, but how long before the planet hackers want to see if they can change the temperature in just one remote, low-population location (something that will be described, no doubt, as "the middle of nowhere")—and then one a little less remote?

The past teaches us that once serious field tests begin, deployment is rarely far behind. Hiroshima and Nagasaki were bombed less than a month after Trinity, the first successful nuclear test—despite the fact that many of the scientists involved in the Manhattan Project thought they were building a nuclear bomb that would be used only as a deterrent. And though slamming the door on any kind of knowledge is always wrenching, it's worth remembering that we have collectively foregone certain kinds of research before, precisely because we understand that the risks are too great. One hundred and sixty-eight nations are party to a treaty banning the development of biological weapons. The same taboos have been attached to research into eugenics because it can so easily become a tool to marginalize and even eliminate whole groups of people. Moreover the U.N. Environmental Modification Convention, which was adopted by governments in the late 1970s, already bans the use of weather modification as a weapon—a prohibition that today's would-be geoengineers are skirting by insisting that their aims are peaceful (even if their work could well feel like an act of war to billions).

Monster Earth

Not all geoengineering advocates dismiss the grave dangers their work could unleash. But many simply shrug that life is full of risks—and just as geoengineering is attempting to fix a problem created by industrialization, some future fix will undoubtedly solve the problems created by geoengineering.

One version of the "we'll fix it later" argument that has gained a good deal of traction comes from the French sociologist Bruno Latour. His argument is that humanity has failed to learn the lessons of the prototypical cautionary story about playing god: Mary Shelley's *Frankenstein*. According to Latour, Shelley's real lesson is not, as is commonly understood, "don't mess with mother nature." Rather it is, don't run away from your technological mess-ups, as young Dr. Frankenstein did when he abandoned the monster to which he had given life. Instead, Latour says we must stick around and continue to care for our "monsters" like the deities that we have become. "The real goal must be to have the same type of patience and commitment to our creations as God the Creator, Himself," he writes, concluding, "From now on, we should stop flagellating ourselves and take up explicitly and seriously what we have been doing all along at an ever-increasing scale."

(British environmentalist Mark Lynas makes a similar, defiantly hubristic argument in calling on us to become "The God Species" in his book of the same name.)[45]

Latour's entreaty to "love your monsters" has become a rallying cry in certain green circles, particularly among those most determined to find climate solutions that adhere to market logic. And the idea that our task is to become more responsible Dr. Frankensteins, ones who don't flee our creations like deadbeat dads, is unquestionably appealing. But it's a terribly poor metaphor for geoengineering. First, "the monster" we are being asked to love is not some mutant creature of the laboratory but the earth itself. We did not create it; it created—and sustains—us. The earth is not our prisoner, our patient, our machine, or, indeed, our monster. It is our entire world. And the solution to global warming is not to fix the world, it is to fix ourselves.

Because geoengineering will certainly monsterize the planet as nothing experienced in human history. We very likely would not be dealing with a single geoengineering effort but some noxious brew of mixed-up techno-fixes—sulfur in space to cool the temperature, cloud seeding to fix the droughts it causes, ocean fertilization in a desperate gambit to cope with acidification, and carbon-sucking machines to help us get off the geo-junk once and for all.

This makes geoengineering the very antithesis·of good medicine, whose goal is to achieve a state of health and equilibrium that requires no further intervention. These technologies, by contrast, respond to the lack of balance our pollution has created by taking our ecosystems even further away from self-regulation. We would require machines to constantly pump pollution into the stratosphere and would be unable to stop unless we invented other machines that could suck existing pollution out of the lower atmosphere, then store and monitor that waste indefinitely. If we sign on to this plan and call it stewardship, we effectively give up on the prospect of ever being healthy again. The earth—our life support system—would itself be put on life support, hooked up to machines 24/7 to prevent it from going full-tilt monster on us.

And the risks are greater still because we might well be dealing with multiple countries launching geoengineering efforts at once, creating unknown and unknowable interactions. In other words, a Frankenstein world, in which we try to solve one problem by making new ones, then pile techno-fixes onto those. And almost no one seems to want to talk about what happens if our geoengineering operations are interrupted for some reason—by war, terrorist attack, mechanical failure, or extreme weather. Or what if, in the middle of simulating the effects of a Mount Pinatubo–like eruption, a real Mount Pinatubo erupts. Would we risk bringing on what David Keith has described as "a worldwide Ice Age, a snowball earth," just because we forgot, yet again, that we are not actually in the driver's seat?[46]

The dogged faith in technology's capacity to allow us to leapfrog out of crisis is born of earlier technological breakthroughs—splitting the atom or putting a man on the moon. And some of the players pushing most aggressively for a techno-fix for climate change were directly involved in those earlier technological triumphs—like Lowell Wood, who helped develop advanced nuclear weaponry, or Gates and Myrhvold, who revolutionized computing. But as longtime sustainability expert Ed Ayres wrote in *God's Last Offer*, the "if we can put a man on the moon" boosterism "glosses over the reality that building rockets and building livable communities are two fundamentally different endeavors: the former required uncanny narrow focus; the latter must engage a holistic view. Building a livable world *isn't* rocket science; it's far more complex than that."[47]

Have We Really Tried Plan A?

On day two of the geoengineering retreat at Chicheley Hall, a spirited debate breaks out about whether the U.N. has any role to play in governing geoengineering experiments. The scientists anxious to get their field tests off the ground are quickest to dismiss the institution, fearing an unwieldy process that would tie their hands. The participants from NGOs are not quite ready to throw out the institution that has been the primary forum for climate governance, flawed as it is.

Just when things are getting particularly heated, there is a commotion outside the glass doors of the lecture hall.

A fleet of brand-new luxury cars has pulled up outside and a retinue of people—noticeably better dressed than the ones in the geoengineering session—pile out, their polished wingtips and high heels crunching noisily on the gravel pathways. One of our hosts from the Royal Society explains that for the rest of the day, another retreat put on by the auto company Audi will also be holding its sessions in the refurbished coach house. I peek outside and notice that several signs bearing Audi's Olympics-like logo have appeared along the driveway.

For the rest of the afternoon, our tense discussions about the ethics of blocking the sun are periodically interrupted by loud cheers coming from next door. The reason for the cheering is, we are told, a corporate secret, but the team from Audi is obviously very happy about something—next season's models, perhaps, or maybe sales figures.

The Royal Society regularly rents out Chicheley Hall for corporate re-treats and Downton Abbey–inspired weddings so the fact that these two meetings are taking place cheek-by-jowl in a country mansion is, of course, pure coincidence. Still, separated by nothing more than a thin sliding wall, it's hard not to feel that the angsty would-be geoengineers and the carefree German car sellers are in conversation with each other—as if, more than anything, the reckless experiments the people in our room are attempting to rationalize are really about allowing the car people in the next room to keep their party going.

The mind has a habit of making connections out of random proximate events, but in this case, it's not entirely random. There is no doubt that some of the people pushing geoengineering see these technologies not as emergency bridges away from fossil fuels, but as a means to keep the fossil fuel frenzy going for as long as possible. Nathan Myhrvold, for one, has even proposed using the mountains of yellow sulfur that are produced as waste in the Alberta tar sands to shield the sun, which would conveniently allow the oil majors to keep digging and drilling indefinitely. "You could put one little pumping facility up there, and with one corner of one of those sulfur moun-tains, you could solve the whole global warming problem for the Northern Hemisphere." And David Keith's start-up company Carbon Engineering has not only Bill Gates as an investor, but also Murray Edwards, whose oil company Canadian Natural Resources is one of the biggest players in the tar sands.[48]

Neither of these is an isolated case. Corporations that either dig up fossil fuels or that, like car companies, are responsible for a disproportionate share of their combustion, have a long track record of promoting geoengineering as a response to climate change, one that they clearly see as preferable to stopping their pollution. This goes as far back as 1992, when the National Academy of Sciences copublished a controversial report titled *Policy Implications of Greenhouse Warming*. To the consternation of many climate scientists, the document included a series of geoengineering options, some of them rather outlandish, from sending fifty thousand mirrors into earth's orbit to putting "billions of aluminized, hydrogen-filled balloons in the stratosphere to provide a reflective screen."[49]

Adding to the controversy was the fact that this chapter of the report was led by Robert A. Frosch, then a vice president at General Motors. As he explained at the time: "I don't know why anybody should feel obligated to reduce carbon dioxide if there are better ways to do it. When you start making deep cuts, you're talking about spending some real money and changing the entire economy. I don't understand why we're so casual about tinkering with the whole way people live on the Earth, but not tinkering a little further with the way we influence the environment."[50]

And notably, it was BP's chief scientist, Steven Koonin, who convened one of the first formal scientific gatherings on geoengineering back in 2008. The gathering produced a report outlining a decade-long research project into climate modification, with a particular focus on Solar Radiation Management. (Koonin left BP to work for the Obama administration as the Department of Energy's under secretary for science.)[51]

It's much the same story at several influential think tanks that are generously funded with fossil fuel dollars. For instance, over a period of years, as it stoked the flames of climate change denial, the American Enterprise Institute (AEI) took millions of dollars in donations from ExxonMobil. It continues to be the top recipient of money from conservative foundations eager to block climate action, bringing in at least $86.7 million from those sources since 2003. And yet, in 2008, the think tank launched a department called the Geoengineering Project. The project has held several conferences, published multiple reports, and sent experts to testify before congressional hearings—all with the consistent message that geoengineering isn't a Plan B should emission cuts fail, but rather a Plan A. Lee Lane, who for several years was AEI's main spokesperson on the subject, explained in 2010, "For those of us who believe that climate change might, at some point, pose a grave threat—and that emissions containment is both costly and politically impractical—climate engineering is beginning to look like the last, best hope."[52]

This position is striking given the think tank's well-documented history of attacks on climate science and concerted efforts to trash virtually every serious attempt to regulate emissions, including mild legislation favoring energy-efficient light bulbs (big government interference in "how we wish to light up our lives," as one AEI researcher put it).[53] Some at the think tank have signaled their openness to a modest or revenue-neutral carbon tax in recent years, which along with geoengineering is an increasingly prominent fetish among non-climate-change-denying Republicans. Still, you would think that turning down the sun for every person on earth is a more intrusive form of big government than asking citizens to change their light bulbs. Indeed you would think that pretty much any policy option would be less intrusive. But that is to miss the point: for the fossil fuel companies and their paid champions, anything is preferable to regulating ExxonMobil, *including* attempting to regulate the sun.

The rest of us tend to see things differently, which is why the fact that geoengineering is being treated so seriously should underline the urgent need for a real Plan A—one based on emission reduction, however economically radical it must be. After all, if the danger of climate change is sufficiently grave and imminent for governments to be considering science-fiction solutions, isn't it also grave and imminent enough for them to consider just plain science-based solutions?

Science tells us we need to keep the vast majority of proven fossil fuel reserves in the ground. It seems reasonable, then, that any government ready to fund experiments into climate alteration should also be willing, at the very least, to put a moratorium on new extreme energy development, while providing sufficient funding for a rapid transition to renewable energy. As the Tyndall Centre's Kevin Anderson points out, "At the moment we're digging out shale gas and tar sands and lots of coal. We're going to be digging under the Arctic. We don't need to concern ourselves too much with geo-engineering for the future, we just need to stop getting fossil fuels out of the ground today."[54]

And how about some other solutions discussed in these pages—like taking far larger shares of the profits from the rogue corporations most responsible for waging war on the climate and using those resources to clean up their mess? Or reversing energy privatizations to regain control over our grids? We have only the briefest window in which this strategy is viable, before we need to get off fossil fuels entirely, so surely it merits discussion.

The Indian author and activist Vandana Shiva, meanwhile, points out that shifting to an agriculture model based on agro-ecological methods would not only sequester large amounts of carbon, it would reduce emissions and increase food security. And unlike geoengineering, "It's not a fifty-year experiment. It's an assured, guaranteed path that has been shown to work."[55] Admittedly, such responses break all the free market rules. Then again, so did bailing out the banks and the auto companies. And they are still not close to as radical as breaking the primordial link between temperature and atmospheric carbon—all to meet our desire for planetary air-conditioning.

If we were staring down the barrel of an imminent and unavoidable climate emergency, the kinds of monstrous calculations implicit in geoengineering—sacrifice part of Latin America in order to save all of China, or save the remaining glaciers and land ice to prevent catastrophic global sea level rise but risk endangering India's food source—might be unavoidable. But even if we acquire enough information to make those kinds of calculations (and it's hard to imagine how we could), we notably are not at that point. We have options, ones that would greatly decrease the chances of ever confronting those impossible choices, choices that indeed deserve to be described as genocidal. To fail to exercise those options—which is exactly what we are collectively doing—knowing full well that eventually the failure could force government to rationalize "risking" turning whole nations, even subcontinents, into sacrifice zones, is a decision our children may judge as humanity's single most immoral act.

The Astronaut's Eye View

There is a photograph from the day Richard Branson launched his $25 million Virgin Earth Challenge that keeps popping into my head at the geoengineering retreat. Branson, dressed in black, has a big grin on his face and he is gleefully tossing a plastic model of Planet Earth into the air as if it were a beach ball. Al Gore, looking unsure about whether this is a good idea, is standing by his side.[56]

This frozen moment strikes me as the perfect snapshot of the first incarnation of the climate movement: a wealthy and powerful man with the whole world literally in his hands, promising to save the fragile blue planet on our behalf. This heroic feat will be accomplished, he has just announced, by harnessing the power of human genius and the desire to get really, really rich.

Pretty much everything is wrong with that picture. The reinvention of a major climate polluter into a climate savior based on little more than good PR. The assumption that dangling enough money can solve any mess we create. And the certainty that the solutions to climate change must come from above rather than below.

But I've begun to think that there is another problem too—it has to do with that pale blue sphere that Branson was tossing skyward. For more than forty years, the view of the Earth from space has been the unofficial logo of the environmental movement—featured on countless T-shirts, pins, and bumper stickers. It is the thing that we are supposed to protect at U.N. climate conferences, and that we are called upon to "save" every Earth Day, as if it were an endangered species, or a starving child far away, or a pet in need of our ministrations. And that idea may be just as dangerous as the Baconian fantasy of the earth as a machine for us to master, since it still leaves us (literally) on top.

When we marvel at that blue marble in all its delicacy and frailty, and resolve to save the planet, we cast ourselves in a very specific role. That role is of a parent, the parent of the earth. But the opposite is the case. It is we humans who are fragile and vulnerable and the earth that is hearty and powerful, and holds us in its hands. In pragmatic terms, our challenge is less to save the earth from ourselves and more to save ourselves from an earth that, if pushed too far, has ample power to rock, burn, and shake us off completely. That knowledge should inform all we do—especially the decision about whether to gamble on geoengineering.

It wasn't supposed to be this way, of course. In the late 1960s, when NASA shared the first photographs of the whole earth from space, there was a great deal of rhapsodizing about how the image would spark a leap in human consciousness. When we were finally able to see our world as an interconnected and holistic entity we at last would understand that this lonely planet is our only home and that it is up to us to be its responsible caretakers.* This was "Spaceship Earth" and the great hope was that being able to see it would cause everyone to grasp what British economist and author Barbara Ward meant when she said in 1966, "This space voyage is totally precarious. We depend upon a little envelope of soil and a rather larger envelope of atmosphere for life itself. And both can be contaminated and destroyed."[57]

So how did we get from that humility before life's precariousness to Branson's game of planet beach ball? One person who saw it all coming was the irascible American novelist Kurt Vonnegut: "Earth is such a pretty blue and pink and white pearl in the pictures NASA sent me," he wrote in *The New York Times Magazine* in 1969. "It looks so *clean*. You can't see all the hungry, angry earthlings down there—and the smoke and the sewage and trash and sophisticated weaponry."[58]

Before those pictures, environmentalism had mostly been intensely local—an earthy thing, not an Earth thing. It was Henry David Thoreau musing on the rows of white bush beans in the soil by Walden Pond. It was Edward Abbey ranging through the red rocks of southern Utah. It was Rachel Carson down in the dirt with DDT-contaminated worms. It was vividly descriptive prose, naturalist sketches, and, eventually, documentary photography and film seeking to awaken and inspire love for specific creatures and places—and, by extension, for creatures and places like them all over the world.

When environmentalism went into outer space, adopting the perspective of the omniscient outsider, things did start getting, as Vonnegut warned, awfully blurry. Because if you are perpetually looking down at the earth from above, rather than up from its roots and soil, it begins to make

---

* Ironically, the most reproduced of the earth-from-space photos was likely taken by Harrison Schmitt, a card-carrying climate change denier, former U.S. senator and a regular speaker at Heartland conferences. He was rather blasé about the experience: "You seen one Earth, you've seen them all," he reportedly said.

a certain kind of sense to shuffle around pollution sources and pollution sinks as if they were pieces on a planet-sized chessboard: a tropical forest to drink up the emissions from a European factory; lower-carbon-fracked gas to replace coal; great fields of corn to displace petroleum; and perhaps in the not too distant future, iron in the oceans and sulfur dioxide in the stratosphere to counter carbon dioxide in the lower atmosphere.

And all the while, just as Vonnegut warned, any acknowledgment of the people way down below the wispy clouds disappears—people with attachments to particular pieces of land with very different ideas about what constitutes a "solution." This chronic forgetfulness is the thread that unites so many fateful policy errors of recent years, from the decision to embrace fracked natural gas as a bridge fuel (failing to notice there were people on those lands who were willing to fight against the shattering of their territory and the poisoning of their water) to cap-and-trade and carbon offsets (forgetting the people once again, the ones forced to breathe the toxic air next to refineries that were being kept open thanks to these backroom deals, as well as the ones locked out of their traditional forests that were being converted into offsets).

We saw the same above-it-all perspective take its toll, tragically, when many of these same players persuaded themselves that biofuels were the perfect low-carbon alternative to oil and gas—only to discover what would have been blindingly obvious if people had figured as prominently in their calculations as carbon: that using prime land to grow fuel puts the squeeze on food, and widespread hunger is the entirely predictable result. And we see the same problems when policymakers ram through industrial-scale wind farms and sprawling desert solar arrays without local participation or consent, only to discover that people are living on those lands with their own inconvenient opinions about how they should be used and who should benefit from their development.

This lethal amensia is once again rearing its head in geoengineering discussions like the one at Chicheley Hall. It is awfully reassuring to imagine that a technological intervention could save Arctic ice from melting but, once again, far too little attention is being paid to the billions of people living in monsoon-fed parts of Asia and Africa who could well pay the price with their suffering, even their lives.

In some cases, the effect of the astronaut's eye view proves particularly extreme. Their minds hovering out in orbit, there are those who begin to imagine leaving the planet for good—saying, "Goodbye Earth!" to quote Princeton physicist Gerard O'Neill, who, in the mid-1970s, started calling for the creation of space colonies to overcome the earth's resource limits. Interestingly, one of O'Neill's most devoted disciples was Stewart Brand, the founder of the *Whole Earth Catalog*, who spent a good chunk of the 1970s arguing that the U.S. government should build space colonies; today he is one of the most vocal proponents of Big Tech fixes to climate change, whether nuclear power or geoengineering.[59]

And he's not the only prominent geoengineering booster nurturing the ultimate escape fantasy. Lowell Wood, co-inventor of the hose-to-the-sky, is an evangelical proponent of terraforming Mars: there is "a 50/50 chance that young children now alive will walk on Martian meadows . . . will swim in Martian lakes," he told an Aspen audience in 2007, describing the technological expertise for making this happen as "kid's stuff."[60]

And then there is Richard Branson, Mr. Retail Space himself. In September 2012, Branson told *CBS This Morning* that, "In my lifetime, I am determined to be part of starting a population on Mars. I think it is absolutely realistic. It will happen." This plan, he said, includes "people inhabiting Mars . . . in sort of giant domes." In another interview, he revealed that he has put a striking amount of thought into who should be invited to this outer space cocktail party: "You're going to want physicians, you're going to want comedians, you're going to want fun people, beautiful people, ugly people, a good cross-section of what happens on Earth on Mars. People have got to be able to get on together, because it's going to be quite confined." Oh and one more person on the list: "It may be a one-way trip. . . . So maybe I'll wait till the last 10 years of my life, and then maybe go, if my wife will let me," Branson said. In explaining his rationale, the Virgin head has invoked physicist Stephen Hawking, who "thinks it's absolutely essential for mankind to colonize other planets because one day, something dreadful might happen to the Earth. And it would be very sad to see years of evolution going to waste."[61]

So said the man whose airlines have a carbon footprint the size of Honduras' and who is pinning his hopes for planetary salvation not on emissions cuts, but on a carbon-sucking machine that hasn't been invented yet.[62] Perhaps this is mere coincidence, but it does seem noteworthy that so many key figures in the geoengineering scene share a strong interest in a planetary exodus. For it is surely a lot easier to accept the prospect of a recklessly high-risk Plan B when you have, in your other back pocket, a Plan C.

The danger is not so much that these visions will be realized; geoengineering the earth is a long shot, never mind terraforming Mars. Yet as Branson's own emissions illustrate so elegantly, these fantasies are already doing real damage in the here and now. As environmental author Kenneth Brower writes, "The notion that science will save us is the chimera that allows the present generation to consume all the resources it wants, as if no generations will follow. It is the sedative that allows civilization to march so steadfastly toward environmental catastrophe. It forestalls the real solution, which will be in the hard, nontechnical work of changing human behavior." And worst of all, it tells us that, "should the fix fail, we have someplace else to go."[63]

We know this escape story all too well, from Noah's Ark to the Rapture. What we need are stories that tell us something very different: that this planet is our only home, and that what goes around comes around (and what goes up, stays up for a very long time, so we'd better be careful what we put there).

Indeed, if geoengineering has anything going for it, it is that it slots perfectly into our most hackneyed cultural narrative, the one in which so many of us have been indoctrinated by organized religion and the rest of us have absorbed from pretty much every Hollywood action movie ever made. It's the one that tells us that, at the very last minute, some of us (the ones that matter) are going to be saved. And since our secular religion is technology, it won't be god that saves us but Bill Gates and his gang of super-geniuses at Intellectual Ventures. We hear versions of this narrative every time a commercial comes on about how coal is on the verge of becoming "clean," about how the carbon produced by the tar sands will soon be sucked out of the air and buried deep underground, and now, about how the mighty sun will be turned down as if it were nothing more than a chandelier on a dimmer. And if one of the current batch of schemes doesn't work, the same story tells us that something else will surely arrive in the nick of time. We are, after all, the super-species, the chosen ones, the God Species. We will triumph in the end because triumphing is what we do.

But after so many of our most complex systems have failed, from BP's deepwater drilling to the derivatives market—with some of our biggest brains failing to foresee these outcomes—there is some evidence that the power of this particular narrative arc is beginning to weaken. The Brookings Institution released a survey in 2012 that found that roughly seven in ten Americans think that trying to turn down the sun will do more harm than good. Only three in ten believe that "scientists would be able to find ways to alter the climate in a way that limits problems" caused by warming. And in a paper published in *Nature Climate Change* in early 2014, researchers analyzed data from interviews and a large online survey conducted in Australia and New Zealand—with the biggest sample size of any geoengineering public opinion study to date. Malcolm Wright, the study's lead author, explained, "The results show that the public has strong negative views towards climate engineering. . . . It is a striking result and a very clear pattern. Interventions such as putting mirrors in space or fine particles into the stratosphere are not well received." Perhaps most interesting of all given the high-tech subject, older respondents were more amenable to geoengineering than younger ones.[64]

And the best news is that the time of astronaut's eye-view environmentalism appears to be passing, with a new movement rising to take its place, one deeply rooted in specific geographies but networked globally as never before. Having witnessed the recent spate of big failures, this generation of activists is unwilling to gamble with the precious and irreplaceable, certainly not based on the reassuring words of overconfident engineers.

This is a movement of many movements, and though utterly undetectable from space, it is beginning to shake the fossil fuel industry to its core.

PART THREE

STARTING ANYWAY

"The day capitalism is forced to tolerate non-capitalist societies in its midst and to acknowledge limits in its quest for domination, the day it is forced to recognize that its supply of raw material will not be endless, is the day when change will come. If there is any hope for the world at all, it does not live in climate-change conference rooms or in cities with tall buildings. It lives low down on the ground, with its arms around the people who go to battle every day to protect their forests, their mountains and their rivers because they know that the forests, the mountains and the rivers protect them.

"The first step towards reimagining a world gone terribly wrong would be to stop the annihilation of those who have a different imagination—an imagination that is outside of capitalism as well as communism. An imagination which has an altogether different understanding of what constitutes happiness and fulfillment. To gain this philosophical space, it is necessary to concede some physical space for the survival of those who may look like the keepers of our past, but who may really be the guides to our future. "

—Arundhati Roy, 2010[1]

"When I started the lawsuit against Chevron in 1993, I thought, 'What we need to do to fight this company and to get justice is we need to unite the Amazon.' And that was a hard challenge. That was a hard task ahead. And now, today, I dare to say that we must unite the entire world. We have to unite the entire world to fight these companies, to fight these challenges."

—Luis Yanza, cofounder, Frente de Defensa de la Amazonía (Amazon Defense Front), 2010[2]

9

BLOCKADIA

The New Climate Warriors

"Where there are threats of serious or irreversible damage, lack of full scientific certainty shall not be used as a reason for postponing cost-effective measures to prevent environmental degradation."

—The United Nations Rio Declaration on Environment and Development, 1992 [1]

"An honest and scrupulous man in the oil business is so rare as to rank as a museum piece."

—U.S. Interior Secretary Harold Ickes, 1936 [2]

"Passport," says the cop, tear gas canisters and grenades hanging off his bulletproof vest like medals of honor. We hand over the passports, along with press passes and other papers attesting that we are nothing more exciting than a vanload of Canadian documentary filmmakers.

The riot cop takes the documents wordlessly, motioning to our translator to get out of the car. He then whispers at length to a colleague whose eyes remain fixed on the enormous biceps bulging from his own crossed arms. Another cop joins the huddle, then another. The last one pulls out a phone and painstakingly reads the names and numbers on each document to whoever is on the other end, occasionally shooting a question to our translator. More uniformed men mill nearby. I count eleven in total.

It's getting dark, the dirt road on which we have been apprehended is a mess and drops off sharply on one side. There are no streetlights.

I have the strong impression we are being deliberately screwed with—

that the whole point of this lengthy document check is to force us to drive this rough road in the dark. But we all know the rules: look pleasant; don't make eye contact; don't speak unless spoken to. Resist the impulse to take pictures of the line of heavily armed cops standing in front of coils of barbed wire (happily it turns out our camera guy was filming through his mesh hat). And Rule No. 1 on encounters with arbitrary power: do *not* show how incredibly pissed off you are.

We wait. Half an hour. Forty minutes. Longer. The sun sets. Our van fills with ravenous mosquitoes. We continue to smile pleasantly.

As far as checkpoints go, I've seen worse. In post-invasion Iraq, everyone had to submit to full pat-downs in order to get in and out of any vaguely official building. Once on the way in and out of Gaza, we were scanned eight different ways and interrogated at length by both the Israeli Defense Forces and Hamas. What's strange about what is happening on this dirt road is that we are not in a war zone, at least not officially. Nor is this a military regime, or an occupied territory, or any other place you might expect to be held and interrogated at length without cause. This is a public road in Greece, a democratic state belonging to the European Union. Moreover this particular road is in Halkidiki, a world-renowned tourist destination that attracts many thousands of visitors every year, drawn to the peninsula's stunning combination of sandy beaches, turquoise waters, olive groves, and old-growth forests filled with four-hundred-year-old beech and oak trees and dotted with waterfalls.

So what's up with all the riot police? The barbed wire? The surveillance cameras strapped to tree branches?

Welcome to Blockadia

What's up is that this area is no longer a Greek vacationland, though the tourists still crowd the white-washed resorts and oceanfront tavernas, with their blue-checked tablecloths and floors sticky with ouzo. This is an outpost of a territory some have taken to calling "Blockadia." Blockadia is not a specific location on a map but rather a roving transnational conflict zone that is cropping up with increasing frequency and intensity wherever extractive projects are attempting to dig and drill, whether for open-pit mines, or gas fracking, or tar sands oil pipelines.

What unites these increasingly interconnected pockets of resistance is the sheer ambition of the mining and fossil fuel companies: the fact that in their quest for high-priced commodities and higher-risk "unconventional" fuels, they are pushing relentlessly into countless new territories, regardless of the impact on the local ecology (in particular, local water systems), as well as the fact that many of the industrial activities in question have neither been adequately tested nor regulated, yet have already shown themselves to be extraordinarily accident-prone.

What unites Blockadia too is the fact the people at the forefront—packing local council meetings, marching in capital cities, being hauled off in police vans, even putting their bodies between the earth-movers and earth—do not look much like your typical activist, nor do the people in one Blockadia site resemble those in another. Rather, they each look like the places where they live, and they look like everyone: the local shop owners, the university professors, the high school students, the grandmothers. (In the quaint seaside Greek village of Ierissos, with its red roofs and lively beach promenade, when an anti-mining rally is called, the owners of the tavernas have to wait tables themselves because their entire staffs are off at the demos.)

Resistance to high-risk extreme extraction is building a global, grassroots, and broad-based network the likes of which the environmental movement has rarely seen. And perhaps this phenomenon shouldn't even be referred to as an environmental movement at all, since it is primarily driven by a desire for a deeper form of democracy, one that provides communities with real control over those resources that are most critical to collective survival—the health of the water, air, and soil. In the process, these place-based stands are stopping real climate crimes in progress.

Seeing those successes, as well as the failures of top-down environmentalism, many young people concerned about climate change are taking a pass on the slick green groups and the big U.N. summits. Instead, they are flocking to the barricades of Blockadia. This is more than a change in strategy; it's a fundamental change in perspective. The collective response to the climate crisis is changing from something that primarily takes place in closed-door policy and lobbying meetings into something alive and unpredictable and very much in the streets (and mountains, and farmers' fields, and forests).

Unlike so many of their predecessors, who've spent years imagining the climate crisis through the astronaut's eye view, these activists have dropped the model globes and are getting lower-case earth under their nails once again. As Scott Parkin, a climate organizer with the Rainforest Action Network, puts it: "People are hungry for climate action that does more than asks you to send emails to your climate-denying congressperson or update your Facebook status with some clever message about fossil fuels. Now, a new antiestablishment movement has broken with Washington's embedded elites and has energized a new generation to stand in front of the bulldozers and coal trucks."[3] And it has taken the extractive industries, so accustomed to calling the shots, entirely by surprise: suddenly, no major new project, no matter how seemingly routine, is a done deal.

In the Skouries forest near Ierissos where our van was stopped, the catalyst was a plan by the Canadian mining company Eldorado Gold to clear-cut a large swath of old-growth forest and reengineer the local water system in order to build a massive open-pit gold and copper mine, along with a processing plant, and a large underground mine.[4] We were pulled over in a part of the forest that will be leveled to make way for a large dam and tailings pond, to be filled with liquid waste from the mining operation. It was like visiting someone who had just been given six months to live.

Many of the people who reside in the villages nearby, who depend on this mountain for freshwater, are adamantly opposed to the mine. They fear for the health of their children and livestock, and are convinced that such a large-scale, toxic industrial operation has no place in a region highly dependent on tourism, fishing, and farming. Locals have expressed their opposition through every means they can think of. In a vacation community like this, that can make for odd juxtapositions: militant marches past miniature amusement parks and heated late night political meetings in thatched-roof bars that specialize in blender drinks. Or a local cheese maker, the pride of the village for his *Guinness Book of World Records* largest ever goat cheese, arrested and held in pretrial detention for weeks. Based on circumstantial evidence, the cheese maker and other villagers were suspects in an incident in which mining trucks and bulldozers were torched by masked intruders.[*][5]

Despite its remote location, the fate of the Skouries forest is a matter of intense preoccupation for the entire country. It is debated in the national parliament and on evening talk shows. For Greece's huge progressive movement, it is something of a cause célèbre: urban activists in Thessaloniki and Athens organize mass demonstrations and travel to the woods for action days and fundraising concerts. "Save Skouries" graffiti can be seen all over the country and the official opposition party, the left-wing Syriza, has pledged that, if elected, it will cancel the mine as one of its first acts in power.

The governing, austerity-enforcing coalition, on the other hand, has also seized on Skouries as a symbol. Greek prime minister Antonis Samaras has announced that the Eldorado mine will go ahead "at all costs," such is the importance of protecting "foreign investment in the country." Invoking Greece's ongoing economic troubles, his coalition has claimed that building the mine, despite the local opposition, is critical to sending a signal to world markets that the country is open for business. That will allow the nation to rapidly move ahead with a slate of other, highly controversial extractive projects currently in the pipeline: drilling for oil and gas in the Aegean and Ionian seas; new coal plants in the north; opening up previously protected beaches to large-scale development; and multiple other mining projects. As one prominent commentator put it, "This is the type of project that the country needs to overcome the economic crisis."[6]

Because of these national stakes, the state has unleashed a level of repression against the anti-mine movement that is unprecedented in Greece since the dark days of dictatorship. The forest has been transformed into a battle zone, with rubber bullets reportedly fired and tear gas so thick it caused older residents to collapse.[7] And of course the checkpoints, which are staggered along all the roads where heavy construction equipment has moved in.

But in this outpost of Blockadia, the police aren't the only ones with

---

* The villagers insist their struggle is committed to nonviolence and blame outsiders or even provocateurs for the arson.

checkpoints: In Ierissos, local residents set up checkpoints at each entrance to their village after over two hundred fully armed riot police marched through the town's narrow streets firing tear gas canisters in all directions; one exploded in the schoolyard, causing children to choke in class.[8] To make sure they are never taken by surprise like this again, the checkpoints are staffed by volunteers around the clock, and when police vehicles are spotted someone runs to the church and rings the bell. In moments the streets are flooded with chanting villagers.

Similar scenes, more reminiscent of civil war than political protest, are un-folding in countless other pieces of contested land around the world, all of which make up Blockadia's multiplying front lines. About eight hun-dred kilometers to the north of the Greek standoff, the farming village of Pungesti, Romania, was gearing up for a showdown against Chevron and its plans to launch the country's first shale gas exploration well.[9] In the fall of 2013, farmers built a protest camp in a field, carted in supplies that could hold them for weeks, dug a latrine, and vowed to prevent Chevron from drilling.

As in Greece, the response from the state was shockingly militarized, es-pecially in such a pastoral environment. An army of riot police with shields and batons charged through the farm fields attacking peaceful demonstra-tors, several of whom were beaten bloody and taken away in ambulances. At one point angry villagers dismantled the fence protecting Chevron's operation, sparking more reprisals. In the village itself, riot police lined the streets like "a kind of occupying army," according to an eyewitness. Mean-while, the roads into town were bisected with police checkpoints and a travel ban was in force, which conveniently prevented media from entering the conflict zone and even reportedly blocked residents from grazing their cattle. For their part, villagers explained that they had no choice but to stop an extraction activity that they were convinced posed a grave threat to their livelihoods. "We live on agriculture here," one local reasoned. "We need clear water. What will our cattle drink if the water gets spoiled?"[10]

Blockadia also stretches into multiple resource hot spots in Canada, my home country. For instance, in October 2013—the same time that Pungesti was in the news—a remarkably similar standoff was playing out in the province of New Brunswick, on land claimed by the Elsipogtog First Nation, a Mi'kmaq community whose roots in what is now eastern Canada go back some ten thousand years. The people of Elsipogtog were leading a blockade against SWN Resources, the Canadian subsidiary of a Texas-based company, as it tried to conduct seismic testing ahead of a possible fracking operation. The land in question has not been handed over by war or treaty and Canada's highest court has upheld the Mi'kmaq's right to continue to access the natural resources of those lands and waters—rights the protesters say would be rendered meaningless if the territory becomes poisoned by fracking toxins.[11]

The previous June, members of the First Nation had announced the lighting of a "sacred fire," a ceremonial bonfire that would burn continuously for days, and invited non-Native Canadians to join them in blockading the gas company's trucks. Many did, and for months demonstrators camped near the seismic testing area, blocking roads and equipment as hand drums pounded out traditional songs. On several occasions, trucks were prevented from working, and at one point a Mi'kmaq woman strapped herself to a pile of seismic testing gear to prevent it from being moved.

The conflict had been mostly peaceful but then on October 17, acting on an injunction filed by the company, the Royal Canadian Mounted Police moved in to clear the road. Once again, a rural landscape was turned into a war zone: more than a hundred police officers—some armed with sniper rifles and accompanied by attack dogs—fired beanbag rounds into the crowd, along with streams of pepper spray and hoses. Elders and children were attacked and dozens were arrested, including the elected chief of the Elsipogtog First Nation. Some demonstrators responded by attacking police vehicles and by the end of the day, five cop cars and one unmarked van had burned. "Native shale-gas protest erupts in violence," read a typical headline.[12]

Blockadia has popped up, too, in multiple spots in the British countryside, where opponents of the U.K. government's "dash for gas" have used a range of creative tactics to disrupt industry activities, from protest picnics blockading the road to a fracking drill site in the tiny hamlet of Balcombe,

West Sussex, to twenty-one activists shutting down a gas power station that towers over the abandoned historical village of West Burton and its beautiful river, the "silver" Trent, as Shakespeare describes it in *Henry IV*. After a daring climb, the group set up camp for more than a week atop two ninety-meter-high water cooling towers, making production impossible (the company was forced to drop a £5-million lawsuit in the face of public pressure). More recently, activists blocked the entrance to a fracking test site near the city of Manchester with a giant wind turbine blade laid on its side.[13]

Blockadia was also aboard the *Arctic Sunrise,* when thirty Greenpeace activists staged a protest in the Russian Arctic to draw attention to the dangers of the rush to drill under the melting ice. Armed Coast Guard officers rappelled onto the vessel from a helicopter, storming it commando-style, and the activists were thrown in jail for two months.[14] Originally facing charges of piracy, which carry sentences of ten to fifteen years, the international activists were all eventually freed and granted amnesty after the Russian government was shamed by a huge international campaign, which included not just demonstrations in at least forty-nine countries but pressure from numerous heads of state and eleven Nobel Peace Prize winners (not to mention Paul McCartney).

The spirit of Blockadia can be seen even in the most repressive parts of China, where herders in Inner Mongolia have rebelled against plans to turn their fossil fuel–rich region into the country's "energy base." "When it's windy, we get covered in coal dust because it's an open mine. And the water level keeps dropping every year," herder Wang Wenlin told the *Los Angeles Times*, adding, "There's really no point living here anymore." With courageous actions that have left several demonstrators dead outside the mines and blockades of coal trucks, locals have staged rolling protests around the region and have been met with ferocious state repression.[15]

It's partly due to this kind of internal opposition to coal mining that China imports increasing amounts of coal from abroad. But many of the places where its coal comes from are in the throes of Blockadia-style uprisings of their own. For instance, in New South Wales, Australia, opposition to new coal mining operations grows more serious and sustained by the month. Beginning in August 2012, a coalition of groups established what they call the "first blockade camp of a coal mine in Australia's history," where for a year and a half (and counting) activists have chained themselves to various entrances of the Maules Creek project—the largest mine under construction in the country, which along with others in the area is set to decimate up to half of the 7,500-hectare (18,500 acre) Leard State Forest and to wield a greenhouse gas footprint representing more than 5 percent of Australia's annual emissions, according to one estimate.[16]

Much of that coal is destined for export to Asia, however, so activists are also gearing up to fight port expansions in Queensland that would hugely increase the number of coal ships sailing from Australia each year, including through the vulnerable ecosystem of the Great Barrier Reef, a World Heritage Site and the earth's largest natural structure made up of living creatures. The Australian Marine Conservation Society describes the dredging of the ocean floor to make way for increased coal traffic as an "unprecedented" threat to the fragile reef, which is already under severe stress from ocean acidification and various forms of pollution runoff.[17]

This is only the barest of sketches of the contours of Blockadia—but no picture would be complete without the astonishing rise of resistance against virtually any piece of infrastructure connected to the Alberta tar sands, whether inside Canada or in the United States.

And none more so than TransCanada's proposed Keystone XL pipeline. Part of the broader Keystone Pipeline System crisscrossing the continent, the first phase of the project, known as Keystone 1, got off to an inauspicious start. In its first year or so of operation, pump stations along the pipeline spilled tar sands oil fourteen times in the U.S. Most spills were small, but two of the biggest forced the entire pipeline to shut down twice in a single month. In one of these cases, a North Dakota rancher woke up to the sight of an oil geyser surging above the cottonwood trees near his farm, remarking that it was "just like in the movies when you strike oil and it's shooting up." If Keystone XL is constructed in full (the southern leg, from Oklahoma to export terminals on the Texas coast, is already up and running), the $7 billion project will add a total of 2,677 kilometers of new pipeline running through seven states and provinces, delivering up to 830,000 barrels per day of mostly tar sands oil to Gulf Coast refineries and export terminals.[18]

It was Keystone that provoked that historic wave of civil disobedience in Washington, D.C., in 2011 (see page 139), followed by what were then the largest protests in the history of the U.S. climate movement (more than 40,000 people outside the White House in February 2013). And it is Keystone that brought together the unexpected alliance of Indigenous tribes and ranchers along the pipeline route that became known as "the Cowboy and Indian alliance" (not to mention unlikely coalitions that brought together vegan activists who think meat is murder with cattle farmers whose homes are decorated with deer heads). In fact the direct-action group Tar Sands Blockade first coined the term "Blockadia" in August 2012, while planning what turned into an eighty-six-day tree blockade challenging Keystone's construction in East Texas. This coalition has used every imaginable method to stop the pipeline's southern leg, from locking themselves inside a length of pipe that had not yet been laid, to creating a complex network of treehouses and other structures along the route.[19]

In Canada, it was the Northern Gateway pipeline, being pushed by the energy company Enbridge, that similarly awoke the sleeping giant of latent ecological outrage. The 1,177-kilometer pipe would begin near Edmonton, Alberta, and carry 525,000 barrels of mostly diluted tar sands oil per day across roughly one thousand waterways, passing through some of the most pristine temperate rainforest in the world (and highly avalanche-prone mountains), finally ending in a new export terminal in the northern British Columbia town of Kitimat. There the oil would be loaded onto supertankers and then navigated through narrow Pacific channels that are often battered by ferocious waves (resorts in this part of B.C. market winter as "storm-watching" season). The sheer audacity of the proposal—putting so much of Canada's most beloved wilderness, fishing grounds, beaches, and marine life at risk—helped give birth to an unprecedented coalition of Canadians who oppose the project, including a historic alliance of Indigenous groups in British Columbia who have vowed to act as "an unbroken wall of opposition from the U.S. border to the Arctic Ocean," to stop any new pipeline that would carry tar sands oil through their collective territory.[20]

The companies at the centers of these battles are still trying to figure out what hit them. TransCanada, for instance, was so sure it would be able to push through the Keystone XL pipeline without a hitch that it went ahead and bought over $1 billion worth of pipe. And why not? President Obama has an "all of the above" energy strategy, and Canadian prime minister Stephen Harper called the project a "no-brainer." But instead of the rubberstamp TransCanada was expecting, the project sparked a movement so large it revived (and reinvented) U.S. environmentalism.[21]

Spend enough time in Blockadia and you start to notice patterns. The slogans on the signs: "Water is life," "You can't eat money," "Draw the line." A shared determination to stay in the fight for the long haul, and to do whatever it takes to win. Another recurring element is the prominent role played by women, who often dominate the front lines, providing not only powerful moral leadership but also some of these movements' most enduring iconography. In New Brunswick, for instance, the image of a lone Mi'kmaq mother, kneeling in the middle of the highway before a line of riot police, holding up a single eagle feather went viral. In Greece, the gesture that captured hearts and minds was when a seventy-four-year-old woman confronted a line of riot police by belting out a revolutionary song that had been sung by the Greek resistance against German occupation. From Romania, the image of an old woman wearing a babushka and holding a knobby walking stick went around the world under the caption: "You know your government has failed when your grandma starts to riot."[22]

The various toxic threats these communities are up against seem to be awakening impulses that are universal, even primal—whether it's the fierce drive to protect children from harm, or a deep connection to land that had been previously suppressed. And though reported in the mainstream press as isolated protests against specific projects, these sites of resistance increasingly see themselves as part of a global movement, one opposing the latest commodities rush wherever it is taking place. Social media in particular has allowed geographically isolated communities to tell their stories to the world, and for those stories, in turn, to become part of a transnational narrative about resistance to a common ecological crisis.

So busloads of anti-fracking and anti-mountaintop-removal activists traveled to Washington, D.C., to protest the Keystone XL pipeline, knowing they are up against a common enemy: the push into ever more extreme and high-risk forms of fossil fuel. Communities in France, upon discovering that their land has been leased to a gas company for something called "hydraulic fracturing"—a previously unknown practice in Europe—got in contact with French-speaking activists in Quebec, who had successfully won a moratorium against the practice (and they, in turn, relied heavily on U.S. activists, in particular the documentary film *Gasland*, which has proved to be a potent global organizing tool).* [23] And eventually the entire global movement came together for a "Global Frackdown" in September 2012, with actions in two hundred communities in more than twenty countries, with even more participating a year later.

Something else unites this network of local resistance: widespread awareness of the climate crisis, and the understanding that these new extraction projects—which produce far more carbon dioxide, in the case of the tar sands, and more methane, in the case of fracking, than their conventional counterparts—are taking the entire planet in precisely the wrong direction. These activists understand that keeping carbon in the ground, and protecting ancient, carbon-sequestering forests from being clear-cut for mines, is a prerequisite for preventing catastrophic warming. So while these conflicts are invariably sparked by local livelihood and safety concerns, the global stakes are never far from the surface.

Ecuadorian biologist Esperanza Martínez, one of the leaders of the movement for an "oil-free Amazon," asks the question at the heart of all of these campaigns: "Why should we sacrifice new areas if fossil fuels should not be extracted in the first place?" Indeed, if the movement has a guiding theory, it is that it is high time to close, rather than expand, the fossil fuel frontier. Seattle-based environmental policy expert KC Golden has called this "the Keystone Principle." He explains, "Keystone isn't simply a pipeline in the sand for the swelling national climate movement." It's an expression of the core principle that before we can effectively solve this crisis, we have to "stop making it worse. Specifically and categorically, we must cease making large, long-term capital investments in new fossil fuel infrastructure that 'locks in' dangerous emission levels for many decades . . . step one for getting out of a hole: Stop digging." [24]

So if Obama's energy policy is "all of the above"—which effectively

---

* Maxime Combes, a French economist and anti-fracking activist, observes, "The scene in the film where landowner Mike Markham ignites gas from a water faucet in his home with a cigarette lighter due to natural gas exploration in the area has had a far greater impact against fracking than any report or speech."

means full steam ahead with fossil fuel extraction, complemented with renewables around the margins—Blockadia is responding with a tough philosophy that might be described as "None of the below." It is based on the simple principle that it's time to stop digging up poisons from the deep and shift, with all speed, to powering our lives from the abundant energies on our planet's surface.

Operation Climate Change

While the scale and connectivity of this kind of anti-extraction activism is certainly new, the movement began long before the fight against Keystone XL. If it's possible to trace this wave back to a time and place, it should probably be the 1990s in what is surely the most oil-ravaged place on the planet: the Niger Delta.

Since the doors to foreign investors were flung open near the end of British colonial rule, oil companies have pumped hundreds of billions of dollars' worth of crude out of Nigeria, most from the Niger Delta, while consistently treating its land, water, and people with undisguised disdain. Wastewater was dumped directly into rivers, streams, and the sea; canals from the ocean were dug willy-nilly, turning precious freshwater sources salty, and pipelines were left exposed and unmaintained, contributing to thousands of spills. In an often cited statistic, an *Exxon Valdez*–worth of oil has spilled in the Delta every year for about fifty years, poisoning fish, animals, and humans.[25]

But none of this compares with the misery that is gas flaring. Over the course of extracting oil, a large amount of natural gas is also produced. If the infrastructure for capturing, transporting, and using that gas were built in Nigeria, it could meet the electricity needs of the entire country. Yet in the Delta, the multinational companies mostly opt to save money by setting it on fire, or flaring it, which sends the gas into the atmosphere in great pillars of polluting fire. The practice is responsible for about 40 percent of Nigeria's total CO_2 emissions (which is why, as discussed, some companies are absurdly trying to collect carbon credits for stopping this practice). Meanwhile, more than half of Delta communities lack electricity and running water, unemployment is rampant, and, in a cruel irony, the region is plagued by fuel shortages.[26]

Since the 1970s, Nigerians living in the Delta have been demanding redress for the damage done to them by multinational oil giants. The fight entered a new phase at the start of the 1990s when the Ogoni—a relatively small Indigenous group in the Niger Delta—organized the Movement for the Survival of the Ogoni People (MOSOP), led by the famed human rights activist and playwright Ken Saro-Wiwa. The group took particular aim at Shell, which had extracted $5.2 billion from Ogoniland between 1958 and 1993.[27]

The new organization did more than beg the government for better conditions, it asserted the rights of the Ogoni people to control the resources under their lands and set about taking those rights back. Not only were oil installations shut down, but as Nigerian political ecologist and environmental activist Godwin Uyi Ojo writes that, on January 4, 1993, "an estimated 300,000 Ogoni, including women and children, staged a historic non-violent protest, and marched against Shell's 'ecological wars.'" That year, Shell was forced to pull out of Ogoni territory, forsaking significant revenues (though the company remains the biggest oil player in other parts of the Delta). Saro-Wiwa stated that the Nigerian state "will have to shoot and kill every Ogoni man, woman and child to take more of their oil."[28]

To this day, oil production has ceased in Ogoniland—a fact that remains one of the most significant achievements of grassroots environmental activism anywhere in the world. Because of Ogoni resistance, carbon has stayed in the ground and out of the atmosphere. In the two decades since Shell withdrew, the land has slowly begun to heal, and there are tentative reports of improved farming output. This represents, according to Ojo, "on a global scale, the most formidable community-wide resistance to corporate oil operations."[29]

But Shell's banishment was not the end of the story. From the start of the protests, the Nigerian government—which relies on oil for 80 percent of its revenues and 95 percent of its export earnings—saw the organized Ogoni as a grave threat. As the region mobilized to take its land back from Shell, thousands of Delta residents were tortured and killed and dozens of Ogoni villages were razed. In 1995, the military regime of General Sani Abacha tried Ken Saro-Wiwa and eight of his compatriots on trumped-up charges. And then all nine men were hanged, fulfilling Saro-Wiwa's prediction that "they are going to arrest us all and execute us. All for Shell."[30]

It was a wrenching blow to the movement, but residents of the Niger Delta fought on. By employing increasingly militant tactics like taking over offshore oil platforms, oil barges, and flow stations, this community-led resistance managed to shut down roughly twenty oil installations, significantly reducing production.[31]

A key and little examined chapter in the Niger Delta's fossil fuel resistance took place at the tail end of 1998. Five thousand young people belonging to the Ijaw Nation, one of the largest ethnic groups in Nigeria, held a gathering in Kaiama, a town in a southern province of the Delta. There, the Ijaw Youth Council drafted the Kaiama Declaration, which asserted that 70 pecent of the government's oil revenues came from Ijaw land and that, "Despite these huge contributions, our reward from the Nigerian State remains avoidable deaths resulting from ecological devastation and military repression." The declaration—endorsed by a huge cross-section of Delta society—stated: "All land and natural resources (including mineral resources) within the Ijaw territory belong to Ijaw communities and are the basis of our survival," and went on to demand "Self Government and resource control."[32]

But it was Clause 4 that commanded the most attention: "We, therefore, demand that all oil companies stop all exploration and exploitation activities in the Ijaw area. . . . Hence, we advise all oil companies staff and contractors to withdraw from Ijaw territories by the 30th December, 1998 pending the resolution of the issue of resource ownership and control in the Ijaw area of the Niger Delta."[33]

The Ijaw Youth Council voted unanimously to call their new offensive Operation Climate Change. "The idea was: we are going to change our world," Isaac Osuoka, one of the movement's organizers, told me. "There was an understanding of the link that the same crude oil that impoverishes us, also impoverishes the Earth. And that a movement to change the wider world can begin from changing our own world." This was, in other words, an attempt at another kind of climate change—an effort by a group of people whose lands had been poisoned and whose future was imperiled to change their political climate, their security climate, their economic climate, and even their spiritual climate.[34]

As promised, on December 30 the youth took to the streets in the thousands. The leadership instructed participants not to carry weapons and not to drink. The demonstrations—called *Ogeles*, which are traditional Ijaw processions—were nonviolent and dramatic. Many participants wore black, held candles, sang, danced, and drummed. Several oil platforms were occupied, not with arms but through the sheer numbers of bodies that overwhelmed security guards. "Sometimes," Osuoka recalled in a phone interview, "a person will have worked for a short time for the oil companies, so they knew which valve was the one to turn off."

The Nigerian government's response was overwhelming. An estimated fifteen thousand troops were mobilized, warships were sent, as were fleets of tanks. In some regions the government declared a state of emergency and imposed a curfew. According to Osouka, "In village after village, soldiers deployed by the state opened fire on unarmed citizens." In the towns of "Kaiama, Mbiama, and Yenagoa people were killed in the streets and women and young girls were raped in their homes as the state unleashed mayhem, ostensibly to defend oil installations."[35]

The confrontations continued for about a week. By the end, as many as 200 or possibly more lives were reported lost, and dozens of houses had been burned to the ground. In at least one case, the soldiers who conducted lethal raids flew into the area on a helicopter taken from a Chevron operation. (The oil giant claimed it had no choice but to allow the equipment to be used by the military, since it came from a joint venture with the Nigerian government, though as Human Rights Watch noted, "The company did not issue any public protest at the killings; nor has it stated that it will take any steps to avoid similar incidents in the future.")[36]

Brutal events like these go a long way toward explaining why many young people in the Niger Delta today have lost their faith in nonviolence. And why, by 2006, the area was in the throes of a full-blown armed insurgency, complete with bombings of oil infrastructure and government targets, rampant pipeline vandalism, ransom kidnapping of oil workers (designated as "enemy combatants" by the militants), and, more recently, amnesty deals that offered cash for guns. Godwin Uyi Ojo writes that, as the armed conflict wore on, "grievance was soon mingled with greed and violent crimes."[37] In the process, the original goals of the movement—to stop the ecological plunder, and take back control over the region's resource—became harder to decipher.

And yet it is worth looking back to the 1990s when the aims were clear. Because what is evident in the original struggles of the Ogoni and Ijaw is that the fight against violent resource extraction and the fight *for* greater community control, democracy, and sovereignty are two sides of the same coin. The Nigerian experience also had a huge and largely uncredited influence on other resource-rich regions in the Global South that found themselves facing off against multinational oil giants.

The most important such exchange took place in 1995, immediately after the killing of Ken Saro-Wiwa, when activists from Environmental Rights Action in Nigeria formed an alliance with a similar organization in Ecuador, called Acción Ecológica. At that time Acción Ecológica was neck deep in an environmental and human health disaster that Texaco had left behind in a northeastern region of the country, an incident that became known as the "Rainforest Chernobyl." (Chevron, after acquiring Texaco, was later ordered to pay $9.5 billion in damages by the Ecuadorian supreme court; the legal battles are still ongoing).[38] These frontline activists in two of the worst oil-impacted regions on the planet formed an organization called Oilwatch International, which has been at the forefront of the global movement to "leave the oil in the soil" and whose influence can be felt throughout Blockadia.

———

As the experiences in Nigeria and Ecuador make clear, anti-extraction activism is not a new phenomenon. Communities with strong ties to the land have always, and will always, defend themselves against businesses that threaten their ways of life. And fossil fuel resistance has a long history in the United States, most notably against mountaintop removal coal mining in Appalachia. Moreover, direct action against reckless resource extraction has been a part of the environmental movement for a very long time and has succeeded in protecting some of the planet's most biologically diverse lands and waters. Many of the specific tactics being used by Blockadia activists today—tree-sits and equipment lockdowns in particular—were developed by Earth First! in the 1980s, when the group fought "wars in the woods" against clear-cut logging.

What has changed in recent years is largely a matter of scale, which is itself a reflection of the dizzying ambitions of the extractive project at this point in history. The rise of Blockadia is, in many ways, simply the flip side of the carbon boom. Thanks to a combination of high commodity prices, new technologies, and depleted conventional reserves, the industry is going further on every front. It is extracting more, pushing into more territory, and relying on more risky methods. Each of these factors is fueling the backlash, so it's worth looking at each in turn.

All in the Sacrifice Zone

Though there are certainly new and amplified risks associated with our era of extreme energy (tar sands, fracking for both oil and gas, deepwater drilling, mountaintop removal coal mining), it's important to remember that these have never been safe or low-risk industries. Running an economy on energy sources that release poisons as an unavoidable part of their extraction and refining has always required sacrifice zones—whole subsets of humanity categorized as less than fully human, which made their poisoning in the name of progress somehow acceptable.

And for a very long time, sacrifice zones all shared a few elements in common. They were poor places. Out-of-the-way places. Places where residents lacked political power, usually having to do with some combination of race, language, and class. And the people who lived in these condemned places knew they had been written off. To quote Paula Swearengin, an activist from a coal mining family near Beckley, West Virginia, a landscape ravaged by mountaintop-removal coal mining: "We live in the land of the lost."[39]

Through various feats of denialism and racism, it was possible for privileged people in North America and Europe to mentally cordon off these unlucky places as hinterlands, wastelands, nowheres—or unluckiest of all, as in the case of Nauru, middle of nowheres. For those fortunate enough to find ourselves outside those condemned borders, myself among them, it seemed as if our places—the ones where we live and to which we escape for pleasure (the assumed somewheres, the centers, or best of all, the centers of everywhere)—would not be sacrificed to keep the fossil fuel machine going.

And up until quite recently, that has held up as the grand bargain of the carbon age: the people reaping the bulk of the benefits of extractivism pretend not to see the costs of that comfort so long as the sacrifice zones are kept safely out of view.

But in less than a decade of the extreme energy frenzy and the commodity boom, the extractive industries have broken that unspoken bargain. In very short order, the sacrifice zones have gotten a great deal larger, swallowing ever more territory and putting many people who thought they were safe at risk. Not only that, but several of the largest zones targeted for sacrifice are located in some of the wealthiest and most powerful countries in the world. For instance, Daniel Yergin, energy industry consultant (and author of *The Prize*), euphorically described the newfound capacity to extract oil from "tight rock" formations—usually shale—as being akin to discovering whole new petrostates: "This is like adding another Venezuela or Kuwait by 2020, except these tight oil fields are in the United States."[40]

And of course it's not just the communities next to these new oil fields that are asked to sacrifice. So much oil is now being extracted in the U.S. (or "Saudi America," as some market watchers call it) that the number of rail cars carrying oil has increased by *4111 percent* in just five years, from 9,500 cars in 2008 to an estimated 400,000 in 2013. (Little wonder that significantly more oil spilled in U.S. rail incidents in 2013 than spilled in the previous forty years combined—or that trains engulfed in smoking fireballs have become increasingly frequent sights on the nightly news.) In practice this means that hundreds if not thousands of towns and cities suddenly find themselves in the paths of poorly maintained, underregulated "oil bomb" trains—towns like Quebec's Lac-Mégantic, where, in July 2013, a train carrying seventy-two tank cars of fracked Bakken oil (more flammable than the regular kind) exploded, killing forty-seven people and flattening half of its picturesque downtown. (Former North Dakota governor George Sinner said the oil trains posed a "ridiculous threat" shortly after one blew up near his native town of Casselton.)[41]

The Alberta tar sands, meanwhile, are growing so fast that the industry will soon be producing more of its particular brand of high-carbon oil than current pipeline capacity can handle—which is why it is so determined to push projects like Keystone XL through the U.S. and Northern Gateway through British Columbia. "If there was something that kept me up at night," said Alberta's (then) energy minister Ron Liepert in June 2011, "it would be the fear that before too long we're going to be landlocked in bitumen. We're not going to be an energy superpower if we can't get the oil out of Alberta."[42] But building those pipelines, as we have seen, impacts a huge number of communities: the ones living along thousands of kilometers of proposed pipe, as well as those who live along vast stretches of coastline that would see their waters crowded with oil tankers, courting disaster.

No place, it seems, is off limits, and no extractive activity has set its sights on more new land than hydraulic fracturing for natural gas. To quote Chesapeake Energy's then-CEO Aubrey McClendon, in 2010, "In the last few years we have discovered the equivalent of two Saudi Arabias of oil in the form of natural gas in the United States. Not one, but two."[43] Which is why the industry is fighting to frack wherever it can. The Marcellus Shale, for instance, spans parts of Pennsylvania, Ohio, New York, West Virginia, Virginia, and Maryland. And it is just one of many such massive blankets of methane-rich rock.

The endgame, according to Republican politician Rick Santorum, is to "drill everywhere"—and it shows. As *The Guardian*'s Suzanne Goldenberg reports, "Energy companies have fracked wells on church property, school grounds and in gated developments. Last November, an oil company put a well on the campus of the University of North Texas in nearby Denton, right next to the tennis courts and across the road from the main sports stadium and a stand of giant wind turbines." Fracking now covers so much territory that, according to a 2013 *Wall Street Journal* investigation, "more than 15 million Americans live within a mile of a well that has been drilled and fracked since 2000."[44]

In Canada, the ambitions are just as aggressive. "As of mid-2012, the entire underground subsoil of Montréal, Laval, and Longueuil (three of the main cities in Québec) had been claimed by gas and petrol companies," reports Kim Cornelissen, a former politician turned anti-fracking campaigner in the province. (So far, Quebec's residents have managed to fend off the gas companies with a moratorium.) In Britain, the area under consideration for fracking adds up to about half the entire island. And in July 2013, residents of the northeast of England were enraged to hear their region described as "uninhabited and desolate" in the House of Lords—and therefore eminently deserving of sacrifice. "Certainly in part of the northeast where there's plenty of room for fracking, well away from anybody's residence where we could conduct [it] without any kind of threat to the rural environment," said Lord Howell, who had been an energy advisor to David Cameron's government.[45]

This is coming as a rude surprise to a great many historically privileged people who suddenly find themselves feeling something of what so many frontline communities have felt for a very long time: how is it possible that a big distant company can come to my land and put me and my kids at risk—and never even ask my permission? How can it be legal to put chemicals in the air right where they know children are playing? How is it possible that the state, instead of protecting me from this attack, is sending police to beat up people whose only crime is trying to protect their families?

This unwelcome awakening has made the fossil fuel sector a whole lot of enemies out of onetime friends. People like South Dakota cattle rancher John Harter, who went to court to try to stop TransCanada from burying a portion of the Keystone XL pipeline on his land. "I've never considered myself a bunny hugger," he told a reporter, "but I guess if that's what I've got to be called now, I'm OK with it." The industry has also alienated people like Christina Mills, who worked as an auditor for oil companies in Oklahoma for much of her career. But when a gas company started fracking in her middle-class North Texas subdivision, her views of the sector changed. "They made it personal here, and that's when I had a problem. . . . They came into the back of our neighbourhood, 300ft from the back fence. That is so intrusive."[46]

And fracking opponents could only laugh when, in February 2014, it emerged that none other than Exxon CEO Rex Tillerson had quietly joined a lawsuit opposing fracking-related activities near his $5 million Texas home, claiming it would lower property values. "I would like to officially welcome Rex to the 'Society of Citizens Really Enraged When Encircled by Drilling' (SCREWED)," wrote Jared Polis, a Democratic Congressman from Colorado, in a sardonic statement. "This select group of everyday citizens has been fighting for years to protect their property values, the health of their local communities, and the environment. We are thrilled to have the CEO of a major international oil and gas corporation join our quickly multiplying ranks."[47]

In 1776, Tom Paine wrote in his rabble-rousing pamphlet *Common Sense*, "It is the good fortune of many to live distant from the scene of sorrow."[48] Well, the distance is closing, and soon enough no one will be safe from the sorrow of ecocide. In a way, the name of the company at the center of Greece's anti-mining movement says it all: Eldorado—a reference to the legendary "lost city of gold" that drove the conquistadors to some of their bloodiest massacres in the Americas. This kind of pillage used to be reserved for non-European countries, with the loot returned to the motherland in Europe. But as Eldorado's activities in northern Greece make clear, today the conquistadors are pillaging on their home turf as well.

That may prove to have been a grave strategic error. As Montana-based environmental writer and activist Nick Engelfried puts it, "Every fracking well placed near a city's water supply and every coal train rolling through a small town gives some community a reason to hate fossil industries. And by failing to notice this, oil, gas and coal companies may be digging their political graves."[49]

None of this means that environmental impacts are suddenly evenly distributed. Historically marginalized people in the Global South, as well as communities of color in the Global North, are still at far greater risk of living downstream from a mine, next door to a refinery, or next to a pipeline, just as they are more vulnerable to the impacts of climate change. But in the era of extreme energy, there is no longer the illusion of discreet sacrifice zones anymore. As Deeohn Ferris, formerly with the Lawyers' Committee for Civil Rights Under Law, aptly put it, "we're all in the same sinking boat, only people of color are closest to the hole."[50]

Another boundary breaker is, of course, climate change. Because while there are still plenty of people who are fortunate enough to live somewhere that is not (yet) directly threatened by the extreme energy frenzy, no one is exempt from the real-world impacts of increasingly extreme weather, or from the simmering psychological stress of knowing that we may very well grow old—and our young children may well grow up—in a climate significantly more treacherous than the one we currently enjoy. Like an oil spill that spreads from open water into wetlands, beaches, riverbeds, and down to the ocean floor, its toxins reverberating through the lifecycles of countless species, the sacrifice zones created by our collective fossil fuel dependence are creeping and spreading like great shadows over the earth. After two centuries of pretending that we could quarantine the collateral damage of this filthy habit, fobbing the risks off on others, the game is up, and we are all in the sacrifice zone now.

Choked in Enemy Territory

The fossil fuel industry's willingness to break the sacrifice bargain in order to reach previously off-limits pools of carbon has galvanized the new climate movement in several important ways. For one, the scope of many new extraction and transportation projects has created opportunities for people whose voices are traditionally shut out of the dominant conversation to form alliances with those who have significantly more social power. Tar sands pipelines have proven to be a particularly potent silo buster in this regard, and something of a gift to political organizing.

Beginning in northern Alberta, in a region where the worst impacts are being felt by Indigenous people, and often ending in places where the worst health impacts are felt by urban communities of color, these pipelines pass a whole lot of other places in between. After all, the same piece of infrastructure will travel through multiple states or provinces (or both); through the watersheds of big cities and tiny towns; through farmlands and fishing rivers; through more lands claimed by Indigenous people and through land occupied by the upper middle class. And despite their huge differences, everyone along the route is up against a common threat and therefore are potential allies. In the 1990s, it was trade deals that brought huge and un-likely coalitions together; today it is fossil fuel infrastructure.

Before the most recent push into extreme energy, Big Oil and Big Coal had grown accustomed to operating in regions where they are so eco-nomically omnipotent that they pretty much ran the show. In places like Louisiana, Alberta, and Kentucky—not to mention Nigeria and, until the Chávez era, Venezuela—the fossil fuel companies treat politicians as their unofficial PR wings and the judiciaries as their own personal legal depart-ments. With so many jobs, and such a large percentage of the tax base on the line, regular people put up with an awful lot too. For instance, even after the Deepwater Horizon disaster, many Louisianans wanted higher safety standards and a bigger share of the royalties from offshore oil wealth—but most didn't join calls for a moratorium on deepwater drilling, despite all they had suffered.[51]

This is the Catch-22 of the fossil fuel economy: precisely because these activities are so dirty and disruptive, they tend to weaken or even destroy other economic drivers: fish stocks are hurt by pollution, the scarred land-scape becomes less attractive to tourists, and farmland becomes unhealthy. But rather than spark a popular backlash, this slow poisoning can end up strengthening the power of the fossil fuel companies because they end up being virtually the only game in town.

As the extractive industries charge into territories previously considered out of bounds, however, they are suddenly finding themselves up against people who are far less compromised. In many of the new carbon frontiers, as well as in territories through which fossil fuel companies must move their product, the water is still relatively clean, the relationship to the land is still strong—and there are a great many people willing to fight very hard to protect ways of life that they view as inherently incompatible with toxic extraction.

For instance, one of the natural gas industry's biggest strategic mistakes was deciding it wanted to frack in and around Ithaca, New York—a liberal college town with a vibrant economic localization movement and blessed with breathtaking gorges and waterfalls. Faced with a direct threat to its idyllic community, Ithaca became not just a hub for anti-fracking activ-ism but a center for serious academic research into the unexplored risks:

it's likely no coincidence that researchers at Cornell University, based in Ithaca, produced the game-changing study on methane emissions linked to fracking, whose findings became an indispensable tool for the global resistance movement. And it was the industry's great misfortune that famed biologist and author Sandra Steingraber, a world-renowned expert on the link between industrial toxins and cancer, had recently taken up a post at Ithaca College. Steingraber threw herself into the fracking fight, providing expert testimony before countless audiences and helping to mobilize tens of thousands of New Yorkers. This work contributed to not just keeping the frackers out of Ithaca but to a total of nearly 180 fracking bans or moratoria adopted by cities and towns across the state.[52]

The industry badly miscalculated again when it began construction on a 12,260-horsepower compressor station carrying Pennsylvania's fracked gas smack in the middle of the town of Minisink, New York. Many homes were within half a mile of the facility, including one just 180 meters away. And the town's residents weren't the only ones whose health was threatened by the station. The surrounding area is prized agricultural land dotted with small family farms, orchards, and vineyards growing organic and artisanal produce for New York's farmer's markets and locavore restaurants. So Millennium Pipeline—the company behind the compressor—found itself up against not just a bunch of angry, local farmers but also a whole lot of angry New York City hipsters, celebrity chefs, and movie stars like Mark Ruffalo, calling not just for an end to fracking but for the state to shift to 100 percent renewables.[53]

And then there was the almost unfathomably stupid idea of trying to open up some of Europe's first major fracking operations nowhere other than the South of France. When residents of the Department of Var—known for its olives, figs, sheep, and for the famed beaches of Saint-Tropez—discovered that several of their communities were in line for gas fracking, they organized furiously. Economist and activist Maxime Combes describes scenes around southern France at the inception of the movement, where "the halls of the town-meetings in impacted communities were packed to overflowing, and very often, there were more participants in these meetings than inhabitants in the villages." Var, Combes wrote, would soon experience "the largest citizen's mobilization seen in the history of a Department that is usually on the right of the political spectrum." As a result of the industry's French folly, it ended up not just losing the right to frack near the Riviera (at least for now), but in 2011 France became the first country to adopt a nationwide fracking ban.[54]

Even something as routine as getting heavy machinery up to northern Alberta to keep the tar sands mines and upgraders running has ignited new resistance movements. In keeping with the mammoth scale of everything associated with the largest industrial project on earth, the machines being transported, which are manufactured in South Korea, can be about as long and heavy as a Boeing 747, and some of the "heavy hauls," as they are called, are three stories high. The shipments are so large, in fact, that these behemoths cannot be trucked normally. Instead, oil companies like Exxon-Mobil have to load them onto specialty trailers that take up more than two lanes of highway, and are too high to make it under most standard overpasses.[55]

The only roads that meet the oil companies' needs are located in distinctly hostile territory. For instance, communities in Montana and Idaho have led a fierce multi-year campaign to prevent the rigs from traveling along the scenic but narrow Highway 12. They object to the human costs of having their critical roadway blocked for hours so that the huge machines can pass, as well as to the environmental risks of a load toppling on one of many hairpin turns and ending up in a stream or river (this is fly-fishing country and locals are passionate about their wild rivers).

In October 2010, a small crew of local activists took me on a drive along the part of Highway 12 that the so-called big rigs would have to travel. We went past groves of cedar and Douglas fir and glowing, golden-tipped larch, past signs for moose crossings and under towering rock outcroppings. As we drove, with fall leaves rushing downstream in Lolo Creek next to the road, my guides scouted locations for an "action camp" they were planning. It would bring together anti–tar sands activists from Alberta, ranchers, and Indigenous tribes all along the proposed route of the Keystone XL pipeline, and locals interested in stopping the big rigs on Highway 12. They discussed a friend who had offered to set up a mobile kitchen and the logistics of camping in early winter. Marty Cobenais, then the pipeline campaigner for the Indigenous Environmental Network, explained how all the campaigns are connected. "If they can stop the rigs here then it affects the [production] capacity in the tar sands to get the oil to put in the pipelines." Then he smiles. "That's why we are building a Cowboys and Indians alliance."[56]

Following a long fight, the rigs were ultimately barred from this section of Highway 12 after the Nez Perce tribe and the conservation group Idaho Rivers United filed a joint lawsuit. "They made a huge mistake trying to go through western Montana and Idaho," Alexis Bonogofsky, a Billings, Montana, based goat rancher and activist, told me. "It's been fun to watch."[57]

An alternate route for the huge trucks was eventually found, this one taking them through eastern Oregon. Another bad move. When the first load made its way through the state in December 2013, it was stopped several times by activist lockdowns and blockades. Members of the Confederated Tribes of the Umatilla Indian Reservation, objecting to the loads crossing their ancestral lands, led a prayer ceremony near the second shipment in Pendleton, Oregon. And though local concerns about the safety of the big rigs were real, many participants were clear that they were primarily motivated by fears over what these machines were helping to do to our climate once they arrived at their destination. "This has gone too far," said one Umatilla blockader before she was arrested. "Our children are going to die from this."[58]

Indeed, the oil and coal industries are no doubt cursing the day that they ever encountered the Pacific Northwest—Oregon, Washington State, and British Columbia. There the sector has had to confront a powerful combination of resurgent Indigenous Nations, farmers, and fishers whose livelihoods depend on clean water and soil, and a great many relative newcomers who have chosen to live in that part of the world because of its natural beauty. It is also, significantly, a region where the local environmental movement never fully succumbed to the temptations of the corporate partnership model, and where there is a long and radical history of land-based direct action to stop clear-cut logging and dirty mining.

This has meant fierce opposition to tar sands pipelines, as we have seen. And the deep-seated ecological values of the Pacific Northwest have also become the bane of the U.S. coal industry in recent years. Between grassroots resistance to building new coal-fired plants, and pressure to shut down old ones, as well as the rapid rise of natural gas, the market for coal in the United States has collapsed. In a span of just four years, between 2008 and 2012, coal's share of U.S. electricity generation plummeted from about 50 percent to 37 percent. That means that if the industry is to have a future, it needs to ship U.S. coal to parts of the world that still want it in large quantities. That means Asia. (It's a strategy that global energy expert and author Michael T. Klare has compared to the one tobacco companies began to employ a few decades ago: "Just as health officials now condemn Big Tobacco's emphasis on cigarette sales to poor people in countries with inadequate health systems," he writes, "so someday Big Energy's new 'smoking' habit will be deemed a massive threat to human survival.") The problem for the coal companies is that U.S. ports along the Pacific Coast are not equipped for such large coal shipments, which means that the industry needs to build new terminals. It also needs to dramatically increase the number of trains carrying coal from the massive mines of the Powder River Basin, in Wyoming and Montana, to the Northwest.[59]

As with the tar sands pipelines and the heavy hauls, the greatest obstacle to the coal industry's plans to reach the sea has been the defiant refusal of residents of the Pacific Northwest to play along. Every community in Washington State and Oregon that was slated to become the new home of a coal export terminal rose up in protest, fueled by health concerns about coal dust, but also, once again, by larger concerns about the global impact of burning all that coal.

This was expressed forcefully by KC Golden, who has helped to usher in many of the most visionary climate policies in Washington State, when he wrote: "The great Pacific Northwest is not a global coal depot, a pusher for fossil fuel addiction, a logistics hub for climate devastation. We're the last place on Earth that should settle for a tired old retread of the false choice between jobs and the environment. Coal export is fundamentally inconsistent with our vision and values. It's not just a slap in the face to 'green' groups. It's a moral disaster and an affront to our identity as a community."[60] After all, what is the point of installing solar panels and rainwater barrels if they are going to be coated in coal dust?

What these campaigns are discovering is that while it's next to impossible to win a direct fight against the fossil fuel companies on their home turf, the chances of victory greatly increase when the battleground extends into a territory where the industry is significantly weaker—places where nonextractive ways of life still flourish and where residents (and politicians) are less addicted to petro and coal dollars. And as the corroded tentacles of extreme energy reach out in all directions like a giant metal spider, the industry is pushing into a whole lot of those kinds of places.

Something else is going on too. As resistance to the extractive industries gains ground along these far-flung limbs, it is starting to spread back to the body of carbon country—lending new courage to resist even in those places that the fossil fuel industry thought it had already conquered.

The city of Richmond, California, across the bay from San Francisco, provides a glimpse of how quickly the political landscape can change. Predominantly African American and Latino, the city is a rough-edged, working-class pocket amidst the relentless tech-fuelled gentrification of the Bay Area. In Richmond, the big employer isn't Google, it's Chevron, whose huge refinery local residents blame for myriad health and safety problems, from elevated asthma rates to frequent accidents at the hulking facility (including a massive fire in 1999 that sent hundreds to hospital). And yet as the city's largest business and employer Chevron still had the power to call the shots.[61]

No more. In 2009 community members successfully blocked a plan by Chevron to significantly expand its oil refinery, which could have allowed the plant to process heavier, dirtier crudes such as bitumen from the tar sands. A coalition of environmental justice groups challenged the expansion in the streets and in the courts, arguing that it would further pollute Richmond's air. In the end, a superior court ruled against Chevron, citing a wholly inadequate environmental impact report (which "fails as an informational document," the judge tartly remarked). Chevron appealed, but in 2010 it lost again. "This is a victory for the grassroots, and the people who have been suffering the health impacts of the refinery for the past 100 years," said Asian Pacific Environmental Network senior organizer Torm Nompraseurt.[62]

Richmond is not the only place dominated by Big Oil finding new reserves of courage to fight back. As the anti–tar sands movement spreads through North America and Europe, Indigenous communities in the belly of the beast—the ones who were raising the alarm about the dangers of the tar sands long before large environmental groups showed any interest in the issue—have also been emboldened to go further than ever. They've launched new lawsuits for violations of their land rights, with potentially grave ramifications for industry's access to carbon reserves, and delegations from deeply impacted First Nations communities are now constantly traveling the globe to alert more people to the devastation of their territories in the hopes that more arteries will be severed. One of these activists is Melina Laboucan-Massimo, a mesmerizing speaker with an understated courage who has spent much of her early thirties on the road, showing ugly slides of oil spills and ravaged landscapes and describing the silent war the oil and gas industry is waging on her people, the Lubicon Lake First Nation. "People are listening now," she told me, with tears in her eyes in the summer of 2013. "But it took a long time for people to get to that place." And this, she said, means that "there is hope. But it can be pretty dire sometimes in Alberta."[63]

What is clear is that fighting a giant extractive industry on your own can seem impossible, especially in a remote, sparsely populated location. But being part of a continent-wide, even global, movement that has the industry surrounded is a very different story.

This networking and cross-pollinating is usually invisible—it's a mood, an energy that spreads from place to place. But for a brief time in September 2013, Blockadia's web of inspiration was made visible. Five carvers from the Lummi Nation in Washington State—the coastal tribe that is leading the fight against the largest proposed coal export terminal on a contested piece of the West Coast—showed up in Otter Creek, Montana. They had traveled roughly 1,300 kilometers from their home territory of mountainous temperate rainforest and craggy Pacific beaches to southeastern Montana's parched grasses and gentle hills, carrying with them a twenty-two-foot cedar totem pole, strapped to a flatbed truck. Otter Creek is the site of a planned massive coal mine and the Lummi visitors stood on that spot, which until recently had been written off as doomed, with more than a hundred people from the nearby Northern Cheyenne Reservation, as well as a group of local cattle ranchers. Together, they explored the ways in which they had been brought together by the ambitions of the carbon frenzy.

If the Otter Creek mine were built in the Powder River Basin, it would compromise the water and air for the ranchers and the Northern Cheyenne, and the railway transporting the coal to the west coast could disturb the Cheyenne's ancient burial grounds. The export port, meanwhile, was set to be built on one of the Lummi's ancient burial grounds, and the coal would then be carried on barges that would disrupt their fishing areas and potentially threaten many livelihoods.

The group stood in the valley by the banks of Otter Creek, under a sunny sky with hawks flying overhead, and blessed the totem pole with pipe smoke, vowing to fight together to keep the coal under their feet in the ground, and to keep both the railway and port from being built. The Lummi carvers then strapped the totem pole—which they had named Kwel hoy' or "We Draw the Line"—back onto the truck and took it on a sixteen-day journey to eight other communities, all of whom found themselves in the path of coal trains, big rigs, or tar sands pipelines and oil tankers. There were ceremonies at every stop, as the visitors and their hosts—both Native and non-Native—together drew connections among their various local battles against the extractive industries. The journey ended on Tsleil-Waututh land in North Vancouver, a pivotal community in the fight against increased oil tanker traffic. There the totem pole was permanently planted, looking out at the Pacific.

While in Montana, Lummi master carver Jewell Praying Wolf James explained the purpose of the long journey: "We're concerned about protecting the environment as well as people's health all the way from the Powder River to the West Coast. . . . We're traveling across the country to help unify people's voices. It doesn't matter who you are, where you are at or what race you are—red, black, white or yellow—we're all in this together."[64]

———

This kind of alliance building among the various outposts of Blockadia has proven the movement's critics wrong time and time again. When the campaign against the Keystone XL pipeline began to gather momentum, several high-profile pundits insisted that it was all a waste of valuable time and energy. The oil would get out through another route regardless, and in the grand scheme of things the carbon it would carry represented little more than "a rounding error," as Jonathan Chait wrote in *New York* magazine. Better, they argued, to fight for a carbon tax, or for stronger EPA regulations, or for a reincarnation of cap-and-trade. *New York Times* columnist Joe Nocera went so far as to call the strategy "utterly boneheaded," and accused James Hansen, whose congressional testimony launched the modern climate movement, of "hurting the very cause he claims to care so much about."[65]

What we now know is that Keystone was always about much more than a pipeline. It was a new fighting spirit, and one that is contagious. One battle doesn't rob from another but rather causes battles to multiply, with each act of courage, and each victory, inspiring others to strengthen their resolve.

The BP Factor: No Trust

Beyond the fossil fuel industry's pace of expansion, and its forays into hostile territory, something else has propelled this movement forward in recent years. That is the widespread conviction that today's extractive activities are significantly higher risk than their predecessors: tar sands oil is unquestionably more disruptive and damaging to local ecosystems than conventional crude. Many believe it to be more dangerous to transport, and once spilled harder to clean up. A similar risk escalation is present in the shift to fracked oil and gas; in the shift from shallow to deepwater drilling (as the BP disaster showed); and most dramatically, in the move from warm water to Arctic drilling. Communities in the path of unconventional energy projects are convinced they are being asked to risk a hell of a lot, and much of the time they are being offered very little in return for their sacrifice, whether lasting jobs or significant royalties.

Industry and government, for their part, have been extremely reluctant to acknowledge, let alone act upon, the stepped-up risks of extreme energy. For years, rail companies and officials have largely treated fracked oil from the Bakken as if it were the same as conventional crude—never mind the mounting evidence that it is significantly more volatile. (After announcing some mostly voluntary new safety measures beginning in early 2014 that were generally deemed inadequate, U.S. regulators claim to be in the process of developing a variety of tougher rules for oil-by-rail transport.)[66]

Similarly, government and industry are pushing the vast expansion of pipelines carrying oil from the Alberta tar sands despite a paucity of reliable, peer-reviewed research assessing whether dilbit, as diluted bitumen is called, is more prone to spill than conventional oil. But there is good reason for concern. As a joint 2011 report published by the Natural Resources Defense Council, the Sierra Club, and others notes, "There are many indications that dilbit is significantly more corrosive to pipeline systems than conventional crude. For example, the Alberta pipeline system has had approximately sixteen times as many spills due to internal corrosion as the U.S. system. Yet, the safety and spill response standards used by the United States to regulate pipeline transport of bitumen are designed for conventional oil."[67]

Meanwhile, there are huge gaps in our knowledge about how spilled tar sands oil behaves in water. Over the last decade, there have been few studies published on the subject, and almost all were commissioned by the oil industry. However, a recent investigation by Environment Canada contained several disturbing findings, including that diluted tar sands oil sinks in saltwater "when battered by waves and mixed with sediments" (rather than floating on the ocean surface where it can be partially recovered) and that dispersants like those used during BP's Deepwater Horizon disaster have only "a limited effect," according to a report in The Globe and Mail. And there has been virtually no formal research at all on the particular risks of transporting tar sands oil via truck or rail.[68]

Similarly, large knowledge gaps exist in our understanding of the ecological and human health impact of the Alberta tar sands themselves, with their enormous open-pit mines, dump trucks that can reach up to five stories high, and roaring upgraders. In huge swaths of country surrounding Fort McMurray, ground zero of Canada's bitumen boom, the boreal forest—once a verdant, spongy bog—has been sucked dry of life. Every few minutes, the rancid air is punctured by the sound of booming cannons, meant to keep migrating birds from landing on the strange liquid silver surface of the huge tailing ponds.* [69] In Alberta the centuries-old war to control nature is not a metaphor; it is a very real war, complete with artillery.

The oil companies, of course, say that they are using the safest methods of environmental protection; that the vast tailings ponds are secure; that water is still safe to drink (though workers stick to bottled); that the land will soon be "reclaimed" and returned to moose and black bears (if any are still around). And despite years of complaints from First Nations communities like the Athabasca Chipewyan, situated downstream from the mines along the Athabasca River, industry and government continued to insist that whatever organic contaminants are found in the river are "naturally occurring"—this is an oil-rich region after all.

To anyone who has witnessed the scale of the tar sands operation, the assurances seem implausible. The government has yet to establish a genuinely independent, comprehensive system for monitoring mining impacts on the surrounding watersheds—in an industrial project whose total worth is approaching $500 billion. After it announced a flashy new federal-provincial monitoring program in 2012, the PR effort quickly spiraled out of its control. Referring to new findings from government and independent researchers, Bill Donahue, an environmental scientist with an advisory role in the program, said in February 2014 that "not only are those tailings ponds leaking, but it looks like it is flowing pretty much from those tailings ponds, through the ground and into the Athabasca River." He added: "So, there goes . . . that message we've been hearing about. 'These tailings ponds are safe, they don't leak,' and so on." In a separate incident, a team of government scientists with Environment Canada corroborated outside research on widespread contamination of snow around tar sands operations, though

---

* In 2008, 1,600 ducks died after they landed in these dangerous waters during a storm; another incident led to the deaths of over five hundred more two years later. (A biologist investigating the later incident for the Alberta government explained that it was not industry's fault that the ducks were forced to land during a violent storm—then pointed out, without apparent irony, that such storms will become more frequent as a result of climate change.)

the Harper administration did its best to keep the researchers from speaking to the press.[70]

And there are still *no* comprehensive studies on the impacts of this pollution on human health. On the contrary, some who have chosen to speak out have faced severe reprisals. Most notable had been the experience of John O'Connor, a gentle, gray-bearded family doctor who still speaks with an accent from his native Ireland. In 2003, O'Connor began to report that, while treating patients in Fort Chipewyan, he was coming across alarming numbers of cancers, including extremely rare and aggressive bile-duct malignancies. He quickly found himself under fire from federal health regulators, who filed several misconduct charges against him with the College of Physicians and Surgeons of Alberta (including raising "undue alarm"). "I don't know, personally, of any situation where a doctor has had to go through what I've gone through," O'Connor has said of the reputational smears and the years spent fighting the allegations. He was, eventually, cleared of all charges and a subsequent investigation of cancer rates vindicated several of his warnings.[71]

But before that happened the message to other doctors was sent: a report commissioned by the Alberta Energy Regulator recently found a "marked reluctance to speak out" in the medical community about the health impact of the tar sands, with several interviewees pointing to Dr. O'Connor's experience. ("Physicians are quite frankly afraid to diagnose health conditions linked to the oil and gas industry," concluded the toxicologist who authored the report.) It has become routine, moreover, for the federal government to prevent senior environmental and climate scientists from speaking to journalists about any environmentally sensitive subjects. ("I'm available when media relations says I'm available," as one scientist told Postmedia.)[72]

And this is just one facet of what has become known as Prime Minister Stephen Harper's "war on science," with environmental monitoring budgets relentlessly slashed, covering everything from oil spills and industrial air pollution to the broader impacts of climate change. Since 2008, more than two thousand scientists have lost their jobs as a result of the cuts.[73]

This is, of course, a strategy. Only by systematically failing to conduct basic research, and silencing experts who are properly tasked to investigate health and environmental concerns, can industry and government continue to make absurdly upbeat claims about how all is under control in the oil patch.* [74]

A similar willful blindness pervades the rapid spread of hydraulic fracking. For years the U.S. gas industry responded to reports of contaminated water wells by insisting that there was no scientific proof of any connection between fracking and the fact that residents living near gas drilling suddenly found they could set their tap water on fire. But the reason there was no evidence was because the industry had won an unprecedented exemption from federal monitoring and regulation—the so-called Halliburton Loophole, ushered in under the administration of George W. Bush. The loophole exempted most fracking from regulations of the Safe Drinking Water Act, helping to ensure that companies did not have to report any of the chemicals they were injecting underground to the Environmental Protection Agency, while shielding their use of the riskiest chemicals from EPA oversight. [75] And if no one knows what you are putting into the ground, it's tough to make a definitive link when those toxins start coming out of people's taps.

And yet as more evidence emerges, it is coming down hard on one side. A growing body of independent, peer-reviewed studies is building the case that fracking puts drinking water, including aquifers, at risk. In July 2013, for instance, a Duke University–led paper analyzed dozens of drinking water wells in northeastern Pennsylvania's Marcellus Shale region. The researchers found that the level of contamination from methane, ethane, and propane closely correlated with proximity to wells for shale gas. The industry response is that this is just natural leakage in regions rich in gas (the same line that tar sands operators in Alberta used when organic pollutants are

---

* And their claims are indeed absurd: according to an independent study published in 2014 in the *Proceedings of the National Academy of Sciences*, for example, emissions of potentially toxic pollutants from the tar sands "are two to three orders-of-magnitude larger than those reported" by companies to their regulators. The discrepancy is evident in actual measurements of these pollutants in the air near tar sands activities. The study's coauthor, Frank Wania, an environmental scientist at the University of Toronto, described the official estimates as "inadequate and incomplete" and made the commonsense observation, "Only with a complete and accurate account of the emissions is it actually possible to make a meaningful assessment of the environmental impact and of the risk to human health."

found in the water there). But this study found that while methane was present in most of the sampled water wells, the concentration was *six times higher* in those within a kilometer of a gas well. In a study not yet published, the Duke team also analyzed water wells in Texas that had been previously declared safe. There, they found that contrary to assurances from government and industry, methane levels in many wells exceeded the minimum safety level set by the U.S. Geological Survey.[76]

The links between fracking and small earthquakes are also solidifying. In 2012, a University of Texas research scientist analyzed seismic activity from November 2009 to September 2011 over part of the huge Barnett Shale region in Texas, which lies under Fort Worth and parts of Dallas, and found the epicenters of sixty-seven small earthquakes.[77] The most reliably located earthquakes were within two miles of an injection well. A July 2013 study in the *Journal of Geophysical Research* linked fracking-related waste injection to 109 small earthquakes that took place in a single year around Youngstown, Ohio, where an earthquake had not been previously recorded since monitoring began in the eighteenth century. The lead researcher of a similar study, published in *Science*, explained, "The fluids [in wastewater injection wells] are driving the faults to their tipping point."[78]

All of this illustrates what is so unsettling about unconventional extraction methods. Conventional oil and gas drilling, as well as underground coal mining, are destructive, to be sure. But comparatively speaking, they are the fossil fuel equivalent of the surgeon's scalpel—the carbon is extracted with relatively small incisions. But extreme, or unconventional extraction takes a sledgehammer to the whole vicinity. When the sledgehammer strikes the surface of the land—as in the case of mountaintop coal removal and open-pit tar sands—the violence can be seen with the naked eye. But with fracking, deepwater drilling, and underground ("in situ") tar sands extraction, the sledgehammer aims deep underground. At first this can seem more benign, since the impacts are less visible. Yet over and over again, we are catching glimpses of how badly we are breaking critical parts of our ecosystems that our best experts have no idea how to fix.

Educated by Disaster

In Blockadia outposts around the world, the initials "BP" act as a kind of mantra or invocation—shorthand for: whatever you do, take no extractive company at its word. The initials mean that passivity and trust in the face of assurances about world-class technology and cutting-edge safety measures are recipes for flammable water in your faucet, an oil slick in your backyard, or a train explosion down the street.

Indeed, many Blockadia activists cite the 2010 BP disaster in the Gulf of Mexico as either their political awakening, or the moment they realized they absolutely had to win their various battles against extreme energy. The facts of that case are familiar but bear repeating. In what became the largest accidental marine oil spill in history, a state-of-the-art offshore oil rig exploded, killing eleven workers, while oil gushed from the ruptured Macondo wellhead about one and a half kilometers below the surface. What made the strongest impression on the horrified public was not the tar-coated tourist beaches in Florida or the oil-soaked pelicans in Louisiana. It was the harrowing combination of the oil giant's complete lack of preparedness for a blowout at those depths, as it scrambled for failed fix after failed fix, and the cluelessness of the government regulators and responders. Not only had regulators taken BP at its word about the supposed safety of the operation, but government agencies were so ill-equipped to deal with the scale of the disaster that they allowed BP—the perpetrator—to be in charge of the cleanup. As the world watched, the experts were clearly making it up as they went along.

The investigations and lawsuits that followed revealed that a desire to save money had played an important role in creating the conditions for the accident. For instance, as Washington raced to reestablish lost credibility, an investigation by a U.S. Interior Department agency found "BP's cost or time saving decisions without considering contingencies and mitigation were contributing causes of the Macondo blowout." A report from the specially created Presidential Oil Spill Commission similarly found, "Whether purposeful or not, many of the decisions that BP, [and its contractors] Halliburton and Transocean, made that increased the risk of the Macondo blowout clearly saved those companies significant time (and money)." Jackie

Savitz, a marine scientist and a vice president at the conservation group Oceana, was more direct: BP "put profits before precautions. They let dollar signs drive a culture of risk-taking that led to this unacceptable outcome."[79]

And any notion that this was a problem unique to BP was quickly dispelled when—only ten days after crews stopped the gush of oil into the Gulf of Mexico—an Enbridge pipeline burst in Michigan, causing the largest onshore oil spill in U.S. history. The pipe ruptured in a tributary of the Kalamazoo River and quickly contaminated more than fifty-five kilometers of waterways and wetlands with over one million gallons of oil, which left swans, muskrats, and turtles coated in black gunk. Homes were evacuated, local residents sickened, and onlookers watched "an alarming brown mist rise as river water the shade of a dark chocolate malt tumbled" over a local dam, according to one report.[80]

Like BP, it seemed that Enbridge had put profits before basic safety, while regulators slept at the switch. For instance, it turned out that Enbridge had known as early as 2005 that the section of pipeline that failed was corroding, and by 2009 the company had identified 329 other defects in the line stretching through southern Michigan that were serious enough to require immediate repair under federal rules. The $40 billion company was granted an extension, and applied for a second one just ten days before the rupture—the same day an Enbridge VP told Congress that the company could mount an "almost instantaneous" response to a leak. In fact it took them seventeen hours to close the valve on the leaking pipeline. Three years after the initial disaster, about 180,000 gallons of oil were still sitting on the bottom of the Kalamazoo.[81]

As in the Gulf, where BP had been drilling at depths unheard of just a few years earlier, the Kalamazoo disaster was also linked to the new era of extreme, higher-risk fossil fuel extraction. It took a while, however, before that became clear. For more than a week Enbridge did not share with the public the very pertinent fact that the substance that had leaked was not conventional crude; it was diluted bitumen, piped from the Alberta tar sands through Michigan. In fact in the early days, Enbridge's then CEO, Patrick Daniel, flatly denied that the oil came from the tar sands and was later forced to backtrack. "What I indicated is that it was not what we have traditionally referred to as tar sands oil," Daniel claimed of bitumen that certainly had come from the tar sands. "If it is part of the same geological formation, then I bow to that expert opinion."[82]

In the fall of 2010, with many of these disasters still under way, Marty Cobenais of the Indigenous Environmental Network told me that the summer of spills was having a huge impact on communities in the path of new infrastructure projects, whether big rigs, pipelines, or tankers. "The oil industry always says there is 0 percent chance of their oil hitting the shores, but with BP, we saw that it did. Their projections are always wrong," he said, adding, "They are always talking about 'fail-proof' but with Kalamazoo we saw they couldn't turn it off for hours."[83]

In other words, a great many people are no longer believing what the industry experts tell them; they are believing what they see. And over the last few years we have all seen a whole lot. Unforgettable images from the bizarre underwater "spill-cam" showing BP's oil gushing for three long months into the Gulf merge seamlessly with shocking footage of methane-laced tap water being set on fire in fracking country, which in turn meld with the grief of Quebec's Lac-Mégantic after the horrific train explosion, with family members searching through the rubble for signs of their loved ones, which in turn fade to memories of 300,000 people in West Virginia being told they could not drink or bathe in their tap water for up to ten days after it had been contaminated by chemicals used in coal mining. And then there was the spectacle in 2012 of Shell's first foray into the highest-risk gambit of all: Arctic drilling. Highlights included one of Shell's giant drill rigs breaking free from its tow and running aground on the coast of Sitkalidak Island; another rig slipping its anchorage; and an oil spill containment dome being "crushed like a beer can," according to a U.S. Bureau of Safety and Environmental Enforcement official.[84]

If it seems like there are more such spills and accidents than before, that's because there are. According to a months-long investigation by EnergyWire, in 2012 there were more than six thousand spills and "other mishaps" at onshore oil and gas sites in the U.S. "That's an average of more than 16 spills a day. And it's a significant increase since 2010. In the 12 states where comparable data were available, spills were up about 17 percent." There is also evidence that companies are doing a poorer job of cleaning up their messes: in an investigation of pipeline leaks of hazardous liquids (mostly petroleum-related), *The New York Times* found that in 2005 and 2006 pipeline operators reported "recovering more than 60 percent of liquids spilled"; between 2007 and 2010 "operators recovered less than a third."[85]

It's not just the engineering failures that are feeding widespread mistrust. As with BP and Enbridge, it's the constant stream of revelations about the role that greed—fully liberated by lax regulation and monitoring—seems to have played in stacking the deck. For example, Shell's Arctic rig ran aground when it braved fierce weather in an apparent attempt by the company to get out of Alaska in time to avoid paying additional taxes in the state.[86]

And Montreal, Maine & Atlantic (MM&A), the rail company behind the Lac-Mégantic disaster, had received, one year before the accident, government permission to cut the number of staff on its trains to a single engineer. Until the 1980s, trains like the one that derailed were generally staffed by five employees, all sharing the duties of operating safely. Now it's down to two—but for MM&A, that was still too much. According to one of the company's former railway workers, "It was all about cutting, cutting, cutting." Compounding these risks, according to a four-month *Globe and Mail* investigation, "companies often don't test their oil shipments for explosiveness before sending the trains." Little wonder then that within a year of Lac-Mégantic, several more oil-laden trains went up in flames, including one in Casselton, North Dakota, one outside a village in northwest New Brunswick, and one in downtown Lynchburg, Virginia.[87]

In a sane world, this cluster of disasters, layered on top of the larger climate crisis, would have prompted significant political change. Caps and moratoriums would have been issued, and the shift away from extreme energy would have begun. The fact that nothing of the sort has happened, and that permits and leases are still being handed out for ever more dangerous extractive activities, is at least partly due to old-fashioned corruption—of both the legal and illegal varieties.

A particularly lurid episode was revealed a year and a half before the BP disaster. An internal U.S. government report pronounced that what was then called the Minerals Management Service—the division of the U.S. Interior Department charged with collecting royalty payments from the oil and gas industry—suffered from "a culture of ethical failure." Not only had officials repeatedly accepted gifts from oil industry employees but, according to a report by the department's inspector general, several officials "frequently consumed alcohol at industry functions, had used cocaine and marijuana, and had sexual relationships with oil and gas company representatives." For a public that had long suspected that their public servants were in bed with the oil and gas lobby, this was pretty graphic proof.[88]

Little wonder, then, that a 2013 Harris poll found that a paltry 4 percent of U.S. respondents believe oil companies are "honest and trustworthy" (only the tobacco industry fared worse). That same year, Gallup polled Americans about their opinions of twenty-five industries, including banking and government. No industry was more disliked than the oil and gas sector. A 2012 poll in Canada, meanwhile, asked Canadians to rate each of eleven groups on their trustworthiness on "energy issues." Oil and gas firms and energy executives took the bottom two slots, well below academics (the most trusted group), as well as environmental and community groups (which also rated positively). And in an EU-wide survey that same year, participants were polled on their impressions of eleven different sectors and asked if they "make efforts to behave responsibly towards society"—along with finance and banking, mining and oil and gas companies again came in last place.[89]

Hard realities like these have posed a challenge to the highly paid spin doctors employed by the extractive industries, the ones who had grown accustomed to being able to gloss over pretty much any controversy with sleek advertising showing blond children running through fields and multiracial actors in lab coats expressing concern about the environment. These days that doesn't cut it. No matter how many millions are spent on advertising campaigns touting the modernity of the tar sands or the cleanliness of natural gas, it's clear that a great many people are no longer being persuaded. And those proving most resistant are the ones whose opinions matter most: the people living on lands that the extractive companies need to access in order to keep their astronomical profits flowing.

The Return of Precaution

For decades, the environmental movement spoke the borrowed language of risk assessment, diligently working with partners in business and government to balance dangerous levels of pollution against the need for profit and economic growth. These assumptions about acceptable levels of risk were taken so deeply for granted that they formed the basis of the official climate change discussion. Action necessary to save humanity from the very real risk of climate chaos was coolly balanced against the risk such action would pose to GDPs, as if economic growth still has a meaning on a planet convulsing in serial disasters.

But in Blockadia, risk assessment has been abandoned on the barricaded roadside, replaced by a resurgence of the precautionary principle—which holds that when human health and the environment are significantly at risk, perfect scientific certainty is not required before taking action. Moreover the burden of proving that a practice is safe should not be placed on the public that could be harmed.

Blockadia is turning the tables, insisting that it is up to industry to prove that its methods are safe—and in the era of extreme energy that is something that simply cannot be done. To quote the biologist Sandra Steingraber, "Can you provide an example of an ecosystem on which was laid down a barrage of poisons, and terrible and unexpected consequences for human beings were not the result?"[90]

The fossil fuel companies, in short, are no longer dealing with those Big Green groups that can be silenced with a generous donation or a conscience-clearing carbon offset program. The communities they are facing are, for the most part, not looking to negotiate a better deal—whether in the form of local jobs, higher royalties, or better safety standards. More and more, these communities are simply saying "No." No to the pipeline. No to Arctic drilling. No to the coal and oil trains. No to the heavy hauls. No to the export terminal. No to fracking. And not just "Not in My Backyard" but, as the French anti-fracking activists say: *Ni ici, ni ailleurs*—neither here, nor elsewhere. In other words: no new carbon frontiers.

Indeed the trusty slur NIMBY has completely lost its bite. As Wendell Berry says, borrowing words from E. M. Forster, conservation "turns on affection"—and if each of us loved our homeplace enough to defend it, there would be no ecological crisis, no place could ever be written off as a sacrifice zone.[91] We would simply have no choice but to adopt nonpoisonous methods of meeting our needs.

This sense of moral clarity, after so many decades of chummy green partnerships, is the real shock for the extractive industries. The climate movement has found its nonnegotiables. This fortitude is not just building a large and militant resistance to the companies most responsible for the climate crisis. As we will see in the next chapter, it is also delivering some of the most significant victories the environmental movement has seen in decades.

LOVE WILL SAVE THIS PLACE

Democracy, Divestment, and the Wins So Far

"I believe that the more clearly we can focus our attention on the wonders and realities of the universe about us, the less taste we shall have for destruction."

—Rachel Carson, 1954[1]

"What good is a mountain just to have a mountain?"

—Jason Bostic, Vice President of the West Virginia Coal Association, 2011[2]

On a drizzly British Columbia day in April 2012, a twenty-seven-seat turboprop plane landed at the Bella Bella airport, which consists of a single landing strip leading to a clapboard building. The passengers descending from the blue-and-white Pacific Coastal aircraft included the three members of a review panel created by the Canadian government. They had made the 480-kilometer journey from Vancouver to this remote island community, a place of deep fjords and lush evergreen forests reaching to the sea, to hold public hearings about one of the most contentious new pieces of fossil fuel infrastructure in North America: Enbridge's proposed Northern Gateway pipeline.

Bella Bella is not directly on the oil pipeline's route (that is 200 kilometers even further north). However, the Pacific ocean waters that are its front yard are in the treacherous path of the oil tankers that the pipeline would load up with diluted tar sands oil—up to 75 percent more oil in some supertankers than the *Exxon Valdez* was carrying in 1989 when it spilled in Alaska's Prince William Sound, devastating marine life and fisheries across the region.[3] A spill in these waters could be even more damaging, since the remoteness would likely make reaching an accident site difficult, especially during winter storms.

The appointed members of the Joint Review Panel—one woman and two men, aided by support staff—had been holding hearings about the pipeline impacts for months now and would eventually present the federal government with their recommendation on whether the project should go ahead. Bella Bella, whose population is roughly 90 percent Heiltsuk First Nation, was more than ready for them.

A line of Heiltsuk hereditary chiefs waited on the tarmac, all dressed in their full regalia: robes embroidered with eagles, salmon, orcas, and other creatures of these seas and skies; headdresses adorned with animal masks and long trails of white ermine fur, as well as woven cedar basket hats. They greeted the visitors with a welcome dance, noisemakers shaking in their hands and rattling from the aprons of their robes, while a line of drummers and singers backed them up. On the other side of the chain link fence was a large crowd of demonstrators carrying anti-pipeline signs and canoe paddles.

Standing a respectful half step behind the chiefs was Jess Housty, a slight twenty-five-year-old woman who had helped to galvanize the community's engagement with the panel (and would soon be elected to the Heiltsuk Tribal Council as its youngest member). An accomplished poet who created Bella Bella's first and only library while she was still a teenager, Housty described the scene at the airport as "the culmination of a huge planning effort driven by our whole community."[4]

And it was young people who had led the way, turning the local school into a hub of organizing. Students had worked for months in preparation for the hearings. They researched the history of pipeline and tanker spills, including the 2010 disaster on the Kalamazoo River, noting that Enbridge, the company responsible, was the same one pushing the Northern Gateway pipeline. The teens were also keenly interested in the *Exxon Valdez* disaster since it took place in a northern landscape similar to their own. As a community built around fishing and other ocean harvesting, they were alarmed to learn about how the salmon of Prince William Sound had become sick in the years after the spill, and how herring stocks had completely collapsed (they are still not fully recovered, more than two decades later).

The students contemplated what such a spill would mean on their coast. If the sockeye salmon, a keystone species, were threatened, it would have a cascade effect—since they feed the killer whales and white-sided dolphins whose dorsal fins regularly pierce the water's surface in nearby bays, as well as the seals and sea lions that bark and sunbathe on the rocky outcroppings. And when the fish return to the freshwater rivers and streams to spawn, they feed the eagles, the black bears, the grizzlies, and the wolves, whose waste then provides the nutrients to the lichen that line the streams and riverbanks, as well as to the great cedars and Douglas firs that tower over the temperate rainforest. It's the salmon that connect the streams to the rivers, the river to the sea, the sea back to the forests. Endanger salmon and you endanger the entire ecosystem that depends on them, including the Heiltsuk people whose ancient culture and modern livelihood is inseparable from this intricate web of life.

Bella Bella's students wrote essays on these themes, prepared to present testimony, and painted signs to greet the panel members. Some went on a forty-eight-hour hunger strike to dramatize the stakes of losing their food source. Teachers observed that no issue had ever engaged the community's young people like this—some even noticed a decline in depression and drug use. That's a very big deal in a place that not long ago suffered from a youth suicide epidemic, the legacy of scarring colonial policies, including generations of children—the great-grandparents, grandparents, and sometimes the parents of today's teens and young adults—being taken from their families and placed in church-run residential schools where abuse was rampant.

Housty recalls, "As I stood behind our chiefs [on the tarmac], I remember thinking how the community had grown around the issue from the first moment we heard rumblings around Enbridge Northern Gateway. The momentum had built and it was strong. As a community, we were prepared to stand up with dignity and integrity to be witnesses for the lands and waters that sustained our ancestors—that sustain us—that we believe should sustain our future generations."

After the dance, the panel members ducked into a white minivan that took them on the five-minute drive into town. The road was lined with hundreds of residents, including many children, holding their handmade poster-board signs. "Oil Is Death," "We Have the Moral Right to Say No," "Keep Our Oceans Blue," "Our Way of Life <u>Cannot</u> Be Bought!," "I Can't Drink Oil." Some held drawings of orcas, salmon, even kelp. Many of the signs simply said: "No Tankers." One man thought the panel members weren't bothering to look out the window, so he thumped the side of the van as it passed and held his sign up to the glass.

By some counts, a third of Bella Bella's 1,095 residents were on the street that day, one of the largest demonstrations in the community's history.[5] Others participated in different ways: by harvesting and preparing food for the evening feast, where the panel members were to be honored guests. It was part of the Heiltsuk's tradition of hospitality but it was also a way to show the visitors the foods that would be at risk if just one of those super-tankers were to run into trouble. Salmon, herring roe, halibut, oolichan, crab, and prawns were all on the menu.

Similar scenes had played out everywhere the panel traveled in British Columbia: cities and towns came out in droves, voicing unanimous or near unanimous opposition to the project. Usually First Nations were front and center, reflecting the fact that the province is home to what is arguably the most powerful Indigenous land rights movement in North America, evidenced by the fact that roughly 80 percent of its land remains "unceded," which means that it has never been relinquished under any treaty nor has it ever been claimed by the Canadian state through an act of war.[6]

Yet there was clearly something about the passion of Bella Bella's greeting that unnerved the panel members. The visitors refused the invitation to the feast that evening, and Chief Councilor Marilyn Slett was put in the unenviable position of having to take the microphone and share a letter she had just received from the Joint Review Panel. It stated that the pipeline hearings for which the assembled crowd had all been preparing for months were canceled. Apparently the demonstration on the way from the airport had made the visitors feel unsafe and, the letter stated, "The Panel cannot be in a situation where it is unsure that the crowd will be peaceful." It later emerged that the sound of that single man thumping the side of the van had somehow been mistaken for gunfire. (Police in attendance asserted that the demonstrations had been nonviolent and that there was never any security threat.)[7]

Housty said the news of the cancellation had a "physical impact. We had done everything according to our teachings, and to feel the back of someone's hand could hardly have been more of an insult." In the end, the hearings went ahead but a day and a half of promised meeting time was lost, depriving many community members of their hope of being heard in person.*8

What shocked many of Bella Bella's residents was not just the weird and false accusation of violence; it was the extent to which the entire spirit of their actions seemed to have been misunderstood. When the panel members looked out the van window, they evidently saw little more than a stereotypical mob of angry Indians, wanting to vent their hatred on anyone associated with the pipeline. But to the people on the other side of the glass, holding their paddles and fish paintings, the demonstration had not primarily been about anger or hatred. It had been about love—a collective and deeply felt expression of love for their breathtaking part of the world.

As the young people of this community explained when they finally got the chance, their health and identity were inextricably bound up in their ability to follow in the footsteps of their forebears—fishing and paddling in the same waters, collecting kelp in the same tidal zones in the outer coastal islands, hunting in the same forests, and collecting medicines in the same meadows. Which is why Northern Gateway was seen not simply as a threat to the local fishery but as the possible undoing of all this intergenerational healing work. And therefore as another wave of colonial violence.

When Jess Housty testified before the Enbridge Gateway review panel (she had to travel for a full day to Terrace, British Columbia to do it), she put this in unequivocal terms.

When my children are born, I want them to be born into a world where hope and transformation are possible. I want them to be born into a world where stories still have power. I want them to grow up able to be Heiltsuk in every sense of the word. To practice the customs and understand the identity that has made our people strong for hundreds of generations.

---

* When a make-up hearing was scheduled by the Joint Review Panel months later, it was held in a predominantly white community elsewhere in the province.

That cannot happen if we do not sustain the integrity of our territory, the lands and waters, and the stewardship practices that link our people to the landscape. On behalf of the young people in my community, I respectfully disagree with the notion that there is any compensation to be made for the loss of our identity, for the loss of our right to be Heiltsuk.[9]

The power of this ferocious love is what the resource companies and their advocates in government inevitably underestimate, precisely because no amount of money can extinguish it. When what is being fought for is an identity, a culture, a beloved place that people are determined to pass on to their grandchildren, and that their ancestors may have paid for with great sacrifice, there is nothing companies can offer as a bargaining chip. No safety pledge will assuage; no bribe will be big enough. And though this kind of connection to place is surely strongest in Indigenous communities where the ties to the land go back thousands of years, it is in fact Blockadia's defining feature.

I saw it shine brightly in Halkidiki, Greece, in the struggle against the gold mine. There, a young mother named Melachrini Liakou—one of the movement's most tireless leaders—told me with unswerving confidence that the difference between the way she saw the land, as a fourth-generation farmer, and the way the mining company saw the same patch of earth, was that, "I am a part of the land. I respect it, I love it and I don't treat it as a useless object, as if I want to take something out of it and then the rest will be waste. Because I want to live here this year, next year, and to hand it down to the generations to come. In contrast, Eldorado, and any other mining company, they want to devour the land, to plunder it, to take away what is most precious for themselves."[10] And then they would leave behind, she said, "a huge chemical bomb for all mankind and nature."

Alexis Bonogofsky (who had told me what a "huge mistake" the oil companies made in trying to bring their big rigs along Highway 12) speaks in similar terms about the fight to protect southeastern Montana from mining companies like Arch Coal. But for Bonogofsky, a thirty-three-year-old goat rancher and environmentalist who does yoga in her spare time, it's less about farming than deer hunting. "It sounds ridiculous but there's this one spot where I can sit on the sandstone rock and you know that the mule deer are coming up and migrating through, you just watch these huge herds come through, and you know that they've been doing that for thousands and thousands of years. And you sit there and you feel connected to that. And sometimes it's almost like you can feel the earth breathe." She adds: "That connection to this place and the love that people have for it, that's what Arch Coal doesn't get. They underestimate that. They don't understand it so they disregard it. And that's what in the end will save that place. Is not the hatred of the coal companies, or anger, but love will save that place."[11]

This is also what makes Blockadia conflicts so intensely polarized. Because the culture of fossil fuel extraction is—by both necessity and design—one of extreme rootlessness. The workforce of big rig drivers, pipefitters, miners, and engineers is, on the whole, highly mobile, moving from one worksite to the next and very often living in the now notorious "man camps"—self-enclosed army-base-style mobile communities that serve every need from gyms to movie theaters (often with an underground economy in prostitution).

Even in places like Gillette, Wyoming, or Fort McMurray, Alberta, where extractive workers may stay for decades and raise their kids, the culture remains one of transience. Almost invariably, workers plan to leave these blighted places as soon as they have saved enough money—enough to pay off student loans, to buy a house for their families back home, or, for the really big dreamers, enough to retire. And with so few well-paying blue-collar jobs left, these extraction jobs are often the only route out of debt and poverty. It's telling that tar sands workers often discuss their time in northern Alberta as if it were less a job than a highly lucrative jail term: there's "the three-year plan" (save $200,000, then leave); "the five-year-plan" (put away half a million); "the ten-year-plan" (make a million and retire at thirty-five). Whatever the details (and however unrealistic, given how much money disappears in the city's notorious party scene), the plan is always pretty much the same: tough it out in Fort Mac (or Fort McMoney as it is often called), then get the hell out and begin your real life. In one survey, 98 percent of respondents in the tar sands area said they planned to retire somewhere else.[12]

There is a real sadness to many of these choices: beneath the bravado of the bar scene are sky-high divorce rates due to prolonged separations and intense work stress, soaring levels of addiction, and a great many people wishing to be anywhere but where they are. This kind of disassociation is part of what makes it possible for decent people to inflict the scale of damage to the land that extreme energy demands. A coalfield worker in Gillette, Wyoming, for instance, told me that to get through his workdays, he had trained himself to think of the Powder River Basin as "another planet."[13] (The moonscape left behind by strip mining no doubt made this mental trick easier).

These are perfectly understandable survival strategies—but when the extractive industry's culture of structural transience bumps up against a group of deeply rooted people with an intense love of their homeplace and a determination to protect it, the effect can be explosive.

Love and Water

When these very different worlds collide, one of the things that seems to happen is that, as in Bella Bella, communities begin to cherish what they have—and what they stand to lose—even more than before the extractive threat arrived. This is particularly striking because many of the people waging the fiercest anti-extraction battles are, at least by traditional measures, poor. But they are still determined to defend a richness that our economy has not figured out how to count. "Our kitchens are filled with homemade jams and preserves, sacks of nuts, crates of honey and cheese, all produced by us," Doina Dediu, a Romanian villager protesting fracking, told a reporter. "We are not even that poor. Maybe we don't have money, but we have clean water and we are healthy and we just want to be left alone."[14]

So often these battles seem to come to this stark choice: water vs. gas. Water vs. oil. Water vs. coal. In fact, what has emerged in the movement against extreme extraction is less an anti–fossil fuels movement than a pro-water movement.

I was first struck by this in December 2011 when I attended a signing ceremony for the Save the Fraser Declaration, the historic Indigenous people's declaration pledging to prevent the Northern Gateway pipeline and any other tar sands project of its kind from accessing British Columbia territory. More than 130 First Nations have signed, along with many nonIndigenous endorsers. The ceremony was held at the Vancouver Public Library, with several chiefs present to add their names. Among those addressing the bank of cameras that day was Marilyn Baptiste, then elected chief of Xeni Gwet'in, one of the communities of the Tsilhqot'in First Nation. She introduced herself, her people, and their stake in the fight by naming interconnected bodies of water: "We are at the headwaters of Chilko, which is one of the largest wild salmon runs, that is also part of the Taseko, that drains into the Chilko, the Chilko into the Chilcotin, and into the Fraser. It's common sense for all of our people to join together."[15]

The point of drawing this liquid map was clear to all present: of course all of these different nations and groups would join together to fight the threat of an oil spill—they are all already united by water; by the lakes and rivers, streams and oceans that drain into one another. And in British Columbia, the living connection among all of these waterways is the salmon, that remarkably versatile traveler, which moves through fresh- and saltwater and back again during its life cycle. That's why the declaration that was being signed was not called the "Stop the Tankers and Pipelines Declaration" but rather the "Save the Fraser Declaration"—the Fraser, at almost 1,400 kilometers, being the longest river in B.C. and home to its most productive salmon fishery. As the declaration states: "A threat to the Fraser and its headwaters is a threat to all who depend on its health. We will not allow our fish, animals, plants, people, and ways of life to be placed at risk. . . . We will not allow the proposed Enbridge Northern Gateway Pipelines, or similar Tar Sands projects, to cross our lands, territories and watersheds, or the ocean migration routes of Fraser River salmon."[16]

If the tar sands pipeline threatens to become an artery of death, carrying poison across an estimated one thousand waterways, then these interconnected bodies of water that Chief Baptiste was mapping are arteries of life, flowing together to bind all of these disparate communities in common purpose.[17]

The duty to protect water doesn't just unite opposition to this one pipeline; it is the animating force behind every single movement fighting extreme extraction. Whether deepwater drilling, fracking, or mining;

whether pipelines, big rigs, or export terminals, communities are terrified about what these activities will do to their water systems. This fear is what binds together the southeastern Montana cattle ranchers with the Northern Cheyenne with the Washington State communities fighting coal trains and export terminals. Fear of contaminated drinking water is what kick-started the anti-fracking movement (and when a proposal surfaced that would allow the drilling of roughly twenty thousand fracking wells in the Delaware River Basin—the source of freshwater for fifteen million Americans—it is what kicked the movement squarely into the U.S. mainstream).[18]

The movement against Keystone XL would, similarly, never have resonated as powerfully as it did had TransCanada not made the inflammatory decision to route the pipeline through the Ogallala Aquifer—a vast underground source of freshwater beneath the Great Plains that provides drinking water to approximately two million people and supplies roughly 30 percent of the country's irrigation groundwater.[19]

In addition to the contamination threats, almost all these extractive projects also stand out simply for how much water they require. For instance, it takes 2.3 barrels of water to produce a single barrel of oil from tar sands mining—much more than the 0.1 to 0.3 barrels of water needed for each barrel of conventional crude. Which is why the tar sands mines and upgrading plants are surrounded by those giant tailings "ponds" visible from space. Fracking for both shale gas and "tight oil" similarly requires far more water than conventional drilling and is much more water-intensive than the fracking methods used in the 1990s. According to a 2012 study, modern fracking "events" (as they are called) use an average of five million gallons of water—"70 to 300 times the amount of fluid used in traditional fracking." Once used, much of this water is radioactive and toxic. In 2012, the industry created 280 billion gallons of such wastewater in the U.S. alone— "enough to flood all of Washington DC beneath a 22ft deep toxic lagoon," as The Guardian noted.[20]

In other words, extreme energy demands that we destroy a whole lot of the essential substance we need to survive—water—just to keep extracting more of the very substances threatening our survival and that we can power our lives without.

This is coming, moreover, at a time when freshwater sources are imperiled around the world. Indeed, the water used in extraction operations often comes from aquifers that are already depleted from years of serial droughts, as is the case in southern California, where prospectors are eyeing the enormous Monterey Shale, and in Texas, where fracking has skyrocketed in recent years. Meanwhile, the Karoo—an arid and spectacular region of South Africa that Shell is planning to frack—literally translates as "land of the great thirst." Which helps explain why Oom Johannes Willemse, a local spiritual leader, says, "Water is so holy. If you don't have water, you don't have anything worth living for." He adds, "I will fight to the death. I won't allow this water to be destroyed."[21]

The fight against pollution and climate change can seem abstract at times; but wherever they live, people will fight for their water. Even die for it.

"Can we live without water?" the anti-fracking farmers chant in Pungesti, Romania.

"No!"

"Can we live without Chevron?"

"Yes!"[22]

These truths emerge not out of an abstract theory about "the commons" but out of lived experience. Growing in strength and connecting communities in all parts of the world, they speak to something deep and unsettled in many of us. We know that we are trapped within an economic system that has it backward; it behaves as if there is no end to what is actually finite (clean water, fossil fuels, and the atmospheric space to absorb their emissions) while insisting that there are strict and immovable limits to what is actually quite flexible: the financial resources that human institutions manufacture, and that, if imagined differently, could build the kind of caring society we need. Anni Vassiliou, a youth worker who is part of the struggle against the Eldorado gold mine in Greece, describes this as living in "an upside down world. We are in danger of more and more floods. We are in danger of never, here in Greece, never experiencing spring and fall again. And they're telling us that we are in danger of exiting the Euro. How crazy is that?"[23] Put another way, a broken bank is a crisis we can fix; a broken Arctic we cannot.

Early Wins

It's not yet clear which side will win many of the struggles outlined in these pages—only that the companies in the crosshairs are up against far more than they bargained for. There have, however, already been some solid victories, too many to fully catalogue here.

For instance, activists have won fracking bans or moratoria in dozens of cities and towns and in much larger territories too. Alongside France, countries with moratoria include Bulgaria, the Netherlands, the Czech Republic, and South Africa (though South Africa has since lifted the ban). Moratoria or bans are also in place in the states and provinces of Vermont, Quebec, as well as Newfoundland and Labrador (as of early 2014, New York's contentious moratorium still held but it looked shaky). This track record is all the more remarkable considering that so much local anti-fracking activism has not received foundation funding, and is instead financed the old-fashioned way: by passing the hat at community events and with countless volunteer hours.

And some victories against fossil fuel extraction receive almost no media attention, but are significant nonetheless. Like the fact that in 2010 Costa Rica passed a landmark law banning new open-pit mining projects anywhere in the country. Or that in 2012, the residents of the Colombian archipelago of San Andrés, Providencia, and Santa Catalina successfully fended off government plans to open the waters around their beautiful islands to offshore oil drilling. The region is home to one of the largest coral reefs in the Western Hemisphere and as one account of the victory puts it, what was established was the fact that coral is "more important than oil." [*][24]

And then there is the wave of global victories against coal. Under mounting pressure, the World Bank as well as other large international funders have announced that they will no longer offer financing to coal

---

[*] Sadly, this pristine UNESCO Biosphere Reserve is once again at risk after an international court ruling declared the waters surrounding the Caribbean islands to be legally owned by the government of Nicaragua (though the islands themselves remain part of Colombia). And Nicaragua has stated its intention to drill.

projects except in exceptional circumstances, which could turn out to be a severe blow to the industry if other financiers follow suit. In Gerze, Turkey, a major proposed coal plant on the Black Sea was scuttled under community pressure. The Sierra Club's hugely successful "Beyond Coal" campaign has, along with dozens of local partner organizations, succeeded in retiring 170 coal plants in the United States and prevented over 180 proposed plants since 2002.[25]

The campaign to block coal export terminals in the Pacific Northwest has similarly moved from strength to strength. Three of the planned terminals—one near Clatskanie, Oregon, another in Coos Bay, Oregon, and another in Hoquiam, Washington—have already been nixed, the result of forceful community activism, much of it organized by the Power Past Coal coalition. Several port proposals are still pending but resistance is fierce, particularly to the largest of the bunch, just outside Bellingham, Washington. "It's not a fun time to be in the coal industry these days," said Nick Carter, president and chief operating officer of the U.S. coal company Natural Resource Partners. "It's not much fun to get up every day, go to work and spend your time fighting your own government."[26]

In comparison, the actions against the various tar sands pipelines have not yet won any clear victories, only a series of very long delays. But those delays matter a great deal because they have placed a question mark over the capacity of Alberta's oil patch to make good on its growth projections. And if there is one thing billion-dollar investors hate, it's political uncertainty. If Alberta's landlocked oil patch can't guarantee its investors a reliable route to the sea where bitumen can be loaded onto tankers, then, as the province's former minister of energy Ron Liepert put it, "the investment is going to dry up." The head of one of the largest oil companies in the tar sands confirmed this in January 2014. "If there were no more pipeline expansions, I would have to slow down," Cenovus CEO Brian Ferguson said. He clearly considered this some kind of threat, but from a climate perspective it sounded like the best news in years.[27]

Even if these tactics succeed only in slowing expansion plans, the delays will buy time for clean energy sources to increase their market share and to be seen as more viable alternatives, weakening the power of the fossil fuel lobby. And, even more significantly, the delays give residents of the largest markets in Asia a window of opportunity to strengthen their own demands for a clean energy revolution.

Already, these demands are spreading so rapidly that it isn't at all clear how long the market for new coal-fired plants and extra-dirty gasoline in Asia will continue to expand. In India, Blockadia-style uprisings have been on full display in recent years, with people's movements against coal-fired power plants significantly slowing the rush to dirty energy in some regions. The southeastern state of Andhra Pradesh has been the site of several iconic struggles, like one in the village of Kakarapalli, surrounded by rice patties and coconut groves, where local residents can be seen staffing a semipermanent checkpoint under a baobab tree at the entrance to town. The encampment chokes off the only road leading to a half-built power plant where construction was halted amidst protests in 2011. In nearby Sompeta, another power plant proposal was stopped by a breakthrough alliance of urban middle-class professionals and subsistence farmers and fishers who united to protect the nearby wetlands. After police charged a crowd of protesters in 2010, shooting dead at least two people, a national uproar forced the National Environment Appellate Authority to revoke the permit for the project.[28] The community remains vigilant, with a daily rotating hunger strike entering its 1,500th day at the beginning of 2014.

China, meanwhile, is in the midst of a very public and emotional debate about its crisis levels of urban air pollution, in large part the result of the country's massive reliance on coal. There have been surprisingly large and militant protests against the construction of new coal-fired plants, most spectacularly in Haimen, a small city in Guangdong Province. In December 2011, as many as thirty thousand residents surrounded a government building and blocked a highway to protest plans to expand a coal-fired power plant. Citing concerns about cancer and other health problems blamed on the existing plant, the demonstrators withstood days of attacks by police, including tear gas and reported beatings with batons. They were there to send the message, as one protester put it, that, "This is going to affect our future generations. They still need to live." The plant expansion was suspended.[29]

Chinese peasants who rely on traditional subsistence activities like agriculture and fishing have a history of militant uprisings against industrial projects that cause displacement and disease, whether toxic factories, high-ways, or mega-dams. Very often these actions attract severe state repres-sion, including deaths in custody of protest leaders. The projects usually go ahead regardless of the opposition, though there have been some notable successes.

What has changed in China in recent years—and what is of paramount concern to the ruling party—is that the country's elites, the wealthy winners in China's embrace of full-throttle capitalism, are increasingly distressed by the costs of industrialization. Indeed, Li Bo, who heads Friends of Nature, the oldest environmental organization in China, describes urban air pollu-tion as "a superman for Chinese environment issues," laughing at the irony of an environmentalist having "to thank smog." The reason, he explains, is that the elites had been able to insulate themselves from previous environ-mental threats, like baby milk and water contamination, because "the rich, the powerful, have special channels of delivery, safer products [delivered] to their doorsteps." But no matter how rich you are, there is no way to hide from the "blanket" of toxic air. "Nobody can do anything for special [air] delivery," he says. "And that's the beauty of it."[30]

To put the health crisis in perspective, the World Health Organization sets the guideline for the safe presence of fine particles of dangerous air pol-lutants (known as $PM_{2.5}$) at 25 micrograms or less per cubic meter; 250 is considered hazardous by the U.S. government. In January 2014, in Beijing, levels of these carcinogens hit 671. The ubiquitous paper masks haven't been enough to prevent outbreaks of respiratory illness, or to protect chil-dren as young as eight from being diagnosed with lung cancer. Shanghai, meanwhile, has introduced an emergency protocol in which kindergartens and elementary schools are automatically shut down and all large-scale out-door gatherings like concerts and soccer games are canceled when the levels of particulate matter in the air top 450 micrograms per cubic meter. No wonder Chen Jiping, a former senior Communist Party official, now retired, admitted in March 2013 that pollution is now the single greatest cause of social unrest in the country, even more than land disputes.[31]

China's unelected leaders have long since deflected demands for democ-racy and human rights by touting the ruling party's record of delivering gal-loping economic growth. As Li Bo puts it, the rhetoric was always, "We get rich first, we deal with the environment problems second." That worked for a long time, but now, he says, "their argument has all of a sudden suffocated in the smog."

The pressure for a more sustainable development path has forced the government to cut its targeted growth to a rate lower than China had experienced in more than a decade, and to launch huge alternative energy programs. Many dirty-energy projects, meanwhile, have been canceled or delayed. In 2011, a third of the Chinese coal-fired power plants that had been approved for construction "were stalled and investments in new coal plants weren't even half the level they were in 2005," according to Justin Guay, associate director of the Sierra Club's International Climate Program. "Even better, China actually closed down over 80 gigawatts of coal plants between 2001–2010 and is planning to phase out another 20 GW. To put that in perspective that's roughly the size of *all* electricity sources in Spain, home to the world's 11th largest electricity sector." (In an effort to reduce smog, the government is also exploring the potential for natural gas fracking, but in an earthquake-prone country with severe water shortages, it's a plan unlikely to quell unrest.)[32]

All this pushback from within China is of huge significance to the broader fossil-fuel resistance, from Australia to North America. It means that if tar sands pipelines and coal export terminals can be held off for just a few more years, the market for the dirty products the coal and oil companies are trying to ship to Asia could well dry up. Something of a turning point took place in July 2013 when the multinational investment banking firm Goldman Sachs published a research paper titled, "The Window for Thermal Coal Investment Is Closing." Less than six months later, Goldman Sachs sold its 49 percent stake in the company that is developing the largest of the proposed coal export terminals, the one near Bellingham, Washington, having apparently concluded that window had already closed.[33]

These victories add up: they have kept uncountable millions of tons of carbon and other greenhouse gases out of the atmosphere. Whether or not climate change has been a primary motivator, the local movements behind them deserve to be recognized as unsung carbon keepers, who, by protecting their beloved forests, mountains, rivers, and coastlines, are helping to protect all of us.

Fossil Free: The Divestment Movement

Climate activists are under no illusion that shutting down coal plants, blocking tar sands pipelines, and passing fracking bans will be enough to lower emissions as rapidly and deeply as science demands. There are just too many extraction operations already up and running and too many more being pushed simultaneously. And oil multinationals are hyper-mobile—they move wherever they can dig.

With this in mind, discussions are under way to turn the "no new fossil frontiers" principle behind these campaigns into international law. Proposals include a Europe-wide ban on fracking (in 2012, more than a third of the 766 members of the European Parliament cast votes in favor of an immediate moratorium).[34] There is a growing campaign calling for a worldwide ban on offshore drilling in the sensitive Arctic region, as well as in the Amazon rainforest. And activists are similarly beginning to push for a global moratorium on tar sands extraction anywhere in the world, on the grounds that it is sufficiently carbon-intensive to merit transnational action.

Another tactic spreading with startling speed is the call for public interest institutions like colleges, faith organizations, and municipal governments to sell whatever financial holdings they have in fossil companies. The divestment movement emerged organically out of various Blockadia-style attempts to block carbon extraction at its source—specifically, out of the movement against mountaintop removal coal mining in Appalachia, which was looking for a tactic to put pressure on coal companies that had made it clear that they were indifferent to local opinion. Those local activists were later joined by a national and then international campaign spearheaded by 350.org, which extended the divestment call to include all fossil fuels, not just coal. The idea behind the tactic was to target not just individual unpopular projects but the logic that is driving this entire wave of frenetic, high-risk extraction.

The divestment campaign is based on the idea—outlined so compellingly by Bill McKibben—that anyone with a basic grasp of arithmetic can look at how much carbon the fossil fuel companies have in their reserves, subtract how much carbon scientists tell us we can emit and still keep global warming below 2 degrees Celsius, and conclude that the fossil fuel companies have every intention of pushing the planet beyond the boiling point.

These simple facts have allowed the student-led divestment movement to put the fossil fuel companies' core business model on trial, arguing that they have become rogue actors whose continued economic viability relies on radical climate destabilization—and that, as such, any institution claiming to serve the public interest has a moral responsibility to liberate itself from these odious profits. "What the fossil fuel divestment movement is saying to companies is your fundamental business model of extracting and burning carbon is going to create an uninhabitable planet. So you need to stop. You need a new business model," explains Chloe Maxmin, coordinator of Divest Harvard.[35] And young people have a special moral authority in making this argument to their school administrators: these are the institutions entrusted to prepare them for the future; so it is the height of hypocrisy for those same institutions to profit from an industry that has declared war on the future at the most elemental level.

No tactic in the climate wars has resonated more powerfully. Within six months of the campaign's official launch in November 2012, there were active divestment campaigns on over three hundred campuses and in more than one hundred U.S. cities, states, and religious institutions. The demand soon spread to Canada, Australia, the Netherlands, and Britain. At the time of publication, thirteen U.S. colleges and universities had announced their intention to divest their endowments of fossil fuel stocks and bonds, and the leaders of more than twenty-five North American cities had made similar commitments, including San Francisco and Seattle. Around forty religious institutions had done the same. The biggest victory to date came in May 2014 when Stanford University—with a huge endowment worth $18.7 billion—announced it would be selling its coal stocks.[36]

Critics have been quick to point out that divestment won't bankrupt Exxon; if Harvard, with its nearly $33 billion endowment, sells its stock, someone else will snap it up. But this misses the power of the strategy: every time students, professors, and faith leaders make the case for divestment, they are chipping away at the social license with which these companies operate. As Sara Blazevic, a divestment organizer at Swarthmore College, puts it, the movement is "taking away the hold that the fossil fuel industry has over our political system by making it socially unacceptable and morally unacceptable to be financing fossil fuel extraction." And Cameron Fenton, one of the leaders of the divestment push in Canada, adds, "No one is thinking we're going to bankrupt fossil fuel companies. But what we can do is bankrupt their reputations and take away their political power."[37]

The eventual goal is to confer on oil companies the same status as tobacco companies, which would make it much easier to make other important demands—like bans on political donations from fossil fuel companies and on fossil fuel advertising on television (for the same public health reasons that we ban broadcast cigarette ads). Crucially, it might even create the space for a serious discussion about whether these profits are so illegitimate that they deserve to be appropriated and reinvested in solutions to the climate crisis. Divestment is just the first stage of this delegitimization process, but it is already well under way.

None of this is a replacement for major policy changes that would regulate carbon reduction across the board. But what the emergence of this networked, grassroots movement means is that the next time climate campaigners get into a room filled with politicians and polluters to negotiate, there will be many thousands of people outside the doors with the power to amp up the political pressure significantly—with heightened boycotts, court cases, and more militant direct action should real progress fail to materialize. And that is a very significant shift indeed.

Already, the rise of Blockadia and the fossil fuel divestment movement is having a huge impact on the mainstream environmental community, particularly the Big Green groups that had entered into partnerships with fossil fuel companies (never mind The Nature Conservancy, with its own Texas oil and gas operation . . .). Not surprisingly, some of the big pro-corporate green groups view this new militancy as an unwelcome intrusion on their territory. When it comes to fracking in particular, groups like the Environmental Defense Fund have pointedly not joined grassroots calls for drilling bans and a rapid shift to 100 percent renewables, but have instead positioned themselves as brokers, offering up "best practices"—developed with industry groups—that will supposedly address local environmental concerns. (Even when locals make it abundantly clear that the only best practice they are interested in is an unequivocal ban on fracking.) "We fear that those who oppose all natural gas production everywhere are, in effect, making it harder for the U.S. economy to wean itself from dirty coal," charged EDF chief counsel Mark Brownstein.[38]

Predictably, these actions have provoked enormous tensions, with grassroots activists accusing the EDF of providing cover for polluters and undercutting their efforts.[*][39]

But not all the Big Greens are reacting this way. Some—like Food & Water Watch, 350.org, Greenpeace, Rainforest Action Network, and Friends of the Earth—have been a central part of this new wave of anti-fossil fuel activism from the beginning. And for others that were more ambivalent, the rapid spread of a new, take-no-prisoners climate movement appears to have been a wake-up call; a reminder that they had strayed too far from first principles. This shift has perhaps been clearest at the Sierra Club, which, under the leadership of its former executive director, Carl Pope, had attracted considerable controversy with such corporate-friendly actions as lending its logo to a line of "green" cleaning products owned by Clorox. Most damaging, Pope had been an enthusiastic supporter of natural gas and had appeared publicly (even lobbying on Capitol Hill) to sing the praises of the fossil fuel alongside Aubrey McClendon, then CEO of Chesapeake Energy—a company at the forefront of the hydraulic fracking explosion. Many local chapters, neck deep in battles against fracking, had been livid. And it would later emerge that the Sierra Club was, in this same period, secretly receiving many millions in donations from Chesapeake—one of the biggest controversies to hit the movement in decades.[†][40]

---

* For instance, in May 2013, sixty-eight groups and individuals—including Friends of the Earth, Greenpeace, and Robert Kennedy Jr.—signed a letter that directly criticized the EDF and and its president Fred Krupp for their role in creating the industry-partnered Center for Sustainable Shale Development (CSSD). "CSSD bills itself as a collaborative effort between 'diverse interests with a common goal,' but our goals as a nation are not, and cannot, be the same as those of Chevron, Consol Energy, EQT Corporation, and Shell, all partners in CSSD," the letter states. "These corporations are interested in extracting as much shale gas and oil as possible, and at a low cost. We are interested in minimizing the extraction and consumption of fossil fuels and in facilitating a rapid transition to the real sustainable energy sources—the sun, the wind, and hydropower."

† Reached by email, Carl Pope, who had not previously commented on the controversy, explained his actions as follows: "Climate advocates were at war with the coal industry, and at that moment

A great deal has changed at the organization in the years since. The Sierra Club's new executive director, Michael Brune, put an end to the secret arrangement with Chesapeake and canceled the Clorox deal. (Though the money was replaced with a huge donation from Michael Bloomberg's foundation, which—though this was not known at the time—is significantly invested in oil and natural gas.) Brune was also arrested outside the White House in a protest against the construction of Keystone XL tar sands pipeline, breaking the organization's longtime ban on engaging in civil disobedience. Perhaps most significantly, the Sierra Club has joined the divestment movement. It now has a clear policy against investing in, or taking money from, fossil fuel companies and affiliated organizations.[41]

In April 2014, the Natural Resources Defense Council announced that it had helped create "the first equity global index tool that will exclude companies linked to exploration, ownership or extraction of carbon-based fossil fuel reserves. This new investment tool will allow investors who claim to be socially conscious, including foundations, universities, and certain pension groups, to align their investments with their missions." The rigor of this new tool remains to be tested (and I have my doubts) but it represents a shift from a year earlier, when the NRDC admitted that its own portfolio was invested in mutual funds and other mixed assets that did not screen for fossil fuels.[42]

The divestment movement is even (slowly) being embraced by some of the foundations that finance environmental activism. In January 2014, seventeen foundations pledged to divest from fossil fuels and invest in clean energy. While none of the Big Green donors—the Hewlett and Packard Foundations or the Walton Family Foundation, for example, not to mention Ford or Bloomberg—were on board, several smaller ones were, including the Wallace Global Fund and the Park Foundation, both major funders of anti–fossil fuel activism.[43]

Chesapeake was willing to ally with us. I understand the concerns of those who thought that alliance was a bad idea—but it is likely that without it about 75 of the pending 150 new coal fired power plants we stopped would have been built instead." He added, "What I do regret is the failure at the time to understand the scale and form that the shale gas and oil revolution would take, which led us to make inadequate investments in getting ready for the assault that would soon be coming at states like Pennsylvania, West Virginia and Colorado. That was a significant, and costly, failure of vision."

———

Up until quite recently, there was a widely shared belief that the big oil companies had such a fail-safe profit-making formula that none of this—not the divestment campaigns, not the on-the-ground resistance—would make any kind of dent in their power and wealth. That attitude needed some re-adjusting in January 2014 when Shell—which raked in more revenue than any company in the world in 2013—announced fourth-quarter profits that blindsided investors. Rather than the previous year's $5.6 billion quarter, Shell's new CEO, Ben van Beurden, announced that the company was now expecting just $2.9 billion, a jarring 48 percent drop.[44]

No single event could take the credit, but the company's various troubles were clearly adding up: its Arctic misadventures, the uncertainty in the tar sands, the persistent political unrest in Nigeria, and the growing chatter about a "carbon bubble" inflating its stock. Reacting to the news, the financial research company Sanford C. Bernstein & Co. noted that the plummet was "highly unusual for an integrated oil company" and admitted that it was "a bit shellshocked."[45]

The Democracy Crisis

As the anti–fossil fuel forces gain strength, extractive companies are beginning to fight back using a familiar tool: the investor protection provisions of free trade agreements. As previously mentioned, after the province of Quebec successfully banned fracking, the U.S.-incorporated oil and gas company Lone Pine Resources announced plans to sue Canada for at least $230 million under the North American Free Trade Agreement's rules on expropriation and "fair and equitable treatment." In arbitration documents, Lone Pine complained that the moratorium imposed by a democratically elected government amounted to an "arbitrary, capricious, and illegal revocation of the Enterprise's valuable right to mine for oil and gas under the St. Lawrence River." It also claimed (rather incredibly) that this occurred "with no cognizable public purpose"—not to mention "without a penny of compensation."[46]

It's easy to imagine similar challenges coming from any company whose extractive dreams are interrupted by a democratic uprising. And indeed after the Keystone XL pipeline was delayed yet again in April 2014, Canadian and TransCanada officials began hinting of a possible challenge to the U.S. government under NAFTA.

In fact, current trade and investment rules provide legal grounds for foreign corporations to fight virtually any attempt by governments to restrict the exploitation of fossil fuels, particularly once a carbon deposit has attracted investment and extraction has begun. And when the aim of the investment is explicitly to *export* the oil, gas, and coal and sell it on the world market—as is increasingly the case—successful campaigns to block those exports could well be met with similar legal challenges, since imposing "quantitative restrictions" on the free flow of goods across borders violates a fundamental tenet of trade law.[47]

"I really do think in order to combat the climate crisis, fundamentally we need to strip the power out of the fossil fuel industry, which raises enormous investment challenges in the trade context," says Ilana Solomon, the Sierra Club's trade expert. "As we begin to regulate the fossil fuel industry, for example in the United States, the industry may increasingly respond by seeking to export raw materials, whether it's coal, or natural gas, and under trade law it is literally illegal to stop the exports of those resources once they're mined. So it's very hard to stop."[48]

It is unsurprising, then, that as Blockadia victories mount, so do the corporate trade challenges. More investment disputes are being filed than ever before, with a great many initiated by fossil fuel companies—as of 2013, a full sixty out of 169 pending cases at the World Bank's dispute settlement tribunal had to do with the oil and gas or mining sectors, compared to a mere seven extraction cases throughout the entire 1980s and 1990s. According to Lori Wallach, director of Public Citizen's Global Trade Watch, of the more than $3 billion in compensation already awarded under U.S. free trade agreements and bilateral investment treaties, more than 85 percent "pertains to challenges against natural resource, energy, and environmental policies."[49]

None of this should be surprising. Of course the richest and most powerful companies in the world will exploit the law to try to stamp out real and perceived threats and to lock in their ability to dig and drill wherever they wish in the world. And it certainly doesn't help that many of our governments seem determined to hand out even more lethal legal weapons in the form of new and expanded trade deals, which companies, in turn, will use against governments' own domestic laws.

There may, however, be an unexpected upside to the aggressive use of trade law to quash environmental wins: after a decade lull when few seemed to be paying attention to the arcane world of free trade negotiations, a new generation of activists is once again becoming attuned to the democratic threat these treaties represent. Indeed there is now more public scrutiny and debate about trade agreements than there has been in years.

The point of this scrutiny, however, should not be to throw up our hands in the face of yet another obstacle standing in the way of sensible action on climate. Because while it is true that the international legal architecture of corporate rights is both daunting and insidious, the well-kept secret behind these deals is that they are only as powerful as our governments allow them to be. They are filled with loopholes and workarounds so any government that is serious about adopting climate polices that reduce emissions in line with science could certainly find a way to do so, whether by aggressively challenging trade rulings that side with polluters, or finding creative policy tweaks to get around them, or refusing to abide by rulings and daring reprisals (since these institutions cannot actually force governments to change laws), or attempting to renegotiate the rules. Put another way, the real problem is not that trade deals are allowing fossil fuel companies to challenge governments, it's that governments are not fighting back against these corporate challenges. And that has far less to do with any individual trade agreement than it does with the profoundly corrupted state of our political systems.

Beyond Fossilized Democracies

The process of taking on the corporate-state power nexus that underpins the extractive economy is leading a great many people to face up to the underlying democratic crisis that has allowed multinationals to be the authors of the laws under which they operate—whether at the municipal, state/provincial, national, or international level. It is this corroded state of our political systems—as fossilized as the fuel at the center of these battles—that is fast turning Blockadia into a grassroots pro-democracy movement.

Having the ability to defend one's community's water source from danger seems to a great many people like the very essence of self-determination. What is democracy if it doesn't encompass the capacity to decide, collectively, to protect something that no one can live without?

The insistence on this right to have a say in critical decisions relating to water, land, and air is the thread that runs through Blockadia. It's a sentiment summed up well by Helen Slottje, a former corporate lawyer who has helped around 170 New York towns to adopt anti-fracking ordinances: "Are you kidding me? You think you can just come into my town and tell me you're going to do whatever you want, wherever you want, whenever you want it, and I'm going to have no say? Who do you think you are?" I heard much the same from Marily Papanikolaou, a wavy-haired Greek mountain-bike guide who had been perfectly happy raising her toddlers and leading tourists through forest trails, but now spends her spare time at anti-mine demonstrations and meetings. "I can't let anyone come in my village and try to do this and not ask me for my permission. *I* live here!" And you can hear something awfully similar from Texas landowners, irate that a Canadian pipeline company tried to use the law of eminent domain to gain access to their family land. "I just don't believe that a Canadian organization that appears to be building a pipeline for their financial gain has more right to my land than I do," said Julia Trigg Crawford, who has challenged TransCanada in court over its attempt to use her 650-acre ranch near Paris, Texas, which her grandfather purchased in 1948.[50]

And yet the most jarring part of the grassroots anti-extraction uprising has been the rude realization that most communities do appear to lack this power; that outside forces—a far-off central government, working hand-in-glove with transnational companies—are simply imposing enormous health and safety risks on residents, even when that means overturning local laws. Fracking, tar sands pipelines, coal trains, and export terminals are being proposed in many parts of the world where a clear majority of the population has made its opposition unmistakable, at the ballot box, through official consultation processes, and in the streets.

And yet consent seems beside the point. Again and again, after failing to persuade communities that these projects are in their genuine best interest, governments are teaming up with corporate players to roll over the opposition, using a combination of physical violence and draconian legal tools reclassifying peaceful activists as terrorists.[*][51]

Nongovernmental organizations of all kinds find themselves under increasing surveillance, both by security forces and by corporations, often working in tandem. Pennsylvania's Office of Homeland Security hired a private contractor to gather intelligence on anti-fracking groups, which it proceeded to share with major shale gas companies. The same phenomenon is unfolding in France, where the utility EDF was convicted in 2011 of unlawfully spying on Greenpeace. In Canada, meanwhile, it was revealed that Chuck Strahl, then chair of the committee overseeing the country's spy agency, the Canadian Security Intelligence Service, was registered as a lobbyist for Enbridge, the company behind the hugely controversial Northern Gateway tar sands pipeline. That was a problem because the National Energy Board had directed the agency to assess the security threats to pipeline projects, which was thinly veiled code for spying on environmentalists and First Nations.[52]

Strahl's dual role raised the question of whether Enbridge could also gain access to the information gleaned. Then it came out that Strahl wasn't the only one who seemed to be working for the government and the fossil fuel companies simultaneously. As the CBC reported, "Half of the other

---

* This reached truly absurd levels in December 2013 when two twentysomething antifracking activists were charged with staging a "terrorism hoax" after they unfurled cloth protest banners at the headquarters of Devon Energy in Oklahoma City. Playing on the *Hunger Games* slogan, one of the banners said: "THE ODDS ARE NEVER IN OUR FAVOR." Standard, even benign activist fare— except for one detail. According to Oklahoma City Police captain Dexter Nelson, as the banner was lowered it shed a "black powder substance" that was meant to mimic a "biochemical assault," as the police report put it. That nefarious powder, the captain stated, was "later determined to be glitter." Never mind that the video of the event showed absolutely no concern about the falling glitter from the assembled onlookers. "I could have swept it up in two minutes if they gave me a broom," said Stefan Warner, one of those charged and facing the prospect of up to ten years in jail.

Harper government appointees keeping an eye on the spies also have ties to the oil business"—including one member who sits on the board of Enbridge Gas NB, a wholly owned regional subsidiary of the pipeline company, and another who had been on TransCanada's board. Strahl resigned amid the controversy; the others did not.[53]

The collusion between corporations and the state has been so boorishly defiant that it's almost as if the communities standing in the way of these projects are viewed as little more than "overburden"—that ugliest of words used by the extractive industries to describe the "waste earth" that must be removed to access a tar sands or mineral deposit. Like the trees, soil, rocks, and clay that the industry's machines scrape up, masticate, and pile into great slag heaps, democracy is getting torn into rubble too, chewed up and tossed aside to make way for the bulldozers.

That was certainly the message when the three-person Joint Review Panel that had been so scared by the Heiltsuk community's welcome in Bella Bella finally handed down its recommendation to Canada's federal government. The Northern Gateway pipeline should go ahead, the panel announced. And though it enumerated 209 conditions that should be met before construction—from submiting caribou habitat protection plans to producing an updated inventory of waterway crossings "in both Adobe PDF and Microsoft Excel spreadsheet formats"—the ruling was almost universally interpreted as a political green light.[54]

Only two out of the over one thousand people who spoke at the panel's community hearings in British Columbia supported the project. One poll showed that 80 percent of the province's residents opposed having more oil tankers along their marine-rich coastline. That a supposedly impartial review body could rule in favor of the pipeline in the face of this kind of overwhelming opposition was seen by many in Canada as clear evidence of a serious underlying crisis, one far more about money and power than the environment. "Sadly, today's results are exactly what we expected," said anti-pipeline campaigner Torrance Coste, "proof that our democratic system is broken."[55]

In a sense, these are merely local manifestations of the global democratic crisis represented by climate change itself. As Venezuelan political scientist Edgardo Lander aptly puts it, "The total failure of climate negotiation serves to highlight the extent to which we now live in a post-democratic society. The interests of financial capital and the oil industry are much more important than the democratic will of people around the world. In the global neoliberal society profit is more important than life." Or, as George Monbiot, *The Guardian*'s indispensable environmental columnist, put it on the twenty-year anniversary of the Rio Earth Summit, "Was it too much to have asked of the world's governments, which performed such miracles in developing stealth bombers and drone warfare, global markets and trillion-dollar bailouts, that they might spend a tenth of the energy and resources they devoted to these projects on defending our living planet? It seems, sadly, that it was." Indeed, the failure of our political leaders to even attempt to ensure a safe future for us represents a crisis of legitimacy of almost unfathomable proportions.[56]

And yet a great many people have reacted to this crisis not by abandoning the promise of genuine self-government, but rather by attempting to make good on that promise in the spheres where they still have real influence. It's striking, for instance, that even as national governments and international agencies fail us, cities are leading the way on climate action around the world, from Bogotá to Vancouver. Smaller communities are also taking the lead in the democratic preparation for a climate-changed future. This can be seen most clearly in the fast-growing Transition Town movement. Started in 2006 in Totnes—an ancient market town in Devon, England, with a bohemian reputation—the movement has since spread to more than 460 locations in at least forty-three countries worldwide. Each Transition Town (and this may be an actual town or a neighborhood in a larger city) undertakes to design what the movement calls an "energy descent action plan"—a collectively drafted blueprint for lowering its emissions and weaning itself off fossil fuels. The process opens up rare spaces for participatory democracy, with neighbors packing consultation meetings at city halls to share ideas about everything from how to increase their food security through increased local agriculture to building more efficient affordable housing.[57]

Nor is it all dry planning meetings. In Totnes, the local Transition group organizes frequent movie nights, public lectures, and discussions, as well as street festivals to celebrate each landmark toward greater sustainability.

This too is part of responding to the climate crisis, as critical as having secure food supplies and building sturdy seawalls. Because a key determinant in how any community survives an extreme weather event is its connective tissue—the presence of small local businesses and common spaces where neighbors can get to know one another and make sure that elderly people aren't forgotten during crushing heat waves or storms. As the environmental writer and analyst David Roberts has observed, "the ingredients of resilience" are "overlapping social and civic circles, filled with people who, by virtue of living in close proximity and sharing common spaces, know and take care of each other. The greatest danger in times of stress or threat is *isolation*. Finding ways of expanding public spaces and nurturing civic involvement is not just some woolly-headed liberal project—it's a survival strategy."[58]

The intimacy of local politics is also what has turned this tier of government into an important site of resistance to the carbon extraction frenzy—whether it's cities voting to take back control over a coal-burning utility that won't switch to renewables (as so many citizens are doing in Germany), or municipalities adopting policies to divest city holdings of fossil fuels, or towns passing anti-fracking ordinances. And these are not mere symbolic expressions of dissent. Commenting on the stakes of his client's court challenge to local anti-fracking ordinances, Thomas West, a lawyer for Norse Energy Corporation USA, told *The New York Times*, "It's going to decide the future of the oil and gas industry in the state of New York."[59]

———

Local ordinances are not the only—or even the most powerful—unconventional legal tools that may help Blockadia to extend its early victories. This became apparent when the panel reviewing Enbridge's Northern Gateway pipeline announced its recommendations. The news that it had greenlighted the federal government to approve the much loathed tar sands project was not, for the most part, greeted with despair. Instead, a great many Canadians remained convinced that the pipeline would never go ahead and that the British Columbia coast would be saved—no matter what the panel said or what the federal government did.

"The federal cabinet needs First Nations' approval and social license from British Columbians, and they have neither," said Sierra Club BC campaigns director Caitlyn Vernon. And referring to the Save the Fraser Declaration signed by Chief Baptiste and so many others, she added, "First Nations have formally banned pipelines and tankers from their territories on the basis of Indigenous law."[60] It was a sentiment echoed repeatedly in news reports: that the legal title of the province's First Nations was so powerful that even if the federal government did approve the pipeline (which it eventually did in June 2014), the project would be successfully stopped in the courts through Indigenous legal challenges, as well as in the forests through direct action.

Is it true? As the next chapter will explore, the historical claims being made by Indigenous peoples around the world as well as by developing countries for an honoring of historical debts indeed have the potential to act as counterweights to increasingly undemocratic and intransigent governments. But the outcome of this power struggle is by no means certain. As always, it depends on what kind of movement rallies behind these human rights and moral claims.

YOU AND WHAT ARMY?

Indigenous Rights and the Power of Keeping Our Word

"I never thought I would ever see the day that we would come together. Relationships are changing, stereotypes are disappearing, there's more respect for one another. If anything, this Enbridge Northern Gateway has unified British Columbia."

—Geraldine Thomas-Flurer, coordinator of the Yinka Dene Alliance,
a First Nations coalition opposing the Enbridge Northern Gateway pipeline, 2013[1]

"There is never peace in West Virginia because there is never justice."

—Labor organizer Mary Harris "Mother" Jones, 1925[2]

The guy from Standard & Poor's was leafing through the fat binder on the round table in the meeting room, brow furrowed, skimming and nodding.

It was 2004 and I found myself sitting in on a private meeting between two important First Nations leaders and a representative of one of the three most powerful credit rating agencies in the world. The meeting had been requested by Arthur Manuel, a former Neskonlith chief in the interior of British Columbia, now spokesperson for the Indigenous Network on Economies and Trade.

Arthur Manuel, who comes from a long line of respected Native leaders, is an internationally recognized thinker on the question of how to force belligerent governments to respect Indigenous land rights, though you might not guess it from his plainspoken manner or his tendency to chuckle midsentence. His theory is that nothing will change until there is a credible threat that continuing to violate Native rights will carry serious financial costs, whether for governments or investors. So he has been looking for different ways to inflict those costs.

That's why he had initiated a correspondence with Standard & Poor's, which routinely blesses Canada with a AAA credit rating, a much coveted indicator to investors that the country is a safe and secure place in which to sink their money. In letters to the agency, Manuel had argued that Canada did not deserve such a high rating because it was failing to report a very important liability: a massive unpaid debt that takes the form of all the wealth that had been extracted from unceded Indigenous land, without consent—since 1846.[3] He further explained the various Supreme Court cases that had affirmed that Aboriginal and Treaty Rights were still very much alive.

After much back-and-forth, Manuel had managed to get a meeting with Joydeep Mukherji, director of the Sovereign Ratings Group, and the man responsible for issuing Canada's credit rating. The meeting took place at S&P's headquarters, a towering building just off Wall Street. Manuel had invited Guujaaw, the charismatic president of the Haida Nation, to help him make the case about those unpaid debts, and at the last minute had asked me to come along as a witness. Unaware that, post-9/11, official ID is required to get into all major Manhattan office buildings, the Haida leader had left his passport in his hotel room; dressed in a short-sleeved checked shirt and with a long braid down his back, Guujaaw almost didn't make it past security. But after some negotiation with security (and intervention from Manuel's contact upstairs), we made it in.

At the meeting, Manuel presented the Okanagan writ of summons, and explained that similar writs had been filed by many other First Nations. These simple documents, asserting land title to large swaths of territory, put the Canadian government on notice that these bands had every intention of taking legal action to get the economic benefits of lands being used by resource companies without their consent. These writs, Manuel explained, represented trillions of dollars' worth of unacknowledged liability being carried by the Canadian state.

Guujaaw then solemnly presented Mukherji with the Haida Nation's registered statement of claim, a seven-page legal document that had been filed before the Supreme Court of British Columbia seeking damages and reparations from the provincial government for unlawfully exploiting and degrading lands and waters that are rightfully controlled by the Haida. Indeed, at that moment, the case was being argued before the Supreme Court of Canada, challenging both the logging giant Weyerhaeuser and the provincial government of British Columbia over a failure to consult before logging the forests on the Pacific island of Haida Gwaii. "Right now the Canadian and British Columbia governments are using our land and our resources—Aboriginal and Treaty Rights—as collateral for all the loans they get from Wall Street," Manuel said. "We are in fact subsidizing the wealth of Canada and British Columbia with our impoverishment."[4]

Mukherji and an S&P colleague listened and silently skimmed Manuel's documents. A polite question was asked about Canada's recent federal elections and whether the new government was expected to change the enforcement of Indigenous land rights. It was clear that none of this was new to them—not the claims, not the court rulings, not the constitutional language. They did not dispute any of the facts. But Mukherji explained as nicely as he possibly could that the agency had come to the conclusion that Canada's First Nations did not have the power to enforce their rights and therefore to collect on their enormous debts. Which meant, from S&P's perspective, that those debts shouldn't affect Canada's stellar credit rating. The company would, however, continue to monitor the situation to see if the dynamics changed.

And with that we were back on the street, surrounded by New Yorkers clutching iced lattes and barking into cell phones. Manuel snapped a few pictures of Guujaaw underneath the Standard & Poor's sign, flanked by security guards in body armor. The two men seemed undaunted by what had transpired; I, on the other hand, was reeling. Because what the men from S&P were really saying to these two representatives of my country's original inhabitants was: "We know you never sold your land. But how are you going to make the Canadian government keep its word? You and what army?"

At the time, there did not seem to be a good answer to that question. Indigenous rights in North America did not have powerful forces marshaled behind them and they had plenty of powerful forces standing in opposition. Not just government, industry, and police, but also corporate-owned media that cast them as living in the past and enjoying undeserved special rights, while those same media outlets usually failed to do basic public education about the nature of the treaties our governments (or rather their British predecessors) had signed. Even most intelligent, progressive thinkers paid little heed: sure they supported Indigenous rights in theory, but usually as part of the broader multicultural mosaic, not as something they needed to actively defend.

However, in perhaps the most politically significant development of the rise of Blockadia-style resistance, this dynamic is changing rapidly—and an army of sorts is beginning to coalesce around the fight to turn Indigenous land rights into hard economic realities that neither government nor industry can ignore.

The Last Line of Defense

As we have seen, the exercise of Indigenous rights has played a central role in the rise of the current wave of fossil fuel resistance. The Nez Perce were the ones who were ultimately able to stop the big rigs on Highway 12 in Idaho and Montana; the Northern Cheyenne continue to be the biggest barrier to coal development in southeastern Montana; the Lummi present the greatest legal obstacle to the construction of the biggest proposed coal export terminal in the Pacific Northwest; the Elsipogtog First Nation managed to substantially interfere with seismic testing for fracking in New Brunswick; and so on. Going back further, it's worth remembering that the struggles of the Ogoni and Ijaw in Nigeria included a broad demand for self-determination and resource control over land that both groups claimed was illegitimately taken from them during the colonial formation of Nigeria. In short, Indigenous land and treaty rights have proved a major barrier for the extractive industries in many of the key Blockadia struggles.

And through these victories, a great many non-Natives are beginning to understand that these rights represent some of the most robust tools available to prevent ecological crisis. Even more critically, many non-Natives are also beginning to see that the ways of life that Indigenous groups are protecting have a great deal to teach about how to relate to the land in ways that are not purely extractive. This represents a true sea change over a very short period of time. My own country offers a glimpse into the speed of this shift.

The Canadian Constitution and the Canadian Charter of Rights and Freedoms acknowledge and offer protection to "aboriginal rights," including treaty rights, the right to self-government, and the right to practice traditional culture and customs. There was, however, a widespread perception among Canadians that treaties represented agreements to fully surrender large portions of lands in exchange for the provision of public services and designated rights on much smaller reserves. Many Canadians also assumed that in the lands not covered by any treaty (which is a great deal of the country, 80 percent of British Columbia alone), non-Natives could pretty much do what they wished with the natural resources. First Nations had rights on their reserves, but if they once had rights off them as well, they had surely lost them by attrition over the years. Finders keepers sort of thing, or so the thinking went.[5]

All of this was turned upside down in the late 1990s when the Supreme Court of Canada handed down a series of landmark decisions in cases designed to test the limits of Aboriginal title and treaty rights. First came *Delgamuukw v. British Columbia* in 1997, which ruled that in those large parts of B.C. that were not covered by any treaty, Aboriginal title over that land had never been extinguished and still needed to be settled. This was interpreted by many First Nations as an assertion that they still had full rights to that land, including the right to fish, hunt, and gather there. Chelsea Vowel, a Montréal-based Métis educator and Indigenous legal scholar, explains the shockwave caused by the decision. "One day, Canadians woke up to a legal reality in which millions of acres of land were recognized as never having been acquired by the Crown," which would have "immediate implications for other areas of the country where no treaties ceding land ownership were ever signed."[6]

Two years later, in 1999, the ruling known as the *Marshall* decision affirmed that when the Mi'kmaq, Maliseet, and Passamaquoddy First Nations, largely based in New Brunswick and Nova Scotia, signed "peace and friendship" treaties with the British Crown in 1760 and 1761, they did not—as so many Canadians then assumed—agree to give up rights to their ancestral lands. Rather they were agreeing to *share* them with settlers on the condition that the First Nations could continue to use those lands for traditional activities like fishing, trading, and ceremony. The case was sparked by a single fisherman, Donald Marshall Jr., catching eels out of season and without a license; the court ruled that it was within the rights of the Mi'kmaq and Maliseet to fish year-round enough to earn a "moderate livelihood" where their ancestors had fished, exempting them from many of the rules set by the federal government for the non-Native fishing fleet.[7]

Many other North American treaties contained similar resource-sharing provisions. Treaty 6, for instance, which covers large parts of the Alberta tar sands region, contains clear language stating that "Indians, shall have right to pursue their avocations of hunting and fishing throughout the tract surrendered"—in other words, they surrendered only their *exclusive* rights to the territory and agreed that the land would be used by both parties, with settlers and Indigenous peoples pursuing their interests in parallel.[8]

But any parallel, peaceful coexistence is plainly impossible if one party is irrevocably altering and poisoning that shared land. And indeed, though it is not written in the text of the treaty, First Nations elders living in this region contend that Indigenous negotiators gave permission for the land to be used by settlers only "to the depth of a plow"—considerably less than the cavernous holes being dug there today. In the agreements that created modern-day North America such land-sharing provisions form the basis of most major treaties.

In Canada, the period after the Supreme Court decisions was a tumultuous one. Federal and provincial governments did little or nothing to protect the rights that the judges had affirmed, so it fell to Indigenous people to go out on the land and water and assert them—to fish, hunt, log, and build ceremonial structures, often without state permission. The backlash was swift. Across the country non-Native fishers and hunters complained that the "Indians" were above the law, that they were going to empty the oceans and rivers of fish, take all the good game, destroy the woods, and on and on. (Never mind the uninterrupted record of reckless resource mismanagement by all levels of the Canadian government.)

Tensions came to a head in the Mi'kmaq community of Burnt Church, New Brunswick. Enraged that the *Marshall* decision had empowered Mi'kmaq people to exercise their treaty rights and fish outside of government-approved seasons, mobs of non-Native fishermen launched a series of violent attacks on their Native neighbors. In what became known as the Burnt Church Crisis, thousands of Mi'kmaq lobster traps were destroyed, three fish-processing plants were ransacked, a ceremonial arbor was burned to the ground, and several Indigenous people were hospitalized after their truck was attacked. And it wasn't just vigilante violence. As the months-long crisis wore on, government boats staffed with officials in riot gear rammed into Native fishing boats, sinking two vessels and forcing their crews to jump to safety in the water. The Mi'kmaq fishers did their best to defend themselves, with the help of the Mi'kmaq Warrior Society, but they were vastly outnumbered and an atmosphere of fear prevailed for years. The racism was so severe that at one point a non-Native fisherman put on a long-haired wig and performed a cartoonish "war dance" on the deck of his boat in front of delighted television crews.

That was 2000. In 2013, a little more than an hour's drive down the coast from Burnt Church, the same Mi'kmaq Warrior Society was once again in the news, this time because it had joined with the Elsipogtog First Nation to fend off the Texas company at the center of the province's fracking showdown. But the mood and underlying dynamics could not have been more different. This time, over months of protest, the warriors helped to light a series of ceremonial sacred fires and explicitly invited the non-Native community to join them on the barricades "to ensure that the company cannot resume work to extract shale gas via fracking." A statement explained, "This comes as part of a larger campaign that reunites Indigenous, Acadian & Anglo people." (New Brunswick has a large French-speaking Acadian population, with its own historical tensions with the English-speaking majority.)[9]

Many heeded the call and it was frequently noted that protests led by the Elsipogtog First Nation were remarkably diverse, drawing participants from all of the province's ethnic groups, as well as from First Nations across the country. As one non-Native participant, Debbi Hauper, told a video crew, "It's just a real sense of togetherness. We are united in what is most important. And I think we're seeing more and more of government and industries' methods of trying to separate us. And let's face it, these methods have worked for decades. But I think we're waking up."[10]

There were attempts to revive the old hatreds, to be sure. A police officer was overheard saying "Crown land belongs to the government, not to fucking Natives." And after the conflict with police turned violent, New Brunswick premier David Alward observed, "Clearly, there are those who do not have the same values we share as New Brunswickers." But the community stuck together and there were solidarity protests in dozens of cities and towns across the country: "This is not just a First Nations campaign. It's actually quite a historic moment where all the major peoples of this province—English, French and Aboriginal—come together for a common cause," said David Coon, head of the Green Party in New Brunswick. "This is really a question of justice. They want to protect their common lands, water and air from destruction." [11]

By then many in the province had come to understand that the Mi'kmaq's rights to use their traditional lands and waters to hunt and fish—the same rights that had sparked race riots a dozen years earlier— represented the best hope for the majority of New Brunswickers who opposed fracking. [12] And new tools were clearly required. Premier Alward had been a fracking skeptic before he was elected in 2010 but once in office, he promptly changed his tune, saying the revenue was needed to pay for social programs and to create jobs—the sort of flip flop that breeds cynicism about representative democracy the world over.

Indigenous rights, in contrast, are not dependent on the whims of politicians. The position of the Elsipogtog First Nation was that no treaty gave the Canadian government the authority to radically alter their ancestral lands. The right to hunt and fish, affirmed by the *Marshall* decision, was violated by industrial activity that threatened the fundamental health of the lands and waters (since what good is having the right to fish, for instance, when the water is polluted?). Gary Simon of the Elsipogtog First Nation explains, "I believe our treaties are the last line of defense to save the clean water for future generations." [13]

It's the same position the Lummi have taken against the coal export terminal near Bellingham, Washington, arguing that the vast increase in tanker traffic in the Strait of Georgia, as well as the polluting impacts of coal dust, violates their treaty-protected right to fish those waters. (The Lower Elwha Klallam tribe in Washington State made similar points when its leaders fought to remove two dams on the Elwha River. They argued, successfully, that by interfering with salmon runs the dam violated their treaty rights to fish.) And when the U.S. State Department indicated, in February 2014, that it might soon be offering its blessing to the Keystone XL pipeline, members of the Lakota Nation immediately announced that they considered the pipeline construction illegal. As Paula Antoine, an employee of the Rosebud tribe's land office, explained, because the pipeline passes through Lakota treaty-protected traditional territory, and very close to reservation land, "They aren't recognizing our treaties, they are violating our treaty rights and our boundaries by going through there. Any ground disturbance around that proposed line will affect us."[14]

These rights are real and they are powerful, all the more so because many of the planet's largest and most dangerous unexploded carbon bombs lie beneath lands and waters to which Indigenous peoples have legitimate legal claims. No one has more legal power to halt the reckless expansion of the tar sands than the First Nations living downstream whose treaty-protected hunting, fishing, and trapping grounds have already been fouled, just as no one has more legal power to halt the rush to drill under the Arctic's melting ice than Inuit, Sami, and other northern Indigenous tribes whose livelihoods would be jeopardized by an offshore oil spill. Whether they are able to exercise those rights is another matter.

This power was on display in January 2014 when a coalition of Alaskan Native tribes, who had joined forces with several large green groups, won a major court victory against Shell's already scandal-plagued Arctic drilling adventures. Led by the Native village of Point Hope, the coalition argued that when the U.S. Interior Department handed out drilling permits to Shell and others in the Chukchi Sea, it failed to take into account the full risks, including the risks to Indigenous Inupiat ways of life, which are inextricably entwined with a healthy ocean. As Port Hope mayor Steve Oomittuk explained when the lawsuit was launched, his people "have hunted and depended on the animals that migrate through the Chukchi Sea for thousands of years. This is our garden, our identity, our livelihood. Without it we would not be who we are today. . . . We oppose any activity that will endanger our way of life and the animals that we greatly depend on." Faith Gemmill, executive director of Resisting Environmental Destruction on Indigenous Lands, one of the groups behind the lawsuit, notes that for the Inupiat who rely on the Chukchi Sea, "you cannot separate environmental impacts from subsistence impacts, for they are the same."[15]

A federal appeals court ruled in the coalition's favor, finding that the Department of the Interior's risk assessments were based on estimates that were "arbitrary and capricious," or presented "only the *best* case scenario for environmental harm."[16] Rather like the shoddy risk assessments that set the stage for BP's Deepwater Horizon disaster.

John Sauven, executive director of Greenpeace U.K., described the ruling as "a massive blow to Shell's Arctic ambitions." Indeed just days later, the company announced that it was putting its Arctic plans on indefinite hold. "This is a disappointing outcome, but the lack of a clear path forward means that I am not prepared to commit further resources for drilling in Alaska in 2014," said Shell CEO Ben van Beurden. "We will look to relevant agencies and the Court to resolve their open legal issues as quickly as possible." Without Indigenous groups raising the human rights stakes in this battle, it's a victory that might never have taken place.[17]

Worldwide, companies pushing for vast new coal mines and coal export terminals are increasingly being forced to similarly reckon with the unique legal powers held by Indigenous peoples. For instance, in Western Australia in 2013 the prospect of legal battles over native title was an important factor in derailing a planned $45 billion LNG (liquefied natural gas) processing plant and port, and though the state government remains determined to force gas infrastructure and fracking on the area, Indigenous groups are threatening to assert their traditional ownership and procedural rights in court. The same is true of communities facing coal bed methane development in New South Wales.[18]

Meanwhile, several Indigenous groups in the Amazon have been steadfastly holding back the oil interests determined to sacrifice new swaths of the great forests, protecting both the carbon beneath the ground and the carbon-capturing trees and soil above those oil and gas deposits. They have asserted their land rights with increasing success at the Inter-American Court of Human Rights, which has sided with Indigenous groups against governments in cases involving natural resource and territorial rights.[19] And the U'wa, an isolated tribe in Colombia's Andean cloud forests—where the tree canopy is perpetually shrouded in mist—have made history by resisting repeated attempts by oil giants to drill in their territory, insisting that stealing the oil beneath the earth would bring about the tribe's destruction. (Though some limited drilling has taken place.)

As the Indigenous rights movement gains strength globally, huge advances are being made in recognizing the legitimacy of these claims. Most significant was the United Nations Declaration on the Rights of Indigenous Peoples, adopted by the General Assembly in September 2007 after 143 member states voted in its favor (the four opposing votes—United States, Canada, Australia, and New Zealand—would each, under domestic pressure, eventually endorse it as well). The declaration states that, "Indigenous peoples have the right to the conservation and protection of the environment and the productive capacity of their lands or territories and resources." And further that they have "the right to redress" for the lands that "have been confiscated, taken, occupied, used or damaged without their free, prior and informed consent." Some countries have even taken the step of recognizing these rights in revised constitutions. Bolivia's constitution, approved by voters in 2009, states that Indigenous peoples "are guaranteed the right to prior consent: obligatory consultation by the government, acting in good faith and in agreement, prior to the exploitation of non-renewable natural resources in the territory they inhabit." A huge, hard-won legal victory.[20]

Might vs. Rights

And yet despite growing recognition of these rights, there remains a tremendous gap between what governments say (and sign) and what they do—and there is no guarantee of winning when these rights are tested in court. Even in countries with enlightened laws as in Bolivia and Ecuador, the state still pushes ahead with extractive projects without the consent of the Indigenous people who rely on those lands.[21] And in Canada, the United States, and Australia, these rights are not only ignored, but Indigenous people know that if they try to physically stop extractive projects that are clearly illegal, they will in all likelihood find themselves on the wrong side of a can of pepper spray—or the barrel of a gun. And while the lawyers argue the intricacies of land title in court, buzzing chainsaws proceed to topple trees that are four times as old as our countries, and toxic fracking fluids seep into the groundwater.

The reason industry can get away with this has little to do with what is legal and everything to do with raw political power: isolated, often impoverished Indigenous peoples generally lack the monetary resources and social clout to enforce their rights, and anyway, the police are controlled by the state. Moreover the costs of taking on multinational extractive companies in court are enormous. For instance in the landmark "Rainforest Chernobyl" case in which Ecuador's highest court ordered Chevron to pay $9.5 billion in damages, a company spokesman famously said: "We're going to fight this until hell freezes over—and then we'll fight it out on the ice." (And indeed, the fight still drags on.)[22]

I was struck by this profound imbalance when I traveled to the territory of the Beaver Lake Cree Nation in northern Alberta, a community that is in the midst of one of the highest-stakes legal battles in the tar sands. In 2008, the band filed a historic lawsuit charging that by allowing its traditional territories to be turned into a latticework of oil and gas infrastructure, and by poisoning and driving away the local wildlife, the provincial and federal governments, as well as the British Crown, had infringed no fewer than fifteen thousand times on the First Nation's treaty rights to continue to hunt, fish, and trap on their territory.[23] What set the case apart was that it was not about one particular infringement, but an entire model of poisonous, extractive development, essentially arguing that this model itself constituted a grave treaty violation.

"The Governments of Canada and Alberta have made a lot of promises to our people and we intend to see those promises kept," said Al Lameman, the formidable chief of the Beaver Lake Cree Nation at the time the lawsuit was filed (Lameman had made history before, filing some of the first Indigenous human rights challenges against the Canadian government). Against the odds, the case has proceeded through the Canadian court system, and in March 2012 an Alberta court flatly rejected government efforts to have the case dismissed as "frivolous," an "abuse of the Court's process," and "unmanageable."[24]

A year after that ruling, I met Al Lameman, now retired, and his cousin Germaine Anderson, an elected band councilor, as well as the former chief's niece, Crystal Lameman, who has emerged as one of the most compelling voices against the tar sands on the international stage. These are three of the people most responsible for moving the lawsuit forward, and Germaine Anderson had invited me to a family barbecue to discuss the case.

It was early July and after a long dark winter it was as if a veil had lifted: the sun was still bright at 10 p.m. and the northern air had a thin, baked quality. Al Lameman had aged considerably in recent years and slipped in and out of the conversation. Anderson, almost painfully shy, had also struggled with her health. The spot where the family met for this gathering was where she spent the summer months: a small trailer in a clearing in the woods, without running water or electricity, entirely off the grid.

I knew the Beaver Lake Cree were in a David and Goliath struggle. But on that endless summer evening, I suddenly understood what this actually meant: some of the most marginalized people in my country—many of them, like all the senior members of the Lameman clan, survivors of the intergenerational trauma of abusive residential schools—are taking on some of the wealthiest and most powerful forces on the planet. Their heroic battles are not just their people's best chance of a healthy future; if court challenges like Beaver Lake's can succeed in halting tar sands expansion, they could very well be the best chance for the rest of us to continue enjoying a climate that is hospitable to human life.

That is a huge burden to bear and that these communities are bearing it with shockingly little support from the rest of us is an unspeakable social injustice.

A few hours north, a different Indigenous community, the Athabasca Chipewyan First Nation (ACFN), recently launched another landmark lawsuit, this one taking on Shell and the Canadian government over the approval of a huge tar sands mine expansion. The band is also challenging another Shell project, the proposed Pierre River Mine, which it says "would significantly impact lands, water, wildlife and the First Nation's ability to utilize their traditional territory." Once again the mismatch is staggering. The ACFN, with just over one thousand members and an operating budget of about $5 million, is battling both the Canadian government and Shell, with its 92,000 employees across more than seventy countries and 2013 global revenues of $451.2 billion. Many communities see odds like these and, understandably, never even get in the ring.[25]

It is this gap between rights and resources—between what the law says and what impoverished people are able to force vastly more powerful entities to do—that government and industry have banked on for years.

"Honour the Treaties"

What is changing is that many non-Native people are starting to realize that Indigenous rights—if aggressively backed by court challenges, direct action, and mass movements demanding that they be respected—may now represent the most powerful barriers protecting all of us from a future of climate chaos.

Which is why, in many cases, the movements against extreme energy extraction are becoming more than just battles against specific oil, gas, and coal companies and more, even, than pro-democracy movements. They are opening up spaces for a historical reconciliation between Indigenous peoples and non-Natives, who are finally understanding that, at a time when elected officials have open disdain for basic democratic principles, Indigenous rights are not a threat, but a tremendous gift. Because the original Indigenous treaty negotiators in much of North America had the foresight to include language protecting their right to continue living off their traditional lands, they bequeathed to all residents of these and many other countries the legal tools to demand that our governments refrain from finishing the job of flaying the planet.

And so, in communities where there was once only anger, jealousy, and thinly veiled racism, there is now something new and unfamiliar. "We're really thankful for our First Nations partners in this struggle," said Lionel Conant, a property manager whose home in Fort St. James, British Columbia, is within sight of the proposed Northern Gateway pipeline. "[They've] got the legal weight to deal with [the pipeline] . . . because this is all unceded land." In Washington State, anti-coal activists talk about the treaty rights of the Lummi as their "ace in the hole" should all other methods of blocking the export terminals fail. In Montana, the Sierra Club's Mike Scott told me bluntly, "I don't think people understand the political power Natives have as sovereign nations, often because they lack the resources to exercise that power. They can stop energy projects in a way we can't." [26]

In New Brunswick, Suzanne Patles, a Mi'kmaq woman involved in the anti-fracking movement, described how non-Natives "have reached out to the Indigenous people to say 'we need help.'" [27] Which is something of a turnaround from the saviorism and pitying charity that have poisoned relationships between Indigenous peoples and well-meaning liberals for far too long.

It was in the context of this gradual shift in awareness that Idle No More burst onto the political scene in Canada at the end of 2012 and then spread quickly south of the border. North American shopping centers—from the enormous West Edmonton Mall to Minnesota's Mall of America—were suddenly alive with the sounds of hand drums and jingle dresses as Indigenous people held flash mob round dances across the continent at the peak of the Christmas shopping season. In Canada, Native leaders went on hunger strikes, and youths embarked on months-long spiritual walks and blockaded roads and railways.

The movement was originally sparked by a series of attacks by the Canadian government on Indigenous sovereignty, as well as its all-out assault on existing environmental protections, particularly for water, to pave the way for rapid tar sands expansion, more mega-mines, and projects like Enbridge's Northern Gateway pipeline. The attacks came in the form of two omnibus budget bills passed in 2012 that gutted large parts of the country's environmental regulatory framework. As a result, a great many industrial activities were suddenly exempt from federal environmental reviews, which along with other changes, greatly reduced opportunities for community input and gave the intractable right-wing government of Stephen Harper a virtual free hand to ram through unpopular energy and development projects. The omnibus bills also overhauled key provisions of the Navigable Waters Protection Act that protect species and ecosystems from damage. Previously, virtually 100 percent of the country's water bodies had been covered by these protections; under the new order, that was slashed to less than 1 percent, with pipelines simply exempted. (Documents later revealed that the latter change had been specifically requested by the pipeline industry.)[28]

Canadians were in shock at the extent and speed of the regulatory overhaul. Most felt powerless, and with good reason: despite winning only 39.6 percent of the popular vote, the Harper government had a majority in Parliament and could apparently do as it pleased.[29] But the First Nations' response was not to despair; it was to launch the Idle No More movement from coast to coast. These laws, movement leaders said, were an attack on Indigenous rights to clean water and to maintain traditional ways of life. Suddenly, the arguments that had been made in local battles were being taken to the national level, now used against sweeping federal laws. And for a time Idle No More seemed to change the game, attracting support from across Canadian society, from trade unions to university students, to the opinion pages of mainstream newspapers.

These coalitions of rights-rich-but-cash-poor people teaming up with (relatively) cash-rich-but-rights-poor people carry tremendous political potential. If enough people demand that governments honor the legal commitments made to the people on whose land colonial nations were founded, and do so with sufficient force, politicians interested in reelection won't be able to ignore them forever. And the courts, too—however much they may claim to be above such influences—are inevitably shaped by the values of the societies in which they function. A handful of courageous rulings notwithstanding, if an obscure land right or treaty appears to be systematically ignored by the culture as a whole, it will generally be treated tentatively by the courts. If, however, the broader society takes those commitments seriously, then there is a far greater chance that the courts will follow.*

As Idle No More gained steam, many investors took notice. "For the

---

* Indeed, it may be no coincidence that in June 2014, the Supreme Court of Canada issued what may be its most significant indigenous rights ruling to date when it granted the Tsilhqot'in Nation a declaration of Aboriginal title to 1,750 square kilometers of land in British Columbia. The unanimous decision laid out that ownership rights included the right to use the land, to decide how the land should be used by others, and to derive economic benefit from the land. Government, it also stated, must meet certain standards before stepping in, and seek not only consultation with First Nations, but consent from them. Many commented that it would make the construction of controversial projects like tar sands pipelines—rejected by local First Nations—significantly more difficult.

first time in six years, Canadian provinces failed to top the list of the best mining jurisdictions in the world in a 2012/13 survey," Reuters reported in March 2013. "Companies that participated in the survey said they were concerned about land claims." The article quoted Ewan Downie, chief executive of Premier Gold Mines, which owns several projects in Ontario: "I would say one of the big things that is weighing on mining investment in Canada right now is First Nations issues."[30]

Writing in *The Guardian*, journalist and activist Martin Lukacs observed that Canadians seemed finally to be grasping that

> implementing Indigenous rights on the ground, starting with the United Nations Declaration on the Rights of Indigenous Peoples, could tilt the balance of stewardship over a vast geography: giving Indigenous peoples much more control, and corporations much less. Which means that finally honoring Indigenous rights is not simply about paying off Canada's enormous legal debt to First Nations: it is also our best chance to save entire territories from endless extraction and destruction. In no small way, the actions of Indigenous peoples—and the decision of Canadians to stand alongside them—will determine the fate of the planet.
>
> This new understanding is dawning on more Canadians. Thousands are signing onto educational campaigns to become allies to First Nations. . . . Sustained action that puts real clout behind Indigenous claims is what will force a reckoning with the true nature of Canada's economy—and the possibility of a transformed country. That is the promise of a growing mass protest movement, an army of untold power and numbers.[31]

In short, the muscle able to turn rights into might that Standard & Poor's had been looking for in that meeting with Arthur Manuel and Guujaaw back in 2004 may have finally developed.

The power of this collaboration received another boost in January 2014 when the rock legend Neil Young kicked off a cross-Canada tour called "Honour the Treaties." He had visited the tar sands several months earlier and been devastated by what he saw, saying (to much controversy) that the region "looks like Hiroshima." While in the region, he had met with Chief

Allan Adam of the Athabasca Chipewyan and heard about the lawsuits opposing Shell's tar sands expansions, as well as the health impacts current levels of oil production are already having on the community. "I was sitting with the chief in the teepee, on the reserve. I was hearing the stories. I saw that the cancer rate was up among all the tribes. This is not a myth. This is true," Young said.[32]

And he concluded that the best way he could contribute to the fight against the tar sands was to help the Athabasca Chipewyan First Nation exercise its rights in court. So he went on a concert tour, donating 100 percent of the proceeds to the court challenges. In addition to raising $600,000 for their legal battles within two months, the tour attracted unprecedented national attention to both the local and global impacts of runaway tar sands development. The prime minister's office fought back by attacking one of Canada's most beloved icons, but it was a losing battle. Prominent Canadians spoke up to support the campaign, and polls showed that even in Alberta a majority were taking Young's side in the dispute.[33]

Most importantly, the Honour the Treaties tour sparked a national discussion about the duty to respect First Nation legal rights. "It's up to Canadians all across Canada to make up their own minds about whether their integrity is threatened by a government that won't live up to the treaties that this country is founded on," Young said. And the country heard directly from Chief Allan Adam, who described the treaties his ancestors signed as "not just pieces of paper but a last line of defense against encroaching reckless tar sands development that my people don't want and that we are already suffering from."[34]

The Moral Imperative of Economic Alternatives

Making the most of that last line of defense is a complex challenge involving much more than rock concerts and having cash in hand to pay lawyers. The deeper reason why more First Nations communities aren't taking on companies like Shell has to do with the systematic economic and social disenfranchisement that makes doing business with heavily polluting oil or mining companies seem like the only way to cover basic human needs. Yes, there is a desire to protect the rivers, streams, and oceans for traditional fishing. But in Canada, according to a 2011 government report, the water systems in 25 percent of First Nations communities are so neglected and underfunded that they pose a "high overall risk" to health, while thousands of residents of Native reserves are living without sewage or running water at all. If you are the leader of one such community, getting those basic services taken care of, no matter the cost, is very likely going to supersede all other priorities. [35]

And ironically, in many cases, climate change is further increasing the economic pressure on Indigenous communities to make quick-and-dirty deals with extractive industries. That's because disruptive weather changes, particularly in northern regions, are making it much harder to hunt and fish (for example when the ice is almost never solid, communities in the far north become virtually trapped, unable to harvest food for months on end). All this makes it extremely hard to say no to offers of job training and re-source sharing when companies like Shell come to town. Members of these communities know that the drilling will only make it harder to engage in subsistence activities—there are real concerns about the effects of oil de-velopment on the migration of whales, walruses, and caribou—and that's without the inevitable spills. But precisely because the ecology is already so disrupted by climate change, there often seems no other option.

The paucity of good choices is perhaps best on display in Greenland, where receding glaciers and melting ice are revealing a vast potential for new mines and offshore oil exploration. The former Danish colony gained home rule in 1979, but the Inuit nation still relies on an annual infusion of more than $600 million (amounting to a full third of the economy) from Denmark. A 2008 self-governance referendum gave Greenland still more control over its own affairs, but also put it firmly on the path of drilling and mining its way to full independence. "We're very aware that we'll cause more climate change by drilling for oil," a top Greenlandic official, then heading the Office of Self-Governance, said in 2008. "But should we not? Should we not when it can buy us our independence?" Currently, Green-land's largest industry is fishing, which of course would be devastated by a major spill. And it doesn't bode well that one of the companies selected to begin developing Greenland's estimated fifty billion barrels of offshore oil and gas is none other than BP.[36]

Indeed the melancholy dynamic strongly recalls BP's "vessels of opportunity" program launched in the midst of the Deepwater Horizon disaster. For months, virtually the entire Louisiana fishing fleet was docked, unable to make a living for fear that the seafood was unsafe. That's when BP offered to convert any fishing vessel into a cleanup boat, providing it with booms to (rather uselessly) mop up some oil. It was tremendously difficult for local shrimpers and oystermen to take work from the company that had just robbed them of their livelihood—but what choice did they have? No one else was offering to help pay the bills. This is the way the oil and gas industry holds on to power: by tossing temporary life rafts to the people it is drowning.

That many Indigenous people would view the extractive industries as their best of a series of bad options should not be surprising. There has been almost no other economic development in most Native communities, no one else offering jobs or skills training in any quantity. So in virtually every community on the front lines of extractive battles, some faction invariably makes the argument that it's not up to Indigenous people to sacrifice to save the rest of the world from climate change, that they should concentrate instead on getting better deals from the mining and oil companies so that they can pay for basic services and train their young people in marketable skills. Jim Boucher, chief of the Fort McKay First Nation, whose lands have been decimated by the Alberta tar sands, told an oil-industry-sponsored conference in 2014, "There is no more opportunity for our people to be employed or have some benefits except the oil sands"—going so far as to call the mines the "new trap line," a reference to the fur trade that once drove the economics of the region.[37]

Sadly, this argument has created rancorous divisions and families are often torn apart over whether to accept industry deals or to uphold traditional teachings. And as the offers from industry become richer (itself a sign of Blockadia's growing power), those who are trying to hold the line too often feel they have nothing to offer their people but continued impoverishment. As Phillip Whiteman Jr., a traditional Northern Cheyenne storyteller and longtime opponent of coal development, told me, "I can't keep asking my people to suffer with me."[38]

These circumstances raise troubling moral questions for the rising Blockadia movement, which is increasingly relying on Indigenous people to be the legal barrier to new, high-carbon projects. It's fine and well to laud treaty and title rights as the "last line of defense" against fossil fuel extraction. But if non-Native people are going to ask some of the poorest, most systematically disenfranchised people on the planet to be humanity's climate saviors, then, to put it crassly, what are we going to do for them? How can this relationship not be yet another extractive one, in which non-Natives use hard-won Indigenous rights but give nothing or too little in return? As the experience with carbon offsets shows, there are plenty of examples of new "green" relationships replicating old patterns. Large NGOs often use Indigenous groups for their legal standing, picking up some of the costs for expensive legal battles but not doing much about the underlying issues that force so many Indigenous communities to take these deals in the first place. Unemployment stays sky high. Options, for the most part, stay bleak.

If this situation is going to change, then the call to Honour the Treaties needs to go a whole lot further than raising money for legal battles. Non-Natives will have to become the treaty and land-sharing partners that our ancestors failed to be, making good on the full panoply of promises they made, from providing health care and education to creating economic opportunities that do not jeopardize the right to engage in traditional ways of life. Because the only people who will be truly empowered to say no to dirty development over the long term are people who see real, hopeful alternatives. And this is true not just within wealthy countries but between the countries of the wealthy postindustrial North and the fast-industrializing South.

SHARING THE SKY

The Atmospheric Commons and the Power of Paying Our Debts

"The forest is already 'developed,' the forest is life."

—Franco Viteri, Sarayaku leader, Ecuador[1]

"How, in the North, could anything like this ever be possible? How, given the madness that has come upon the wealthy countries, one in which ideologues and elites have cast a mythology of 'debt crisis' and 'bitter medicine' and 'austerity' over all claims to the commonwealth, could the North ever accept the necessity of large-scale financial and technological investments in a climate mobilisation, including massive support to the South? . . . How, given the North's fear of a rising Asia, and its stubborn insistence that the South is both unwilling and unable to restrain its own emissions, will the North ever come to see the implacability of the logic—the fear of a foreclosed future—which most deeply animates the South's negotiators? And how, given that the North's blindness on these points is an almost perfect, ready-made excuse for its own continued free-riding, can there be any path to rapidly increased global ambition that does not begin in the North?"

—Sivan Kartha, Tom Athanasiou, and Paul Baer, climate researchers, 2012[2]

I saw this new kind of partnership in action while reporting on one of the highest-stakes fronts of the fossil fuel wars, in southeastern Montana. There, underneath the rolling hills dotted with cattle, horses, and otherworldly sandstone rock formations, sits a whole lot of coal. So much that you can see it in seams on the side of the road. The region has enough coal to supply current levels of U.S. consumption for nearly two hundred years.[3] Indeed, much of the coal that the industry plans to export to China would come from mines presently planned for this part of the world, all of it impacting, in one form or another, the Northern Cheyenne. The industry wants the coal under and near their reservation, and, as discussed, it wants to build a railroad skirting their reservation to get that coal out—which, together with the mine, would threaten the Tongue River, a key water source.

The Northern Cheyenne have been fighting off the mining companies since the early 1970s, in part due to an important Sweet Medicine prophecy that is often interpreted to mean that digging up the "black rock" would bring on a kind of madness and the end of Cheyenne culture. But when I first visited the reserve in 2010, the region was in the throes of the fossil fuel frenzy, getting hit from every direction—and it wasn't clear how long the community's anti-coal forces were going to be able to hold out.

After an ugly battle, coal opponents had just lost a crucial vote at the State Land Board about the proposed new coal mine at Otter Creek, just outside the boundaries of the Northern Cheyenne Reservation (this was the site visited by the Lummi carvers on their ceremonial totem pole journey). Otter Creek was the biggest new coal mine under consideration in the United States, and at this point it seemed certain to go ahead. Attention was shifting to opposing the construction of the artery needed to get the coal out of the mine, the proposed Tongue River Railroad—which would likely impact Cheyenne burial grounds. Like the tar sands pipelines, it had become a "chokepoint" battle: without the railroad, there would be no hope of getting the coal out, and therefore no point in building the new mine.

But back in 2010, the railroad fight hadn't succeeded in galvanizing the Northern Cheyenne in opposition and it also looked likely to go ahead. Meanwhile, on the neighboring Crow Reservation, there was a plan to build a coal-to-liquids plant, a noxious process that turns coal rock into a highly polluting form of liquid fuel, which emits twice as much carbon as regular gasoline when burned. The Australian company behind the plant, which called the project "Many Stars," had commissioned a famous Crow artist to create its logo: two tepees against a starry sky.[4]

The Sierra Club's Mike Scott described his work to me as "triage"—

racing to try to stop or slow one terrible idea after another. His partner, Alexis Bonogofsky, told me at the time, "There is so much going on, people don't know what to fight."[5] From their goat ranch outside Billings, the pair would go off in different directions every day, trying to beat back yet another offensive in the fossil fuel frenzy.

Bonogofsky's official job tittle is "tribal lands program manager" for the National Wildlife Federation, which in practice meant helping Indigenous tribes exercise their legal rights in order to protect the land, air, and water. The tribe with whom she was working most closely was the Northern Cheyenne, both because they were in the bull's-eye of new coal development and because they had a long history of using the law for land stewardship. For instance, the Northern Cheyenne broke legal ground by arguing that their right to enjoy a traditional way of life included the right to breathe clean air. In 1977, the EPA agreed and granted the Northern Cheyenne Reservation the highest possible classification for its air quality (called Class I under the Clean Air Act). This seemingly bureaucratic technicality allowed the tribe to argue in court that polluting projects as far away as Wyoming were a violation of its treaty rights, since the pollutants could travel to the Northern Cheyenne Reservation and potentially compromise its air and water quality.

Bonogofsky, in her familiar plaid flannel shirts and cowboy boots, was spending many hours a week in her white pickup truck, driving from the ranch to Lame Deer, the small, scrubby town at the center of the Northern Cheyenne Reservation. There, more often than not, she would end up in the converted Mormon church that houses the tribe's environmental protection offices, meeting and scheming with the department's tough and tireless director, Charlene Alden.

Alden has been an anchor of the Northern Cheyenne's long battle against coal and she has won some major victories, like stopping the dumping of untreated wastewater from coal bed methane directly into the Tongue River. But when we met, she wasn't sure how much longer she could hold off the pro-mining forces.

The problems were as much internal as external. The tribe had elected a former coal miner as tribal president and he was determined to open up the lands to the extractive industries. On the day I arrived, pink flyers had appeared on the community bulletin board announcing that in the elections coming up in ten days, tribal members would be polled on their views about developing the reservation's coal and methane.

Charlene Alden was furious about the flyers. The wording was biased, and she said the process violated several election rules. But she also knew why some of her people were tempted to take the money. Tribal unemployment was as high as 62 percent, and by some estimates significantly higher. Substance abuse was ravaging the reservation (a mural at the center of town depicted crystal meth as an evil-eyed green snake being fought off with sacred arrows). And these problems had been plaguing the community for a very long time. In 1995, Alden made a video that aired on the ABC current affairs program *Day One*, cohosted by Diane Sawyer, which for its time was a breakthrough in Indigenous representation on network television. Formatted as a video journal, the piece was a meditation on historical trauma that showed shocking footage of Alden's own sister drinking the toxic cleaning liquid Lysol out of a plastic jug; "Cheyenne champagne," it was called.[6]

It was desperation like that which made it possible for mining companies like Arch and Peabody to gain a ready audience when they sailed into town promising jobs and money to fund new social programs. "People say we have high unemployment, we have no tax base. If we go ahead and do this we can have good schools, a good waste system," Alden told me. And there is no question that "the tribal government has no money at all." But she worried that sacrificing the health of the land for coal dollars would only further alienate Cheyenne people from their culture and traditions, very likely causing more depression and abuse, not less. "In Cheyenne, the word for water is the same as the word for life," she said. "We know that if we start messing around too much with coal, it destroys life."[7]

The only way to break the deadlock, Alden had come to believe, was to prove to the next generation of Cheyenne leaders that there is another path out of poverty and hopelessness—one that does not involve handing over the land for which their ancestors paid so dearly.

And she saw no end to the possibilities. As we were speaking, a colleague popped into Alden's office and told her that someone had broken into the building the night before and stolen an electric heater. Alden was not surprised. It was fall, the nighttime temperatures were dropping, and rez houses are notoriously drafty, most of them having been built from government-issue kits in the 1940s and 1950s. You can see the hinges connecting the walls. Residents blast the heat (even turn on their ovens for backup) only to have it fly through the cracks in the walls, windows, and doors. As a result, heating bills are staggeringly high—$400 a month is average, but I met people with bills that topped $1,000 in winter. And since the heat was coming from coal and propane, it was contributing to the climate crisis that was already hitting this region hard, with persistent droughts and massive wildfires.

To Alden, everything was wrong with that picture—the expensive bills, the crummy housing, the dirty energy source. And it all pointed to the tremendous untapped opportunities for models of development that respect, rather than violate, Cheyenne values. For instance, the converted church where we were sitting had just been fitted with new windows as part of an energy conservation program and Alden was thrilled with the results: the new windows saved heating costs, let in more natural light, and installing them had created jobs for community members. But the scale was so small. Why couldn't there be a program to install windows like these in all the houses on the reservation?

An NGO had come in a few years ago and built a handful of model homes made of straw bale, an ancient form of architecture that keeps buildings cool in summer and warm in winter. Today, Alden reported with some amazement, those families had minuscule electricity bills—"$19 a month instead of $400!" But she couldn't see why the tribe needed outsiders to build homes that are based on Indigenous knowledge. Why not train tribal members to design and build them and get funding to do it across the reservation? There would be a green building boom in no time and the trainees could use their skills elsewhere too. Montana, meanwhile, provides excellent conditions for both wind and rooftop solar.

This takes money though, and money is what the Northern Cheyenne do not have. There had been hopes that President Obama would significantly increase funding for green jobs in disadvantaged communities, but those plans had mostly been shelved in the wake of the economic crisis. Bonogofsky, however, was convinced that finding ways to help the Northern Cheyenne fulfill their aspirations for real economic alternatives to coal was just as important as helping to pay for their anti-coal lawsuits. So she and Alden got to work.

About a year after my visit to the church, Bonogofsky called to say that they had managed to scrape together some money—from the Environmental Protection Agency and from her own NGO—for an exciting new project. Henry Red Cloud, a Lakota social entrepreneur who had won awards for his work bringing wind and solar power to the Pine Ridge Reservation in South Dakota, was going to teach a group of about a dozen Northern Cheyenne to install solar heaters on their reservation homes. The heaters are worth $2,000 each and they would be going in for free, cutting heating bills by as much as half. Did I want to come back to Montana?

The Sun Comes Out

My return trip could not have been more different from the first. It was spring of 2011 and those gentle hills around the reservation were now covered in tiny yellow wildflowers that somehow made the grass look video-game green. The trainings were already under way and about fifteen people had gathered on the lawn of a home to learn how a simple box made mostly of dark glass could capture enough heat to warm an entire house.

Red Cloud, a natural leader with a gift for making his courses feel like a meeting of friends, effortlessly wove technical lessons about passive solar systems with meditations about how "solar power was always part of Natives' way of life. Everything followed the *anpetuwi tawonawaka*, the life-giving force of the sun. It ties in with our culture, our ceremony, our language, our songs."[8]

Each installation began with Red Cloud walking around the house holding up a pocket Solar Pathfinder that told him where the sun would hit on every day of the year. The solar boxes are placed on the sides of the buildings and need at least six hours of daily radiant sunlight to run effectively. A couple of houses were nestled too tightly against trees and mountains to make them strong candidates. For these, roof panels might be used instead, or another power source entirely.

Red Cloud, a former metalworker who used to earn his living on large industrial sites, clearly enjoyed this flexible aspect to renewable power; he called the fiddling and adapting he does "Indianizing" and reminisced about building his first wind turbine out of a 1978 Chevy Blazer that was rusting on the rez. Watching him pace around these homes, a twinkle in his eye, it struck me that this need to adapt to nature is what drives some people mad about renewables: even at a very large scale, they require a humility that is the antithesis of damming a river, blasting bedrock for gas, or harnessing the power of the atom. They demand that we adapt ourselves to the rhythms of natural systems, as opposed to bending those systems to our will with brute force engineering. Put another way, if extractive energy sources are NFL football players, bashing away at the earth, then renewables are surfers, riding the swells as they come, but doing some pretty fancy tricks along the way.

It was precisely this need to adapt ourselves to nature that James Watt's steam engine purportedly liberated us from in the late 1770s, when it freed factory owners from having to find the best waterfalls, and ship captains from worrying about the prevailing winds. As Andreas Malm writes, the first commercial steam engine "was appreciated for having no ways or places of its own, no external laws, no residual existence outside that brought forth by its proprietors; it was absolutely, indeed *ontologically* subservient to those who owned it."[9]

It is this powerfully seductive illusion of total control that a great many boosters of extractive energy are so reluctant to relinquish. Indeed at the climate change denial conference hosted by the Heartland Institute, renewables were derided as "sunbeams and friendly breezes"—the subtext was clear: real men burn coal.[10] And there is no doubt that moving to renewables represents more than just a shift in power sources but also a fundamental shift in power *relations* between humanity and the natural world on which we depend. The power of the sun, wind, and waves can be harnessed, to be sure, but unlike fossil fuels, those forces can never be fully possessed by us. Nor do the same rules work everywhere.

So now we find ourselves back where we started, in dialogue with nature. Proponents of fossil and nuclear energy constantly tell us that renewables are not "reliable," by which they mean that they require us to think closely about where we live, to pay attention to things like when the sun shines and when the wind blows, where and when rivers are fierce and where they are weak.* And it's true: renewables, at least the way Henry Red Cloud sees them, require us to unlearn the myth that we are the masters of nature—the "God Species"—and embrace the fact that we are in relationship with the rest of the natural world. But ours is a new level of relationship, one based on an understanding of nature that far surpasses anything our pre–fossil fuel ancestors could have imagined. We know enough to know how much we will never know, yet enough to find ingenious ways to amplify the systems provided by nature in what feminist historian Carolyn Merchant has described as a "partnership ethic."[11]

It is this collaborative quality that resonated most powerfully with Red Cloud's students. Landon Means, a recent college grad who had just moved back to the reservation, told me that he saw in solar energy a shift in worldview that was about "working synergistically" with the earth, "instead of just using it." This insight seemed to hit hardest with the young Cheyenne men who had spent time working in the coal industry and were tired of suppressing core parts of their identity to earn a paycheck. During the lunch break on the first day of the training, Jeff King, one of the Cheyenne students, confessed that he was still working in Gillette, Wyoming—ground zero of the Powder River Basin coal boom. He described it balefully as the "carbon capital of the world" and clearly wanted out. He hadn't intended to drive trucks to coal mines for a living; a decade ago he had been one of the most promising Cheyenne students of his generation and had gone to Dartmouth on scholarship to study art, which he describes as "a calling." But the coal boom sucked him in. Now, he said, he wasn't sure how he could go back to Gillette. He huddled with a couple of friends to discuss starting their own solar company to serve the reservation.[12]

One of the last houses to get a solar air heater was on a busy street in downtown Lame Deer. As Red Cloud's students measured, drilled, and hammered, they started to draw a crowd. Kids gathered to watch the action.

---

* Renewables are, in fact, much more reliable than power based on extraction, since those energy models require continuous new inputs to avoid a crash, whereas once the initial investment has been made in renewable energy infrastructure, nature provides the raw materials for free.

Old women asked what was going on. "Half the cost of electricity? Really? How do I get one?"

Red Cloud smiled. This is his marketing strategy for building a solar revolution in Indian Country. The first step, he says, it to get "a few solar panels over on Grandma's house. Everyone sees Grandma and says, 'What is that? I want that too.'" Alexis Bonogofsky, meanwhile, beamed from the sidelines. "This has been probably the best week I have ever had at this job—it felt different," she told me as the training wrapped up. "It feels like something has changed."[13]

———

In the coming months, several members of the initial group continued to train with Red Cloud and others joined them, making pilgrimages to his school, the Red Cloud Renewable Energy Center on the Pine Ridge Reservation. Jeff King quit his coal job in Gillette and set about getting a solar business off the ground. The money wasn't as good but, he said, "I have a direction now."

One of Red Cloud's star students turned out to be a twenty-nine-year-old woman named Vanessa Braided Hair, who more than held her own with the power tools in the mostly male class. She worked seasonally as a firefighter for the Bureau of Indian Affairs and in the summer of 2012 battled an unprecedented wildfire that burned over 230 square kilometers (90 square miles) and destroyed nineteen homes on the Northern Cheyenne Reservation alone. (As the Associated Press reported at the time, the fire "ripped through as if the land had been doused with gasoline.") Braided Hair did not need anyone to tell her that climate change was an existential crisis and welcomed the chance to contribute to the solution. But it went deeper than that. Solar power, she said, embodied the worldview in which she had been raised, one in which "You don't take and take and take. And you don't consume and consume and consume. You take what you need and then you put back into the land."[14]

Red Cloud tells his students that deriving energy in a way that heals and protects the natural world is not just about employment. It's a continuation of "what the ancestors shed blood for, always fought for—the earth."

And he says he is training them not just to be technicians but to be "solar warriors."[15]

I confess that when I first heard that I thought it was another example of Red Cloud's marketing flair. But in the months and years that followed, I watched his prediction come true in the lives of the young people he had taught. In 2012, with his training still under way, the fight against the mines and the coal train on the Northern Cheyenne Reservation—the fight that had seemed all but lost in 2010—sprang back to life. Suddenly there was no shortage of Cheyennes willing to hold protests, demand meetings with regulators, or make impassioned speeches at hearings. And Red Cloud's solar warriors were front and center, dressed in red "Beyond Coal" T-shirts and declaring themselves "Idle No More," a reference to the movement that had started in Canada and had swept through Indigenous communities across the continent.

At a technical hearing for the proposed massive coal mine at Otter Creek, Vanessa Braided Hair pulled no punches: "I want you to know that many people do not see any difference between your agency and Arch Coal," she told a panel of squirming officials, including the head of Montana's Department of Environmental Quality. Lucas King, twenty-eight, and another Red Cloud student, told a different hearing on the Otter Creek mine, "This is Cheyenne country. It has been for a long time, longer than any dollar has ever lasted. I don't expect you guys to understand us. You don't. And I'm not saying I understand you. But I know you guys understand 'no.'" He concluded his remarks by saying: "Please go back and tell whoever you have to that we don't want it. It's not for us. Thank you." The room broke out in applause. A new generation of warriors had been born.[16]

Today, the mood among dirty-energy opponents in southeastern Montana is positively jubilant. They speak about "when" they will stop the railroad, not if. Which, if true, means the Otter Creek mine cannot proceed. And there is far less talk of mining on the Cheyenne Reservation itself. The plan for a coal-to-liquid plant on the Crow Reservation is also dead. Mike Scott from the Sierra Club has been working with Crow members on building a wind farm.

What this part of the world has clearly shown is that there is no more potent weapon in the battle against fossil fuels than the creation of real alternatives. Just the glimpse of another kind of economy can be enough to energize the fight against the old one. There are also powerful precedents for this: in two of the countries with the largest commitment to decentralized, community-controlled renewable power—Denmark and Germany—these energy victories trace their roots back to the antinuclear movement. In both countries, communities were forcefully opposed to the risks associated with nuclear power plants, but they knew that to win, they needed an alternative. So instead of just saying no, they demanded government policies that would allow the communities in question to generate their own clean power and earn revenue in the process. Large-scale victories like these, however, are hard to achieve when the communities lack political power. It's clear from the European examples that renewable energy can be a viable alternative to extraction for Indigenous people around the world; it can provide skills training, jobs, and steady revenue streams for impoverished communities. But opportunities are consistently lost.

For instance, the Black Mesa Water Coalition, founded in 2001 by a group of Navajo and Hopi youth in Arizona, won a pivotal battle in 2005 when it helped shut down the notoriously polluting Mohave Generating Station as well as the Black Mesa Mine. But coal mining and coal-power generation continue on Navajo territory, helping to light up and pump water to large stretches of Arizona, including Phoenix, along with parts of Nevada and California. The mining puts the water supply at risk but the Black Mesa activists know that there is no hope of shutting it all down until they are able to provide tangible alternatives to their people. So in 2010, they came up with a highly detailed proposal to convert land that the mining industry had abandoned, land still likely contaminated and depleted, and use it to host vast solar arrays that could power not just their reservation but also large urban centers. Since the infrastructure and transmission lines are already in place, thanks to the coal industry, it would be just a matter of converting the power source. As Jihan Gearon, executive director of the coalition, puts it, "Why not turn those lands into something positive that could bring in monetary income to the people who live in that region and begin that transition away from coal?" But under this plan, the Navajos—not an outside multinational energy company—would be the owners of the power they produced and sold to the grid. And the money generated would be able to support traditional economies, such as Navajo weaving. That is what made the plan different: this time, the arrangement would be non-extractive in every sense—the poisons would stay in the ground, and the money and skills would stay in the community.[17]

Yet half a decade later, this elegant plan is still struggling to get off the ground. As always, a major barrier has been funding. And that is a problem not only for Black Mesa but for everyone concerned about climate change—because if the Navajo cannot show that clean energy can provide a route out of poverty and toward real self-determination, then the coal mining will continue, to everyone's detriment. Part of the job of the climate movement, then, is to make the moral case that the communities who have suffered most from unjust resource relationships should be *first* to be supported in their efforts to build the next, life-based economy now.

And that means a fundamentally new relationship, in which those communities have full control over resource projects, so that they become opportunities for skills training, jobs, and steady revenues (rather than one-off payments). This point needs to be stressed because far too many large-scale renewable energy projects are being imposed on Native lands without proper consultation and consent, replicating old colonial patterns in which profits (and skills and jobs) go to outsiders. The shift from one power system to another must be more than a mere flipping of a switch from underground to aboveground. It must be accompanied by a power correction in which the old injustices that plague our societies are righted once and for all. That's how you build an army of solar warriors.

––––––

The need to provide tangible economic alternatives to extraction is not only pressing in Native communities, of course. The impossible choices faced by the Navajo Nation and the Northern Cheyenne are intensified versions of the same nonchoices offered to a great many low-income communities where the present is so difficult and the pressures to provide the basics of life are so great that focusing on the future can seem like an impossible luxury. Holding on to family farms in the face of fierce competition from Big Ag, for instance, is so tough that there is never any shortage of farmers and ranchers willing to make some extra money by leasing land to fracking or pipeline companies—even if that means going to war with their neighbors who oppose these practices, and even if it means imperiling their own water supply and livestock. Desperate people do desperate things.

The same goes for many of the workers who want to build those pipelines, frack that gas, or work in polluting refineries. Manufacturing in North America is as battered as family farming, which means that well-paying union jobs are so scarce that people will fight for whatever jobs are on offer, no matter how dangerous, precarious, or polluting to themselves, their families, or their own communities. The solution, as the more visionary sectors of the labor movement understand, is to fight for policies that do not force workers to make those kinds of choices.

For instance, a 2012 study from the Canadian Centre for Policy Alternatives compared the public value from a $5 billion pipeline—the rough cost of Enbridge's Northern Gateway—and the value that could be derived from investing the same amount in green economic alternatives. It found that if $5 billion is spent on a pipeline, it produces mostly short-term construction jobs, big private sector profits, and heavy public costs for future environmental damage. But if $5 billion is spent on public transit, building retrofits, and renewable energy, economies can gain, at the very least, three times as many jobs in the short term, while simultaneously helping to reduce the chances of catastrophic warming in the long term. In fact, the number of jobs could be many times more than that, according to the institute's modeling. At the highest end, green investment could create *thirty-four times* more jobs than just building another pipeline.[18]

The problem, of course, is that while companies like Enbridge are putting dollars on the table to build pipelines, governments are unwilling to make comparable sums available for these alternatives. And yet in Canada a minimal national carbon tax of $10 a ton would raise $5 billion a year, the sum in question—and unlike a one-off pipeline investment, it would do so year after year.[19] If policy options like that were on the table, the jobs vs. environment dichotomy would all but evaporate.

Which is another reason why today's climate movement does not have the luxury of simply saying no without simultaneously fighting for a series of transformative yeses—the building blocks of our next economy that can provide good clean jobs, as well as a social safety net that cushions the hardships for those inevitably suffering losses.

Don't Just Divest, Reinvest

As discussed, the resources for this just transition must ultimately come from the state, collected from the profits of the fossil fuel companies in the brief window left while they are still profitable. But until a shift in the political tides makes that necessity a reality, there are ways to start funneling much needed resources to the next economy right now. This is shaping up to be the most exciting aspect of the growing fossil fuel divestment movement: increasingly, participants aren't just calling on public interest institutions like colleges and municipalities to sell their holdings in the companies that are wrecking the planet, they are also asking them to reinvest that money in entities that have a clear vision for the healing process.

Dan Apfel, former head of the Responsible Endowments Coalition, and a key advisor to the movement, argues that "our colleges, other charities, pension funds, and foundations must be the ones to lead the way." He points out, "Five percent of the money in these public purpose and publicly related institutions adds up to about $400 billion. Four hundred billion dollars in new investments could stimulate real climate solutions, help create the market for further investments, encourage policy change and sustain financial returns long into the future."[20]

Already, the group of foundations and wealthy individuals that has joined the fossil fuel divestment movement (see page 357) has taken the additional step of moving the funds that had been profiting from fossil fuel companies and reinvesting them in the clean tech sector (the intiative has become known as "Divest-Invest"). Some colleges are taking a similar approach. As economic analysts Jeremy Brecher, Brendan Smith, and Kristen Sheeran note, "Duke University in North Carolina has invested $8 million in the Self-Help Credit Union, in part to fund affordable green housing. Carleton College in Minnesota and Miami University in Florida are directing investment into renewable energy funds."[21]

These big investors are taking a solid first step; even better would be if they dedicated a share of their investments to projects that go deeper: not just switching from brown energy to green energy but supporting cutting-edge projects that are designed to bolster local economies, improve public transit, and otherwise strengthen the starved public sphere. Critically, smart reinvestment strategies can even give the communities at the front lines of fossil fuel extraction the economic tools they need to resist carbon pollution at source—like the Black Mesa Water Coalition's plan for a municipal-scale solar utility, or the solar co-ops employing growing numbers of African American and Latino workers in Richmond, California, who might otherwise see no option besides the Chevron refinery. Brecher, Smith, and Sheeran elaborate on these kinds of creative possibilities for how the divestment movement can "leverage its power to build a new sustainable economy for both the planet and local communities":

> Institutions should think in much more positive terms: How can their money maximize the transition to a new sustainable economy? Here's one place to start: There are hundreds of community investment funds, socially oriented banks and credit unions, union pension funds, and other financial vehicles that have long experience in investing for social purposes. There are thousands of co-ops, worker- and community-owned businesses, non-profits, municipal initiatives, and other enterprises that are engaged on a small scale in creating a new economy.
>
> These are the elements of a growing sector of enterprises devoted to public purposes with augmented control by workers and employees. They are insulating and solarizing buildings; expanding public transportation; developing low-carbon equipment and techniques for schools and hospitals; developing new recycling systems for handling waste. They are thereby also creating community-based economies that provide economic security, empower local and workplace democracy, and ward off the running away of jobs. But this is a sector that is generally starved for capital. Expanding the resources to grow this sector as rapidly as possible should be a priority for divestors.[22]

The main power of divestment is not that it financially harms Shell and Chevron in the short term but that it erodes the social license of fossil fuel companies and builds pressure on politicians to introduce across-the-board emission reductions. That pressure, in turn, increases suspicions in the investment community that fossil fuel stocks are overvalued. The benefit of an accompanying reinvestment strategy, or a visionary investment strategy from the start, is that it has the potential to turn the screws on the industry much tighter, strengthening the renewable energy sector so that it is better able to compete directly with fossil fuels, while bolstering the frontline land defenders who need to be able to offer real economic alternatives to their communities.

All of this points to something else that sets Blockadia apart from many previous social movements of its kind. In the past, people committed to social change often believed they had to choose between fighting the system and building alternatives to it. So in the 1960s, the counterculture splintered between those who stayed in cities to try to stop wars and bash away at inequalities and those who chose to drop out and live their ecological values among like-minded people on organic farms or in manageable-sized cities like Bellingham, Washington. The activists and the exodus.

Today's activists don't have the luxury of these choices even if they wanted them. During these times of continual economic stress and exclusion, the communities on the front lines of saying no to dirty energy have discovered that they will never build the base they need unless they can simultaneously provide economic alternatives to the projects they are opposing. So after three years of just saying no to the Keystone XL pipeline, a group of farmers in Nebraska came up with just such a strategy: they built a barn, powered by wind and solar, in the pipeline's path. And they pointed out that the power generated from just that one barn would bring more energy to the region than the oil in the pipeline that was headed for the export terminal in Texas.[23] On one level, the Build Our Energy Barn was just PR: the farmers were daring President Obama to tear down a renewable energy installation to make way for dirty oil. But it also showed their neighbors that, if the right policies are in place, there is another way to earn some much needed extra income without putting their land at risk.

Similarly, after the British village of Balcombe in West Sussex was the site of huge anti-fracking protests and angry clashes in 2013, a new power company formed called REPOWERBalcombe. Its goal is "to supply the equivalent of 100% of Balcombe's electricity demand through community owned, locally generated renewable energy"—with financing coming from residents buying shares in the energy co-operative. The fracking fight continues to play out in the courts, but solar panels are already on their way and residents who were originally in favor of oil and gas drilling are joining the co-op, attracted to the promise of self-sufficiency and cost-saving.[24] A comparable process is underway in Pungesti, the Romanian farming community fighting fracking. The claims by Chevron's supporters that gas extraction is the only option for jobs in this poor region of the country forced fracking opponents to put forward proposals of their own—like a community wind farm, a processing plant for the vegetables grown locally, and an abattoir for their livestock, all of which would add value to the livelihoods that are the region's heritage.

In short, some of the most tangible responses to the ecological crisis today come not from utopian dropout projects, but rather are being forged in the flames of resistance, by communities on the front lines of the battles against extreme extraction. And at the same time, many of those who, decades ago, built alternatives at the local level are finding themselves forced back to the barricades. That's because many of the most idyllic pockets where the sixties-era dropouts went to build their utopias are suddenly under siege: oil and coal tankers threaten their shores, oil and coal trains threaten their downtowns, and frackers want their land.

And even in places that are lucky enough to have been spared (so far) all these threats, climate change is demolishing the idea that any countercultural pocket can provide a safe haven. In August 2011, that became clear to the organic farmers in Vermont who had pioneered one of the most advanced and sustainable local agriculture systems in North America. Probably most famous is the Intervale, a network of urban farms in Burlington that supplies roughly 10 percent of the city's fresh food, while at the same time composting its waste and sustainably generating a significant portion of its power. But when Hurricane Irene descended on the state, floodwaters destroyed not only historic covered bridges but as Bill McKibben, a Vermonter and staunch supporter of food localization, said to me shortly after, "It washed away huge amounts of that beautiful local agriculture. The Intervale in Burlington is suddenly under five feet of water. Nothing gets harvested there. There are tons of farms where the beautiful, rich topsoil is now just covered with feet of sand from the river." He took away from that experience the fact that "If we can't solve the climate problem, then all the rest of this is for naught."[25]

I witnessed something similar, if on a smaller scale, in New York City one year later in the immediate aftermath of Superstorm Sandy. While visiting Red Hook, Brooklyn, one of the hardest hit neighborhoods, I stopped by the Red Hook Community Farm—an amazing place that teaches kids from nearby housing projects how to grow healthy food, provides composting for a huge number of residents, hosts a weekly farmer's market, and runs a Community Supported Agriculture (CSA) program, getting all kinds of produce to people who need it. Not only was the farm improving the lives of people in the neighborhood, it was also doing everything right from a climate perspective—reducing food miles; staying away from petroleum inputs; sequestering carbon in the soil; reducing landfill by composting. But when the storm came, none of that mattered. The entire fall harvest was lost. And the urban farmers I met there—still in shock from seeing so much collective work gone to waste—were preoccupied with the fear that the water that had inundated the fields had been so toxic that they would need to bring in new soil.

In short, dropping out and planting vegetables is not an option for this generation. There can be no more green museums because the fossil fuels runaway train is coming for us one way or another. There may have been a time when engaging in resistance against a life-threatening system and building alternatives to that system could be meaningfully separated, but today we have to do both simultaneously: build and support inspiring alternatives like the Red Hook Community Farm—and make sure they have a fighting chance of thriving by trying to change an economic model so treacherous that nowhere is safe. John Jordan, a longtime ecological activist in Britain and France, describes resistance and alternatives as "the twin strands of the DNA of social change. One without the other is useless."[26]

The denizens of Blockadia live and know this. Which is why theirs is neither a movement of negation (no to the miners/drillers/pipe layers/heavy haulers), nor solely of protection (defending cherished but static ways of life). Increasingly, it is also a constructive movement, actively building an alternative economy based on very different principles and values.

They are also learning—in a kind of people's inversion of the shock doctrine—that one of the most opportune times to build that next economy may be in the aftermath of disasters, particularly climate-related disasters. That's because recurring mega-tragedies like Superstorm Sandy and Typhoon Haiyan that kill thousands and cause billions in damages serve dramatically to educate the public about the terrible costs of our current system, driving an argument for radical change that addresses the root, rather than only the symptoms, of the climate crisis. In the outpourings of volunteerism and donations, as well as the rage at any whiff of profiteering, these disasters also activate the latent and broadly shared generosity that capitalism works so hard to deny. Not to mention the fact that, as the disaster capitalists well know, these events result in a whole lot of public money being put on the table—an increasingly rare event during times of relentless economic austerity.

With the right kind of public pressure, that money can be marshaled not just to rebuild cities and communities, but to transform them into models of nonextractive living. This can go far beyond the usual calls for stronger seawalls: activists can demand everything from free, democratically controlled public transit, to more public housing along those transit lines, powered by community-controlled renewable energy—with the jobs created by this investment going to local workers and paying a living wage. And unlike the disaster capitalists who use crises to end-run around democracy, a People's Recovery (as many from the Occupy movement called for post-Sandy) would require new democratic processes, including neighborhood assemblies, to decide how hard-hit communities should be rebuilt. The overriding principle must be to address the twin crises of inequality and climate change at the same time.

One example of this kind of inverted shock doctrine took place in the rural town of Greensburg, Kansas. In 2007, a super tornado ripped through the area, turning about 95 percent of the town into rubble. As a result of an extraordinary, community-led process that began just days after the disaster, with neighbors holding meetings in tents amid the wreckage of their former lives, Greensburg today stands as a model "green town," often described as the greenest in America. The hospital, city hall, and school have all been built to the highest certification level issued by Leadership in Energy and Environmental Design (LEED). And the town has become a destination for hundreds of policy makers, anxious to learn more about its low-energy lighting and its cutting-edge green architecture and waste reduction, as well as the wind turbines that earn municipal revenue by producing more power than local residents need.[27]

Most striking of all, this "living laboratory" is taking place in the heart of an overwhelmingly Republican-voting county, where a great many people are entirely unconvinced that climate change is real. But those debates seem to matter little to residents: the shared experience of tremendous loss, as well as the outpouring of generosity that follow the disaster, have, in Greensburg, rekindled the values of land stewardship and intergenerational responsibility that have deep roots in rural life. "The number one topic at those tent meetings was talking about who we are—what are our values?" recalls Greensburg mayor Bob Dixson, a former postmaster who comes from a long line of farmers. He added, "Sometimes we agreed to disagree, but we were still civil to each other. And let's not forget that our ancestors were stewards of the land. My ancestors lived in the original green homes: sod houses. . . . We learned that the only true green and sustainable things in life are how we treat each other."[28]

Responding to disaster with this kind of soul-searching is profoundly different from the top-down model of the shock doctrine—these are attempts not to exploit crisis, but to harness it to actually solve the underlying problems at their root, and in ways that expand democratic participation rather than the opposite. After Hurricane Katrina, New Orleans became a laboratory for corporate interests intent on capturing and radically shrinking the public sphere, attacking public health and education and leaving the city far more vulnerable to the next disaster. But there is no reason why future disasters cannot be laboratories for those who believe in reviving and reinventing the commons, and in ways that actively reduce the chances that we will all be battered by many more such devastating blows in the future.

From Local to Global Debts

On my first visit to the Northern Cheyenne Reservation, the question of how to finance the kind of healthy economy anti-coal activists were fighting for came up often. At one point, Lynette Two Bulls, who runs an organization that teaches Cheyenne youth about their history, told me that she had heard about something exciting happening in Ecuador. She was talking about the call for the international community to compensate the country for not extracting the oil in the Yasuní rainforest, with the money raised going to social programs and a clean energy transition. It sounded like just what was needed on the reservation and she wanted to know: if Ecuador could be compensated for keeping its oil in the ground, then why couldn't the Northern Cheyenne be compensated for being carbon keepers for their coal?

It was a very good question, and the parallels were striking. Yasuní National Park is an extraordinary swath of Ecuadorian rainforest, home to several Indigenous tribes and a surreal number of rare and exotic animals (it has nearly as many species of trees in 2.5 acres as are native to all of North America). And underneath that riot of life sits an estimated 850 million barrels of crude oil, worth about $7 billion. Burning that oil—and logging the rainforest to get it—would add another 547 million tons of carbon dioxide to the atmosphere. Of course the oil majors want in.[29]

So in 2006, the environmental group Acción Ecológica (the same group that made an early alliance with the anti-oil movement in Nigeria) put forward a counterproposal: the Ecuadorian government should agree not to sell the oil, but it should be supported in this action by the international community, which would benefit collectively from the preservation of biodiversity and from keeping planet-warming gases out of our shared atmosphere. That would mean partially compensating Ecuador for what it would have earned from oil revenues had it opted to drill. As Esperanza Martínez, president of Acción Ecológica, explained, the "proposal establishes a precedent, arguing that countries should be rewarded for not exploiting their oil. . . . Funds gathered would be used for the [renewable] energy transition and could be seen as payments for the ecological debt from North to South, and they should be distributed democratically at the local and global levels." Besides, she writes, surely "the most direct way to reduce emissions of carbon dioxide was to leave fossil fuels in the ground."[30]

The Yasuní plan was based on the premise that Ecuador, like all developing countries, is owed a debt for the inherent injustice of climate change—the fact that wealthy countries had used up most of the atmospheric capacity for safely absorbing CO_2 before developing countries had a chance to industrialize. And since the entire world would reap the benefits of keeping that carbon in the ground (since it would help stabilize the global climate), it is unfair to expect Ecuador, as a poor country whose people had contributed little to the climate crisis, to shoulder the economic burden for giving up those potential petro dollars. Instead, that burden should be shared between Ecuador and the highly industrialized countries most responsible for the buildup of atmospheric carbon. This is not charity, in other words: if wealthy countries do not want poorer ones to pull themselves out of poverty in the same dirty way that we did, the onus is on Northern governments to help foot the bill.

This, of course, is the core of the argument for the existence of a "climate debt"—the same argument that Bolivia's climate negotiator, Angélica Navarro Llanos, had laid out for me in Geneva in 2009, helping me to see how climate change could be the catalyst to attack inequality at its core, the basis for a "Marshall Plan for the Earth."[31] The math behind the argument is simple enough. As discussed, climate change is the result of *cumulative* emissions: the carbon dioxide we emit stays in the atmosphere for approximately one to two centuries, with a portion remaining for a millennium or even more.[32] And since the climate is changing as a result of two-hundred-odd years of such accumulated emissions, that means that the countries that have been powering their economies with fossil fuels since the Industrial Revolution have done far more to cause temperatures to rise than those that just got in on the globalization game in the last couple of decades.[33] Developed countries, which represent less than 20 percent of the world's population, have emitted almost 70 percent of all the greenhouse gas pollution that is now destabilizing the climate. (The United States alone, which comprises less than 5 percent of the global population, now contributes about 14 percent of all carbon emissions.)[34]

And while developing countries like China and India spew large (and rapidly growing) amounts of carbon dioxide, they are not equally responsible for the cost of the cleanup, the argument goes, because they have contributed only a fraction of the two hundred years of cumulative pollution that has caused the crisis. Moreover, not everyone needs carbon for the same sorts of things. For instance, India still has roughly 300 million people living without electricity. Does it have the same degree of responsibility to cut its emissions as, say, Britain, which has been accumulating wealth and emitting industrial levels of carbon dioxide ever since James Watt introduced his successful steam engine in 1776?[35]

Of course not. That is why 195 countries, including the United States, ratified the United Nations Framework Convention on Climate Change in 1992, which enshrines the principle of "common but differentiated responsibilities." That basically means that everyone is responsible for being part of the climate solution but the countries that have emitted more over the past century should be the first to cut and should also help finance poorer countries to switch to clean development models.[36]

Few dispute that climate debt is an argument with justice and international law on its side. And yet Ecuador's attempt to put that principle into practice in the forest has been fraught with difficulties and may well fail. Once again, being right, and even having rights, is not enough on its own to move the rich and powerful.

In 2007, the center-left government of Rafael Correa took up the Yasuní proposal and championed it, albeit briefly, on the world stage. Inside Ecuador, the Yasuní-ITT initiative, as the plan is known (named for the coveted Ishpingo, Tambococha, and Tiputini oil fields inside the park), became a populist rallying cry, a vision for real economic development that did not require sacrificing some of the most cherished parts of the country. A 2011 poll found that 83 percent of Ecuadorians supported leaving the Yasuní's oil in the ground, up from 41 percent in 2008, a measure of how quickly a transformative vision can capture the public imagination. But contributions from developed countries were slow to arrive (only $13 million of a $3.6 billion goal was raised), and in 2013 Correa announced that he was going to allow drilling to begin.[37]

Local supporters of the plan, however, have not given up and Correa's backsliding has opened a new Blockadia front: protestors opposing drilling have already faced arrests and rubber bullets and, in the absence of a political solution, Indigenous groups are likely to resist extraction with their bodies. Meanwhile, in April 2014, a coalition of NGOs and citizen groups collected more than 750,000 signatures calling for the matter to be put to a national referendum (at the time of publication, it seemed that Correa was determined to block the vote and push ahead with drilling). As Kevin Koenig, Ecuador program director at Amazon Watch, wrote in *The New York Times*, "Though the government should be held to account," this is not all Correa's fault. "The stillbirth of Yasuní-ITT is a shared failure."[38]

This setback, moreover, is a microcosm of the broader failure of the international climate negotiations, which have stalled again and again over the central question of whether climate action will reflect the history of who created the crisis. The end result: emissions keep soaring way past safe levels, everyone loses, the poorest lose first and worst.

Giving up on real solutions, like the imaginative one first proposed to save the Yasuní, is therefore not an option. As with Indigenous land rights, if governments are unwilling to live up to their international (and domestic) responsibilities, then movements of people have to step into that leadership vacuum and find ways to change the power equation.

The right, as usual, understands this better than the left, which is why the climate change denial crowd consistently claims that global warming is a socialist conspiracy to redistribute wealth (the Competitive Enterprise Institute's Chris Horner likes to say rich countries are being "extorted" by the poor).[39] Climate debt is not extortion but climate change, when fully confronted, does raise some awfully thorny questions about what we in the wealthy world owe to the countries on the front lines of a crisis they had little hand in creating. At the same time, as elites in countries like China and India grow ever more profligate in their consumption and emissions, traditional North-South categories begin to break down and equally tough questions are raised about the responsibilities of the rich and the rights of the poor wherever they happen to live in the world. Because without facing those questions, there is no hope of getting emissions under control where it counts the most.

As we have seen, emissions in North America and Europe still need to come down dramatically but, thanks largely to the offshoring of production enabled by the free-trade era, they have pretty much stopped growing. It's the fast-rising economies of the Global South—with China, India, Brazil, and South Africa leading the pack—that are mostly responsible for the surge in emissions in recent years, which is why we are racing toward tipping points far more quickly than anticipated.

The reason for this shift in the source of emissions has everything to do with the spectacular success multinational corporations have had in globalizing the high-consumption-based economic model pioneered in wealthy western countries. The trouble is, the atmosphere can't take it. As the atmospheric physicist and mitigation expert Alice Bows-Larkin put it in an interview, "The number of people that went through industrialization the first time around is like a drop in the ocean compared to the number of people going through industrialization this time." And to quote President Obama in late 2013, if China's and India's energy consumption imitates the U.S. model, "we'll be four feet under water."[40]

The truth is—and this is a humbling thing for cultures accustomed to assuming that our actions shape the destiny of the world to accept—the real battle will not be lost or won by us. It will be won or lost by those movements in the Global South that are fighting their own Blockadia-style struggles—demanding their own clean energy revolutions, their own green jobs, their own pools of carbon left in the ground. And they are up against powerful forces within their own countries that insist that it is their "turn" to pollute their way to prosperity and that nothing matters more than economic growth. Indeed, citing the rank unfairness of expecting developing countries to bear the bulk of the burden for humanity avoiding climate catastrophe has become an enormously effective excuse for governments of the Global South to shirk their own responsibilities.

For this reason, if we accept the scientific evidence that we need to act fast to prevent catastrophic climate change, it makes sense to focus our action where it can have the greatest impact. And that's clearly in the Global South. To cite just one example: about a third of all greenhouse gas emissions come from buildings (heating, cooling, and lighting them). Building stock in the Asia Pacific region is projected to grow by a dramatic 47 percent by 2021, while remaining relatively stable in the developed world. This means that, while making existing buildings more energy efficient is important wherever we live, there is nothing more important than helping ensure that new structures in Asia are built to the highest standards of efficiency. Otherwise, we are all—the North, South, East, and West—in for catastrophic emissions growth.[41]

Tipping the Balance

There is, however, much that can be done in the industrialized north to help tip the balance of forces toward a model of development that does not rely on endless growth and dirty fuels. Fighting the pipelines and export terminals that would send fossil fuels to Asia is one piece of the puzzle. So is battling new free-trade deals, reining in our own overconsumption, and sensibly relocalizing our economies, since plenty of the carbon China is burning is going toward making useless stuff for us.

But the most powerful lever for change in the Global South is the same as in the Global North: the emergence of positive, practical, and concrete alternatives to dirty development that do not ask people to choose between higher living standards and toxic extraction. Because if dirty coal is the only way to turn on the lights in India, then that is how those lights will be turned on. And if public transit is a disaster in Delhi, then more and more people will keep choosing to drive cars.

And there are alternatives—models of development that do not require massive wealth stratification, tragic cultural losses, or ecological devastation. As in the Yasuní case, movements in the Global South are fighting hard for these alternative development models—policies that would bring power to huge numbers of people through decentralized renewable energy and revolutionize urban transportation so that public transit is far more desirable than private cars (indeed, as discussed, there have been riots demanding free public transit in Brazil).

One proposal receiving increasing attention is for a "global feed-in tariff," which would create an internationally administered fund to support clean energy transitions throughout the developing world. The architects of this plan—economist Tariq Banuri and climate expert Niclas Hällström—estimate that a $100 billion annual investment for ten to fourteen years "could effectively help 1.5 billion people gain access to energy, while taking decisive steps toward a renewable energy future in time to prevent all our societies from suffering from climate catastrophe."[42]

Sunita Narain, director general of one of the most influential environmental organizations in India, the New Delhi–based Centre for Science and Environment, stresses that the solution is not for the wealthy world to contract its economies while allowing the developing world to pollute its way to prosperity (even if this were possible). It is for developing countries to "develop differently. We do not want to first pollute and then clean up. So we need money, we need technology, to be able to do things differently."[43] And that means the wealthy world must pay its climate debts.

And yet financing a just transition in fast-developing economies has not been a priority of activists in the North. Indeed a great many Big Green groups in the United States consider the idea of climate debt to be politically toxic, since, unlike the standard "energy security" and green jobs arguments that present climate action as a race that rich countries can win, it requires emphasizing the importance of international cooperation and solidarity.

Sunita Narain hears these objections often. "I'm always being told—especially by my friends in America—that . . . issues of historical responsibility are something that we should not talk about. What my forefathers did is not my responsibility." But, she said in an interview, this overlooks the fact that those past actions have a direct bearing on why some countries are rich and others are poor. "Your wealth today has a relationship with the way society has drawn on nature, and overdrawn on nature. That has to be paid back. That's the historical responsibility issues that we need to confront."[44]

These debates are, of course, familiar from other reparations battles. In Latin America, progressive economists have long argued that Western powers owe an "ecological debt" for centuries of colonial land grabs and resource extraction, while Africa and Caribbean governments have, at various points (most notably the 2001 World Conference Against Racism in Durban, South Africa) called for reparations to be paid for transatlantic slavery. After receding for more than a decade after the Durban conference, these claims were back in the news in 2013 when fourteen Caribbean nations banded together to make a formal reparations claim to

Britain, France, the Netherlands, and other European countries that participated in the slave trade. "Our constant search and struggle for development resources is linked directly to the historical inability of our nations to accumulate wealth from the efforts of our peoples during slavery and colonialism," said Baldwin Spencer, prime minister of Antigua and Barbuda, in July 2013. The goal of reparations, he argued, was to break the chains of dependency once and for all.[45]

The rich world, for the most part, pretty much ignores these calls, dismissing it all as ancient history, much as the U.S. government manages to disregard calls for slavery reparations from African Americans (though in the spring of 2014, the calls grew distinctly louder, thanks to breakthrough reporting by *The Atlantic*'s Ta-Nehisi Coates, which once again rekindled the debate).[46] But the case for climate debt is a little different. We can debate the legacy of colonialism, and we can argue about how much slavery shapes modern underdevelopment. But the science of climate change doesn't leave much room for that kind of disagreement. Carbon leaves an unmistakable trail, the evidence etched in coral and ice cores. We can accurately measure how much carbon we can collectively emit into the atmosphere and who has taken up what share of that budget over the past two hundred years or so.

On the other hand, all of these various suppressed and neglected debts are not separate from one another but are better understood as different chapters in the same, continuous story. It was planet-warming coal that powered the textile mills and sugar refineries in Manchester and London that needed to be fed with ever more raw cotton and sugarcane from the colonies, most of it harvested by slave labor. And Eric Williams, the late scholar and Trinidad's first prime minister, famously argued that profits from slavery directly subsidized the growth of industrialization in England, a process that we now know led inextricably to climate change. The details of Williams's claims have long been vigorously debated, but his work received additional vindication in 2013 when researchers at University College London released a database collecting information on the identities and finances of British slave-owners in the mid-nineteenth century.[47]

The research project delved into the fact that when the British Parliament ruled to abolish slavery in its colonies in 1833, it pledged to compensate British slave owners for the loss of their human property—a backward form of reparations for the perpetrators of slavery, not its victims. This led to payouts adding up to £20 million—a figure that, according to *The Independent*, "represented a staggering 40 per cent of the Treasury's annual spending budget and, in today's terms, calculated as wage values, equates to around £16.5bn." Much of that money went directly into the coal-powered infrastructure of the now roaring Industrial Revolution—from factories to railways to steamships. These, in turn, were the tools that took colonialism to a markedly more rapacious stage, with the scars still felt to this day.[48]

Coal didn't create structural inequality—the boats that enabled the transatlantic slave trade and first colonial land grabs were powered by wind, and the early factories powered by water wheels. But the relentless and predictable power of coal certainly supercharged the process, allowing both human labor and natural resources to be extracted at rates previously unimaginable, laying down the bones of the modern global economy.

And now it turns out that the theft did not end when slavery was abolished, or when the colonial project faltered. In fact, it is still in progress, because the emissions from those early steamships and roaring factories were the beginning of the buildup of excess atmospheric carbon. So another way of thinking about this history is that, starting two centuries ago, coal helped Western nations to deliberately appropriate other people's lives and lands; and as the emissions from that coal (and later oil and gas) continually built up in the atmosphere, it gave these same nations the means to inadvertently appropriate their descendants' sky as well, gobbling up most of our shared atmosphere's capacity to safely absorb carbon.

As a direct result of these centuries of serial thefts—of land, labor, and atmospheric space—developing countries today are squeezed between the impacts of global warming, made worse by persistent poverty, and by their need to alleviate that poverty, which, in the current economic system, can be done most cheaply and easily by burning a great deal more carbon, dramatically worsening the climate crisis. They cannot break this deadlock without help, and that help can only come from those countries and corporations that grew wealthy, in large part, as a result of those illegitimate appropriations.

The difference between this reparations claim and older ones is not that the case is stronger. It's that it does not rest on ethics and morality alone: wealthy countries do not just need to help the Global South move to a low-emissions economic path because it's the right thing to do. We need to do it because our collective survival depends on it.

At the same time, we need common agreement that having been wronged does not grant a country the right to repeat the same crime on an even grander scale. Just as having been raped does not bestow the right to rape, or having been robbed the right to steal, having been denied the opportunity to choke the atmosphere with pollution in the past does not grant anyone the right to choke it today. Especially because today's polluters know full well the catastrophic implications of that pollution in ways that early industrialists did not.

So a middle ground must be found. Fortunately, a group of researchers with the think tank EcoEquity and the Stockholm Environment Institute have attempted to do just that: they have developed a detailed and innovative model of what a rigorously fair approach to emission reductions might look like on a global scale. Called the "Greenhouse Development Rights" framework, it is an attempt to better reflect the new realities of wealth and carbon pollution moving to the developing world, while firmly protecting the right to sustainable development and recognizing the West's greater responsibility for cumulative emissions. Such an approach, they believe, is precisely what's needed to break the climate deadlock, since it addresses "the vast disparities found not only between but also *within* countries." Northern countries could be assured that the wealthy of the Global South will do their part now and in the future, while access to what is left of the atmospheric commons would be properly safeguarded for the poor.[49]

With this in mind, each country's fair share of the global carbon-cutting burden is determined by two key factors: responsibility for historical emissions and capacity to contribute, based on the country's level of development. In an illustrative scenario, the United States's share of global emissions cuts needed by the end of the decade might be something like 30 percent (the largest of any single country). But not all of those reductions would need to be done at home—some could be met by financing and otherwise supporting the transition to low-carbon pathways in the South.

And according to the researchers, with every nation's share of the global burden clearly defined and quantified, there would be no need to rely on ineffective and easily gamed market mechanisms like carbon trading.[50]

At a time when rich countries are crying poor and slashing social services for their own people, asking governments to make those kinds of international commitments can sound impossible. We barely do the old kind of aid anymore, let alone some ambitious new justice-based approach. But there are, in the immediate term, plenty of affordable ways for Northern countries to begin to honor our climate debts without going broke—from erasing the foreign debts currently owed by developing countries in exchange for climate action to loosening green energy patents and transferring the associated technological know-how.

Moreover, much of the cost does not need to come from regular taxpayers; it can and should come from the corporations most responsible for driving this crisis. That can take the form of any combination of the polluter-pays measures already discussed, from a financial transaction tax, to eliminating subsidies for fossil fuel companies.

What we cannot expect is that the people least responsible for this crisis will foot all, or even most, of the bill. Because that is a recipe for catastrophic amounts of carbon ending up in our common atmosphere. Like the call to honor our treaties and other land-sharing agreements with Indigenous peoples, climate change is once again forcing us to look at how injustices that many assumed were safely buried in the past are shaping our shared vulnerability to global climate collapse.

With many of the biggest pools of untapped carbon on lands controlled by some of the poorest people on the planet, and with emissions rising most rapidly in what were, until recently, some of the poorest parts of the world, there is simply no credible way forward that does not involve redressing the real roots of poverty.

THE RIGHT TO REGENERATE

Moving from Extraction to Renewal

"Stop calling me resilient. I'm not resilient. Because every time you say, 'Oh, they're resilient,' you can do something else to me."

—Tracie Washington, New Orleans–based civil rights attorney, 2010[1]

"That woman is the first environment is an original instruction. In pregnancy our bodies sustain life. . . . At the breast of women, the generations are nourished. From the bodies of women flows the relationship of those generations both to society and to the natural world. In this way is the earth our mother, the old people tell us. In this way, we as women are earth."

—Katsi Cook, Mohawk midwife, 2007[2]

At the beginning of this book, I wrote about how becoming a mother in an age of extinction brought the climate crisis into my heart in a new way. I had felt the crisis before, of course, as all of us do on some level. But for the most part, my climate fears expressed themselves as low-level melancholy, punctuated by moments of panic, rather than full-blown grief.

At some point about seven years ago, I realized that I had become so convinced that we were headed toward a grim ecological collapse that I was losing my capacity to enjoy my time in nature. The more beautiful and striking the experience, the more I found myself grieving its inevitable loss—like someone unable to fall fully in love because she can't stop imagining the inevitable heartbreak.

Looking out at an ocean bay on British Columbia's Sunshine Coast, a place teeming with life, I would suddenly picture it barren—the eagles, herons, seals, and otters, all gone. It got markedly worse after I covered the BP spill in the Gulf of Mexico: for two years after, I couldn't look at any body of water without imagining it covered in oil. Sunsets were particularly difficult; the pink glow on the waves looked too much like petroleum sheen. And once, while grilling a beautiful piece of fresh sockeye salmon, I caught myself imagining how, as a wizened old woman, I would describe this extraordinary fish—its electric color, its jeweled texture—to a child living in a world where these wild creatures had disappeared.

I called my morbid habit "pre-loss," a variation on the "pre-crimes" committed in the movie *Minority Report*. And I know I'm not the only one afflicted. A few years ago, *The Nation* magazine, where I am a columnist, hosted a one-week cruise to Alaska. The full-page ad that ran in the magazine carried the tag line: "Come see the glaciers before they melt." I called my editor in a fury: How could we joke about melting glaciers while promoting a carbon-spewing holiday? Are we saying that global warming is funny? That we have no role to play in trying to stop it? The ad was pulled, but I realized then that, poor taste aside, this is how a great many of us are consuming wilderness these days—as a kind of nihilistic, final farewell. Gobble it all up before it's gone.

This ecological despair was a big part of why I resisted having kids until my late thirties. For years I joked about giving birth to a Mad Maxian climate warrior, battling alongside her friends for food and fuel. And I was also fully aware that if we were to avoid that future, we would all have to cut down on the number of super-consumers we were producing. It was around the time that I began work on this book that my attitude started to shift. Some of it, no doubt, was standard-issue denial (what does one more kid matter . . .). But part of it was also that immersing myself in the international climate justice movement had helped me imagine various futures that were decidedly less bleak than the post-apocalyptic cli-fi pastiche that had become my unconscious default. Maybe, just maybe, there was a future where replacing our own presence on earth could once again be part of a cycle of creation, not destruction.

And I was lucky: pregnant the first month we started trying. But then, just as fast, my luck ran out. A miscarriage. An ovarian tumor. A cancer scare. Surgery. Month after month of disappointing single pink lines on home pregnancy tests. Another miscarriage.

Then I stepped into the vortex I came to call the fertility factory ("do you have to call it that?" my patient husband pleaded). In its labyrinth of rooms in a downtown office building, drugs, hormones, and day surgeries were dispensed as liberally as toothbrushes at a dentist's office. The working assumption was that any woman who steps through the door will do whatever it takes to land a newborn in her arms, even if that means having three (or five) newborns instead of one. And even if that means seriously compromising her own health with risky drugs and poorly regulated medical procedures in the process.

I did try to be a good patient for a while, but it didn't work. The last straw was a doctor telling me, after my first (and only) round of in vitro fertilization (IVF) that I probably had "egg quality issues" and I should consider an egg donor. Feeling like a supermarket chicken past its best-before date, and with more than a few questions about how much these doctors were driven by a desire to improve their own "live birth" success rates, I stopped going. I tossed the pills, safely disposed of the syringes, and moved on.

Informing friends and family that I had given up on a technological fix to my apparent inability to conceive was surprisingly difficult. People often felt the need to tell me stories about friends and acquaintances who had become parents despite incredible odds. Usually these stories involved people who got pregnant using one of the technologies that I had decided not to try (with the guilt-inducing implication that, by drawing the line where I did, I was clearly not committed enough to procreation). Quite a few were about women who had used every technology available—nine rounds of IVF, egg donors, surrogacy—and then gotten pregnant as soon as they stopped. Common to all these stories was the unquestioned assumption that the body's No never really means no, that there is always a workaround. And, moreover, that there is something wrong with choosing not to push up against biological barriers if the technology is available.

On some level this faith is perfectly understandable. The female reproductive system is amazingly resilient—two ovaries and fallopian tubes when one would do; hundreds of thousands of eggs when all that is really needed are a few dozen good ones; and a generous window of opportunity to conceive spanning ages twelve to fifty (more or less). Yet what I felt my body telling me was that, even with all this ingenious built-in resilience, there is still a wall that can be hit, a place beyond which we cannot push. I felt that wall as a real structure inside my body and slamming against it had left me bruised. I didn't want to keep bashing away.

My resistance to further intervention did not come from some fixed idea about how babies should be conceived "naturally" or not at all. I know that for men and women with clear infertility diagnoses, these technologies are a joyous miracle, and that for gay, lesbian, and trans couples, some form of assisted reproduction is the only route to biological parenthood. And I believe that everyone who wants to become a parent should have the option, regardless of their marital status, sexual orientation, or income (in my view, these procedures should be covered by public health insurance, rather than restricted to those who can afford the astronomical fees).

What made me uneasy at the clinic were, oddly, many of the same things that made me wary of the geoengineers: a failure to address fundamental questions about underlying causes, as well as the fact that we seem to be turning to high-risk technologies not just when no other options are available, but at the first sign of trouble—even as a convenient shortcut ("tick tock," women of a certain age are told). Where I live, for instance, the system makes it significantly less complex to find an egg donor or a surrogate than to adopt a baby.

And then there is the matter of unacknowledged risks. Despite the casual attitude of many practitioners in this more than $10 billion global industry, the risks are real. A Dutch study, for instance, showed that women who had undergone in vitro fertilization were twice as likely to develop "ovarian malignancies"; an Israeli one found that women who had taken the widely prescribed fertility drug clomiphene citrate (which I was on) were at "significantly higher" risk for breast cancer; and Swedish researchers showed that IVF patients in the early stages of pregnancy were seven times more likely to develop a life-threatening blood clot in the lung. Other studies showed various kinds of risks to the children born of these methods.[3]

I did not know about this research when I was going to the clinic; my concerns stemmed from a generalized fear that by taking drugs that dramatically increased the number of eggs available for fertilization each month, I was overriding one of my body's safety mechanisms, forcing something that was better not forced. But there was little space for expressing these doubts at the clinic: conversations with the doctors were as brief as speed dates and questions seemed to be regarded as signs of weakness. Just look at all those joyous birth announcements from grateful couples papering every available surface in the examination rooms and hallways—what could be more important than that?

So why share these experiences and observations in a book about climate change? Partly in the spirit of transparency. The five years it took to research and write this book were the same years that my personal life was occupied with failed pharmaceutical and technological interventions, and ultimately, pregnancy and new motherhood. I tried, at first, to keep these parallel journeys segregated, but it didn't always work. Inevitably, one would escape its respective box to interrupt the other. What I was learning about the ecological crisis informed the responses to my own fertility crisis; and what I learned about fertility began to leave its mark on how I saw the ecological crisis.

Some of the ways in which these two streams in my life intersected were simply painful. For instance, if I was going through a particularly difficult infertility episode, just showing up to a gathering of environmentalists could be an emotional minefield. The worst part were the ceaseless invocations of our responsibilities to "our children" and "our grandchildren." I knew these expressions of intergenerational duty were heartfelt and in no way meant to be exclusionary—and yet I couldn't help feeling shut out. If caring about the future was primarily a function of love for one's descendants, where did that leave those of us who did not, or could not, have children? Was it even possible to be a real environmentalist if you didn't have kids?

And then there was the whole Earth Mother/Mother Earth thing: the idea that women, by right of our biological ability to carry children, enjoy a special connection to that fertile and bountiful matriarch that is the earth herself. I have no doubt that some women experience that bond with nature as a powerful creative force. But it's equally true that some of the most wildly creative and nurturing women (and men) that I know are childless by choice. And where did the equation of motherhood with the earth leave women like me, who wanted to conceive but were not able to? Were we exiles from nature? In my bleaker moments, I battled the conviction that the connection between my body and the cycle of creation had been unnaturally severed, like a dead telephone line.

But along the way, that feeling changed. It's not that I got in touch with my inner Earth Mother; it's that I started to notice that if the earth is indeed our mother, then far from the bountiful goddess of mythology, she is a mother facing a great many fertility challenges of her own. Indeed one of the most distressing impacts of the way in which our industrial activities affect the natural world is that they are interfering with systems at the heart of the earth's fertility cycles, from soil to precipitation. I also began to notice that a great many species besides ours are bashing up against their own infertility walls, finding it harder and harder to successfully reproduce and harder still to protect their young from the harsh new stresses of a changing climate.

On a much more optimistic note, I started to learn that protecting and valuing the earth's ingenious systems of reproducing life and the fertility of all of its inhabitants, may lie at the center of the shift in worldview that must take place if we are to move beyond extractivism. A worldview based on regeneration and renewal rather than domination and depletion.

An Aquatic Miscarriage

Since I had already quit the clinic, I had no idea that I was pregnant when I went to Louisiana to cover the BP spill. A few days after I got home, though, I could tell something was off and did a home pregnancy test. Two lines this time, but the second one was strangely faint. "You can't be just a little bit pregnant," the saying goes. And yet that is what I seemed to be. After going in for more tests, my family doctor called to tell me (in the hope-dampening tone with which I had become familiar) that while I was pregnant my hormone levels were much too low and I would likely miscarry, for the third time.

Instantly my mind raced back to the Gulf. While covering the spill, I had breathed in toxic fumes for days and, at one point, waded up to my waist in contaminated water to get to a secluded beach covered in oil. I searched on the chemicals BP was using in huge quantities, and found reams of online chatter linking them to miscarriages. Whatever was happening, I had no doubt that it was my doing.

After about a week of monitoring, the pregnancy was diagnosed as ectopic, which means that the embryo had implanted itself outside the uterus, most likely in a fallopian tube. I was rushed from the doctor's office to the emergency room. Ectopics are a leading cause of maternal death, particularly in the developing world: if undiagnosed, the embryo keeps growing in its impossible location, causing a rupture and massive internal hemorrhaging. If caught in time, the somewhat creepy treatment is one or more injections of methotrexate, a powerful drug used in chemotherapy to arrest cell development (and carrying many of the same side effects). Once fetal development has stopped, the pregnancy miscarries on its own, but it can take weeks.

It was a tough, drawn-out loss for my husband and me. But it was also a relief to learn that the miscarriage had nothing to do with anything that had happened in the Gulf. Knowing that did make me think a little differently about my time covering the spill, however. As I waited for the pregnancy to "resolve," I thought in particular about a long day spent on the *Flounder Pounder*, a sport fishing boat that a group of us had chartered as we went looking for evidence that the oil had entered the marshlands.

Our guide was Jonathan Henderson, an organizer with the Gulf Restoration Network, a heroic local organization devoted to repairing the damage done to the wetlands by the oil and gas industry. As we navigated through the narrow bayous of the Mississippi River Delta, Henderson kept leaning far over the side of the boat to get a better look at the bright green grass. What concerned him most was not what we were all seeing—fish jumping in fouled water, Roseau cane coated in reddish brown oil—but something much harder to detect, at least without a microscope and sample jars. Spring is the beginning of spawning season on the Gulf Coast and Henderson knew that these marshes were teeming with nearly invisible zooplankton and tiny juveniles that would develop into adult shrimp, oysters, crabs, and fin fish. In these fragile weeks and months, the marsh grass acts as an aquatic incubator, providing nutrients and protection from predators. "Everything is born in these wetlands," he said.[4] Unless, of course, something interferes with the process.

When fish are in their egg and larval phases, they have none of the defensive tools available to more mature animals. These tiny creatures travel where the tides carry them, unable to avoid whatever poison crosses their path. And at this early stage of development, exquisitely fragile membranes offer no protection from toxins; even negligible doses can cause death or mutation.

As far as Henderson was concerned, the prospects for these microscopic creatures did not look good. Each wave brought in more oil and dispersants, sending levels of carcinogenic polycyclic aromatic hydrocarbons (PAHs) soaring. And this was all happening at the absolute worst possible moment on the biological calendar: not only shellfish, but also bluefin tuna, grouper, snapper, mackerel, swordfish, and marlin were all spawning during these same key months. Out in the open water, floating clouds of translucent proto-life were just waiting for one of the countless slicks of oil and dispersants to pass through them like an angel of death. As John Lamkin, a fisheries biologist for the National Oceanic and Atmospheric Administration, put it: "Any larvae that came into contact with the oil doesn't have a chance."[5]

Unlike the oil-coated pelicans and sea turtles, which were being featured on the covers of the world's newspapers that week, these deaths would attract no media attention, just as they would go uncounted in the official assessments of the spill's damage. Indeed, if a certain species of larva was in the process of being snuffed out, we would likely not find out about it for years—until those embryonic life-forms would have normally reached maturity. And then, rather than some camera-ready mass die-off, there would just be . . . nothing. An absence. A hole in the life cycle.

That's what happened to the herring after the *Exxon Valdez* disaster. For three years after the spill, herring stocks were robust. But in the fourth, populations suddenly plummeted by roughly three quarters. The next year, there were so few, and they were so sick, that the herring fishery in Prince William Sound was closed. The math made sense: the herring that were in their egg and larval stages at the peak of the disaster would have been reaching maturity right about then.[6]

This was the kind of delayed disaster that Henderson was worried about as he peered into the marsh grass. When we reached Redfish Bay, usually a sport-fishing paradise, we cut the engine on the *Flounder Pounder* and drifted for a while in silence, taking video of the oily sheen that covered the water's surface.

As our boat rocked in that terrible place—the sky buzzing with Black Hawk helicopters and snowy white egrets—I had the distinct feeling that we were suspended not in water but in amniotic fluid, immersed in a massive multi-species miscarriage. When I learned that I too was in the early stages of creating an ill-fated embryo, I started to think of that time in the marsh as my miscarriage inside a miscarriage.

It was then that I let go of the idea that infertility made me some sort of exile from nature, and began to feel what I can only describe as a kinship of the infertile. It suddenly dawned on me that I was indeed part of a vast biotic community, and it was a place where a great many of us—humans and nonhuman alike—found ourselves engaged in an uphill battle to create new living beings.

A Country for Old Men

For all the talk about the right to life and the rights of the unborn, our culture pays precious little attention to the particular vulnerabilities of children, let alone developing life. When drugs and chemicals are approved for safe use and exposure, risk assessments most often focus on the effects on adults. As biologist Sandra Steingraber has observed, "Entire regulatory systems are premised on the assumption that all members of the population basically act, biologically, like middle-aged men. . . . Until 1990, for example, the reference dose for radiation exposure was based on a hypothetical 5'7" tall white man who weighed 157 pounds." More than three quarters of the mass-produced chemicals in the United States have never been tested for their impacts on fetuses or children. That means they are being released in the environment with no consideration for how they will impact those who weigh, say, twenty pounds, like your average one-year-old girl, let alone a half-pound, like a nineteen-week fetus.[7]

And yet when clusters of infertility and infant illness arise, they very often are the first warning signs of a broader health crisis. For instance, for years it seemed that while there were certainly water and air safety issues associated with fracking, there was no clear evidence that the practice was seriously impacting human health. But in April 2014, researchers with the Colorado School of Public Health and Brown University published a peer-reviewed study looking at birth outcomes in rural Colorado, where a lot of fracking is under way. It found that mothers living in the areas with the most natural gas development were 30 percent more likely to have babies with congenital heart defects than those who lived in areas with no gas wells near their homes. They also found some evidence that high levels of maternal exposure to gas extraction increased the risks of neurological defects.[8]

At around the same time, academics at Princeton, Columbia, and MIT gave a talk at the annual meeting of the American Economic Association, where they presented preliminary findings of a still unpublished study based on Pennsylvania birth records from 2004 to 2011. As Mark Whitehouse of *Bloomberg View* reported (he was one of the few journalists who saw their talk), "They found that proximity to fracking increased the likelihood of low birth weight by more than half, from about 5.6 percent to more than 9 percent. The chances of a low Apgar score, a summary measure of the health of newborn children, roughly doubled."[9]

These kinds of infant health impacts—and much worse—are all too familiar in communities that live in closest proximity to the dirtiest parts of our fossil fuel economy. For instance, the Aamjiwnaang First Nation, which is located just south of the industrial city of Sarnia in southern Ontario, has been the subject of intense scientific scrutiny because of its "lost boys." Up until 1993, the number of boys and girls born to the small Indigenous community was pretty much in keeping with the national average, with slightly more boys than girls. But as people continued living near the petrochemical plants, which had earned the region the nickname "Chemical Valley," that changed. By 2003, the day care was filled with girls and just a handful of boys, and there were years when the community could barely scrape together enough boys to form a baseball or hockey team. Sure enough, a study of birth records confirmed that by the end of the period between 1993 and 2003, twice as many girls as boys had been born on the reserve. Between 1999 and 2003, just 35 percent of Aamjiwnaang's births were boys—"one of the steepest declines ever reported in the ratio of boys to girls," as *Men's Health* magazine revealed in a 2009 exposé. Studies also found that 39 percent of Aamjiwnaang's women had had miscarriages, compared with roughly 20 percent in the general female population. Research published in 2013 showed that hormone-disrupting chemicals may be to blame, since women and children in the area had higher-than-average levels of PCBs in their bodies.[10]

I heard similar fertility horror stories in Mossville, Lousiana, a historic African-American town near Lake Charles. More than half of its two thousand families have left in recent years, fleeing the relentless pollution from their uninvited next-door neighbors: a network of massive industrial plants that convert the oil and gas pumped out of the Gulf into petroleum, plastics, and chemicals. Mossville is a textbook case of environmental racism: founded by freed slaves, it was once a safe haven for its residents, who enjoyed comfortable lives thanks in part to the rich hunting and fishing grounds in the surrounding wetlands. But beginning in the 1930s and 1940s, state politicians aggressively courted petrochemical and other industries with lavish tax breaks, and one giant plant after another set up shop on Mossville's doorstep, some just a few hundred feet from the clapboard homes. Today, fourteen chemical plants and refineries surround the town, including the largest concentration of vinyl production facilities in the U.S. Many of the hulking structures appear to be made entirely of metal pipes; spires in menacing chemical cathedrals. Roaring machinery spews emissions twenty-four hours a day, while floodlights and flares ignite the night sky.[11]

Accidental leaks are commonplace and explosions are frequent. But even when factories are running smoothly, they spew approximately four million pounds of toxic chemicals into the surrounding soil, air, and groundwater each year.[12] Before arriving in Mossville, I had heard about cancer and respiratory illnesses, and I knew that some residents have dioxin levels three times the national average. What I was unprepared for were the stories of miscarriages, hysterectomies, and birth defects.

Debra Ramirez, who after years of struggle was finally forced to abandon her home and move to Lake Charles, described Mossville to me as "a woman's womb of chemicals. And we're dying in that womb." Having just left BP's aquatic miscarriage, I found the idea of a toxic womb particularly chilling. It became more so after Ramirez shared part of her own family's health history. She had undergone a hysterectomy three decades earlier. So had all three of her sisters and her daughter. "It was just repeating from generation to generation," she said. Five hysterectomies in one family might have been bad genetic luck. But then Ramirez showed me footage from a town hall special that CNN's Dr. Sanjay Gupta had hosted on this "toxic town." On camera, Ramirez told the visiting correspondent that she had had a total hysterectomy, "like most young women do in this area." Taken aback, Gupta asked the rest of the women in the room whether they had had hysterectomies—multiple women answered yes, nodding silently. And yet despite the many studies that have sought to document the impact of toxins on human health in Mossville, not one has looked closely at their impact on fertility.[13]

Perhaps this should come as no surprise. As a culture, we do a very poor job of protecting, valuing, or even noticing fertility—not just among humans but across life's spectrum. Indeed vast amounts of money and cutting-edge technology are devoted to practices that actively interfere with the life cycle. We have a global agricultural model that has succeeded in making it illegal for farmers to engage in the age-old practice of saving seeds, the building blocks of life, so that new seeds have to be repurchased each year. And we have a global energy model that values fossil fuels over water, where all life begins and without which no life can survive.

Our economic system, meanwhile, does not value women's reproductive labor, pays caregivers miserably, teachers almost as badly, and we generally hear about female reproduction only when men are trying to regulate it.

BP's Legacy and a "Handful of Nothing"

If we tend to neglect the impact our industrial activities are having on human reproduction, the more vulnerable nonhumans fare significantly worse. A case in point is the risk assessment report that BP produced ahead of the Gulf Coast disaster. Before securing approval to drill in such deep water, the company had to produce a credible plan assessing what would happen to the ecosystem in the event of a spill, and what the company would do to respond. With the risk minimization that is one of the industry's hallmarks, the company confidently predicted that many adult fish and shellfish would be able to survive a spill whether by swimming away or by "metaboliz[ing] hydrocarbons," while marine mammals like dolphins might experience some "stress."[14] Conspicuously absent from the report are the words "eggs," "larvae," "fetus," and "juvenile." In other words, the working assumption, once again, was that we live in a world where all creatures are already fully grown.

That, unsurprisingly, proved to be a tragic assumption. Just as was feared in the early days of the spill, one of the most lasting legacies of the BP disaster may well be an aquatic infertility crisis, one that in some parts of the Gulf could reverberate for decades if not longer. Two years after the spill, Donny Waters, a large-scale fisherman in Pensacola, Florida, who primarily catches red snapper and grouper, reported, "We don't see any significant numbers of small fish"—a reference to the young fish that would have been in their larval stage at the peak of the disaster. That had not yet impacted the commercial catch since small fish are released anyway. But Waters, who holds one of the largest individual fishing quotas in the Pensacola area, worried that when 2016 or 2017 rolls around—when those small fish would normally be reaching maturity—he and his colleagues will be hauling in their lines only to "come up with a handful of nothing."[15]

One year after the spill, shrimpers, crabbers, and oystermen working in some of the most affected parts of Louisiana and Mississippi also began to report sharply reduced catches—and in some areas, that female crabs were relatively scarce, and that many of those caught during spawning season didn't have any eggs. (Some shellfish catches in these areas have shown improvement, but reports of missing or egg-less female crabs have persisted; similar signs of reproductive impairment have been observed in the shrimp and oyster fisheries.)[16]

The precise contribution of the spill to these fertility problems remains unclear as much of the research is still incomplete—but a growing body of scientific data adds weight to anecdotal reports from of fishing crews.

In one study, for instance, researchers sampled oysters after the spill and found alarmingly elevated concentrations of three heavy metals contained in petroleum—with 89 percent of the oysters also displaying a form of metaplasia, or stress-related tissue abnormality that is known to interfere with reproduction. Another study, this one by researchers at the Georgia Institute of Technology, tested the impact of BP oil mixed with Corexit on rotifers—microscopic animals at the bottom of the food web—which "provide food for baby fish, shrimp and crabs in estuaries." It found that even tiny amounts of the mixture "inhibited rotifer egg hatching by 50 percent."[17]

Perhaps most worrying are the findings of Andrew Whitehead, a biology professor at the University of California, Davis, who has conducted a series of studies with colleagues on the impact of BP's oil on one of the most abundant fish in the Gulf marshes, the minnow-sized killifish. He found that when killifish embryos were exposed to sediments contaminated with BP oil (including sediment samples collected over a year after the spill), "these embryos are getting whacked. . . . They're not growing, developing properly, they're not hatching out properly. They've got cardiovascular system developmental problems, their hearts aren't forming properly."[18]

Missing fish don't tend to make the news; for one thing, there are no pictures, just a "handful of nothing," as Waters feared. But that is decidedly not the case when baby dolphins start dying en masse, which is what happened in early 2011. In the month of February alone, NOAA's National Marine Fisheries Service reported that thirty-five dead baby dolphins had been collected on Gulf Coast beaches and in marshes—an eighteen-fold increase from the usual number (only two dead baby dolphins are found in a typical February). By the end of April 2014, 235 baby bottlenose dolphins had been discovered along the Gulf Coast, a staggering figure since scientists estimate that the number of cetacean corpses found on or near shore represents only 2 percent of the "true death toll"; the rest are never found.[19]

After examining the dolphins, NOAA scientists discovered that some of the calves had been stillborn, while others died days after birth. "Something has happened that these animals are now either aborting or the animals are not fit enough to survive," said Moby Solangi, the executive director of the Institute for Marine Mammal Studies (IMMS) in Gulfport, Mississippi, and one of the scientists investigating the incidents.*[20]

The deaths took place during the first birthing season for bottlenose dolphins since the BP disaster. That means that for much of their twelve-month gestation period, these calves were developing inside mothers who very likely swam in waters polluted with oil and chemical dispersant and who may well have inhaled toxic fumes when they surfaced to breathe. Metabolizing hydrocarbons is hard work and could have made the dolphins significantly more vulnerable to bacteria and diseases. Which might explain why, when NOAA-led scientists examined twenty-nine dolphins off the Louisiana coast, they found high levels of lung disease, as well as strikingly low levels of cortisol, an indication of adrenal insufficiency and a severely compromised ability to respond to stress. They also found one dolphin that was pregnant with a five-month-old "nonviable" fetus—an extremely rare occurrence in dolphins, indeed one undocumented in the scientific literature up until this incident. "I've never seen such a high prevalence of very sick animals—and with unusual conditions such as the adrenal hormone abnormalities," said Lori Schwacke, lead author of a paper on these findings that was published in late 2013. Commenting on the study, NOAA warned that the dolphins would "likely" face "reduced survival and ability to reproduce."[21]

The spill wasn't the only added stress these animals faced in this fateful period. The winter of 2010–2011 saw an abnormally heavy snowfall in the region, a phenomenon scientists have linked to climate change. When the huge snowpack melted, it sent torrents of freshwater into the Gulf of Mexico, where it not only dangerously lowered salinity and temperature levels for mammals accustomed to warm saltwater, but likely combined with the oil and dispersant to create an even more dangerous mess for dolphins and other cetaceans. As Ruth Carmichael, senior marine scientist at the Dauphin Island Sea Lab, explains, "These freight trains of cold fresh water may

---

* The dolphin die-off did not restrict itself to the young. By the end of April 2014, over one thousand dead dolphins, of all ages, had been discovered along the Gulf Coast, part of what NOAA termed an "unusual mortality event." Those numbers only scratch the surface of the death toll.

have assaulted [the dolphins], essentially kicking them when they were already down."[22]

This is the one-two punch of an economy built on fossil fuels: lethal when extraction goes wrong and the interred carbon escapes at the source; lethal when extraction goes right and the carbon is successfully released into the atmosphere. And catastrophic when these two forces combine in one ecosystem, as they did that winter on the Gulf Coast.

Disappearing Babies in a Warming World

In species after species, climate change is creating pressures that are depriving life-forms of their most essential survival tool: the ability to create new life and carry on their genetic lines. Instead, the spark of life is being extinguished, snuffed out in its earliest, most fragile days: in the egg, in the embryo, in the nest, in the den.

For sea turtles—an ancient species that managed to survive the asteroid collision that killed the dinosaurs—the problem is that the sand in which females bury their eggs is becoming too hot. In some cases, eggs are reaching lethal temperatures and many eggs aren't hatching at all, or else they are hatching but mostly as females. At least one species of coral is poised for a similar climate-related reproductive crisis: when water temperature reaches above 34 degrees Celsius (93 Fahrenheit), egg fertilization stops. Meanwhile, high temperatures can make reef-building coral so hungry that they reabsorb their own eggs and sperm.[23]

For oysters along the Pacific Coast of Oregon and Washington State, the problem in recent years is that the water is acidifying with such alarming rapidity that larvae are unable to form their tiny shells in the earliest days of life, leading to mass die-offs. Richard Feely, an oceanographer with the National Oceanic and Atmospheric Administration, explains that before the die-offs began, "What we knew at the time was that many organisms as adults are sensitive to acidification. What we did not know is that the larval stages of those organisms are much more sensitive." By 2014, the same problem was leading to a collapse of scallops off British Columbia. One of the largest scallop farming operations on the coast reported that some ten million mollusks had died in its operation alone.[24]

On land, climate change is also hitting the very young first and worst. In West Greenland, for instance, there has been a dramatic decrease in birth and survival rates of caribou calves. It seems that rising temperatures have changed the growing patterns of plants that are the source of critical energy for caribou calves, as well as for their mothers during reproduction and lactation. Populations of songbirds like the pied flycatcher, meanwhile, are collapsing in some parts of Europe because the caterpillars that parents depend upon to feed their young are hatching too early. In Maine, Arctic tern chicks are starving to death for similar reasons: they rely on small fish that have fled for colder waters. Meanwhile, there are reports that around Canada's Hudson Bay, birth dens of polar bears are collapsing in the thawing permafrost, which leaves tiny cubs dangerously exposed.[25]

As I delved into the impacts of climate change on reproduction and youth, I came across many more such examples of bottom-up threats, endangering the youngest members of species ranging from wolverine cubs (whose parents are having trouble storing food in ice) to peregrine falcon chicks (which are catching hypothermia and drowning in unusual downpours) to Arctic ring seal pups (whose snowy birthing dens, like those of polar bears, are threatened).[26] Once this pattern is recognized, it seems obvious: of course the very young are much more vulnerable than adults; of course even the most subtle environmental changes will hurt them more; and of course fertility is one of the first functions to erode when animals are under stress. And yet what struck me most in this research was the frequency with which all this came as a surprise, even to the experts in the field.

In a way, these various oversights make sense. We are used to thinking about extinction as a process that affects a species or cluster of species of every age group—the asteroid that wipes out the dinosaurs, or the way that our ancestors hunted a range of animals until they were all gone. And we still extinguish species that way, of course. But in the age of fossil fuels, we can render the earth less alive by far more stealthy means: by interfering with the capacity of adults to reproduce in the first place, and by making the first days of life simply too difficult to survive. No corpses, just an absence—more handfuls of nothing.

Fallow Time

A few months after I stopped going to the fertility clinic, a friend recommended that I see a naturopathic doctor who had helped several people she knew to get pregnant. This practitioner had her own theories about why so many women without an obvious medical reason were having trouble conceiving, and they were radically different from the ones I had come across so far.

Carrying a baby is one of the hardest physical tasks we can ask of ourselves, she pointed out, and if our bodies decline the task, it is often a sign that they are facing too many other demands—high-stress work that keeps us in a near constant state of "fight or flight," perhaps, or the physical stress of having to metabolize toxins or allergens, or just the stresses of modern life (or some combination of all of the above). With the body in overdrive fending off these real and perceived threats, it can start sending signals that it does not have the excess energy necessary to build and nourish a whole new life.

Most fertility clinics use drugs and technology to override this bodily resistance, and they work for a lot of people. But if they do not (and they often do not) women are frequently left even more stressed, with their hormones more out of whack, than when they began the process. The naturopath proposed an approach that was in every way the mirror opposite: try to figure out what might be overtaxing my system and then remove those things, and hope that a healthier, more balanced endocrine system will start sending some more welcoming signals to babies-to-be.

After a series of tests, I was diagnosed with a whole mess of allergies I didn't know I had, as well as with adrenal insufficiency and low cortisol levels (the same diagnosis, weirdly, that the NOAA scientists made for the Gulf Coast dolphins). The doctor asked me a lot of questions about my lifestyle, including how many hours I had spent in the air over the past year. "Why," I asked warily, knowing that the answer was going to be ugly. "Because of the radiation. There have been some studies done with flight attendants that show it might not be good for fertility." Great. Turns out flying was not just poisoning the atmosphere, it may have been poisoning me too.[27]

I admit that I was far from convinced that this new approach would result in a pregnancy, or even that the science behind it was wholly sound. And I was keenly aware that attributing infertility to female stress has an extensive and inglorious history. "Just relax," women who cannot conceive have long been told (in other words: it's all in your head/all your fault). Then again, the doctors at the fertility factory had clearly been engaged in their own version of highly lucrative guesswork, and after that experience, this doctor was a tonic.*28 Finally someone was trying to figure out *why* my infertility was happening, instead of trying to force my body to do something it clearly was rejecting. As for the downsides, they reminded me of a popular cartoon about global warming: A man stands up at a climate summit and asks, "What if it's a big hoax and we create a better world for nothing?" If all this adrenal stuff turned out to be a big hoax, the worst thing that could happen is that I'd end up healthier and less stressed out.

So I did it all. The yoga, the meditation, the dietary changes (the usual wars on wheat, gluten, dairy, and sugar—as well as various more esoteric odds and ends). I went to acupuncture and drank bitter Chinese herbs, and my kitchen counter became a gallery of powders and supplements. I also left my urban home in Toronto and moved to rural British Columbia, a ferry ride from the nearest city and a twenty-minute drive to the nearest hardware store. This is the part of the world where my parents live, where my grandparents are buried, and where I have always gone to write and rest. I would see what it was like to live there full-time.

Gradually, I learned to identify a half dozen birds by sound, and the sea mammals by the ripples that appeared on the water's surface. I even caught myself appreciating beautiful moments without simultaneously mourning their loss. The golden card in my wallet attesting to my frequent flyer status expired for the first time in a decade, and I was glad.

I still traveled for research, though—and when I did, I often noticed

---

* New research published in May 2014 in the journal *Human Reproduction* shows a strong connection between stress and infertility. The study followed roughly five hundred women in the U.S. as they were trying to conceive, none of them with known fertility problems. It discovered that women whose saliva measured high for alpha-amylase—a biomarker for stress—were twice as likely to be diagnosed as infertile as those with low levels.

parallels between my new doctor's theories about infertility and some of the ideas I was encountering about the changes humanity must make if we are to avoid collapse. Her advice had pretty much boiled down to this: before you can take care of another human being you have to take care of yourself. In a sense she was saying that I had to give myself some fallow time, as opposed to the mechanistic "push harder" approach that dominates Western medicine.

I thought of this advice when I left my hideout and traveled to the Land Institute in Salina, Kansas, one of the most exciting living laboratories for cutting-edge, agro-ecological farming methods. Wes Jackson, the founder and president of the center, says that he is trying to solve what he calls "the 10,000-year-old problem of agriculture."[29] That problem, in essence, is that ever since humans started planting seeds and tilling fields, they have been stripping the soil of its fertility.

Without human interference, plants grow in different varieties next to one another and as perennials, reseeding themselves year after year, with their roots staying put and growing ever longer and deeper. This combination of diversity and perennialism keeps soil healthy, stable, and fertile: the roots hold the soil in place, the plants allow rain water to be more safely slowly absorbed, and different plants provide different fertility functions (some, like legumes and clover, are better at fixing nitrogen, critical to forming the building blocks of plant life), while diversity controls pests and invasive weeds.

It's a self-sustaining cycle, with decomposing plants serving as natural fertilizer for new plants and the life cycle being constantly renewed. Maintaining this cycle, according to the farmer and philosopher Wendell Berry, must be the centerpiece of humanity's relationship with nature. "The problem of sustainability is simple enough to state," he says. "It requires that the fertility cycle of birth, growth, maturity, death, and decay . . . should turn continuously in place, so that the law of return is kept and nothing is wasted."[30] Simple enough: respect fertility, keep it going.

But when humans started planting single crops that needed to be replanted year after year, the problem of fertility loss began. The way industrial agriculture deals with this problem is well known: irrigate heavily to make up for the fact that annual plants do a poor job of retaining moisture (a growing problem as fresh water becomes more scarce), and lay on the chemicals, both to fertilize and ward off invasive pests and weeds.

This in turn creates a host of new environmental and health problems, including massive aquatic dead zones caused by agricultural runoff. In other words, rather than solving the fertility problem in the soil, we have simply moved it, transforming a land-based crisis into an ocean-based one. And the chain of infertility is longer still because some of the chemicals used in industrial farming are endocrine disruptors such as the herb killer atrazine, which research shows causes sterility in amphibians, fish, reptiles, and rats—as well as triggering bizarre spontaneous sex changes in male frogs. And these same chemicals have been linked to increased incidence of birth defects and miscarriages in humans, though the manufacturer of atrazine disputes all these links. Honeybees, meanwhile, our most critical natural pollinators, are under threat around the world—another victim, many experts say, of agriculture's chemical dependency.[31]

Many traditional agricultural societies have developed methods to maintain soil fertility despite planting annual crops. The maize-growing cultures of Mesoamerica, for example, allowed fields to lie fallow so they could regenerate and incorporated nitrogen-fixing legumes such as beans into mixtures of crops grown side by side. These methods, which mimic the way similar plants grow in the wild, have succeeded in keeping land fertile for thousands of years. Healthy soil also has the added bonus of sequestering carbon (helping to control emissions), and polycultures are less vulnerable to being wiped out by extreme weather.[32]

Wes Jackson and his colleagues at the Land Institute are taking this approach one step further: they are trying to remake the way industrial societies produce grains by breeding perennial varieties of wheat, wheatgrass, sorghum, and sunflowers that do not need to be replanted every year—just like the original tall grasses that dominated the prairie landscape before large-scale agriculture began. "Our goal is to fashion an agriculture as sustainable as the native ecosystems it displaced," the institute's literature explains, "to find a way of growing crops that rewards the farmer and the landscape more than the manufacturers of external inputs. We envision an agriculture that not only protects irreplaceable soil, but lessens our dependence on fossil fuels and damaging synthetic chemicals."[33]

And it is beginning to work: when I first visited the institute in 2010, the gift shop had started selling the first batch of flour made from perennial wheatgrass that Jackson and his team have domesticated and dubbed Kernza. When I returned a year later, the Southern Plains were in the grip of a devastating crop-destroying drought. Texas was having its driest year on record, with wheat, corn, and sorghum down 50–60 percent and agricultural losses topping $7 billion;[34] And yet the test sorghum field at the Land Institute was robust and healthy, the plants' long roots able to hold onto even tiny amounts of water. It was the only patch of green for miles around.

———

It was right around then that my son was conceived. For the first few months, the hardest part of the pregnancy was believing that everything really was normal and healthy. No matter how many tests came back with reassuring results, I stayed braced for tragedy. What helped most was hiking, and during the final anxious weeks before the birth, I would calm my nerves by walking for as long as my sore hips would let me on a well-groomed trail along a pristine creek. The stream begins near the top of a snowcapped mountain and the clear water rushes down a waterfall, gathers in dozens of pools, and flows through rapids until it finally empties into the Pacific.

On these hikes, I would keep my eyes open for silvery salmon smolts making their journey to the sea after months of incubation in shallow estuaries. And I would picture the cohos, pinks, and chums swimming with all their might through the rapids and falls, determined to reach the spawning grounds where they were born. This was my son's determination, I would tell myself. He was clearly a fighter, having managed to make his way to me despite the odds; he would find a way to be born safely too.

You can't ask for a better symbol of the tenacity of life than the Pacific salmon. To reach their spawning grounds, cohos will leap up massive waterfalls like deranged kayakers in reverse, dodging eagles and grizzly bears. At the end of their lives, salmon will expend their last life force to complete their mission. Salmon fry must go through a dramatic physical transformation (smolting) to prepare their bodies for the transition from freshwater to the oceans, where they will live out their lives until it is their turn to make the journey upstream.

But these triumphant feats of biology are only one part of the story of regeneration. Because as everyone who lives in salmon country knows, sometimes the autumn streams are eerily empty, filled with nothing but dead leaves, and perhaps one or two mottled fish. The salmon are indeed our Olympic athletes, their determination one of the planet's most powerful expressions of the drive to carry on the life cycle—but they are not invincible. Their strength can be defeated by overfishing, by fish farming operations that spread sea lice that kill young salmon in droves, by warming waters that scientists believe may threaten their food supply, by careless logging operations that leave spawning streams clogged with debris, by concrete dams that defy even the most acrobatic cohos. And of course they can be stopped dead in their tracks by oil spills and other industrial accidents.

All of which is why salmon have disappeared from about 40 percent of their historical range in the Pacific Northwest and several populations of coho, chinook, and sockeye are under perpetual threat and at risk of extinction.[35] To know where these kinds of numbers lead, we need only look to New England and Continental Europe, where commercial runs of Atlantic salmon have disappeared from the rivers where they once were plentiful. Like humans, salmon can overcome an awful lot—but not everything.

Which is why the happy ending to my own story still makes me uneasy, and feels incomplete. I know that for some, my fertility saga seems to reinforce the idea that human resilience will always conquer in the end, but that's not what it feels like. I don't know why this pregnancy succeeded any more than I know why my earlier pregnancies failed—and neither do my doctors, whether of the high-tech or low-tech varieties. Infertility is just one of the many areas in which we humans are confronted with our oceans of ignorance. So mostly I feel lucky—like I could just as easily have failed, no matter how serene my life became, having pushed my system too far. And it's also possible that I could have ended up with a cute baby picture on the wall of the frenetic fertility factory if I had been willing to keep upping the technological ante.

I suppose a part of me is still in that oiled Louisiana marsh, floating in a sea of poisoned larvae and embryos, with my own ill-fated embryo inside me. It's not self-pity that keeps me returning to that sad place. It's the conviction that there is something valuable in the body-memory of slamming up against a biological limit—of running out of second, third, and fourth chances—something that we all need to learn. Hitting the wall didn't dispel my belief in healing and recovery. It just taught me that these gifts require a special kind of nurturing, and a constant vigilance about the limits beyond which life cannot be pushed.

Because the truth is that humans *are* marvelously resilient, capable of adapting to all manner of setbacks. We are built to survive, gifted with adrenaline and embedded with multiple biological redundancies that allow us the luxury of second, third, and fourth chances. So are our oceans. So is the atmosphere. But surviving is not the same as thriving, not the same as living well. And as we have seen, for a great many species it's not the same as being able to nurture and produce new life. Just because biology is full of generosity does not mean its forgiveness is limitless. With proper care, we stretch and bend amazingly well. But we break too—our individual bodies, as well as the communities and ecosystems that support us.

Coming Back to Life

In early 2013, I came across a speech by Mississauga Nishnaabeg writer and educator Leanne Simpson, in which she describes her people's teachings and governance structures like this: "Our systems are designed to promote more life."[36] The statement stopped me in my tracks. It struck me that this guiding purpose was the very antithesis of extractivism, which is based on the premise that life can be drained indefinitely, and which, far from promoting future life, specializes in turning living systems into garbage, whether it's the piles of "overburden" lining the roads in the Alberta tar sands, or the armies of discarded people roving the world looking for temporary work, or the particulates and gases that choke the atmosphere that were once healthy parts of ecosystems. Or, indeed, the cities and towns turned to rubble after being hit by storms made more powerful by the heat those gases are trapping.

After listening to the speech, I wrote to Simpson and asked whether she would be willing to tell me more about what lay behind that statement. When we met at a Toronto cafe, I could tell that Simpson, in a black rocker T-shirt and motorcycle boots, was wary of having her own mind mined by yet another white researcher, having devoted a great deal of her life to collecting, translating, and artistically interpreting her people's oral histories and stories.

We ended up having a long, wide-ranging conversation about the difference between an extractivist mind-set (which Simpson describes bluntly as "stealing" and taking things "out of relationship") and a regenerative one. She described Anishinaabe systems as "a way of living designed to generate life, not just human life but the life of all living things." This is a concept of balance, or harmony, common to many Indigenous cultures and is often translated to mean "the good life." But Simpson told me that she preferred the translation "continuous rebirth," which she first heard from fellow Anishinaabe writer and activist Winona LaDuke.[37]

It's understandable that we associate these ideas today with an Indigenous worldview: it is primarily such cultures that have kept this alternate way of seeing the world alive in the face of the bulldozers of colonialism and corporate globalization. Like seed savers safeguarding the biodiversity of the global seed stock, other ways of relating to the natural world and one another have been safeguarded by many Indigenous cultures, based partly on a belief that a time will come when these intellectual seeds will be needed and the ground for them will become fertile once again.

One of the most important developments in the emergence of what I have been referring to as Blockadia is that, as this movement has taken shape, and as Indigenous people have taken on leadership roles within it, these long protected ways of seeing are spreading in a way that has not occurred for centuries. What is emerging, in fact, is a new kind of reproductive rights movement, one fighting not only for the reproductive rights of women, but for the reproductive rights of the planet as a whole—for the decapitated mountains, the drowned valleys, the clear-cut forests, the fracked water tables, the strip-mined hillsides, the poisoned rivers, the "cancer villages." All of life has the right to renew, regenerate, and heal itself.

Based on this principle, countries like Bolivia and Ecuador—with large

Indigenous populations—have enshrined the "rights of Mother Earth" into law, creating powerful new legal tools that assert the right of ecosystems not only to exist but to "regenerate."*[38] The gender essentialism of the term still makes some people uncomfortable. But it seems to me that the specifically female nature is not of central importance. Whether we choose to see the earth as a mother, a father, a parent, or an ungendered force of creation, what matters is that we are acknowledging that we are not in charge, that we are part of a vast living system on which we depend. The earth, wrote the great ecologist Stan Rowe, is not merely "resource" but "source."

These legal concepts are now being adopted and proposed in non-Indigenous contexts, including in North America and Europe, where increasingly, communities trying to protect themselves from the risks of extreme extraction are passing their own "rights of nature" ordinances. In 2010 the Pittsburgh City Council passed such a law, explicitly banning all natural gas extraction and stating that nature has "inalienable and fundamental rights to exist and flourish" in the city. A similar effort in Europe is attempting to make ecocide a crime under international law. The campaign defines ecocide as "the extensive damage to, destruction of or loss of ecosystem(s) of a given territory, whether by human agency or by other causes, to such an extent that peaceful enjoyment by the inhabitants of that territory has been or will be severely diminished."[39]

As Indigenous-inspired ideas have spread in these somewhat surprising contexts, something else is happening too: many people are remembering their own cultures' stewardship traditions, however deeply buried, and recognizing humanity's role as one of life promotion. The notion that we could

---

* When Ecuador adopted a new constitution in 2008, it became the first country to enshrine the rights of nature in law. Article 71 of the country's constitution states: "Nature or *Pachamama*, where the life is created and reproduced, has as a right that its existence is integrally respected as well as the right of the maintenance and regeneration of its vital cycles, structures, functions and evolutionary processes. Every person, community, people or nationality can demand from the public authority that these rights of nature are fulfilled." Similar principles were enshrined in the "Peoples Agreement" of the World People's Conference on Climate Change and the Rights of Mother Earth, which was adopted by 30,000 members of international civil society gathered in Cochabamba, Bolivia, in April 2010. Noting that, "the regenerative capacity of the planet has been already exceeded," the agreement asserts that the earth has "the right to regenerate its bio-capacity and to continue its vital cycles and processes free of human alteration."

separate ourselves from nature, that we did not need to be in perpetual partnership with the earth around us, is, after all, a relatively new concept, even in the West. Indeed it was only once humans came up with the lethal concept of the earth as an inert machine and man its engineer, that some began to forget the duty to protect and promote the natural cycles of regeneration on which we all depend.

The good news is that not everybody agreed to forget. Another of the more interesting and unexpected side effects of the extreme energy frenzy is that, faced with heightened threats to collective safety, these old ideas are reasserting themselves—cross-pollinating, hybridizing, and finding applications in new contexts.

In Halkidiki, Greece, for instance, as villagers defend their land against open-pit gold mining, the secret weapon has been the intergenerational character of the struggle—teenage girls in skinny jeans and big sunglasses standing side by side with their black-clad grandmothers in orthopedic shoes. This is something new: before the miners threatened the mountain and the streams, many old people had been forgotten, parked at home in front of their TVs, stuffed away like outdated cell phones. But as the villages organized, local young people discovered that, though they were expert at certain things, like flash-mob-style organizing and getting their messages out on social media, their grandparents—who had survived wars and occupations—knew a great deal more about living and working in large groups. Not only could they cook for fifty people (an important skill on the barricades), but they remembered a time when agriculture was done collectively, and were able to help their children and grandchildren believe that it was possible to live well without tearing up the land.

In "young" countries like Canada, the United States, Australia, and New Zealand, which tend to have myths rather than memories, this remembering process is far more complex. For descendants of settlers and newer immigrants, it begins with learning the true histories of where we live—with reading treaties, for instance, and coming to terms with how we ended up with what we have, however painful. And yet Mike Scott, the goat rancher and environmentalist at the forefront of Montana's anti-coal fight, says that the process of Indigenous and non-Indigenous people working closely together "has reawakened a worldview in a lot of people."[40]

The deep sense of interdependence with the natural world that animates rural Blockadia struggles from Greece to coastal British Columbia is, of course, far less obvious in the densely populated cities where so many of us live and work: where our reliance on nature is well hidden by highways, pipes, electrical lines, and overstocked supermarkets. It is only when something in this elaborately insulated system cracks, or comes under threat, that we catch glimpses of how dependent and vulnerable we really are.

And yet these cracks are appearing with greater regularity. At a time when unprecedented wildfires engulf suburban homes in Melbourne, when waters from the rising Thames flood homes in London commuter towns, and when Superstorm Sandy transforms the New York subway into a canal system, the barriers that even the most urban and privileged among us have erected to hold back the natural world are clearly starting to break down.

Sometimes it is extreme extraction that breaks down those barriers, as its tentacles creep into our most modern cities—with fracking in backyards in Los Angeles and proposed tar sands pipelines running through cities like Toronto. Sydney's residents had little reason to think about where their drinking water was coming from—but when it looked like the source of the Australian city's water was going to get fracked, a great many people educated themselves fast. In truth, we never lost our connections with nature—they were always there, in our bodies and under our paved lives. A great many of us just forgot about them for a while.

———

As communities move from simply resisting extractivism to constructing the world that must rise in its rubble, protecting the fertility cycle is at the heart of the most rapidly multiplying models, from permaculture to living buildings to rainwater harvesting. Again and again, linear, one-way relationships of pure extraction are being replaced with systems that are circular and reciprocal. Seeds are saved instead of purchased. Water is recycled. Animal manure, not chemicals, is used as fertilizer, and so on. There are no hard-and-fast formulas, since the guiding principle is that every geography is different and our job, as Wes Jackson says (citing Alexander Pope), is to "consult the genius of the place."[41] There is, however, a recurring pattern:

systems are being created that require minimal external inputs and produce almost no waste—a quest for homeostasis that is the opposite of the Monster Earth that the would-be geoengineers tell us we must learn to love.

And contrary to capitalism's drift toward monopoly and duopoly in virtually every arena, these systems mimic nature's genius for built-in redundancy by amplifying diversity wherever possible, from more seed varieties to more sources of energy and water. The goal becomes not to build a few gigantic green solutions, but to infinitely multiply smaller ones, and to use policies—like Germany's feed-in tariff for renewable energy, for instance—that encourage multiplication rather than consolidation. The beauty of these models is that when they fail, they fail on a small and manageable scale—with backup systems in place. Because if there is one thing we know, it's that the future is going to have plenty of shocks.

Living nonextractively does not mean that extraction does not happen: all living things must take from nature in order to survive. But it does mean the end of the extractivist mindset—of taking without caretaking, of treating land and people as resources to deplete rather than as complex entities with rights to a dignified existence based on renewal and regeneration. Even such traditionally destructive practices as logging can be done responsibly, as can small-scale mining, particularly when the activities are controlled by the people who live where the extraction is taking place and who have a stake in the ongoing health and productivity of the land. But most of all, living nonextractively means relying overwhelmingly on resources that can be continuously regenerated: deriving our food from farming methods that protect soil fertility; our energy from methods that harness the ever-renewing strength of the sun, wind, and waves; our metals from recycled and reused sources.

These processes are sometimes called "resilient" but a more appropriate term might be "regenerative." Because resilience—though certainly one of nature's greatest gifts—is a passive process, implying the ability to absorb blows and get back up. Regeneration, on the other hand, is active: we become full participants in the process of maximizing life's creativity.

This is a far more expansive vision than the familiar eco-critique that stressed smallness and shrinking humanity's impact or "footprint." That is simply not an option today, not without genocidal implications: we are here, we are many, and we must use our skills to act. We can, however, change the nature of our actions so that they are constantly growing, rather than extracting life. "We can build soil, pollinate, compost and decompose," Gopal Dayaneni, a grassroots ecologist and activist with the Oakland, California, based Movement Generation, told me. "We can accelerate, simply though our labor, the restoration and regeneration of living systems, if we engage in thoughtful, concerted action. We are actually the keystone species in this moment so we have to align our strategies with the healing powers of Mother Earth—there is no getting around the house rules. But it isn't about stopping or retreating. It's about aggressively applying our labor toward restoration."[42]

That spirit is already busily at work promoting and protecting life in the face of so many life-negating and life-forgetting threats. It has even reached the creek where I used to take hikes during my pregnancy. When I first discovered the trail, I had thought that the salmon that still swam in the stream were there purely thanks to the species' indomitable will. But as I met and spoke with locals on those walks, I learned that since 1992 the fish had been helped along by a hatchery a few kilometers upstream, as well as by teams of volunteers that worked to clear the water of logging debris and made sure there was enough shade to protect the young fry. Hundreds of thousands of pink, coho, chum, and chinook fry are released into nearby streams each year. It's a partnership of sorts between the fish, the forest, and the people who share this special piece of the world.

So about two months after my son was born, our little family went on a field trip to that hatchery, now being powered through micro turbines and geothermal. Though he was so small he could barely see over the sling, I wanted him to meet some of the baby salmon that had been so important to me before he was born. It was fun: we peered together into the big green tanks where the young fish were being kept safe until they grew strong enough to protect themselves. And we went home with a "salmon alphabet" poster that still hangs in his room ("s" is for smolt).

This was not a fish farm or a fertility factory—nothing was being created from scratch or forced. It was just a helping hand, a boost to keep the fertility cycle going. And it's an expression of the understanding that from here on, when we take, we must not only give back, but we must also take care.

Conclusion

The Leap Years

Just Enough Time for Impossible

"We as a nation must undergo a radical revolution of values. We must rapidly begin the shift from a 'thing-oriented society' to a 'person-oriented society.' When machines and computers, profit motives and property rights, are considered more important than people, the giant triplets of racism, extreme materialism, and militarism are incapable of being conquered."

—Martin Luther King Jr., "Beyond Vietnam," 1967 [1]

"Developed countries have created a global crisis based on a flawed system of values. There is no reason we should be forced to accept a solution informed by that same system."

—Marlene Moses, Ambassador to the U.N. for Nauru, 2009 [2]

In December 2012, Brad Werner—a complex systems researcher with pink hair and a serious expression—made his way through the throng of 24,000 earth and space scientists at the Fall Meeting of the American Geophysical Union in San Francisco. That year's conference had some big-name participants, from Ed Stone of NASA's Voyager project, explaining a new milestone on the path to interstellar space, to the filmmaker James Cameron, discussing his adventures in deep-sea submersibles. But it was Werner's own session that was attracting much of the buzz. It was titled "Is Earth F**ked?" (full title: "Is Earth F**ked? Dynamical Futility of Global Environmental Management and Possibilities for Sustainability via Direct Action Activism").[3]

Standing at the front of the conference room, the University of California, San Diego professor took the crowd through the advanced computer model he was using to answer that rather direct question. He talked about system boundaries, perturbations, dissipation, attractors, bifurcations, and a whole bunch of other stuff largely incomprehensible to those of us uninitiated in complex systems theory. But the bottom line was clear enough: global capitalism has made the depletion of resources so rapid, convenient, and barrier-free that "earth-human systems" are becoming dangerously unstable in response. When a journalist pressed Werner for a clear answer on the "Is Earth f**ked" question, he set the jargon aside and replied, "More or less."[4]

There was one dynamic in the model, however, that offered some hope. Werner described it as "resistance"—movements of "people or groups of people" who "adopt a certain set of dynamics that does not fit within the capitalist culture." According to the abstract for his presentation, this includes "environmental direct action, resistance taken from outside the dominant culture, as in protests, blockades and sabotage by Indigenous peoples, workers, anarchists and other activist groups." Such mass uprisings of people—along the lines of the abolition movement and the civil rights movement—represent the likeliest source of "friction" to slow down an economic machine that is careening out of control.[5]

This, he argued, is clear from history, which tells us that past social movements have "had tremendous influence on . . . how the dominant culture evolved." It stands to reason, therefore, that "if we're thinking about the future of the earth, and the future of our coupling to the environment, we have to include resistance as part of that dynamics." And that, Werner said, is not a matter of opinion, but "really a geophysics problem."[6]

Put another way, only mass social movements can save us now. Because we know where the current system, left unchecked, is headed. We also know, I would add, how that system will deal with the reality of serial climate-related disasters: with profiteering, and escalating barbarism to segregate the losers from the winners. To arrive at that dystopia, all we need to do is keep barreling down the road we are on. The only remaining variable is whether some countervailing power will emerge to block the road, and simultaneously clear some alternate pathways to destinations that are safer. If that happens, well, it changes everything.

The movements explored in these pages—Blockadia's fast multiplying local outposts, the fossil fuel divestment/reinvestment movement, the local laws barring high-risk extraction, the bold court challenges by Indigenous groups and others—are early manifestations of this resistance. They have not only located various choke points to slow the expansion plans of the fossil fuel companies, but the economic alternatives these movements are proposing and building are mapping ways of living within planetary boundaries, ones based on intricate reciprocal relationships rather than brute extraction. This is the "friction" to which Werner referred, the kind that is needed to put the brakes on the forces of destruction and destabilization.

When I despair of the prospects for change, I think back on some of what I have witnessed in the five years of writing of this book. Admittedly, much of it is painful. From the young climate activist breaking down and weeping on my shoulder at the Copenhagen summit, to the climate change deniers at the Heartland Institute literally laughing at the prospect of extinction. From the country manor in England where mad scientists plotted to blot out the sun, to the stillness of the blackened marshes during the BP oil disaster. From the roar of the earth being ripped up to scrape out the Alberta tar sands, to the shock of discovering that the largest green group in the world was itself drilling for oil.

But that's not all I think about. When I started this journey, most of the resistance movements standing in the way of the fossil fuel frenzy either did not exist or were a fraction of their current size. All were significantly more isolated from one another than they are today. North Americans, overwhelmingly, did not know what the tar sands were. Most of us had never heard of fracking. There had never been a truly mass march against climate change in North America, let alone thousands willing to engage together in civil disobedience. There was no mass movement to divest from fossil fuels. Hundreds of cities and towns in Germany had not yet voted to take back control over their electricity grids to be part of a renewable energy revolution. My own province did not have a green energy program that was bold enough to land us in trade court. The environmental news out of China was almost exclusively awful. There was far less top-level research proving that economies powered by 100 percent renewable energy were within our grasp. Only the isolated few dared question the logic of economic growth. And few climate scientists were willing to speak bluntly about the political implications of their work for our frenzied consumer culture. All of this has changed so rapidly as I have been writing that I have had to race to keep up. Yes, ice sheets are melting faster than the models projected, but resistance is beginning to boil. In these existing and nascent movements we now have a clear glimpses of the kind of dedication and imagination demanded of everyone who is alive and breathing during climate change's "decade zero."

Because the carbon record doesn't lie. And what that record tells us is that emissions are still rising: every year we release more greenhouse gases than the year before, the growth rate increasing from one decade to the next—gases that will trap heat for generations to come, creating a world that is hotter, colder, wetter, thirstier, hungrier, angrier. So if there is any hope of reversing these trends, glimpses won't cut it; we will need the climate revolution playing on repeat, all day every day, everywhere.

Werner was right to point out that mass resistance movements have grabbed the wheel before and could very well do so again. At the same time, we must reckon with the fact that lowering global emissions in line with climate scientists' urgent warnings demands changes of a truly daunting speed and scale. Meeting science-based targets will mean forcing some of the most profitable companies on the planet to forfeit trillions of dollars of future earnings by leaving the vast majority of proven fossil fuel reserves in the ground.[7] It will also require coming up with trillions more to pay for zero-carbon, disaster-ready societal transformations. And let's take for granted that we want to do these radical things democratically and without a bloodbath, so violent, vanguardist revolutions don't have much to offer in the way of road maps.

The crucial question we are left with, then, is this: has an economic shift of this kind *ever* happened before in history? We know it can happen during wartime, when presidents and prime ministers are the ones commanding the transformation from above. But has it ever been demanded from below, by regular people, when leaders have wholly abdicated their responsibilities? Having combed through the history of social movements in search of precedents, I must report that the answer to that question is predictably complex, filled with "sort ofs" and "almosts"—but also at least one "yes."

In the West, the most common precedents invoked to show that social movements really can be a disruptive historical force are the celebrated human rights movements of the past century—most prominently, civil, women's, and gay and lesbian rights. And these movements unquestionably transformed the face and texture of the dominant culture. But given that the challenge for the climate movement hinges on pulling off a profound and radical *economic* transformation, it must be noted that for these movements, the legal and cultural battles were always more successful than the economic ones.

The U.S. civil rights movement, for instance, fought not only against legalized segregation and discrimination but also *for* massive investments in schools and jobs programs that would close the economic gap between blacks and whites once and for all. In his 1967 book, *Where Do We Go from Here: Chaos or Community?*, Martin Luther King Jr. pointed out that, "The practical cost of change for the nation up to this point has been cheap. The limited reforms have been obtained at bargain rates. There are no expenses, and no taxes are required, for Negroes to share lunch counters, libraries, parks, hotels and other facilities with whites. . . . The real cost lies ahead. . . . The discount education given Negroes will in the future have to be purchased at full price if quality education is to be realized. Jobs are harder and costlier to create than voting rolls. The eradication of slums housing millions is complex far beyond integrating buses and lunch counters."[8]

And though often forgotten, the more radical wing of the second-wave feminist movement also argued for fundamental challenges to the free market economic order. It wanted women not only to get equal pay for equal work in traditional jobs but to have their work in the home caring for children and the elderly recognized and compensated as a massive unacknowledged market subsidy—essentially a demand for wealth redistribution on a scale greater than the New Deal.

But as we know, while these movements won huge battles against institutional discrimination, the victories that remained elusive were those that, in King's words, could not be purchased "at bargain rates." There would be no massive investments in jobs, schools, and decent homes for African Americans, just as the 1970s women's movement would not win its demand for "wages for housework" (indeed paid maternity leave remains a battle in large parts of the world). Sharing legal status is one thing; sharing resources quite another.

If there is an exception to this rule it is the huge gains won by the labor movement in the aftermath of the Great Depression—the massive wave of unionization that forced owners to share a great deal more wealth with their workers, which in turn helped create a context to demand ambitious social programs like Social Security and unemployment insurance (programs from which the majority of African American and many women workers were notably excluded). And in response to the market crash of 1929, tough new rules regulating the financial sector were introduced at real cost to unfettered profit making. In the same period, social movement pressure created the conditions for the New Deal and programs like it across the industrialized world. These made massive investments in public infrastructure— utilities, transportation systems, housing, and more—on a scale comparable to what the climate crisis calls for today.

If the search for historical precedents is extended more globally (an impossibly large task in this context, but worth a try), then the lessons are similarly mixed. Since the 1950s, several democratically elected socialist governments have nationalized large parts of their extractive sectors and begun to redistribute to the poor and middle class the wealth that had previously hemorrhaged into foreign bank accounts, most notably Mohammad Mosaddegh in Iran and Salvador Allende in Chile. But those experiments were interrupted by foreign-sponsored coups d'état before reaching their potential. Indeed postcolonial independence movements—which so often had the redistribution of unjustly concentrated resources, whether of land or minerals, as their core missions—were consistently undermined through political assassinations, foreign interference, and, more recently, the chains of debt-driven structural adjustment programs (not to mention the corruption of local elites).

Even the stunningly successful battle against apartheid in South Africa suffered its most significant losses on the economic equality front. The country's freedom fighters were not, it is worth remembering, only demanding the right to vote and move freely. They were also, as the African National Congress's official policy platform, the Freedom Charter, made clear, struggling for key sectors of the economy—including the mines and the banks—to be nationalized, with their proceeds used to pay for the social programs that would lift millions in the townships out of poverty. Black South Africans won their core legal and electoral battles, but the wealth accumulated under apartheid remained intact, with poverty deepening significantly in the post-apartheid era.[9]

There have been social movements, however, that have succeeded in challenging entrenched wealth in ways that are comparable to what today's movements must provoke if we are to avert climate catastrophe. These are the movements for the abolition of slavery and for Third World independence from colonial powers. Both of these transformative movements forced ruling elites to relinquish practices that were still extraordinarily profitable, much as fossil fuel extraction is today.

The movement for the abolition of slavery in particular shows us that a transition as large as the one confronting us today has happened before—and indeed it is remembered as one of the greatest moments in human history. The economic impacts of slavery abolition in the mid-nineteenth century have some striking parallels with the impacts of radical emission reduction, as several historians and commentators have observed. Journalist and broadcaster Chris Hayes, in an award-winning 2014 essay titled "The New Abolitionism," pointed out "the climate justice movement is demanding that an existing set of political and economic interests be forced to say goodbye to trillions of dollars of wealth" and concluded that "it is impossible to point to any precedent other than abolition."[10]

There is no question that for a large sector of the ruling class at the time, losing the legal right to exploit men and women in bondage represented a major economic blow, one as huge as the one players ranging from Exxon to Richard Branson would have to take today. As the historian Greg Grandin has put it, "In the realm of economics, the importance of slaves went well beyond the wealth generated from their uncompensated labor. Slavery was the flywheel on which America's market revolution turned—not just in the United States, but in all of the Americas." In the eighteenth century, Caribbean sugar plantations, which were wholly dependent on slave labor, were by far the most profitable outposts of the British Empire, generating revenues that far outstripped the other colonies. In *Bury the Chains*, Adam Hochschild quotes enthusiastic slave traders describing the buying and selling of humans as "the hinge on which all the trade of this globe moves" and "the foundation of our commerce . . . and first cause of our national industry and riches." [11]

While not equivalent, the dependency of the U.S. economy on slave labor—particularly in the Southern states—is certainly comparable to the modern global economy's reliance on fossil fuels.* According to historian Eric Foner, at the start of the Civil War, "slaves as property were worth more than all the banks, factories and railroads in the country put together." Strengthening the parallel with fossil fuels, Hayes points out that "in 1860, slaves represented about 16 percent of the total household assets—that is, all the wealth—in the entire [United States], which in today's terms is a stunning $10 trillion." That figure is very roughly similar to the value of the carbon reserves that must be left in the ground worldwide if we are to have a good chance of keeping warming below 2 degrees Celsius. [12]

But the analogy, as all acknowledge, is far from perfect. Burning fossil fuels is of course not the moral equivalent of owning slaves or occupying countries. (Though heading an oil company that actively sabotages climate science, lobbies aggressively against emission controls while laying claim to enough interred carbon to drown populous nations like Bangladesh and boil sub-Saharan Africa is indeed a heinous moral crime.) Nor were the movements that ended slavery and defeated colonial rule in any way bloodless: nonviolent tactics like boycotts and protests played major roles, but slavery in the Caribbean was only outlawed after numerous slave rebellions were brutally suppressed, and, of course, abolition in the United States came only after the carnage of the Civil War.

Another problem with the analogy is that, though the liberation of mil-

---

* The reliance was certainly not limited to the Southern states: cutting-edge historical research has been exploding long-held perceptions that the North and South of the United States had distinct and irreconcilable economies in this period. In fact, Northern industrialists and Wall Street were far more dependent on and connected to slavery than has often been assumed, and even some crucial innovations in scientific management and accounting can be traced to the American plantation economy.

lions of slaves in this period—some 800,000 in the British colonies and four million in the U.S.—represents the greatest human rights victory of its time (or, arguably, any time), the economic side of the struggle was far less successful. Local and international elites often managed to extract steep payoffs to compensate themselves for their "losses" of human property, while offering little or nothing to former slaves. Washington broke its promise, made near the end of the Civil War, to grant freed slaves ownership of large swaths of land in the U.S. South (a pledge known colloquially as "40 acres and a mule"). Instead the lands were returned to former slave owners, who proceeded to staff them through the indentured servitude of sharecropping. Britain, as discussed, awarded massive paydays to its slave owners at the time of abolition. And France, most shockingly, sent a flotilla of warships to demand that the newly liberated nation of Haiti pay a huge sum to the French crown for the loss of its bonded workforce—or face attack.[13] Reparations, but in reverse.

The costs of these, and so many other gruesomely unjust extortions, are still being paid in lives, from Haiti to Mozambique. The reverse-reparations saddled newly liberated nations and people with odious debts that deprived them of true independence while helping to accelerate Europe's Industrial Revolution, the extreme profitability of which most certainly cushioned the economic blow of abolition. In sharp contrast, a real end to the fossil fuel age offers no equivalent consolation prizes to the major players in the oil, gas, and coal industries. Solar and wind can make money, sure. But by nature of their decentralization, they will never supply the kind of concentrated super-profits to which the fossil fuel titans have become all too accustomed. In other words, if climate justice carries the day, the economic costs to our elites will be real—not only because of the carbon left in the ground but also because of the regulations, taxes, and social programs needed to make the required transformation. Indeed, these new demands on the ultra rich could effectively bring the era of the footloose Davos oligarch to a close.

The Unfinished Business of Liberation

On one level, the inability of many great social movements to fully realize those parts of their visions that carried the highest price tags can be seen as a cause for inertia or even despair. If they failed in their plans to usher in a more equitable economic system, how can the climate movement hope to succeed?

There is, however, another way of looking at this track record: these economic demands—for basic public services that work, for decent housing, for land redistribution—represent nothing less than the unfinished business of the most powerful liberation movements of the past two centuries, from civil rights to feminism to Indigenous sovereignty. The massive global investments required to respond to the climate threat—to adapt humanely and equitably to the heavy weather we have already locked in, and to avert the truly catastrophic warming we can still avoid—is a chance to change all that; and to get it right this time. It could deliver the equitable redistribution of agricultural lands that was supposed to follow independence from colonial rule and dictatorship; it could bring the jobs and homes that Martin Luther King dreamed of; it could bring jobs and clean water to Native communities; it could at last turn on the lights and running water in every South African township. Such is the promise of a Marshall Plan for the Earth.

The fact that our most heroic social justice movements won on the legal front but suffered big losses on the economic front is precisely why our world is as fundamentally unequal and unfair as it remains. Those losses have left a legacy of continued discrimination, double standards, and entrenched poverty—poverty that deepens with each new crisis. But, at the same time, the economic battles the movements *did* win are the reason we still have a few institutions left—from libraries to mass transit to public hospitals—based on the wild idea that real equality means equal access to the basic services that create a dignified life. Most critically, all these past movements, in one form or another, are still fighting today—for full human rights and equality regardless of ethnicity, gender, or sexual orientation; for real decolonization and reparation; for food security and farmers' rights; against oligarchic rule; and to defend and expand the public sphere.

So climate change does not need some shiny new movement that will magically succeed where others failed. Rather, as the furthest-reaching crisis created by the extractivist worldview, and one that puts humanity on a firm and unyielding deadline, climate change can be the force—the grand push—that will bring together all of these still living movements. A rushing river fed by countless streams, gathering collective force to finally reach the sea. "The basic confrontation which seemed to be colonialism versus anticolonialism, indeed capitalism versus socialism, is already losing its importance," Frantz Fanon wrote in his 1961 masterwork, *The Wretched of the Earth*. "What matters today, the issue which blocks the horizon, is the need for a redistribution of wealth. Humanity will have to address this question, no matter how devastating the consequences may be."[14] Climate change is our chance to right those festering wrongs at last—the unfinished business of liberation.

Winning will certainly take the convergence of diverse constituencies on a scale previously unknown. Because, although there is no perfect historical analogy for the challenge of climate change, there are certainly lessons to learn from the transformative movements of the past. One such lesson is that when major shifts in the economic balance of power take place, they are invariably the result of extraordinary levels of social mobilization. At those junctures, activism becomes something that is not performed by a small tribe within a culture, whether a vanguard of radicals or a subcategory of slick professionals (though each play their part), but becomes an entirely normal activity throughout society—it's rent payers associations, women's auxiliaries, gardening clubs, neighborhood assemblies, trade unions, professional groups, sports teams, youth leagues, and on and on. During extraordinary historical moments—both world wars, the aftermath of the Great Depression, or the peak of the civil rights era—the usual categories dividing "activists" and "regular people" became meaningless because the project of changing society was so deeply woven into the project of life. Activists were, quite simply, everyone.

Which brings us back to where we started: climate change and bad timing. It must always be remembered that the greatest barrier to humanity rising to meet the climate crisis is not that it is too late or that we don't know what to do. There is just enough time, and we are swamped with green tech and green plans. And yet the reason so many of us are inclined to answer Brad Werner's provocative question in the affirmative is that we are afraid—with good reason—that our political class is wholly incapable of seizing those tools and implementing those plans, since doing so involves unlearning the core tenets of the stifling free-market ideology that governed every stage of their rise to power.

And it's not just the people we vote into office and then complain about—it's us. For most of us living in postindustrial societies, when we see the crackling black-and-white footage of general strikes in the 1930s, victory gardens in the 1940s, and Freedom Rides in the 1960s, we simply cannot imagine being part of any mobilization of that depth and scale. That kind of thing was fine for them but surely not us—with our eyes glued to smart phones, attention spans scattered by click bait, loyalties split by the burdens of debt and insecurities of contract work. Where would we organize? Who would we trust enough to lead us? Who, moreover, is "we"?

In other words, we are products of our age and of a dominant ideological project. One that too often has taught us to see ourselves as little more than singular, gratification-seeking units, out to maximize our narrow advantage, while simultaneously severing so many of us from the broader communities whose pooled skills are capable of solving problems big and small. This project also has led our governments to stand by helplessly for more than two decades as the climate crisis morphed from a "grandchildren" problem to a banging-down-the-door problem.

All of this is why any attempt to rise to the climate challenge will be fruitless unless it is understood as part of a much broader battle of worldviews, a process of rebuilding and reinventing the very idea of the collective, the communal, the commons, the civil, and the civic after so many decades of attack and neglect. Because what is overwhelming about the climate challenge is that it requires breaking so many rules at once—rules written into national laws and trade agreements, as well as powerful unwritten rules that tell us that no government can increase taxes and stay in power, or say no to major investments no matter how damaging, or plan to gradually contract those parts of our economies that endanger us all.

And yet each of those rules emerged out of the same, coherent worldview. If that worldview is delegitimized, then all of the rules within it become much weaker and more vulnerable. This is another lesson from social movement history across the political spectrum: when fundamental change does come, it's generally not in legislative dribs and drabs spread out evenly over decades. Rather it comes in spasms of rapid-fire lawmaking, with one breakthrough after another. The right calls this "shock therapy"; the left calls it "populism" because it requires so much popular support and mobilization to occur. (Think of the regulatory architecture that emerged in the New Deal period, or, for that matter, the environmental legislation of the 1960s and 1970s.)

So how do you change a worldview, an unquestioned ideology? Part of it involves choosing the right early policy battles—game-changing ones that don't merely aim to change laws but change patterns of thought. That means that a fight for a minimal carbon tax might do a lot less good than, for instance, forming a grand coalition to demand a guaranteed minimum income. That's not only because a minimum income, as discussed, makes it possible for workers to say no to dirty energy jobs but also because the very process of arguing for a universal social safety net opens up a space for a full-throated debate about values—about what we owe to one another based on our shared humanity, and what it is that we collectively value more than economic growth and corporate profits.

Indeed a great deal of the work of deep social change involves having debates during which new stories can be told to replace the ones that have failed us. Because if we are to have any hope of making the kind of civilizational leap required of this fateful decade, we will need to start believing, once again, that humanity is not hopelessly selfish and greedy—the image ceaselessly sold to us by everything from reality shows to neoclassical economics.

Paradoxically, this may also give us a better understanding of our personal climate inaction, allowing many of us to view past (and present) failures with compassion, rather than angry judgment. What if part of the reason so many of us have failed to act is not because we are too selfish to care about an abstract or seemingly far-off problem—but because we are utterly overwhelmed by how much we do care? And what if we stay silent not out of acquiescence but in part because we lack the collective spaces in which to confront the raw terror of ecocide? The end of the world as we know it, after all, is not something anyone should have to face on their own. As the sociologist Kari Norgaard puts it in *Living in Denial*, a fascinating exploration of the way almost all of us suppress the full reality of the climate crisis, "Denial can—and I believe should—be understood as testament to our human capacity for empathy, compassion, and an underlying sense of moral imperative to respond, even as we fail to do so."[15]

Fundamentally, the task is to articulate not just an alternative set of policy proposals but an alternative worldview to rival the one at the heart of the ecological crisis—embedded in interdependence rather than hyper-individualism, reciprocity rather than dominance, and cooperation rather than hierarchy. This is required not only to create a political context to dramatically lower emissions, but also to help us cope with the disasters we can no longer to avoid. Because in the hot and stormy future we have already made inevitable through our past emissions, an unshakable belief in the equal rights of all people and a capacity for deep compassion will be the only things standing between civilization and barbarism.

This is another lesson from the transformative movements of the past: all of them understood that the process of shifting cultural values—though somewhat ephemeral and difficult to quantify—was central to their work. And so they dreamed in public, showed humanity a better version of itself, modeled different values in their own behavior, and in the process liberated the political imagination and rapidly altered the sense of what was possible. They were also unafraid of the language of morality—to give the pragmatic, cost-benefit arguments a rest and speak of right and wrong, of love and indignation.

In *The Wealth of Nations*, Adam Smith made a case against slavery that had little to do with morality and everything to do with the bottom line. Work by paid laborers, he argued, "comes cheaper in the end than that performed by slaves": not only were slave owners responsible for the high costs of the "wear and tear" of their human property but, he claimed, paid laborers had a greater incentive to work hard.[16] Many abolitionists on both sides of the Atlantic would embrace such pragmatic arguments.

However, as the push to abolish the slave trade (and later, slavery itself) ramped up in Britain in the late eighteenth century, much of the movement put considerably more emphasis on the moral travesties of slavery and the corrosive worldview that made it possible. Writing in 1808, British abolitionist Thomas Clarkson described the battle over the slave trade as "a contest between those who felt deeply for the happiness and the honour of their fellow-creatures, and those who, through vicious custom and the impulse of avarice, had trampled under-foot the sacred rights of their nature, and had even attempted to efface all title to the divine image from their minds."[17]

The rhetoric and arguments of American abolitionists could be even starker and more uncompromising. In an 1853 speech, the famed abolitionist orator Wendell Phillips insisted on the right to denounce those who in the harshest terms defended slavery. "Prove to me now that harsh rebuke, indignant denunciation, scathing sarcasm, and pitiless ridicule are wholly and always unjustifiable; else we dare not, in so desperate a case, throw away any weapon which ever broke up the crust of an ignorant prejudice, roused a slumbering conscience, shamed a proud sinner, or changed, in any way, the conduct of a human being. Our aim is to alter public opinion." And indispensable to that goal were the voices of freed slaves themselves, people like Frederick Douglass, who, in his writing and oratory, challenged the very foundations of American patriotism with questions like "What, to the American slave, is your 4th of July?"[18]

This kind of fiery, highly polarizing rhetoric was typical of a battle with so much at stake. As the historian David Brion Davis writes, abolitionists understood that their role was not merely to ban an abhorrent practice but to try to change the deeply entrenched values that had made slavery acceptable in the first place. "The abolition of New World slavery depended in large measure on a major transformation in moral perception—on the emergence of writers, speakers, and reformers, beginning in the mid-eighteenth century, who were willing to condemn an institution that had been sanctioned for thousands of years and who also strove to make human society something more than an endless contest of greed and power."[19]

This same understanding about the need to assert the intrinsic value of life is at the heart of all major progressive victories, from universal suffrage to universal health care. Though these movements all contained economic arguments as part of building their case for justice, they did not win by putting a monetary value on granting equal rights and freedoms. They won by asserting that those rights and freedoms were *too* valuable to be measured and were inherent to each of us. Similarly, there are plenty of solid economic arguments for moving beyond fossil fuels, as more and more patient investors are realizing. And that's worth pointing out. But we will not win the battle for a stable climate by trying to beat the bean counters at their own game—arguing, for instance, that it is more cost-effective to invest in emission reduction now than disaster response later. We will win by asserting that such calculations are morally monstrous, since they imply that there is an acceptable price for allowing entire countries to disappear, for leaving untold millions to die on parched land, for depriving today's children of their right to live in a world teeming with the wonders and beauties of creation.

The climate movement has yet to find its full moral voice on the world stage, but it is most certainly clearing its throat—beginning to put the very real thefts and torments that ineluctably flow from the decision to mock international climate commitments alongside history's most damned crimes. Some of the voices of moral clarity are coming from the very young, who are calling on the streets and increasingly in the courts for intergenerational justice. Some are coming from great social justice movements of the past, like Nobel laureate Desmond Tutu, former archbishop of Cape Town, who has joined the fossil fuel divestment movement with enthusiasm, declaring that "to serve as custodians of creation is not an empty title; it requires that we act, and with all the urgency this dire situation demands."[20] Most of all, those clarion voices are coming from the front lines of Blockadia, from those lives most directly impacted by both high-risk fossil fuel extraction and early climate destabilization.

Suddenly, Everyone

Recent years have been filled with moments when societies suddenly decide they have had enough, defying all experts and forecasters—from the Arab Spring (tragedies, betrayals, and all), to Europe's "squares movement" that saw city centers taken over by demonstrators for months, to Occupy Wall Street, to the student movements of Chile and Quebec. The Mexican journalist Luis Hernández Navarro describes those rare political moments that seem to melt cynicism on contact as the "effervescence of rebellion."[21]

What is most striking about these upwellings, when societies become consumed with the demand for transformational change, is that they so often come as a surprise—most of all to the movements' own organizers. I've heard the story many times: "One day it was just me and my friends dreaming up impossible schemes, the next day the entire country seemed to be out in the plaza alongside us." And the real surprise, for all involved, is that we are so much more than we have been told we are—that we long for more and in that longing have more company than we ever imagined.

No one knows when the next such effervescent moment will open, or whether it will be precipitated by an economic crisis, another natural disaster, or some kind of political scandal. We do know that a warming world will, sadly, provide no shortage of potential sparks. Sivan Kartha, senior scientist at the Stockholm Environment Institute, puts it like this: "What's politically realistic today may have very little to do with what's politically realistic after another few Hurricane Katrinas and another few Superstorm Sandys and another few Typhoon Bophas hit us."[22] It's true: the world tends to look a little different when the objects we have worked our whole lives to accumulate are suddenly floating down the street, or smashed to pieces, turned to garbage.

The world also doesn't look much like it did in the late 1980s. Climate change, as we have seen, landed on the public agenda at the peak of free market, end-of-history triumphalism, which was very bad timing indeed. Its do-or-die moment, however, comes to us at a very different historical juncture. Many of the barriers that paralyzed a serious response to the crisis are today significantly eroded. Free market ideology has been discredited by decades of deepening inequality and corruption, stripping it of much of its persuasive power (if not yet its political and economic power). And the various forms of magical thinking that have diverted precious energy—from blind faith in technological miracles to the worship of benevolent billionaires—are also fast losing their grip. It is slowly dawning on a great many of us that no one is going to step in and fix this crisis; that if change is to take place it will only be because leadership bubbled up from below.

We are also significantly less isolated than many of us were even a decade ago: the new structures built in the rubble of neoliberalism—everything from social media to worker co-ops to farmer's markets to neighborhood sharing banks—have helped us to find community despite the fragmentation of postmodern life. Indeed, thanks in particular to social media, a great many of us are continually engaged in a cacophonous global conversation that, however maddening it is at times, is unprecedented in its reach and power.

Given these factors, there is little doubt that another crisis will see us in the streets and squares once again, taking us all by surprise. The real question is what progressive forces will make of that moment, the power and confidence with which it will be seized. Because these moments when the impossible seems suddenly possible are excruciatingly rare and precious. That means more must be made of them. The next time one arises, it must be harnessed not only to denounce the world as it is, and build fleeting pockets of liberated space. It must be the catalyst to actually build the world that will keep us all safe. The stakes are simply too high, and time too short, to settle for anything less.

———

A year ago, I was having dinner with some newfound friends in Athens. I asked them for ideas about what questions I should put to Alexis Tsipras, the young leader of Greece's official opposition party and one of the few sources of hope in a Europe ravaged by austerity.

Someone suggested, "Ask him: History knocked on your door, did you answer?"

That's a good question, for all of us.

NOTES

N.B.: In the interest of having an endnote section that is shorter than the main text, not every fact in the book has a citation. Facts for which sources are provided include: all quotes, statistics, data points, and facts relating to climate science and carbon mitigation, though often only when the fact first appears in the text and not on repeat references. Facts that do not fall into these categories, but are controversial for some reason, are also sourced.

Sources are not provided for references to uncontroversial facts (usually news events) that can be easily confirmed with a keyword search. Facts that clearly come from the author's personal reporting (but are not quotes) are also generally not sourced.

In cases where there are sources for multiple facts and quotes in a paragraph, one superscript note number appears at the end of the paragraph rather than a number after each individual fact. In the notes section here, the sources are listed in the order in which the facts appear in the paragraph, unless otherwise indicated. This has been done in the interest of achieving a less-cluttered text and further shortening the length of the endnote section.

Quotations that come from interviews conducted by the author or her researchers (usually Rajiv Sicora or Alexandra Tempus) or from the documentary film accompanying this book (directed by Avi Lewis) appear in the endnotes as "personal interview."

If there is a source for a footnote, it is cited in the numbered endnote most closely following the asterisk in the text; such sources are marked FOOTNOTE.

Web addresses for news articles available online are not included because of the transient nature of Web architecture. In cases where a document is available exclusively online, the home page where it appears is cited, not the longer URL for the specific text, once again because links change frequently.

All dollar amounts in the book are in U.S. currency.

EPIGRAPHS
1. "Rebecca Tarbotton," Rainforest Action Network, http://ran.org/becky.
2. Kim Stanley Robinson, "Earth: Under Repair Forever," *OnEarth*, December 3, 2012.

INTRODUCTION
1. Mario Malina et al., "What We Know: The Reality, Risks and Response to Climate Change," AAAS Climate Science Panel, American Association for the Advancement of Science, 2014, pp. 15–16.
2. "Sarah Palin Rolls Out at Rolling Thunder Motorcycle Ride," Fox News, May 29, 2011.
3. Martin Weil, "US Airways Plane Gets Stuck in 'Soft Spot' on Pavement at Reagan National," *Washington Post*, July 7, 2012; "Why Is My Flight Cancelled?" Imgur, http://imgur.com.
4. Weil, "US Airways Plane Gets Stuck in 'Soft Spot' on Pavement at Reagan National."
5. For important sociological and psychological perspectives on the everyday denial of climate change, see: Kari Marie Norgaard, *Living in Denial: Climate Change, Emotions, and Everyday Life* (Cambridge, MA: MIT Press, 2011); Rosemary Randall, "Loss and Climate Change: The Cost of Parallel Narratives," *Ecopsychology* 1.3 (2009): 118-29; and the essays in Sally Weintrobe, ed., *Engaging with Climate Change* (East Sussex: Routledge, 2013).
6. Angélica Navarro Llanos, "Climate Debt: The Basis of a Fair and Effective Solution to Climate

Change," presentation to Technical Briefing on Historical Responsibility, Ad Hoc Working Group on Long-term Cooperative Action, United Nations Framework Convention on Climate Change, Bonn, Germany, June 4, 2009.

7. "British PM Warns of Worsening Floods Crisis," Agence France-Presse, February 11, 2014.

8. "Exponential Growth in Weather Risk Management Contracts," Weather Risk Management Association, press release, June 2006; Eric Reguly, "No Climate-Change Deniers to Be Found in the Reinsurance Business," *Globe and Mail*, November 28, 2013.

9. "Investor CDP 2012 Information Request: Raytheon Company," Carbon Disclosure Project, 2012, https://www.cdp.net.

10. "Who Will Control the Green Economy?" ETC Group, 2011, p. 23; Chris Glorioso, "Sandy Funds Went to NJ Town with Little Storm Damage," NBC News, February 2, 2014.

11. " 'Get It Done: Urging Climate Justice, Youth Delegate Anjali Appadurai Mic-Checks UN Summit," Democracy Now!, December 9, 2011.

12. Corinne Le Quéré et al., "Global Carbon Budget 2013," *Earth System Science Data* 6 (2014): 253; "Greenhouse Gases Rise by Record Amount," Associated Press, November 3, 2011.

13. Sally Weintrobe, "The Difficult Problem of Anxiety in Thinking About Climate Change," in *Engaging with Climate Change*, ed. Sally Weintrobe (East Sussex: Routledge, 2013). 43.

14. For critical scholarship on the history and politics of the 2 degree target, see: Joni Seager, "Death By Degrees: Taking a Feminist Hard Look at the 2 Degrees Climate Policy," *Kvinder, Køn og Foraksning* (Denmark) 18 (2009): 11-22; Christopher Shaw, "Choosing a Dangerous Limit for Climate Change: An Investigation into How the Decision Making Process Is Constructed in Public Discourses," PhD thesis, University of Sussex, 2011, available at http://www.notargets .org.uk; Christopher Shaw, "Choosing a Dangerous Limit for Climate Change: Public Representations of the Decision Making Process," *Global Environmental Change* 23 (2013): 563-571. COPENHAGEN: Copenhagen Accord, United Nations Framework Convention on Climate Change, December 18, 2009, p. 1; "DEATH SENTENCE": "CJN CMP Agenda Item 5 Intervention," speech delivered by activist Sylvia Wachira at Copenhagen climate conference, Climate Justice Now!, December 10, 2009, http://www.climate-justice-now.org; GREENLAND: J. E. Box et al., "Greenland Ice Sheet," Arctic Report Card 2012, National Oceanic and Atmospheric Administration, January 14, 2013; ACIDIFICATION: Bärbel Hönisch et al., "The Geological Record of Ocean Acidification," *Science* 335 (2012): 1058-1063; Adrienne J. Sutton et al., "Natural Variability and Anthropogenic Change in Equatorial Pacific Surface Ocean pCO_2 and pH," *Global Biogeochemical Cycles* 28 (2014): 131-145; PERILOUS IMPACTS: James Hansen et al., "Assessing 'Dangerous Climate Change': Required Reduction of Carbon Emissions to Protect Young People, Future Generations and Nature," *PLOS ONE* 8 (2013): e81648.

15. "Climate Change Report Warns of Dramatically Warmer World This Century," World Bank, press release, November 18, 2012.

16. *Ibid.*; Hans Joachim Schellnhuber et al., "Turn Down the Heat: Why a 4°C Warmer World Must Be Avoided," A Report for the World Bank by the Potsdam Institute for Climate Impact Research and Climate Analytics, November 2012, p. xviii; Kevin Anderson, "Climate Change Going Beyond Dangerous—Brutal Numbers and Tenuous Hope," *Development Dialogue* no. 61, September 2012, p. 29.

17. For general overviews synthesizing scientific research on the likely impacts of a 4 degrees C world, refer to Schellnhuber et al., "Turn Down the Heat," as well as the special theme issue entitled "Four Degrees and Beyond: the Potential for a Global Temperature Increase of Four Degrees and its Implications," compiled and edited by Mark G. New et al., *Philosophical Transactions of The Royal Society* A 369 (2011): 1-241. In 2013, the World Bank released a follow up report exploring the regional impacts of a 4 degree temperature rise, with a focus on Africa and Asia: Hans Joachim Schellnhuber et al., "Turn Down the Heat: Climate Extremes, Regional Impacts, and the Case for Resilience," A Report for the World Bank by the Potsdam Institute for Climate Impact Research and Climate Analytics, June 2013. Even for the most emissions-intensive scenarios that could lead to 4 degrees of warming, IPCC global sea level rise projections are lower than those cited here, but many experts regard them as too conservative. For

examples of research informing this passage, see Schellnhuber et al., "Turn Down the Heat," p. 29; Anders Levermann et al., "The Multimillennial Sea-Level Commitment of Global Warming," *Proceedings of the National Academy of Sciences* 110 (2013): 13748; Benjamin P. Horton et al., "Expert Assessment of Sea-level Rise by AD 2100 and AD 2300," *Quaternary Science Reviews* 84 (2014): 1-6. For more information about the vulnerability of small island nations and coastal areas of Latin America and South and Southeast Asia to sea level rise under "business as usual" and other emissions scenarios (including more optimistic ones), refer to the Working Group II contributions to the 4th and 5th Assessment Reports of the IPCC, both available at http://www .ipcc.ch. See chapters 10, 13, and 16 of M.L. Perry et al., ed., *Climate Change 2007: Impacts, Adaptation and Vulnerability, Contribution of Working Group II to the Fourth Assessment Report of the Intergovernmental Panel on Climate Change* (Cambridge: Cambridge University Press, 2007); and chapters 24, 27, and 29 of V.R. Barros et al., ed., *Climate Change 2014: Impacts, Adaptation, and Vulnerability, Part B: Regional Aspects, Contribution of Working Group II to the Fifth Assessment Report of the Intergovernmental Panel on Climate Change* (Cambridge: Cambridge University Press, 2014). On California and the northeastern United States, see Matthew Heberger et al., "Potential Impacts of Increased Coastal Flooding in California Due to Sea-Level Rise," *Climatic Change* 109, Issue 1 Supplement (2011): 229-249; and Asbury H. Sallenger Jr., Kara S. Doran, and Peter A. Howd, "Hotspot of Accelerated Sea-Level Rise on the Atlantic Coast of North America," *Nature Climate Change* 2 (2012): 884-888. For a recent analysis of major cities that may be particularly threatened by sea level rise, see: Stephane Hallegatte et al., "Future Flood Losses in Major Coastal Cities," *Nature Climate Change* 3 (2013): 802-806.

18. For an overview of regional temperature increases associated with a global rise of 4 degrees C or more, see: M.G. Sanderson, D.L. Hemming and R.A. Betts, "Regional Temperature and Precipitation Changes Under High-end (≥4°C) Global Warming," *Philosophical Transactions of the Royal Society A* 369 (2011): 85-98. See also: "Climate Stabilization Targets: Emissions, Concentrations, and Impacts over Decades to Millennia," Committee on Stabilization Targets for Atmospheric Greenhouse Gas Concentrations, National Research Council, National Academy of Sciences, 2011, p. 31; Schellnhuber et al., "Turn Down the Heat," pp. 37–41. TENS OF THOUSANDS: Jean-Marie Robine et al., "Death Toll Exceeded 70,000 in Europe During the Summer of 2003," *Comptes Rendus Biologies* 331 (2008): 171-78; CROP LOSSES: "Climate Stabilization Targets," National Academy of Sciences, pp. 160–63.

19. ICE-FREE ARCTIC: *Ibid.*, pp. 132–36. VEGETATION: Andrew D. Friend et al., "Carbon Residence Time Dominates Uncertainty in Terrestrial Vegetation Responses to Future Climate and Atmospheric CO_2," *Proceedings of the National Academy of Sciences* 111 (2014): 3280; "4 Degree Temperature Rise Will End Vegetation 'Carbon Sink,'" University of Cambridge, press release, December 17, 2013; WEST ANTARCTICA STUDY: E. Rignot et al., "Widespread, Rapid Grounding Line Retreat of Pine Island, Thwaites, Smith, and Kohler Glaciers, West Antarctica, from 1992 to 2011," *Geophysical Research Letters* 41 (2014): 3502–3509; "APPEARS UNSTOPPABLE": "West Antarctic Glacier Loss Appears Unstoppable," Jet Propulsion Laboratory, NASA, press release, May 12, 2014; "DISPLACE MILLIONS" AND STILL TIME: Eric Rignot, "Global Warming: It's a Point of No Return in West Antarctica. What Happens Next?" *Observer*, May 17, 2014.

20. "World Energy Outlook 2011," International Energy Agency, 2011, p. 40; "World Energy Outlook 2011" (video), Carnegie Endowment for International Peace, November 28, 2011; Timothy M. Lenton et al., "Tipping Elements in the Earth's Climate System," *Proceedings of the National Academy of Sciences* 105 (2008): 1788; "Too Late for Two Degrees?" Low Carbon Economy Index 2012, PricewaterhouseCoopers, November 2012, p. 1.

21. Lonnie G. Thompson, "Climate Change: The Evidence and Our Options," *The Behavior Analyst* 33 (2010): 153.

22. In the U.S., Britain, and Canada, terms for "victory gardens" and "victory bonds" differed between countries and from World War I to World War II; other terms used included "war gardens" and "defense bonds," for example. Ina Zweiniger-Bargielowska, *Austerity in Britain: Rationing, Controls, and Consumption, 1939–1955* (Oxford: Oxford University Press, 2000), 54–55; Amy

Bentley, *Eating for Victory: Food Rationing and the Politics of Domesticity* (Chicago: University of Illinois Press, 1998), 138–39; Franklin D. Roosevelt, "Statement Encouraging Victory Gardens," April 1, 1944, The American Presidency Project, http://www.presidency.ucsb.edu.

23. Pablo Solón, "Climate Change: We Need to Guarantee the Right to Not Migrate," Focus on the Global South, http://focusweb.org.

24. Glen P. Peters et al., "Rapid Growth in CO$_2$ Emissions After the 2008–2009 Global Financial Crisis," *Nature Climate Change* 2 (2012): 2.

25. Spencer Weart, *The Discovery of Global Warming* (Cambridge, MA: Harvard University Press, 2008), 149.

26. Corrine Le Quéré et al., "Trends in the Sources and Sinks of Carbon Dioxide," *Nature Geoscience* 2 (2009): 831, as cited in Andreas Malm, "China as Chimney of the World: The Fossil Capital Hypothesis," *Organization & Environment* 25 (2012): 146; Glen P. Peters et al., "Rapid Growth in CO$_2$ Emissions After the 2008–2009 Global Financial Crisis," *Nature Climate Change* 2 (2012): 2.

27. Kevin Anderson and Alice Bows, "Beyond 'Dangerous' Climate Change: Emission Scenarios for a New World," *Philosophical Transactions of the Royal Society* A 369 (2011): 35; Kevin Anderson, "EU 2030 Decarbonisation Targets and UK Carbon Budgets: Why So Little Science?" Kevin Anderson.info, June 14, 2013, http://kevinanderson.info.

28. Gro Harlem Brundtland et al., "Environment and Development Challenges: The Imperative to Act," joint paper by the Blue Planet Prize laureates, The Asahi Glass Foundation, February 20, 2012, p. 7.

29. "World Energy Outlook 2011," IEA, p. 40; James Herron, "Energy Agency Warns Governments to Take Action Against Global Warming," *Wall Street Journal*, November 10, 2011.

30. Personal interview with Henry Red Cloud, June 22, 2011.

31. Gary Stix, "Effective World Government Will Be Needed to Stave Off Climate Catastrophe," *Scientific American*, March 17, 2012.

32. Daniel Cusick, "Rapid Climate Changes Turn North Woods into Moose Graveyard," *Scientific American*, May 18, 2012; Jim Robbins, "Moose Die-Off Alarms Scientists," *New York Times*, October 14, 2013.

33. Josh Bavas, "About 100,000 Bats Dead After Heatwave in Southern Queensland," ABC News (Australia), January 8, 2014.

34. Darryl Fears, "Sea Stars Are Wasting Away in Larger Numbers on a Wider Scale in Two Oceans," *Washington Post*, November 22, 2013; Amanda Stupi, "What We Know—And Don't Know—About the Sea Star Die-Off," KQED, March 7, 2014.

PART I: BAD TIMING

1. William Stanley Jevons, *The Coal Question: An Inquiry Concerning the Progress of the Nation, and the Probable Exhaustion of Our Coal-Mines* (London: Macmillan and Co., 1865), viii.

2. Hugo's original: "C'est une triste chose de songer que la nature parle et que le genre humain n'écoute pas." Victor Hugo, *Œuvres complètes de Victor Hugo*, Vol. 35, ed. Jeanlouis Cornuz (Paris: Éditions Recontre, 1968), 145.

CHAPTER 1: THE RIGHT IS RIGHT

1. Mario Malina et al., "What We Know: The Reality, Risks and Response to Climate Change," AAAS Climate Science Panel, American Association for the Advancement of Science, 2014, p. 3.

2. Thomas J. Donohue, "Managing a Changing Climate: Challenges and Opportunities for the Buckeye State, Remarks," speech, Columbus, Ohio, May 1, 2008.

3. Session 4: Public Policy Realities (video), 6th International Conference on Climate Change, The Heartland Institute, June 30, 2011.

4. *Ibid.*

5. "Va. Taxpayers Request Records from University of Virginia on Climate Scientist Michael Mann," American Tradition Institute, press release, January 6, 2011; Christopher Horner, "ATI Environmental Law Center Appeals NASA Denial of Request for Dr. James Hansen's Ethics Dis-

closures," American Tradition Institute, press release, March 16, 2011; Session 4: Public Policy Realities (video), The Heartland Institute.

6. Obama for America, "Barack Obama's Plan to Make America a Global Energy Leader," October 2007; personal interview with Patrick Michaels, July 1, 2011; Session 5: Sharpening the Scientific Debate (video), The Heartland Institute; personal interview with Marc Morano, July 1, 2011.

7. Larry Bell, *Climate of Corruption: Politics and Power Behind the Global Warming Hoax* (Austin: Greenleaf, 2011), xi.

8. Peter T. Doran and Maggie Kendall Zimmerman, "Examining the Scientific Consensus on Climate Change," *Eos* 90, (2009): 22-23; William R. L. Anderegg et al., "Expert Credibility in Climate Change," *Proceedings of the National Academy of Sciences*, 107 (2010): 12107-12109.

9. Keynote Address (video), The Heartland Institute, July 1, 2011; Bob Carter, "There IS a Problem with Global Warming . . . It Stopped in 1998," *Daily Telegraph*, April 9, 2006; Willie Soon and David R. Legates, "Avoiding Carbon Myopia: Three Considerations for Policy Makers Concerning Manmade Carbon Dioxide," *Ecology Law Currents* 37 (2010): 3; Willie Soon, "It's the Sun, Stupid!" The Heartland Institute, March 1, 2009, http://heartland.org; Keynote Address (video), The Heartland Institute, June 30, 2011.

10. Personal interview with Joseph Bast, June 30, 2011.

11. In the years following the conference, news coverage rebounded to twenty-nine stories in 2012 and thirty stories in 2013. Douglas Fischer, "Climate Coverage Down Again in 2011," *The Daily Climate*, January 3, 2012; Douglas Fischer, "Climate Coverage Soars in 2013, Spurred by Energy, Weather," *The Daily Climate*, January 2, 2014.

12. Joseph Bast, "Why Won't Al Gore Debate?" The Heartland Institute, press release, June 27, 2007; Will Lester, "Vietnam Veterans to Air Anti-Kerry Ads in W. Va.," Associated Press, August 4, 2004; Leslie Kaufman, "Dissenter on Warming Expands His Campaign," *New York Times*, April 9, 2009; John H. Richardson, "This Man Wants to Convince You Global Warming Is a Hoax," *Esquire*, March 30, 2010; Session 4: Public Policy Realities (video), The Heartland Institute.

13. "Big Drop in Those Who Believe That Global Warming Is Coming," Harris Interactive, press release, December 2, 2009; "Most Americans Think Devastating Natural Disasters Are Increasing," Harris Interactive, press release, July 7, 2011; personal interview with Scott Keeter, September 12, 2011.

14. Lydia Saad, "A Steady 57% in U.S. Blame Humans for Global Warming," Gallup Politics, March 18, 2014; "October 2013 Political Survey: Final Topline," Pew Research Center for the People & the Press, October 9–13, 2013, p. 1; personal email communication with Riley Dunlap, March 29, 2014.

15. DEMOCRATS AND LIBERALS: Aaron M. McCright and Riley E. Dunlap, "The Politicization of Climate Change and Polarization in the American Public's Views of Global Warming 2001–2010," *The Sociological Quarterly* 52 (2011): 188, 193; Saad, "A Steady 57% in U.S. Blame Humans for Global Warming"; REPUBLICANS: Anthony Leiserowitz et al., "Politics and Global Warming: Democrats, Republicans, Independents, and the Tea Party," Yale Project on Climate Change Communication and George Mason University Center for Climate Change Communication, 2011, pp. 3–4; 20 PERCENT: Lawrence C. Hamilton, "Climate Change: Partisanship, Understanding, and Public Opinion," Carsey Institute, Spring 2011, p. 4; OCTOBER 2013 POLL: "Focus Canada 2013: Canadian Public Opinion About Climate Change," The Environics Institute, November 18, 2013, http://www.environicsinstitute.org; AUSTRALIA, U.K., AND WESTERN EUROPE: Bruce Tranter, "Political Divisions over Climate Change and Environmental Issues in Australia," Environmental Politics 20 (2011): 78-96; Ben Clements, "Exploring public opinion on the issue of climate change in Britain," *British Politics* 7 (2012): 183-202; Aaron M. McCright, Riley E. Dunlap, and Sandra T. Marquart-Pyatt, "Climate Change and Political Ideology in the European Union," Michigan State University, working paper, 2014.

16. For a broad, accessible overview of the study of right-wing science denial, see Chris Mooney, *The Republican Brain: The Science of Why They Deny Science—and Reality* (Hoboken, NJ: John Wiley & Sons, 2012). CULTURAL WORLDVIEW: Dan M. Kahan et al., "The Second National Risk

and Culture Study: Making Sense of—and Making Progress in—the American Culture War of Fact," The Cultural Cognition Project at Yale Law School, September 27, 2007, p. 4, available at http://www.culturalcognition.net.

17. Dan Kahan, "Cultural Cognition as a Conception of the Cultural Theory of Risk," in *Handbook of Risk Theory: Epistemology, Decision Theory, Ethics, and Social Implications of Risk*, ed. Sabine Roeser et al. (London: Springer, 2012), 731.

18. Kahan et al., "The Second National Risk and Culture Study," p. 4.

19. Dan Kahan, "Fixing the Communications Failure," *Nature* 463 (2010): 296; Dan Kahan et al., "Book Review—Fear of Democracy: A Cultural Evaluation of Sunstein on Risk," *Harvard Law Review* 119 (2006): 1083.

20. Kahan, "Fixing the Communications Failure," 296.

21. Rebecca Rifkin, "Climate Change Not a Top Worry in U.S.," Gallup, March 12, 2014; "Deficit Reduction Declines as Policy Priority," Pew Research Center for the People & the Press, January 27, 2014; "Thirteen Years of the Public's Top Priorities," Pew Research Center for the People & the Press, January 27, 2014, http://www.people-press.org.

22. Heather Gass, "EBTP at the One Bay Area Agenda 21 Meeting," East Bay Tea Party, May 7, 2011, http://www.theeastbayteaparty.com.

23. For more on the conservative movement's role in climate change denial, see: Riley E. Dunlap and Aaron M. McCright, "Organized Climate Change Denial," in *The Oxford Handbook of Climate Change and Society*, ed. John S. Dryzek, Richard B. Norgaard, and David Schlosberg (Oxford: Oxford University Press, 2011), 144–160; and Aaron M. McCright and Riley E. Dunlap, "Anti-Reflexivity: The American Conservative Movement's Success in Undermining Climate Science and Policy," *Theory, Culture, and Society* 27 (2010): 100–133. DENIAL BOOKS STUDY: Riley E. Dunlap and Peter J. Jacques, "Climate Change Denial Books and Conservative Think Tanks: Exploring the Connection," *American Behavioral Scientist* 57 (2013): 705–706.

24. Bast interview, June 30, 2011.

25. Robert Manne, "How Can Climate Change Denialism Be Explained?" *The Monthly*, December 8, 2011.

26. GORE: "Al Gore Increases His Carbon Footprint, Buys House in Ritzy Santa Barbara Neighborhood," Hate the Media! May 2, 2010; HANSEN: William Lajeunesse, "NASA Scientist Accused of Using Celeb Status Among Environmental Groups to Enrich Himself," Fox News, June 22, 2011; Christopher Horner, "A Brief Summary of James E. Hansen's NASA Ethics File," American Tradition Institute, November 18, 2011; VINDICATED: David Adam, "'Climategate' Review Clears Scientists of Dishonesty over Data," *Guardian*, July 7, 2010; FUELED: James Delingpole, "Climategate: The Final Nail in the Coffin of 'Anthropogenic Global Warming'?" *Daily Telegraph*, November 20, 2009; James Delingpole, "Climategate: FOIA—The Man Who Saved the World," *Daily Telegraph*, March 13, 2013; BILLBOARD CAMPAIGN: Wendy Koch, "Climate Wars Heat Up with Pulled Unabomber Billboards," *USA Today*, May 4, 2012.

27. Personal interview with James Delingpole, July 1, 2011; Bast interview, June 30, 2011.

28. Bast interview, July 1, 2011.

29. "The Rt Hon. Lord Lawson of Blaby," Celebrity Speakers, http://www.speakers.co.uk; Nigel Lawson, *The View from No. 11: Britain's Longest-Serving Cabinet Member Recalls the Triumphs and Disappointments of the Thatcher Era* (New York: Doubleday, 1993), 152–62, 237–40; Tim Rayment and David Smith, "Should High Earners Pay Less Tax," *The Times* (London), September 11, 2011; Nigel Lawson, *An Appeal to Reason: A Cool Look at Global Warming* (New York: Duckworth Overlook, 2008), 101.

30. Naomi Oreskes and Erik M. Conway, *Merchants of Doubt* (New York: Bloomsbury, 2010), 5, 25–26, 82, 135,164; Václav Klaus, "The Climate Change Doctrine Is Part of Environmentalism, Not of Science," *Inaugural Annual GWPF Lecture*, October 19, 2010, http://www.thegwpf.org.

31. Robert J. Brulle, "Institutionalizing Delay: Foundation Funding and the Creation of U.S. Climate Change Counter-Movement Organizations," *Climatic Change* 122 (2014): 681.

32. In addition to questioning whether the concept of "worldview" is truly distinct from political

ideology and possesses unique explanatory power, social scientists have criticized cultural cognition theory for neglecting the structural drivers of the climate change denial movement. For key examples of scholarship focusing on the social, political, and economic dynamics of that movement, see Dunlap and McCright, "Organized Climate Change Denial," and McCright and Dunlap, "Anti-Reflexivity." On the Heartland Institute's funding: according to Greenpeace USA's ExxonSecrets project, the organization "has received $676,500 from ExxonMobil since 1998"; according to Heartland itself, it received a total of $100,000 from the Sarah Scaife Foundation in 1992 and 1993 and $50,000 from the Charles G. Koch Charitable Foundation in 1994; and according to the Conservative Transparency database maintained by the American Bridge 21st Century Foundation, Heartland received an additional total of $42,578 from the Charles G. Koch Charitable Foundation between 1986 and 1989 and in 2011, an additional total of $225,000 from the Sarah Scaife Foundation between 1988 and 1991 and in 1995, a total of $40,000 from the Claude R. Lambe Charitable Foundation (connected to the Koch family) between 1992 and 1999, and a total of $10,000 from The Carthage Foundation (a Scaife foundation) in 1986. See: "Factsheet: Heartland Institute," ExxonSecrets.org, Greenpeace USA, http://www.exxonsecrets.org; Joseph L. Bast, "A Heartland Letter to People for the American Way," The Heartland Institute, August 20, 1996, http://heartland.org; "Heartland Institute," Conservative Transparency, Bridge Project, American Bridge 21st Century Foundation, http://conservativetransparency.org. "MERITS OF OUR POSITIONS": "Reply to Our Critics," The Heartland Institute, http://heartland.org/reply-to-critics; LEAKED DOCUMENTS: "2012 Fundraising Plan," The Heartland Institute, January 15, 2012, pp. 20–21.

33. "Money Troubles: How to Kick-Start the Economy," *Fareed Zakaria GPS*, CNN, August 15, 2010; "Factsheet: Cato Institute," ExxonSecrets.org, Greenpeace USA, http://www.exxonsecrets.org; "Koch Industries Climate Denial Front Group: Cato Institute," Greenpeace USA, http://www.greenpeace.org; "Case Study: Dr. Willie Soon, a Career Fueled by Big Oil and Coal," Greenpeace USA, June 28, 2011, http://www.greenpeace.org.

34. "Factsheet: Committee for a Constructive Tomorrow," ExxonSecrets.org, Greenpeace USA, http://www.exxonsecrets.org; Suzanne Goldenberg, "Secret Funding Helped Build Vast Network of Climate Denial Thinktanks," *Guardian*, February 14, 2013.

35. Lawrence C. Hamilton, "Climate Change: Partisanship, Understanding, and Public Opinion," Carsey Institute, Spring 2011, p. 4; "Vast Majority Agree Climate Is Changing," Forum Research, July 24, 2013, p. 1, http://www.forumresearch.com.

36. Doran and Zimmerman, "Examining the Scientific Consensus on Climate Change," 23; Upton Sinclair, *I, Candidate for Governor: And How I Got Licked* (Berkeley: University of California Press, 1994), 109.

37. Personal email communication with Aaron McCright, September 30, 2011; Aaron McCright and Riley Dunlap, "Cool Dudes: The Denial of Climate Change Among Conservative White Males in the United States," *Global Environmental Change* 21 (2011): 1167, 1171.

38. Session 5: Sharpening the Scientific Debate (video), The Heartland Institute; Chris Hooks, "State Climatologist: Drought Officially Worst on Record," *Texas Tribune*, April 4, 2011; Keynote Address (video), The Heartland Institute, July 1, 2011; "France Heat Wave Death Toll Set at 14,802," Associated Press, September 25, 2003; Keynote Address (video), The Heartland Institute, June 30, 2011.

39. "World Bank Boosts Aid for Horn of Africa Famine," Agence France-Presse, September 24, 2011; "Mankind Always Adapts to Climate, Rep. Barton Says," Republicans on the House and Energy Commerce Committee, press release, March 25, 2009, http://republicans.energycommerce.house.gov.

40. "Turn Down the Heat: Why a 4°C Warmer World Must Be Avoided," Potsdam Institute for Climate Impact Research and Climate Analytics, World Bank, November 2012, p. ix; personal interview with Patrick Michaels, July 1, 2011.

41. "Petition of the Chamber of Commerce of the United States of America for EPA to Conduct Its Endangerment Finding Proceeding on the Record Using Administrative Procedure Act §§ 556

and 557," Attachment 1, "Detailed Review of the Health and Welfare Science Evidence and IQA Petition for Correction," U.S. Chamber of Commerce, 2009, p. 4.

42. Christian Parenti, *Tropic of Chaos: Climate Change and the New Geography of Violence* (New York: Nation Books, 2011).

43. Bryan Walsh, "The Costs of Climate Change and Extreme Weather Are Passing the High-Water Mark," *Time*, July 17, 2013; Suzanne Goldenberg, "Starbucks Concerned World Coffee Supply Is Threatened by Climate Change," *Guardian*, October 13, 2011; Emily Atkin, "Chipotle Warns It Might Stop Serving Guacamole If Climate Change Gets Worse," Climate Progress, March 4, 2014; Robert Kopp et al., "American Climate Prospectus: Economic Risks in the United States," prepared by Rhodium Group for the Risky Business Project, June 2014.

44. "Insurer Climate Risk Disclosure Survey," *Ceres*, March 2013, p. 53, http://www.ceres.org; Eduardo Porter, "For Insurers, No Doubts on Climate Change," *New York Times*, May 14, 2013; "2012 Fundrasing Plan," The Heartland Institute, January 15, 2012, pp. 24–25.

45. Joseph Bast, "About the Center on Finance, Insurance, and Real Estate at the Heartland Institute," Policy Documents, The Heartland Institute, June 5, 2012; personal interview with Eli Lehrer, August 20, 2012.

46. Lehrer interview, August 20, 2012.

47. *Ibid.*

48. John R. Porter et al., "Food Security and Food Production Systems," in *Climate Change 2014: Impacts, Adaptation, and Vulnerability, Part A: Global and Sectoral Aspects, Contribution of Working Group II to the Fifth Assessment Report of the Intergovernmental Panel on Climate Change*, ed. C.B Field et al. (Cambridge: Cambridge University Press, 2014), 20-21; Joan Nymand Larsen et al., "Polar Regions," in *Climate Change 2014: Impacts, Adaptation, and Vulnerability, Part B: Regional Aspects, Contribution of Working Group II to the Fifth Assessment Report of the Intergovernmental Panel on Climate Change*, ed. V.R. Barros et al. (Cambridge: Cambridge University Press, 2014), 20; Julie Satow, "The Generator Is the Machine of the Moment," *New York Times*, January 11, 2013.

49. William Alden, "Around Goldman's Headquarters, an Oasis of Electricity," *New York Times*, November 2, 2012; "How FedEx Survived Hurricane Sandy," KLTV, October 31, 2012; Kimi Yoshino, "Another Way the Rich Are Different: 'Concierge-Level' Fire Protection," *Los Angeles Times*, October 26, 2007; P. Solomon Banda, "Insurance Companies Send Crews to Protect Homes," Associated Press, July 5, 2012.

50. Jim Geraghty, "Climate Change Offers Us an Opportunity," *Philadelphia Inquirer*, August 28, 2011; FOOTNOTE: "House Bill No. 459," 2011 Montana Legislature, February 15, 2011; Brad Johnson, "Wonk Room Interviews Montana Legislator Who Introduced Bill to Declare Global Warming 'Natural,'" ThinkProgress Green, February 17, 2011.

51. FOOTNOTE: "Mission Statement," American Freedom Alliance, http://www.americanfreedomalliance.org; Chris Skates, *Going Green: For Some It Has Nothing to Do with the Environment* (Alachua, FL: Bridge-Logos, 2011).

52. Kurt M. Campbell, Jay Gulledge, J. R. McNeill, et al., "The Age of Consequences: The Foreign Policy National Security Implications of Global Climate Change," Center for Strategic and International Studies and Center for a New American Security, November 2007, p. 85.

53. Lee Fang, "David Koch Now Taking Aim at Hurricane Sandy Victims," *The Nation*, December 22, 2012.

54. "230,000 Join Mail Call to Use Some of the UK's £11billion Foreign Aid Budget to Tackle Floods Crisis," *Daily Mail*, February 14, 2014.

55. Joe Romm, "Krauthammer, Part 2: The Real Reason Conservatives Don't Believe in Climate Science," Climate Progress, June 1, 2008.

56. Spencer Weart, *The Discovery of Global Warming* (Cambridge, MA: Harvard University Press, 2008), 149.

57. Global Carbon Project emissions data, 2013 Budget v2.4 (July 2014), available at http://cdiac.ornl.gov.

58. *Ibid.*; Michael Mann interview, *The Big Picture with Thom Hartmann*, RT America, March 25,

2014; Kevin Anderson, "Why Carbon Prices Can't Deliver the 2°C Target," KevinAnderson. info, August 13, 2013, http://kevinanderson.info.

59. Kahan et al., "The Second National Risk and Culture Study," pp. 5-6.

60. Robert Jay Lifton and Richard Falk, *Indefensible Weapons: The Political and Psychological Case Against Nuclearism* (New York: Basic Books, 1982).

61. Dan Kahan et al., "The Tragedy of the Risk-Perception Commons: Culture Conflict, Rationality Conflict, and Climate Change," Cultural Cognition Project Working Paper No. 89, 2011, pp. 15-16, available at http://culturalcognition.net; Umair Irfan, "Report Finds 'Motivated Avoidance' Plays a Role in Climate Change Politics," *ClimateWire*, December 19, 2011; Irina Feygina, John T. Jost, and Rachel E. Goldsmith, "System Justification, the Denial of Global Warming, and the Possibility of 'System-Sanctioned Change,'" *Personality and Social Psychology Bulletin* 36, (2010): 336.

62. Ted Nordhaus and Michael Shellenberger, "The Long Death of Environmentalism," Breakthrough Institute, February 25, 2011; Michael Shellenberger and Ted Nordhaus, "Evolve," *Orion*, September/October 2011.

63. Scott Condon, "Expert: Win Climate Change Debate by Easing off Science," *Glenwood Springs Post Independent*, July 29, 2010.

64. For an example of how psychologists interested in generational differences have analyzed data from "The American Freshman" survey run out of the University of California, Los Angeles, see: Jean M. Twenge, Elise C. Freeman, and W. Keith Campbell, "Generational Differences in Young Adults' Life Goals, Concern for Others, and Civic Orientation, 1966–2009," *Journal of Personality and Social Psychology* 102 (2012): 1045-1062. Alternative explanations such as the rising cost of education (itself a product of the neoliberal era) could help explain the shift in materialist attitudes. For the 1966 and 2013 survey data, see: Alexander W. Astin, Robert J. Panos, and John A. Creager, "National Norms for Entering College Freshmen—Fall 1966," Ace Research Reports, Vol. 2, No. 7, 1967, p. 21; Kevin Eagan et al., "The American Freshman: National Norms Fall 2013," Cooperative Institutional Research Program at the Higher Education Research Institute, University of California, Los Angeles, 2013, p. 40. THATCHER QUOTE: Ronald Butt, "Mrs Thatcher: The First Two Years," *Sunday Times* (London), May 3, 1981.

65. John Immerwahr, "Waiting for a Signal: Public Attitudes Toward Global Warming, the Environment, and Geophysical Research," Public Agenda, American Geophysical Union, April 15, 1999, pp. 4–5.

66. Yuko Heath and Robert Gifford, "Free-Market Ideology and Environmental Degradation: The Case of Belief in Global Climate Change," *Environment and Behavior* 38 (2006): 48–71; Tim Kasser, "Values and Ecological Sustainability: Recent Research and Policy Possibilities," in *The Coming Transformation: Values to Sustain Human and Natural Communities*, eds. Stephen R. Kellert and James Gustave Speth, Yale School of Forestry & Environmental Studies, 2009, pp. 180–204; Tim Crompton and Tim Kasser, *Meeting Environmental Challenges: The Role of Human Identity* (Surrey: WWF-UK, 2009), 10.

67. Milton Friedman and Rose D. Friedman, *Two Lucky People: Memoirs* (Chicago: University of Chicago Press, 1998), 594.

68. Rebecca Solnit, *A Paradise Built in Hell: The Extraordinary Communities That Arise in Disaster* (New York: Penguin Books, [2009] 2010).

CHAPTER 2: HOT MONEY

1. Ken Burns, *The Dust Bowl*, PBS, 2012.

2. Marlene Moses, "The Choice Is Ours," Planet B, Rio + 20 Special Edition, June 2012, p. 80.

3. US CHALLENGE TO CHINA: "China—Measures Concerning Wind Power Equipment," Request for Consultations by the United States, World Trade Organization, December 22, 2010, p. 1; CHINA CHALLENGE TO EU: "European Member States—Certain Measures Affecting the Renewable Energy Generation Sector," Request for Consultations by China, World Trade Organization, November 7, 2012, p. 1; CHINA THREATENS US: "Announcement No. 26 of 2012 of the Ministry of Commerce of the People's Republic of China on the Preliminary

Investigation Conclusion on the U.S. Policy Support and Subsidies for Its Renewable Energy Sector," Ministry of Commerce, People's Republic of China, May 27, 2012, http://english.mof com.gov; US CHALLENGE TO INDIA: "India—Certain Measures Relating to Solar Cells and Solar Modules," Request for Consultations by the United States, World Trade Organization, February 11, 2013, pp. 1–2; CONTEMPLATING CLOSURE: Chandra Bhushan, "Who Is the One Not Playing by the Rules—India or the US?" Centre for Science and Environment, February 8, 2013; personal interview with Chandra Bhushan, Deputy Director General, Centre for Science and Environment, May 10, 2013; INDIA RESPONSE: "Certain Local Content Requirements in Some of the Renewable Energy Sector Programs," Questions by India to the United States, World Trade Organization, April 17, 2013, p. 1; "Subsidies," Questions Posed by India to the United States Under Article 25.8 of the Agreement on Subsidies and Countervailing Measures—State Level Renewable Energy Sector Subsidy Programmes with Local Content Requirements, World Trade Organization, April 18, 2013.

4. Personal interview with Paolo Maccario, January 9, 2014.
5. Green Energy and Green Economy Act, 2009, S.O. 2009, c.12—Bill 150, Government of Ontario, 2009.
6. Jenny Yuen, "Gore Green with Envy," *Toronto Star*, November 25, 2009; "International Support for Ontario's Green Energy Act," Government of Ontario, Ministry of Energy, June 24, 2009.
7. "Feed-in Tariff Program: FIT Rules Version 1.1," Ontario Power Authority, September 30, 2009, p. 14.
8. Michael A. Levi, "The Canadian Oil Sands: Energy Security vs. Climate Change," Council on Foreign Relations, 2009, p. 12; Gary Rabbior, "Why the Canadian Dollar Has Been Bouncing Higher," *Globe and Mail*, October 30, 2009.
9. GAS: Mississauga Power Plant Cancellation Costs, Special Report, Office of the Auditor General of Ontario, April 2013, pp. 7–8; Oakville Power Plant Cancellation Costs, Special Report, Office of the Auditor General of Ontario, October 2013, pp. 7–8; WIND: Dave Seglins, "Ont. Couple Seeks Injunction to Stop Wind-Farm Expansion," CBC News, September 11, 2012; SOLAR: "Ontario Brings More Clean Solar Power Online, Creates Jobs," Government of Ontario, Ministry of Energy, press release, July 31, 2012; ONE COAL PLANT: "Ontario—First Place in North America to End Coal-Fired Power," Government of Ontario, Office of the Premier, November 21, 2013; JOBS: "Progress Report 2014: Jobs and Economy," Government of Ontario, May 1, 2014, http://www.ontario.ca.
10. "Wayne Wright, Silfab Solar" (video), BlueGreen Canada, YouTube, June 2, 2011.
11. "Canada—Certain Measures Affecting the Renewable Energy Generation Sector," Request for Consultations by Japan, World Trade Organization, September 16, 2010, pp. 2–3.
12. "Canada—Certain Measures Affecting the Renewable Energy Generation Sector; Canada—Measures Relating to the Feed-in Tariff Program," Reports of the Appellate Body, World Trade Organization, May 6, 2013; "Ontario to Change Green Energy Law After WTO Ruling," Canadian Press, May 29, 2013; "Ontario Lowering Future Energy Costs," Government of Ontario, Ministry of Energy, press release, December 11, 2013.
13. Elizabeth Bast et al., "Low Hanging Fruit: Fossil Fuel Subsidies, Climate Finance, and Sustainable Development," Oil Change International for the Heinrich Böll Stiftung North America, June 2012, p. 16; Nicholas Stern, *The Economics of Climate Change: The Stern Review* (Cambridge: Cambridge University Press, [2006] 2007), xviii.
14. "Facts About Wind Power: Facts and Numbers," Danish Energy Agency, http://www.ens.dk; "Renewables Now Cover More than 40% of Electricity Consumption," Danish Energy Agency, press release, September 24, 2012; Greg Pahl, *The Citizen-Powered Energy Handbook: Community Solutions to a Global Crisis* (White River Junction, VT: Chelsea Green, 2007), 69; Shruti Shukla and Steve Sawyer (Global Wind Energy Council), *30 Years of Policies for Wind Energy: Lessons from 12 Wind Energy Markets* (Abu Dhabi, UAE: International Renewable Energy Agency, 2012), 55.
15. Scott Sinclair, "Negotiating from Weakness," Canadian Centre for Policy Alternatives, April 2010, p. 11.

16. Aaron Cosbey, "Renewable Energy Subsidies and the WTO: The Wrong Law and the Wrong Venue," *Subsidy Watch* 44 (2011): 1.
17. "Multi-Association Letter Regarding EU Fuel Quality Directive," Institute for 21st Century Energy, May 20, 2013, http://www.energyxxi.org; "Froman Pledges to Preserve Jones Act, Criticizes EU Clean Fuel Directive," *Inside US Trade*, September 20, 2013; "Non-paper on a Chapter on Energy and Raw Materials in TTIP," Council of the European Union, May 27, 2014, http://www.scribd.com; Lydia DePillis, "A Leaked Document Shows Just How Much the EU Wants a Piece of America's Fracking Boom," *Washington Post*, July 8, 2014.
18. The quote is from an interview conducted by Victor Menotti, executive director of the International Forum on Globalization, in 2005. Victor Menotti, "G8 'Climate Deal' Ducks Looming Clash with WTO," International Forum on Globalization, July 2007, http://www.ifg.org.
19. "Notice of Arbitration Under the Arbitration Rules of the United Nations Commission on International Trade Law and Chapter Eleven of the North American Free Trade Agreement," Lone Pine Resources, September 6, 2013.
20. "U.S. Solar Market Insight Report: 2013 Year-in-Review," Executive Summary, GTM Research, Solar Energy Industries Association, p. 4; Bhushan, "Who Is the One Not Playing by the Rules— India or the US?"; Bhushan interview, May 10, 2013; Maccario interview, January 9, 2014; "Climate Change, China, and the WTO," March 30, 2011 (video), panel discussion, Columbia Law School.
21. Personal interview with Steven Shrybman, October 4, 2011.
22. Oceanographer Roger Revelle, who led the team that wrote on atmospheric CO2 in the report for President Johnson, had used similar language describing carbon emissions as a "geophysical experiment" as early as 1957, in a landmark climate science paper co-authored with chemist Hans Suess: Roger Revelle and Hans E. Suess, "Carbon Dioxide Exchange Between Atmosphere and Ocean and the Question of an Increase of Atmospheric CO_2 during the Past Decades," *Tellus* 9 (1957): 19–20. For in-depth histories of climate science and politics, see: Spencer Weart, *The Discovery of Global Warming* (Cambridge, MA: Harvard University Press, 2008); Joshua P. Howe, *Behind the Curve: Science and the Politics of Global Warming* (Seattle: University of Washington Press, 2014). HISTORY: Weart, *The Discovery of Global Warming*, 1–37; JOHNSON REPORT: Roger Revelle et al., "Atmospheric Carbon Dioxide," in *Restoring the Quality of Our Environment*, Report of the Environmental Pollution Panel, President's Science Advisory Committee, The White House, November 1965, Appendix Y4, pp. 126–27.
23. "Statement of Dr. James Hansen, Director, NASA Goddard Institute for Space Studies," presented to United States Senate, June 23, 1988; Philip Shabecoff, "Global Warming Has Begun, Expert Tells Senate," *New York Times*, June 24, 1988; Weart, *The Discovery of Global Warming*, 150–51.
24. Thomas Sancton, "Planet of the Year: What on EARTH Are We Doing?," *Time*, January 2, 1989.
25. *Ibid.*
26. President R. Venkataraman, "Towards a Greener World," speech at WWF-India, New Delhi, November 3, 1989, in *Selected Speeches, Volume I: July 1987–December 1989* (New Delhi: Government of India, 1991), 612.
27. Daniel Indiviglio, "How Americans' Love Affair with Debt Has Grown," *The Atlantic*, September 26, 2010.
28. One bold proposal imagines future restrictions on trade in all goods produced with fossil fuels, arguing that once the green transition is underway and industries have begun to decarbonize, such measures could be introduced and ramped up gradually: Tilman Santarius, "Climate and Trade: Why Climate Change Calls for Fundamental Reforms in World Trade Policies," German NGO Forum on Environment and Development, Heinrich Böll Foundation, pp. 21–23. U.N. CLIMATE AGREEMENT: United Nations Framework Convention on Climate Change, United Nations, 1992, Article 3, Principle 5; "PIVOTAL MOMENT": Robyn Eckersley, "Understanding the Interplay Between the Climate and Trade Regimes," in *Climate and Trade Policies in a Post-2012 World*, United Nations Environment Programme, p. 17.

29. Martin Khor, "Disappointment and Hope as Rio Summit Ends," in *Earth Summit Briefings* (Penang: Third World Network, 1992), p. 83.

30. Steven Shrybman, "Trade, Agriculture, and Climate Change: How Agricultural Trade Policies Fuel Climate Change," Institute for Agriculture and Trade Policy, November 2000, p. 1.

31. Sonja J. Vermeulen, Bruce M. Campbell, and John S.I. Ingram, "Climate Change and Food Systems," *Annual Review of Environment* 37 (2012): 195; personal email communication with Steven Shrybman, April 23, 2014.

32. "Secret Trans-Pacific Partnership Agreement (TPP)—Environment Consolidated Text," WikiLeaks, January 15, 2014, https://wikileaks.org; "Summary of U.S. Counterproposal to Consolidated Text of the Environment Chapter," released by RedGE, February 17, 2014, http://www.redge.org.pe.

33. Traffic refers to containerized port traffic, measured by twenty-foot equivalent units (TEUs). From 1994 to 2013 global containerized port traffic increased from 128,320,326 to an estimated 627,930,960 TEUs, an increase of 389.4 percent: United Nations Conference on Trade and Development, "Review of Maritime Transport," various years, available at http://unctad.org. For years 2012 and 2013, port traffic was projected based on industry estimates from Drewry: "Container Market Annual Review and Forecast 2013/14," Drewry, October 2013. NOT ATTRIBUTED: "Emissions from Fuel Used for International Aviation and Maritime Transport (International Bunker Fuels)," United Nations Framework Convention on Climate Change, http://unfccc.int; SHIPPING EMMISSIONS: Øyvind Buhaug et al., "Second IMO GHG Study 2009," International Maritime Organization, 2009, p. 1.

34. "European Union CO_2 Emissions: Different Accounting Perspectives," European Environmental Agency Technical Report No. 20/2013, 2013, pp. 7–8.

35. Glen P. Peters et al., "Growth in Emission Transfers via International Trade from 1990 to 2008," *Proceedings of the National Academy of Sciences* 108 (2011): 8903-4.

36. Corrine Le Quéré et al., "Global Budget 2013," *Earth System Science Data* 6 (2014): 252; Corrine Le Quéré et al., "Trends in the Sources and Sinks of Carbon Dioxide," *Nature Geoscience* 2 (2009): 831; Ross Garnaut et al., "Emissions in the Platinum Age: The Implications of Rapid Development for Climate-Change Mitigation," *Oxford Review of Economic Policy* 24 (2008): 392; Glen P. Peters et al., "Rapid Growth in CO_2 Emissions After the 2008–2009 Global Financial Crisis," *Nature Climate Change* 2 (2012): 2; "Technical Summary," in O. Edenhofer et al., ed., *Climate Change 2014: Mitigation of Climate Change, Contribution of Working Group III to the Fifth Assessment Report of the Intergovernmental Panel on Climate Change* (Cambridge: Cambridge University Press), 15.

37. Andreas Malm, "China as Chimney of the World: The Fossil Capital Hypothesis," *Organization & Environment* 25 (2012): 146, 165; Yan Yunfeng and Yang Laike, "China's Foreign Trade and Climate Change: A Case Study of CO_2 Emissions," *Energy Policy* 38 (2010): 351; Ming Xu et al., "CO_2 Emissions Embodied in China's Exports from 2002 to 2008: A Structural Decomposition Analysis," *Energy Policy* 39 (2011): 7383.

38. Personal interview with Margrete Strand Rangnes, March 18, 2013.

39. Malm, "China as Chimney of the World," 147, 162.

40. Elisabeth Rosenthal, "Europe Turns Back to Coal, Raising Climate Fears," *New York Times*, April 23, 2008; Personal email communication with IEA Clean Coal Centre, March 19, 2014.

41. Jonathan Watts, "Foxconn offers pay rises and suicide nets as fears grow over wave of deaths," *Guardian*, May 28, 2010; Shahnaz Parveen, "Rana Plaza factory collapse survivors struggle one year on," BBC News, April 23, 2014.

42. Mark Dowie, *Losing Ground: American Environmentalism at the Close of the Twentieth Century* (Cambridge, MA: MIT Press, 1996), 185-86; Keith Schneider, "Environment Groups Are Split on Support for Free-Trade Pact," *New York Times*, September 16, 1993.

43. Dowie, *Losing Ground*, 186–87; Gilbert A. Lewthwaite, "Gephardt Declares Against NAFTA; Democrat Cites Threat to U.S. Jobs," *Baltimore Sun*, September 22, 1993; John Dillin, "NAFTA Opponents Dig In Despite Lobbying Effort," *Christian Science Monitor*, October 12, 1993;

Mark Dowie, "The Selling (Out) of the Greens; Friends of Earth–or Bill?" *The Nation*, April 18, 1994.

44. Bill Clinton, "Remarks on the Signing of NAFTA (December 8, 1993)," Miller Center, University of Virginia.

45. Stan Cox, "Does It Really Matter Whether Your Food Was Produced Locally?" *Alternet*, February 19, 2010.

46. Solomon interview, August 27, 2013.

47. Kevin Anderson, "Climate Change Going Beyond Dangerous—Brutal Numbers and Tenuous Hope," *Development Dialogue* no. 61, September 2012, pp. 16-40.

48. The "8 to 10 percent" range relies on interviews with Anderson and Bows-Larkin as well as their published work. For the underlying emissions scenarios, refer to pathways C+1, C+3, C+5, and B6 3 in: Kevin Anderson and Alice Bows, "Beyond 'Dangerous' Climate Change: Emission Scenarios for a New World," *Philosophical Transactions of the Royal Society* A 369 (2011): 35. See also: Kevin Anderson, "EU 2030 Decarbonisation Targets and UK Carbon Budgets: Why So Little Science?" KevinAnderson.info, June 14, 2013, http://kevinanderson.info. HUGELY DAMAGING: Anderson, "Climate Change Going Beyond Dangerous," pp. 18–21; DE BOER: Alex Morales, "Kyoto Veterans Say Global Warming Goal Slipping Away," Bloomberg, November 4, 2013.

49. Stern, *The Economics of Climate Change*, 231–32.

50. *Ibid.*, 231; Global Carbon Project emissions data, 2013 Budget v2.4 (July 2014), available at http://cdiac.ornl.gov; Carbon Dioxide Information Analysis Center emissions data, available at http://cdiac.ornl.gov.

51. Kevin Anderson and Alice Bows, "A 2°C Target? Get Real, Because 4°C Is on Its Way," *Parliamentary Brief* 13 (2010): 19; FOOTNOTE: Anderson and Bows, "Beyond 'Dangerous' Climate Change," 35; Kevin Anderson, "Avoiding Dangerous Climate Change Demands De-growth Strategies from Wealthier Nations," KevinAnderson.info, November 25, 2013, http://kevinanderson.info.

52. Anderson and Bows-Larkin have based their analysis on the commitment made by governments at the 2009 U.N. climate summit in Copenhagen that emission cuts should be done on the basis "of equity" (meaning rich countries must lead so that poor countries have room to develop). Some argue that rich countries don't have to cut quite so much. Even if that were true, however, the basic global picture still suggests that the necessary reductions are incompatible with economic growth as we have known it. As Tim Jackson shows in *Prosperity Without Growth*, global annual emission cuts of as little as 4.9 percent cannot be achieved simply with green tech and greater efficiencies. Indeed he writes that to meet that target, with the world population and income per capita continuing to grow at current rates, the carbon intensity of economic activity would need to go down "almost ten times faster than it is doing right now." And by 2050, we would need to be twenty-one times more efficient than we are today. So, even if Anderson and Bows-Larkin have vastly overshot, they are still right on their fundamental point: we need to change our current model of growth. See: Tim Jackson, *Prosperity Without Growth: Economics for a Finite Planet* (London: Earthscan, 2009): 80, 86.

53. Anderson and Bows, "A New Paradigm for Climate Change," 640.

54. Kevin Anderson, "Romm Misunderstands Klein's and My View of Climate Change and Economic Growth," KevinAnderson.info, September 24, 2013.

55. Clive Hamilton, "What History Can Teach Us About Climate Change Denial," in *Engaging with Climate Change: Psychoanalytic and Interdisciplinary Perspectives*, ed. Sally Weintrobe (East Sussex: Routledge, 2013), 18.

56. For the foundational scenario modeling work on a "Great Transition" to global sustainability, led by researchers at the Tellus Institute and the Stockholm Environment Institute, see: Paul Raskin et al., "Great Transition: The Promise and Lure of the Times Ahead," Report of the *Global Scenario Group*, Stockholm Environment Institute and Tellus Institute, 2002. This research has continued as part of Tellus' Great Transition Initiative, available at: "Great Transition Initiative:

Toward a Transformative Vision and Praxis," Tellus Institute, http://www.greattransition.org. For parallel work at the U.K.'s New Economics Foundation, see: Stephen Spratt, Andrew Simms, Eva Neitzert, and Josh Ryan-Collins, "The Great Transition," The New Economics Foundation, June 2010.

57. Bows interview, January 14, 2013.
58. Rebecca Willis and Nick Eyre, "Demanding Less: Why We Need a New Politics of Energy," Green Alliance, October 2011, pp. 9, 26.
59. FOOTNOTE: "EP Opens Option for a Common Charger for Mobile Phones," European Commission, press release, March 13, 2014; Adam Minter, *Junkyard Planet* (New York: Bloomsbury, 2013), 6–7, 67, 70.
60. This quote has been clarified slightly at Anderson's request. Paul Moseley and Patrick Byrne, "Climate Expert Targets the Affluent," BBC, November 13, 2009.
61. Phaedra Ellis-Lamkins, "How Climate Change Affects People of Color," *The Root*, March 3, 2013.
62. Tim Jackson, "Let's Be Less Productive," *New York Times*, May 26, 2012.
63. John Stutz, "Climate Change, Development and the Three-Day Week," Tellus Institute, January 2, 2008, pp. 4-5. See also: Juliet B. Schor, *Plenitude: The New Economics of True Wealth* (New York: Penguin Press, 2010); Kyle W. Knight, Eugene A. Rosa, and Juliet B. Schor, "Could Working Less Reduce Pressures on the Environment? A Cross-National Panel Analysis of OECD Countries, 1970–2007," *Global Environmental Change* 23 (2013): 691-700.
64. Alyssa Battistoni, "Alive in the Sunshine," *Jacobin* 13 (2014): 25.

CHAPTER 3: PUBLIC AND PAID FOR

1. Sunita Narain, "Come Out and Claim the Road," *Business Standard*, November 10, 2013.
2. George Orwell, *The Lion and the Unicorn: Socialism and the English Genius* (London: Secker & Warburg, [1941] 1962), 64.
3. Anna Leidreiter, "Hamburg Citizens Vote to Buy Back Energy Grid," World Future Council Climate and Energy Commission, September 25, 2013; personal email communication with Hans Thie, economic policy advisor, German Bundestag (Left Party), March 14, 2014.
4. Personal email communication with Wiebke Hansen, 20 March 2014.
5. The German data, measuring renewable electricity supply as a share of gross electricity consumption, differs slightly from the U.S. data, which measures the wind and solar share of net electricity generation: "Renewable Energy Sources in Germany—Key Information 2013 at a Glance," German Federal Ministry for Economic Affairs and Energy, Working Group on Renewable Energy-Statistics (AGEE-Stat), http://www.bmwi.de; "Table 1.1.A. Net Generation from Renewable Sources: Total (All Sectors), 2004–April 2014," Electric Power Monthly, U.S. Energy Information Administration, http://www.eia.gov; "Table 1.1. Net Generation by Energy Source: Total (All Sectors), 2004–April 2014," Electric Power Monthly, U.S. Energy Information Administration. FRANKFURT AND MUNICH: "City of Frankfurt 100% by 2050," Go 100% Renewable Energy, http://www.go100percent.org; "City of Munich," Go 100% Renewable Energy.
6. "Factbox—German Coalition Agrees on Energy Reforms," Reuters, November 27, 2013.
7. Leidreiter, "Hamburg Citizens Vote to Buy Back Energy Grid."
8. Nicholas Brautlecht, "Hamburg Backs EU2 Billion Buyback of Power Grids in Plebiscite," Bloomberg, September 23, 2013; personal interview with Elisabeth Mader, spokesperson, German Association of Utilities, March 20, 2014.
9. "Energy Referendum: Public Buy-Back of Berlin Grid Fails," *Spiegel Online*, November 4, 2013; personal email communication with Arwen Colell, cofounder, BürgerEnergie Berlin (Citizen Energy Berlin), March 20, 2014.
10. Personal interview with Steve Fenberg, March 19, 2014.
11. "Campaign for Local Power" (video), New Era Colorado, YouTube, September 1, 2013; "Boulder and Broomfield Counties' Final 2011 Election Results," *Daily Camera*, November 1, 2011.
12. "Campaign for Local Power" (video), YouTube.

13. NETHERLANDS: International Energy Agency, *Energy Policies of IEA Countries: The Netherlands; 2008 Review* (Paris: International Energy Agency and the Organisation for Economic Co-operation and Development, 2009), 9–11, 62–64; AUSTRIA: International Energy Agency, *Energy Policies of IEA Countries; Austria: 2007 Review* (Paris: International Energy Agency and the Organisation for Economic Co-operation and Development, 2008), 11–16; NORWAY: International Energy Agency, *Renewable Energy: Medium-Term Market Report 2012; Market Trends and Projections to 2017* (Paris: International Energy Agency and the Organisation for Economic Co-operation and Development, 2012), 71–76; AUSTIN: "Climate Protection Resolution No. 20070215-023," 2013 Update, Office of Sustainability, City of Austin, p. 3, http://www.austin texas.gov; SACRAMENTO: "Our Renewable Energy Portfolio," Sacramento Municipal Utility District, https://www.smud.org; "Greenhouse Gas Reduction," Sacramento Municipal Utility District, https://www.smud.org.; "LOBBY AS HARD AS WE CAN": Personal interview with John Farrell, March 19, 2014.

14. Translation provided by Tadzio Mueller. "Unser Hamburg, Unser Netz," Hamburger Energienetze in die Öffentliche Hand!, http://unser-netz-hamburg.de.

15. "Energy Technology Perspectives 2012: Pathways to a Clean Energy System," International Energy Agency, 2012, p. 149.

16. David Hall et al., "Renewable Energy Depends on the Public Not Private Sector," Public Services International Research Unit, University of Greenwich, June 2013, p. 2.

17. *Ibid.*, pp. 2, 3–5.

18. Mark Z. Jacobson and Mark A. Delucchi, "A Plan to Power 100 Percent of the Planet with Renewables," *Scientific American*, November 2009, pp. 58-59; Mark Z. Jacobson and Mark A. Delucchi, "Providing All Global Energy with Wind, Water, and Solar Power, Part I: Technologies, Energy Resources, Quantities and Areas of Infrastructure, and Materials," *Energy Policy* 39 (2011): 1154–69, 1170–90.

19. Matthew Wright and Patrick Hearps, "Zero Carbon Australia 2020: Stationary Energy Sector Report—Executive Summary" (2nd ed.), University of Melbourne Energy Research Institute and Beyond Zero Emissions, August 2011, pp. 2, 6.

20. As of July 2014, the NOAA researchers had presented the results of their 5-year study and expected to publish in the near future. Alexander MacDonald and Christopher Clack, "Low Cost and Low Carbon Emission Wind and Solar Energy Systems are Feasible for Large Geographic Domains," presentation at Sustainable Energy and Atmospheric Sciences seminar, Earth System Research Laboratory, U.S. National Oceanic and Atmospheric Administration, May 27, 2014; personal email communication with Alexander MacDonald, director, ESRL, and Christopher Clack, research scientist, ESRL, July 28, 2014.

21. M. M. Hand et al., "Renewable Electricity Futures Study—Volume 1: Exploration of High-Penetration Renewable Electricity Futures," National Renewable Energy Laboratory, 2012, pp. xvii–xviii.

22. Mark Z. Jacobson et al., "Examining the Feasibility of Converting New York State's All-Purpose Energy Infrastructure to One Using Wind, Water, and Sunlight," *Energy Policy* 57 (2013): 585; Elisabeth Rosenthal, "Life After Oil and Gas," *New York Times*, March 23, 2013.

23. Louis Bergeron, "The World Can Be Powered by Alternative Energy, Using Today's Technology, in 20–40 Years, Says Stanford Researcher Mark Z. Jacobson," Stanford Report, January 26, 2011; Elisabeth Rosenthal, "Life After Oil and Gas," *New York Times*, March 23, 2013.

24. Personal interview with Nastaran Mohit, November 10, 2012.

25. Steve Kastenbaum, "Relief from Hurricane Sandy Slow for Some," CNN, November 3, 2012.

26. Johnathan Mahler, "How the Coastline Became a Place to Put the Poor," *New York Times*, December 3, 2012; personal interview with Aria Doe, executive director, Action Center for Education and Community Development, February 3, 2013.

27. Sarah Maslin Nir, "Down to One Hospital, Rockaway Braces for Summer Crowds," *New York Times*, May 20, 2012; personal email communication with Nastaran Mohit, March 28, 2014; Mohit interview, November 10, 2012.

28. *Ibid.*; FOOTNOTE: Greg B. Smith, "NYCHA Under Fire for Abandoning Tenants in Hurricane Sandy Aftermath," New York *Daily News*, November 19, 2012.

29. Mohit interview, November 10, 2012.

30. *Ibid.*

31. Andrew P. Wilper et. al., "Health Insurance and Mortality in U.S. Adults," *American Journal of Public Health* 99 (2009): 2289–95; Mohit, November 10, 2012.

32. Doe interview, February 3, 2013.

33. John Aglionby, Mark Odell, and James Pickford, "Tens of Thousands Without Power After Storm Hits Western Britain," *Financial Times*, February 13, 2014; Tom Bawden, "St. Jude's Day Storm: Four Dead After 99mph Winds and Night of Destruction—But at Least We Saw It Coming," *The Independent* (London), October 29, 2013.

34. Alex Marshall, "Environment Agency Cuts: Surviving the Surgeon's Knife," *The ENDS Report*, January 3, 2014; Damian Carrington, "Massive Cuts Risk England's Ability to Deal with Floods, MPs Say," *Guardian*, January 7, 2014; Damian Carrington, "Hundreds of UK Flood Defence Schemes Unbuilt Due to Budget Cuts," *Guardian*, July 13, 2012.

35. Dave Prentis, "Environment Agency Workers Are Unsung Heroes," UNISON, January 6, 2014.

36. EM-DAT, International Disaster Database, Centre for Research on the Epidemiology of Disasters (advanced searches), http://www.emdat.be/database; personal email communication with Michael Mann, March 27, 2014.

37. "Billion-Dollar Weather/Climate Disasters," National Climatic Data Center, http://www.ncdc .noaa.gov; "Review of Natural Catastrophes in 2011: Earthquakes Result in Record Loss Year," Munich RE, press release, January 4, 2012.

38. Personal interview with Amy Bach, September 18, 2012.

39. "Climate Change: Impacts, Vulnerabilities and Adaptation in Developing Countries," UNFCCC, 2007, pp. 18–26, 29–38; "Agriculture Needs to Become 'Climate-Smart,'" Food and Agriculture Organization of the UN, October 28, 2010.

40. "World Economic and Social Survey 2011: The Great Green Technological Transformation," United Nations Department of Economic and Social Affairs, 2011, pp. xxii, 174.

41. The oil and gas sector was either the most represented or tied for the most represented sector in the top 20 of Fortune's Global 500 rankings for 2012 and 2013: "Fortune Global 500," CNN Money, 2013, http://money.cnn.com; "Fortune Global 500," CNN Money, 2012, http://money .cnn.com. BLOCKED PROGRESS: James Hoggan with Richard Littlemore, *Climate Cover-Up: The Crusade to Deny Global Warming* (Vancouver: Greystone Books, 2009); $900 BILLION: Daniel J. Weiss, "Big Oil's Lust for Tax Loopholes," Center for American Progress, January 31, 2011; 2011 EARNINGS: "2011 Summary Annual Report," ExxonMobil, p. 4; 2012 EARNINGS: "2012 Summary Annual Report," ExxonMobil, p. 4; "Exxon's 2012 Profit of $44.9B Misses Record," Associated Press, February 1, 2013.

42. BP, for instance, pledged $8 billion for alternative energy in 2005. Saaed Shah, "BP Looks 'Beyond Petroleum' with $8bn Renewables Spend," *The Independent* (London), November 29, 2005; BEYOND PETROLEUM: Terry Macalister and Eleanor Cross, "BP Rebrands on a Global Scale," *Guardian*, July 24, 2000; HELIOS MARK: "BP Amoco Unveils New Global Brand to Drive Growth," press release, July 24, 2000; BROWNE: Terry Macalister and Eleanor Cross, "BP Rebrands on a Global Scale," *Guardian*, July 24, 2000; CHEVRON: "We Agree: Oil Companies Should Support Renewable Energy" (video), Chevron, YouTube, 2010; 2009 STUDY: Daniel J. Weiss and Alexandra Kougentakis, "Big Oil Misers," Center for American Progress, March 31, 2009; EXECUTIVE PAY: James Osborne, "Exxon Mobil CEO Rex Tillerson Gets 15 Percent Raise to $40.3 Million," *Dallas Morning News*, April 12, 2013.

43. Antonia Juhasz, "Big Oil's Lies About Alternative Energy," *Rolling Stone*, June 25, 2013; Ben Elgin, "Chevron Dims the Lights on Green Power," Bloomberg Businessweek, May 29, 2014; Ben Elgin, "Chevron Backpedals Again on Renewable Energy," Bloomberg Businessweek, June 9, 2014

44. Brett Martel, "Jury Finds Big Tobacco Must Pay $590 Million for Stop-Smoking Programs," Associated Press, May 21, 2004; Bruce Alpert, "U.S. Supreme Court Keeps Louisiana's $240 Mil-

lion Smoking Cessation Program Intact," *Times-Picayune*, June 27, 2011; Sheila McNulty and Ed Crooks, "BP Oil Spill Pay-outs Hit $5bn Mark," *Financial Times*, August 23, 2011; Lee Howell, "Global Risks 2013," World Economic Forum, 2013, p. 19.

45. Marc Lee, "Building a Fair and Effective Carbon Tax to Meet BC's Greenhouse Gas Targets," Canadian Centre for Policy Alternatives, August 2012.

46. U.S. Department of Defense emissions were calculated using the federal Greenhouse Gas Inventory for fiscal year 2011 (total Scope 1 emissions, excluding biogenic). "Fiscal Year 2011 Greenhouse Gas Inventory," U.S. Department of Energy, Office of Energy Efficiency and Renewable Energy, June 14, 2013, http://energy.gov; "Greenhouse Gas 100 Polluters Index," Political Economy Research Institute, University of Massachusetts Amherst, June 2013, http://www.peri.umass.edu.

47. Borgar Aamaas, Jens Borken-Kleefeld, and Glen P. Peters, "The Climate Impact of Travel Behavior: A German Case Study with Illustrative Mitigation Options," *Environmental Science & Policy* 33 (2013): 273, 276.

48. Thomas Piketty, *Capital in the Twenty-First Century*, trans. Arthur Goldhammer (Cambridge, MA: Harvard University Press, 2014); Gar Lipow, *Solving the Climate Crisis through Social Change: Public Investment in Social Prosperity to Cool a Fevered Planet* (Santa Barbara: Praeger, 2012), 56; Stephen W. Pacala, "Equitable Solutions to Greenhouse Warming: On the Distribution of Wealth, Emissions and Responsibility Within and Between Nations," presentation to International Institute for Applied Systems Analysis, November 2007, p. 3.

49. "Innovative Financing at a Global and European Level," European Parliament, resolution, March 8, 2011, http://www.europarl.europa.eu.

50. "Revealed: Global Super-Rich Has at Least $21 Trillion Hidden in Secret Tax Havens," Tax Justice Network, press release, July 22, 2012.

51. "World Economic and Social Survey 2012: In Search of New Development Finance," United Nations Department of Economic and Social Affairs, 2012, p. 44.

52. Sam Perlo-Freeman, et. al., "Trends in World Military Expenditure, 2012," Stockholm International Peace Research Institute, April 2013 http://sipri.org.

53. "Mobilizing Climate Finance: A Paper Prepared at the Request of G20 Finance Ministers," World Bank Group, October 6, 2011, p.15, http://www.imf.org.

54. "Governments Should Phase Out Fossil Fuel Subsidies or Risk Lower Economic Growth, Delayed Investment in Clean Energy and Unnecessary Climate Change Pollution," Oil Change International and Natural Resources Defense Council, June 2012, p. 2.

55. For a more in-depth, U.S.-focused discussion of raising climate funds from these kinds of sources, see Lipow, *Solving the Climate Crisis through Social Change*, 55-61.

56. For much more on rationing, climate change, and environmental and economic justice, see: Stan Cox, *Any Way You Slice It: The Past, Present, and Future of Rationing* (New York: The New Press, 2013). 16 PERCENT: Ina Zweiniger-Bargielowska, *Austerity in Britain: Rationing, Controls, and Consumption, 1939–1955* (Oxford: Oxford University Press, 2000), 55, 58.

57. Nicholas Timmins, "When Britain Demanded Fair Shares for All," *The Independent* (London), July 27, 1995; Martin J. Manning and Clarence R. Wyatt, *Encyclopedia of Media and Propaganda in Wartime America*, Vol. 1 (Santa Barbara, CA: ABC-CLIO: 2011), 533; Terrence H. Witkowski, "The American Consumer Home Front During World War II," *Advances in Consumer Research* 25 (1998).

58. *Rationing, How and Why?* pamphlet, Office of Price Administration, 1942, p. 3.

59. Donald Thomas, *The Enemy Within: Hucksters, Racketeers, Deserters and Civilians During the Second World War* (New York: New York University Press, 2003), 29; Hugh Rockoff, *Drastic Measures: A History of Wage and Price Controls in the United States* (Cambridge: Cambridge University Press, 1984), 166–67.

60. Jimmy Carter, "Crisis of Confidence" speech (transcript), *American Experience*, PBS.

61. "The Pursuit of Progress" (video), *Richard Heffner's Open Mind*, PBS, February 10, 1991.

62. Eleanor Taylor, "British Social Attitudes 28," Chapter 6, Environment, NatCen Social Research, p. 104.

63. Will Dahlgreen, "Broad Support for 50P Tax," YouGov, January 28, 2014; "Nine in Ten Canadians Support Taxing the Rich 'More' (88%) and a Potential 'Millionaire's Tax' (89%)," Ipsos, May 30, 2013; Anthony Leiserowitz et al., "Public Support for Climate and Energy Policies in November 2013," Yale Project on Climate Change Communication and George Mason University Center for Climate Change Communication, November 2013; "Voter Attitudes Toward Pricing Carbon and a Clean Energy Refund" (memo), Public Opinion Strategies, April 21, 2010.

64. "Americans Support Limits on CO_2," Yale Project on Climate Change Communication, April 2014.

CHAPTER 4: PLANNING AND BANNING

1. John Berger, *Keeping a Rendezvous* (New York: Pantheon, 1991), 156.

2. James Gustave Speth, *The Bridge at the End of the World: Capitalism, the Environment, and Crossing from Crisis to Sustainability* (New Haven: Yale University Press, 2008), 178.

3. "The Second McCain-Obama Presidential Debate" (transcript), Commission on Presidential Debates, October 7, 2008.

4. Sam Gindin, "The Auto Crisis: Placing Our Own Alternative on the Table," Bullet/Socialist Project, E-Bulletin No. 200, April 9, 2009.

5. Ricardo Fuentes-Nieva and Nick Galasso, "Working for the Few," Oxfam, January 20, 2014, p. 2; FOOTNOTE: Jason Walsh, "European Workers Rebel as G-20 Looms," *Christian Science Monitor*, April 1, 2009; Rupert Hall, "Swansea Factory Workers Start Production at Former Remploy Site," Wales Online, October 14, 2013; Alejandra Cancino, "Former Republic Windows and Doors Workers Learn to Be Owners," *Chicago Tribune*, November 6, 2013.

6. According to U.S. Bureau of Labor Statistics data, the net loss in manufacturing jobs between January 2008 and January 2014 was 114,500; "Employment, Hours, and Earnings from the Current Employment Statistics Survey (National)," U.S. Bureau of Labor Statistics, http://data.bls.gov.

7. Michael Grunwald, *The New New Deal: The Hidden Story of Change in the Obama Era* (New York: Simon & Schuster, 2012), 10–11, 163–168; "Expert Reaction to Two New Nature Papers on Climate," Science Media Centre, December 4, 2011.

8. Roger Lowenstein, "The Nixon Shock," *Bloomberg Businessweek Magazine*, August 4, 2011; Bruce Bartlett, "Keynes and Keynesianism," *New York Times*, May 14, 2013.

9. The 3.7 million jobs estimate comes from the Apollo Alliance Project, which merged with the BlueGreen Alliance in 2011. "Make It in America: The Apollo Clean Transportation Manufacturing Action Plan," Apollo Alliance, October 2010; Smart Growth America, "Recent Lessons from the Stimulus: Transportation Funding and Job Creation," February 2011, p. 2.

10. "Working Towards Sustainable Development: Opportunities for Decent Work and Social Inclusion in a Green Economy," International Labour Organization, May 2012.

11. "More Bang for Our Buck," BlueGreen Canada, November 2012; Jonathan Neale, "Our Jobs, Our Planet: Transport Workers and Climate Change," A report originally written for the European Transport Workers Federation, October 2011, p. 49; "About," One Million Climate Jobs, http://www.climatejobs.org.

12. Will Dahlgreen, "Nationalise Energy and Rail Companies, Say Public," YouGov, November 4, 2013.

13. "2011 Wind Technologies Market Report," U.S. Department of Energy, August 2012, p. iii; Matthew L. Wald, "New Energy Struggles on Its Way to Markets," *New York Times*, December 27, 2013.

14. Personal interview with Ben Parfitt, September 21, 2013.

15. Michelle Kinman and Antonia Juhasz, ed., "The True Cost of Chevron: An Alternative Annual Report," True Cost of Chevron Network, May 2011, pp. 12, 18, 22, 43; Patrick Radden Keefe, "Reversal of Fortune," *The New Yorker*, January 9, 2012; Pierre Thomas et al., "B.P.'s Dismal Safety Record," ABC News, May 27, 2010; Alan Levin, "Oil Companies Fought Stricter Regulation," *USA Today*, May 20, 2010; Chip Cummins et al., "Five Who Laid Groundwork for Historic Spike in Oil Market," *Wall Street Journal*, December 20, 2005.

16. Seth Klein, "Moving Towards Climate Justice: Overcoming Barriers to Change," Canadian Centre for Policy Alternatives, April 2012.

17. Lucia Kassai, "Brazil to Boost Oil Exports as Output Triples, IEA Says," Bloomberg, November 12, 2013; Jeffrey Jones, "Statoil, PTTEP Deal to Test Tighter Oil Sands Rules," *Globe and Mail*, January 30, 2014; "PetroChina Buys Entire Alberta Oilsands Project," Canadian Press, January 3, 2012.

18. David Bollier, *Think Like a Commoner: A Short Introduction to the Life of the Commons* (Gabriola Island, BC: New Society Publishers, 2014).

19. Personal interview with Hans Thie, economic policy advisor, German Bundestag (Left Party), March 20, 2014; "Solarstrombranche (Photovoltaik)," Statistische Zahlen der deutschen, BSW Solar, April 2014, p. 1; "Status Des Windenergieausbasus An Land In Deutschland," Deutsche WindGuard, 2013, p. 1; "Flyer: Renewably Employed!" Federal Ministry for the Environment, Nature Conservation, Building and Nuclear Safety, August 2013.

20. Hans Thie, "The Controversial Energy Turnaround in Germany: Successes, Contradictions, Perspectives," Vienna Theses, July 2013.

21. "Danish Key Figures," Facts and Figures, Danish Energy Agency, 2010, http://www.ens.dk/en; personal email communication with Carsten Vittrup, strategic energy consultant, Energinet.dk, March 20, 2014.

22. Russ Christianson, "Danish Wind Co-ops Can Show Us the Way," Wind-Works, August 3, 2005.

23. Personal interview with Dimitra Spatharidou, May 20, 2013.

24. Andrea Stone, "Family Farmers Hold Keys to Agriculture in a Warming World," *National Geographic*, May 2, 2014.

25. Calogero Carletto, Sara Savastano, and Alberto Zezza, "Fact or Artifact: The Impact of Measurement Errors on the Farm Size–Productivity Relationship," *Journal of Development Economics* 103 (2013): 254-261; "Typhoon Haiyan Exposes the Reality of Climate Injustice," La Via Campesina, press release, December 4, 2013; Raj Patel, *Stuffed and Starved: The Hidden Battle for the World Food System* (Brooklyn: Melville House, 2012), 6-7.

26. De Schutter's analysis has been echoed by mainstream development bodies including the U.N. Conference on Trade and Development (UNCTAD) and the International Assessment of Agricultural Knowledge, Science and Technology for Development (IAASTD), both of which have issued reports in recent years that have held up small-scale agroecological farming, particularly when land is controlled by women, as key solutions to both the climate crisis and to persistent poverty. See: "Trade and Environment Review 2013: Wake Up Before It Is Too Late," United Nations Conference on Trade and Development, 2013; "Agriculture at a Crossroads: Synthesis Report," International Assessment of Agriculture Knowledge, Science and Technology for Development, 2009. "LARGE SEGMENT": "Eco-Farming Can Double Food Production in 10 Years, Says New UN Report," United Nations, Office of the High Commissioner for Human Rights, press release, March 8, 2011.

27. Verena Seufert, Navin Ramankutty, and Jonathan A. Foley, "Comparing the Yields of Organic and Conventional Agriculture," *Nature* 485 (2012): 229-232; "Eco-Farming Can Double Food Production in 10 Years, Says New UN Report," United Nations.

28. Personal email communication with Raj Patel, June 6, 2014.

29. *Ibid.*

30. Thie interview, March 20, 2014; "Greenhouse Gas Emissions Rise Slightly Again in 2013, by 1.2 Percent," German Federal Environment Agency (UBA), press release, March 10, 2014.

31. Thie interview, March 20, 2014; Helen Pidd, "Germany to Shut All Nuclear Reactors," *Guardian*, May 30, 2011; Peter Friederici, "WW II-Era Law Keeps Germany Hooked on 'Brown Coal' Despite Renewables Shift," InsideClimate News, October 1, 2013.

32. Mark Z. Jacobson and Mark A. Delucchi, "A Plan to Power 100 Percent of the Planet with Renewables," *Scientific American*, November 2009, pp. 58-59; Mark Z. Jacobson, "Nuclear Power Is Too Risky," CNN, February 22, 2010; *Real Time with Bill Maher*, HBO, episode 188, June 11, 2010.

33. Net electricity generation from nuclear was 11.9% of the world total in 2011, the most recent year

with complete data from the U.S. Energy Information Administration. "International Energy Statistics," U.S. Energy Information Administration, http://www.eia.gov.

34. Sven Teske, "Energy Revolution: A Sustainable EU 27 Energy Outlook," Greenpeace International and the European Renewable Energy Council, 2012, p. 11.

35. Thie interview, March 20, 2014; Andreas Rinke, "Merkel Signals Support for Plan to Lift Carbon Prices," Reuters, October 16, 2013.

36. Personal email communication with Tadzio Mueller, March 14, 2014.

37. "Development of Baseline Data and Analysis of Life Cycle Greenhouse Gas Emissions of Petroleum-Based Fuels," U.S. Department of Energy, National Energy Technology Laboratory, DOE/ NETL-2009/1346, 2008, p. 13.

38. Bill McKibben, "Join Us in Civil Disobedience to Stop the Keystone XL Tar-Sands Pipeline," Grist, June 23, 2011.

39. James Hansen, "Game Over for the Climate," New York Times, May 9, 2012.

40. Barack Obama, "Barack Obama's Remarks in St. Paul" speech, St. Paul, Minnesota, New York Times, June 3, 2008.

41. "Remarks by the President on Climate Change," speech, Washington, D.C., June 25, 2013, White House Office of the Press Secretary.

42. Jackie Calmes and Michael D. Shear, "Interview with President Obama," New York Times, July 27, 2013.

43. "Presidential Memorandum—Power Sector Carbon Pollution Standards," White House Office of the Press Secretary, June 25, 2013, http://www.whitehouse.gov; Mark Hertsgaard, "A Top Obama Aide Says History Won't Applaud the President's Climate Policy," Harper's, June 2, 2014.

44. Keynote Address (video), 6th International Conference on Climate Change, The Heartland Institute, June 30, 2011.

45. Robert W. Howarth, Renee Santoro, and Anthony Ingraffea, "Methane and the Greenhouse-Gas Footprint of Natural Gas from Shale Formations," Climatic Change 106 (2011): 679–90.

46. Ibid., 681-85, 687; Gunnar Myhre et al., "Anthropogenic and Natural Radiative Forcing," in Climate Change 2013: The Physical Science Basis. Contribution of Working Group I to the Fifth Assessment Report of the Intergovernmental Panel on Climate Change, ed. T. F. Stocker et al. (Cambridge: Cambridge University Press, 2013), 714.

47. Ibid.; personal interview with Robert Howarth, April 10, 2014.

48. Howarth provides a helpful overview of subsequent research on methane emissions from shale gas, arguing that this work has bolstered the key conclusions of the 2011 paper, in: Robert W. Howarth, "A Bridge to Nowhere: Methane Emissions and the Greenhouse Gas Footprint of Natural Gas," Energy Science & Engineering 2(2014): 47–60. FIRST PEER-REVIEWED: Howarth, Santoro, and Ingraffea, "Methane and the Greenhouse-Gas Footprint of Natural Gas from Shale Formations," 687; QUICK TO VOLUNTEER: Bryan Schutt, "Methane Emissions 'Achilles' Heel' of Shale Gas, Cornell Professor Contends," SNL Daily Gas Report, May 23, 2011; LACK OF TRANSPARENCY: Robert W. Howarth, Renee Santoro, and Anthony Ingraffea, "Venting and Leaking of Methane from Shale Gas Development: Response to Cathles et al.," Climatic Change 113 (2012): 539–40; FOOTNOTE: "U.S. Energy-Related Carbon Dioxide Emissions, 2012," U.S. Energy Information Administration, October 2013, p. ii; Scot M. Miller et al., "Anthropogenic emissions of methane in the United States," Proceedings of the National Academy of Sciences 110 (2013): 20018–20022; "Changing the Game? Emissions and Market Implications of New Natural Gas Supplies," Energy Modeling Forum, Stanford University, EMF Report No. 26, Vol. 1, September 2013, p. vii; Shakeb Afsah and Kendyl Salcito, "Us Coal Exports Erode All CO_2 Savings from Shale Gas," CO_2 Scorecard Group, March 24, 2014, http://www.co2scorecard.org.

49. Stefan Wagstyl, "German Coal Use at Highest Level Since 1990," Financial Times, January 7, 2014; Stefan Nicola and Ladka Bauerova, "In Europe, Dirty Coal Makes a Comeback," Bloomberg Businessweek, February 27, 2014.

50. Chester Dawson and Carolyn King, "Exxon Unit Seeks Canada Approval for Oil-Sands Project," Wall Street Journal, December 17, 2013; "Environmental Responsibility," Kearl, Operations,

Imperial, http://www.imperialoil.ca; "Fuel for Thought: The Economic Benefits of Oil Sands Investment for Canada's Regions," Conference Board of Canada, October 2012, pp. 3, 9.

51. Leila Coimbra and Sabrina Lorenzi, "BG to Spend $30 Billion on Brazil Offshore Oil by 2025," Reuters, May 24, 2012; "Chevron Announces $39.8 Billion Capital and Exploratory Budget for 2014," Chevron, press release, December 11, 2013; "Gorgon Project Overview," Chevron, May 2014, pp. 1-2, http://www.chevronaustralia.com; Andrew Callus, "Record-Breaking Gas Ship Launched, Bigger One Planned," Reuters, December 3, 2013; "A Revolution in Natural Gas Production," Shell Global, http://www.shell.com.

52. "Gorgon Project Overview," p. 1; "Prelude FLNG in Numbers," Shell Global, http://www.shell.com; "Operations: Kearl Oil Sands Project," Overview, Imperial Oil, http://www.imperialoil.ca; "Sunrise Energy Project," Husky Energy, http://www.huskyenergy.com; Kevin Anderson and Alice Bows, "Beyond 'Dangerous' Climate Change: Emission Scenarios for a New World," *Philosophical Transactions of the Royal Society* A 369 (2011): 35.

53. "Reserve-Replacement Ratio," Investopedia Dictionary, http://www.investopedia.com.

54. Fred Pals, "Shell Lagged Behind BP in Replacing Reserves in 2008," Bloomberg, March 17, 2009; "Royal Dutch Shell Plc Strategy Update 2009—Final," Fair Disclosure Wire, March 17, 2009; Robin Pagnamenta, "Anger as Shell Cuts Back on Its Investment in Renewables," *The Times* (London), March 18, 2009; "Royal Dutch Shell Plc Updates on Strategy to Improve Performance and Grow," Royal Dutch Shell, press release, March 16, 2010; Robert Perkins, "Shell Eyes 2012 Output of 3.5 Million Boe/d," *Platts Oilgram Price Report*, March 17, 2010.

55. "World Energy Outlook 2013," International Energy Agency, 2013, pp. 471–72.

56. "Exxon Mobil Corporation Announces 2011 Reserves Replacement," ExxonMobil, press release, February 23, 2012.

57. The figures given in this passage can vary, as there are different estimates of the carbon budget for limiting warming to 2°C. The original Carbon Tracker report relied on a seminal *Nature* paper published in 2009: James Leaton, "Unburnable Carbon: Are the World's Financial Markets Carrying a Carbon Bubble?" Carbon Tracker Initiative, 2011, pp. 6–7; Malte Meinshausen et al., "Greenhouse-Gas Emission Targets for Limiting Global Warming to 2°C," *Nature* 458 (2009): 1161. For an updated analysis from Carbon Tracker, refer to: James Leaton et al., "Unburnable Carbon 2013: Wasted Capital and Stranded Assets," Carbon Tracker Initiative, 2013. For 2 degree carbon budget estimates in the latest IPCC Assessment Report, see: "Summary for Policymakers," in *Climate Change 2013: The Physical Science Basis. Contribution of Working Group I to the Fifth Assessment Report of the Intergovernmental Panel on Climate Change*, ed. T. F. Stocker et al. (Cambridge: Cambridge University Press, 2013), 27–28. "THE THING TO NOTICE": Bill McKibben, speech, New York City, Do the Math Tour, 350.org, November 16, 2012.

58. John Fullerton, "The Big Choice," Capital Institute, July 19, 2011; Leaton, "Unburnable Carbon," p. 6.

59. Total oil and gas industry lobby spending in 2013 was $144,878,531, according to the Center for Responsive Politics: "Oil & Gas," OpenSecrets.org, Center for Responsive Politics, https://www.opensecrets.org; ELECTION SPENDING: "Oil and Gas: Long-Term Contribution Trends," OpenSecrets.org, Center for Responsive Politics, https://www.opensecrets.org.

60. Daniel Cayley-Daoust and Richard Girard, "Big Oil's Oily Grasp: The Making of Canada as a Petro-State and How Oil Money is Corrupting Canadian Politics," Polaris Institute, December 2012, p. 3; Damian Carrington, "Energy Companies Have Lent More Than 50 Staff to Government Departments," *Guardian*, December 5, 2011.

61. Google Finance historical price records for ExxonMobil, Chevron, Royal Dutch Shell, ConocoPhillips, BP, Anglo American, and Arch Coal between the dates of December 1, 2009 and December 31, 2009, particularly December 18.

62. Suzanne Goldenberg, "Exxon Mobil Agrees to Report on Climate Change's Effect on Business Model," *Guardian*, March 20, 2014; "Energy and Carbon—Managing the Risks," ExxonMobil, 2014, pp. 1, 8, 16.

63. Personal email communication with John Ashton, March 20, 2014.

64. Mark Dowie, *Losing Ground: American Environmentalism at the Close of the Twentieth Century* (Cambridge, MA: MIT Press, 1996), 25.

65. Yotam Marom, "Confessions of a Climate Change Denier," Waging Nonviolence, July 30, 2013.

66. "Paxman vs. Brand—Full Interview" (video), *BBC Newsnight*, October 23, 2013.

67. "System Change—Not Climate Change," A People's Declaration from Klimaforum09, December 2009.

68. Miya Yoshitani, "Confessions of a Climate Denier in Tunisia," Asian Pacific Environment Network, May 8, 2013.

69. Nick Cohen, "The Climate Change Deniers Have Won," *The Observer*, March 22, 2014.

70. Philip Radford, "The Environmental Case for a Path to Citizenship," Huffington Post, March 14, 2013; Anna Palmer and Darren Samuelsohn, "Sierra Club Backs Immigration Reform," *Politico*, April 24, 2013; "Statement on Immigration Reform," BlueGreen Alliance, http://www.bluegreenalliance.org; May Boeve, "Solidarity with the Immigration Reform Movement," 350.org, March 22, 2013, http://350.org.

71. Pamela Gossin, *Encyclopedia of Literature and Science* (Westport, CT: Greenwood, 2002), 208; William Blake, "And Did Those Feet in Ancient Time," poem in *The Complete Poetry and Prose of William Blake* (Berkeley: University of California Press, 2008), 95.

72. Personal interview with Colin Miller, March 14, 2011; Simon Romero, "Bus-Fare Protests Hit Brazil's Two Biggest Cities," *New York Times*, June 13, 2013; Larry Rohter, "Brazil's Workers Take to Streets in One-Day Strike," *New York Times*, July 11, 2013.

CHAPTER 5: BEYOND EXTRACTIVISM

1. Steve Stockman, Twitter post, March 21, 2013, 2:33 p.m. ET, https://twitter.com.

2. Ben Dangl, "Miners Just Took 43 Police Officers Hostage in Bolivia," *Vice*, April 3, 2014.

3. Rodrigo Castro et al., "Human-Nature Interaction in World Modeling with Modelica," prepared for the Proceedings of the 10th International Modelica Conference, March 10–12, 2014, http://www.ep.liu.se.

4. Personal interview with Nerida-Ann Steshia Hubert, March 30, 2012.

5. Hermann Joseph Hiery, *The Neglected War: The German South Pacific and the Influence of World War I* (Honolulu: University of Hawai'i Press, 1995), 116–25, 241; "Nauru," New Zealand Ministry of Foreign Affairs and Trade, updated December 9, 2013, http://wwww.mfat.govt.nz; "Nauru" (video), NFSA Australia, NFSA Films.

6. Charles J. Hanley, "Tiny Pacific Isle's Citizens Rich, Fat and Happy—Thanks to the Birds," Associated Press, March 31, 1985; Steshia Hubert interview, March 30, 2012.

7. "Country Profile and National Anthem," Permanent Mission of the Republic of Nauru to the United Nations, United Nations, http://www.un.int; Jack Hitt, "The Billion-Dollar Shack," *New York Times Magazine*, December 10, 2000.

8. Hiery, *The Neglected War*, 116–25, 241; "Nauru," New Zealand Ministry of Foreign Affairs and Trade.

9. Hitt, "The Billion-Dollar Shack"; David Kendall, "Doomed Island," *Alternatives Journal*, January 2009.

10. "Nauru" (video), NFSA Films.

11. Philip Shenon, "A Pacific Island Is Stripped of Everything," *New York Times*, December 10, 1995.

12. Hitt, "The Billion-Dollar Shack"; Robert Matau, "Road Deaths Force Nauru to Review Traffic Laws," Islands Business, July 10, 2013; "The Fattest Place on Earth" (video), *Nightline*, ABC, January 3, 2011; Steshia Hubert interview, March 30, 2012.

13. Hitt, "The Billion-Dollar Shack"; "Nauru," Country Profile, U.N. Data, http://data.un.org.

14. "Nauru," Overview, Rand McNally, http://education.randmcnally.com; Tony Thomas, "The Naughty Nation of Nauru," *The Quadrant*, January/February 2013; Andrew Kaierua et al., "Nauru," in *Climate Change in the Pacific*, Scientific Assessment and New Research, Volume 2:

Country Reports, Australian Bureau of Meteorology and CSIRO, 2011, pp. 134, 140; "Fresh Water Supplies a Continual Challenge to the Region," Applied Geoscience and Technology Division, Secretariat of the Pacific Community, press release, January 18, 2011.

15. Glenn Albrecht, "The Age of Solastalgia," *The Conversation*, August 7, 2012.

16. Kendall, "Doomed Island."

17. "Nauru: Phosphate Roller Coaster; Elections with Tough Love Theme," August 13, 2007, via WikiLeaks, http://www.wikileaks.org.

18. Nick Bryant, "Will New Nauru Asylum Centre Deliver Pacific Solution?" *BBC News*, June 20, 2013; Rob Taylor, "Ruling Clouds Future of Australia Detention Center," *Wall Street Journal*, January 30, 2014; "Nauru Camp a Human Rights Catastrophe with No End in Sight," Amnesty International, press release, November 23, 2012; "What We Found on Nauru," Amnesty International, December 17, 2012; "Hundreds Continue 11-Day Nauru Hunger Strike," ABC News (Australia), November 12, 2012.

19. Bryant, "Will New Nauru Asylum Centre Deliver Pacific Solution?"; Oliver Laughland, "Nauru Immigration Detention Centre—Exclusive Pictures," *Guardian*, December 6, 2013; "Hundreds Continue 11-Day Nauru Hunger Strike," ABC News (Australia); "Police Attend Full-Scale Riot at Asylum Seeker Detention Centre on Nauru," ABC News (Australia), July 20, 2013.

20. "Nauru Camp a Human Rights Catastrophe with No End in Sight," Amnesty International, press release, November 23, 2012; "UNHCR Monitoring Visit to the Republic of Nauru, 7 to 9 October 2013," United Nations High Commissioner for Refugees, November 26, 2013; Mark Isaacs, *The Undesirables* (Richmond, Victoria: Hardie Grant Books, 2014), 99; Deborah Snow, "Asylum Seekers: Nothing to Lose, Desperation on Nauru," *Sydney Morning Herald*, March 15, 2014.

21. "The Middle of Nowhere," *This American Life*, December 5, 2003, http://www.thisamericanlife.org; Mitra Mobasherat and Ben Brumfield, "Riot on a Tiny Island Highlights Australia Shutting a Door on Asylum," CNN, July 20, 2013; Rosamond Dobson Rhone, "Nauru, the Richest Island in the South Seas," *National Geographic* 40 (1921): 571, 585.

22. Marcus Stephen, "On Nauru, a Sinking Feeling," *New York Times*, July 18, 2011.

23. Francis Bacon, *De Dignitate et Augmentis Scientiarum, Works*, ed. James Spedding, Robert Leslie Ellis, and Douglas Devon Heath, Vol. 4 (London: Longmans Green, 1870), 296.

24. William Derham, *Physico-Theology: or, A demonstration of the Being and Attributes of God, from His Works of Creation* (London: Printed for Robinson and Roberts, 1768), 110.

25. Barbara Freese, *Coal: A Human History* (New York: Penguin, 2004), 44.

26. Emphasis in original. Many of the sources in this recounting were originally cited in Andreas Malm, "The Origins of Fossil Capital: From Water to Steam in the British Cotton Industry," *Historical Materialism* 21 (2013): 31.

27. J. R. McCulloch [unsigned], "Babbage on Machinery and Manufactures," *Edinburgh Review* 56 (January 1833): 313–32; François Arago, *Historical Eloge of James Watt*, trans. James Patrick Muirhead (London: J. Murray, 1839), 150.

28. C. H. Turner, *Proceedings of the Public Meeting Held at Freemasons' Hall, on the 18th June, 1824, for Erecting a Monument to the Late James Watt* (London: J. Murray, 1824), pp. 3–4, as cited in Andreas Malm, "Steam: Nineteenth-Century Mechanization and the Power of Capital," in *Ecology and Power: Struggles over Land and Material Resources in the Past, Present, and Future*, eds. Alf Hornborg, Brett Clark, and Kenneth Hermele (London: Routledge, 2013), 119.

29. M. A. Alderson, *An Essay on the Nature and Application of Steam: With an Historical Notice of the Rise and Progressive Improvement of the Steam-Engine* (London: Sherwood, Gilbert and Piper, 1834), 44.

30. Asa Briggs, *The Power of Steam: An Illustrated History of the World's Steam Age* (Chicago: University of Chicago Press, 1982), 72.

31. Jackson J. Spielvogel, *Western Civilization: A Brief History, Volume II: Since 1500*, 8th ed. (Boston: Wadsworth, 2014), 445.

32. Herman E. Daly and Joshua Farley, *Ecological Economics: Principles and Applications* (Washington, D.C.: Island Press, 2011), 10.

33. Rebecca Newberger Goldstein, "What's in a Name? Rivalries and the Birth of Modern Science," in *Seeing Further: The Story of Science, Discovery, and the Genius of the Royal Society*, ed. Bill Bryson (London: Royal Society, 2010), 120.

34. Ralph Waldo Emerson, *The Conduct of Life* (New York: Thomas Y. Crowell, 1903), 70.

35. Clive Hamilton, "The Ethical Foundations of Climate Engineering," in *Climate Change Geoengineering: Philosophical Perspectives, Legal Issues, and Governance Frameworks*, ed. Wil C. G. Burns and Andrew L. Strauss (New York: Cambridge University Press, 2013), 58.

36. Esperanza Martínez, "The Yasuní—ITT Initiative from a Political Economy and Political Ecology Perspective," in Leah Temper et al., "Towards a Post-Oil Civilization: Yasunization and Other Initiatives to Leave Fossil Fuels in the Soil," EJOLT Report No. 6, May 2013, p. 12.

37. Jean-Paul Sartre, *Critique of Dialectical Reason*, trans. Alan Sheridan-Smith (London: Verso, 2004), 154; Tim Flannery, *Here on Earth: A Natural History of the Planet* (New York: Grove), 185.

38. Karl Marx, *Capital*, Vol. 3, as cited in John Bellamy Foster, *Marx's Ecology: Materialism and Nature* (New York: Monthly Review Press, 2000), 155.

39. "Yearly Emissions: 1987," CAIT database, World Resources Institute, http://cait.wri.org; Nicholas Stern, *The Economics of Climate Change: The Stern Review* (Cambridge: Cambridge University Press, [2006] 2007), 231; Judith Shapiro, *Mao's War Against Nature: Politics and the Environment in Revolutionary China* (Cambridge: Cambridge University Press, 2001); Mara Hvistendahl, "China's Three Gorges Dam: An Environmental Catastrophe?" *Scientific American*, March 25, 2008; Will Kennedy and Stephen Bierman, "Free Khodorkovsky to Find Oil Industry Back in State Control," Bloomberg, December 20, 2013; Tom Metcalf, "Russian Richest Lost $13 Billion as Global Stocks Fell," Bloomberg, March 4, 2014.

40. ROUGHLY 81 PERCENT: "Stockholm Action Plan for Climate and Energy, 2012–2015: With an Outlook to 2030," Stockholm Environment and Health Administration, p. 12; MAJORITY STATE-OWNED: "Annual Report on Form 20-F," Statoil, 2013, p. 117, http://www.statoil.com; TAR SANDS: "Oil Sands," Statoil, http://www.statoil.com; ARCTIC: "Large-Scale Arctic Oil and Gas Drilling Decades Away," Reuters, November 29, 2013; "Statoil Stepping Up in the Arctic," Statoil, press release, August 28, 2012; IRAQ: "Iraq," Our Operations, Annual Report 2011, Statoil, http://www.statoil.com; Stephen A. Carney, "Allied Participation in Operation Iraqi Freedom," Center of Military History, United States Army, 2011, http://www.history.army.mil.

41. Instituto de Pesquisa Econômica Aplicada (IPEA) data, http://www.ipeadata.gov.br; Mark Weisbrot and Jake Johnston, "Venezuela's Economic Recovery: Is It Sustainable?" Center for Economic and Policy Research, September 2012, p. 26; "Ecuador Overview," Ecuador, World Bank, http://www.worldbank.org; "Population Below National Poverty Line, Urban, Percentage," Millenium Development Goals Database, U.N. Data, http://data.un.org.

42. "Bolivia: Staff Report for the 2013 Article IV Consultation," International Monetary Fund, February 2014, p. 6.

43. Luis Hernández Navarro, "Bolivia Has Transformed Itself by Ignoring the Washington Consensus," *Guardian*, March 21, 2012.

44. ECUADOR: Nick Miroff, "In Ecuador, Oil Boom Creates Tensions," *Washington Post*, February 16, 2014; BOLIVIA AND VENEZUELA: Dan Luhnow and José de Córdoba, "Bolivia Seizes Natural-Gas Fields in a Show of Energy Nationalism," *Wall Street Journal*, May 2, 2006; ARGENTINA: "Argentine Province Suspends Open-Pit Gold Mining Project Following Protests," MercoPress, January 31, 2012; "GREEN DESERTS": "The Green Desert," *The Economist*, August 6, 2004; BRAZIL: "The Rights and Wrongs of Belo Monte," *The Economist*, May 4, 2013; RAW RESOURCES: Exports of Primary Products as Percentage of Total Exports, "Statistical Yearbook for Latin America and the Caribbean," Economic Commission for Latin America and the Caribbean, United Nations, 2012, p. 101; CHINA: Joshua Schneyer and Nicolás Medina Mora Pérez, "Special Report: How China Took Control of an OPEC Country's Oil," Reuters, November 26, 2013.

45. Eduardo Gudynas, "Buen Vivir: Today's Tomorrow," *Development* 54 (2011): 442–43; Martínez in

Temper et al., "Towards a Post-Oil Civilization," p.17; Eduardo Gudynas, "The New Extractivism of the 21st Century: Ten Urgent Theses About Extractivism in Relation to Current South American Progressivism," Americas Program Report, Washington, D.C.: Center for International Policy, January 21, 2010.

46. Personal interview with Alexis Tsipras, May 23, 2013.

47. Patricia Molina, "The 'Amazon Without Oil' Campaign: Oil Activity in Mosetén Territory," in Temper et al., "Towards a Post-Oil Civilization," p. 75.

48. William T. Hornaday, *Wild Life Conservation in Theory and Practice* (New Haven: Yale University Press, 1914), v–vi.

49. "Who Was John Muir?" Sierra Club, http://www.sierraclub.org; John Muir, *The Yosemite* (New York: Century, 1912), 261–62.

50. Bradford Torrey, ed., *The Writings of Henry David Thoreau: Journal, September 16, 1851–April 30, 1852* (New York: Houghton Mifflin, 1906), 165; Aldo Leopold, *A Sand County Almanac* (Oxford: Oxford University Press, 1949), 171; FOOTNOTE: Henry David Thoreau, *Walden* (New York: Thomas Y. Crowell, 1910), 393–94.

51. Leopold, *A Sand Counrty Almanac*, 171; Jay N. Darling to Aldo Leopold, November 20, 1935, Aldo Leopold Archives, University of Wisconsin Digital Collections.

52. Rachel Carson, *Silent Spring* (New York: Houghton Mifflin, 1962), 57, 68, 297.

53. *Ibid.*, 297.

54. Christian Parenti, "'The Limits to Growth': A Book That Launched a Movement," *The Nation*, December 5, 2012.

PART II: MAGICAL THINKING

1. William Barnes and Nils Gilman, "Green Social Democracy or Barbarism: Climate Change and the End of High Modernism," in *The Deepening Crisis: Governance Challenges After Neoliberalism*, ed. Craig Calhoun and Georgi Derluguian (New York: New York University Press, 2011), 50.

2. Christine MacDonald, *Green, Inc.: An Environmental Insider Reveals How a Good Cause Has Gone Bad* (Guilford, CT: Lyons Press, 2008), 236.

CHAPTER 6: FRUITS, NOT ROOTS

1. Barry Commoner, "A Reporter at Large: The Environment," *New Yorker*, June 15, 1987.

2. Eric Pooley, *The Climate War* (New York: Hyperion, 2010), 351–52.

3. Valgene W. Lehmann, "Attwater's Prairie Chicken—Its Life History and Management," *North American Fauna* 57, U.S. Fish and Wildlife Service, Department of the Interior, 1941, pp. 6–7; "Attwater's Prairie-Chicken Recovery Plan," Second Revision, U.S. Fish and Wildlife Service, 2010, p. 5.

4. "Texas Milestones," The Nature Conservancy, http://www.nature.org.

5. Joe Stephens and David B. Ottaway, "How a Bid to Save a Species Came to Grief," *Washington Post*, May 5, 2003; "Texas City Prairie Preserve," Nature Conservancy, http://www.nature.org, version saved by the Internet Archive Wayback Machine on February 8, 2013, http://web.archive.org.

6. Richard C. Haut et al., "Living in Harmony—Gas Production and the Attwater's Prairie Chicken," prepared for presentation at the Society of Petroleum Engineers Annual Technical Conference and Exhibition, Florence, Italy, September 19–22, 2010, pp. 5, 10; Oil and Gas Lease, Nature Conservancy of Texas, Inc. to Galveston Bay Resources, Inc., March 11, 1999, South 1,057 Acres; Stephens and Ottaway, "How a Bid to Save a Species Came to Grief"; personal interview with Aaron Tjelmeland, April 15, 2013.

7. Janet Wilson, "Wildlife Shares Nest with Profit," *Los Angeles Times*, August 20, 2002; Stephens and Ottaway, "How a Bid to Save a Species Came to Grief."

8. *Ibid.*

9. *Ibid.*

10. *Ibid.*

11. "Nature Conservancy Changes," *Living on Earth*, Public Radio International, June 20, 2003.

12. Personal email communications with Vanessa Martin, associate director of marketing and communications, Texas chapter, The Nature Conservancy, May 16 and 21 and June 24, 2013.

13. Outside of the original 1999 well and its replacement well drilled on the same pad in 2007, an additional two wells were drilled under Nature Conservancy leases, both in 2001: a gas well that was plugged and abandoned in 2004, and another well that turned out to be a dry hole. Haut et al., "Living in Harmony," p. 5; personal email communications with Vanessa Martin, April 24 and May 16, 2013.

14. Oil and Gas Lease, Nature Conservancy of Texas, Inc. to Galveston Bay Resources, Inc., pp. 3–5; Martin email communications, May 21 and June 24, 2013; "Attwater's Prairie Chicken Background," The Nature Conservancy, provided on April 24, 2013, p. 3; personal interview with James Petterson, July 31, 2014.

15. NOVEMBER 2012: Personal email communication with Mike Morrow, wildlife biologist, Attwater Prairie Chicken National Wildlife Refuge, April 17, 2013; "NONE THAT WE KNOW ABOUT": Tjelmeland interview, April 15, 2013; "BIGGEST" AND THIRTY-FIVE COUNTRIES: D.T. Max, "Green is Good," The New Yorker, May 12, 2014; MEMBERS: "About Us: Learn More About the Nature Conservancy," Nature Conservancy, http://www.nature.org; ASSETS: "Consolidated Financial Statements," Nature Conservancy, June 30, 2013, p. 3; MILLIONS: Stephens and Ottaway, "How a Bid to Save a Species Came to Grief"; WEBSITE: "Texas City Prairie Preserve," Nature Conservancy, http://www.nature.org.

16. DONATIONS FROM SHELL AND BP TO CF, CI, AND TNC, AND FROM AEP TO CF: Christine MacDonald, Green, Inc.: An Environmental Insider Reveals How a Good Cause Has Gone Bad (Guilford, CT: Lyons Press, 2008), 25; SUPPORT FROM AEP TO TNC: Ibid., 139; WWF AND SHELL: Alexis Schwarzenbach, Saving the World's Wildlife: WWF—The First 50 Years (London: Profile, 2011), 145–48, 271; "The Gamba Complex—Our Solutions," World Wildlife Fund Global, http://wwf.panda.org; WRI AND SHELL FOUNDATION: "WRI's Strategic Relationships," World Resources Institute, http://www.wri.org; CI PARTNERHSIPS: "Corporate Partners," Conservation International, http://www.conservation.org; $2 MILLION: Joe Stephens, "Nature Conservancy Faces Potential Backlash from Ties with BP," Washington Post, May 24, 2010; FOOTNOTE: "Undercover with Conservation International" (video), Don't Panic, May 8, 2011; Tom Zeller Jr., "Conservation International Duped by Militant Greenwash Pitch," Huffington Post, May 17, 2011; Peter Seligmann, "Partnerships for the Planet: Why We Must Engage Corporations," Huffington Post, May 19, 2011.

17. John F. Smith Jr., a former CEO and later chairman of General Motors, and E. Linn Draper Jr., formerly the CEO and chairman of American Electric Power, both served on The Nature Conservancy's board of directors: "Past Directors of The Nature Conservancy," Nature Conservancy, http://www.nature.org; David B. Ottaway and Joe Stephens, "Nonprofit Land Bank Amasses Billions," Washington Post, May 4, 2003. BUSINESS COUNCIL: "Working with Companies: Business Council," Nature Conservancy, http://www.nature.org; BOARD OF DIRECTORS: "About Us: Board of Directors," Nature Conservancy, http://www.nature.org.

18. "Consolidated Financial Statements," Nature Conservancy, June 30, 2012, pp. 20-21; "Consolidated Financial Statements," Nature Conservancy, June 30, 2013, p. 21; Naomi Klein, "Time for Big Green to Go Fossil Free," The Nation, May 1, 2013; FOOTNOTE: Mark Tercek email communication to senior managers, August 19, 2013.

19. Shell has agreed to pay out $15.5 million to settle a case involving such human rights abuse claims, although it continues to deny wrongdoing, as does Chevron: Jad Mouawad, "Shell to Pay $15.5 Million to Settle Nigerian Case," New York Times, June 8, 2009; Michelle Kinman and Antonia Juhasz, ed., "The True Cost of Chevron: An Alternative Annual Report," True Cost of Chevron network, May 2011, p. 46; "Bowoto v. Chevron," EarthRights International, http://www.earthrights.org. SIERRA CLUB: Bryan Walsh, "How the Sierra Club Took Millions from the Natural Gas Industry—and Why They Stopped," Time, February 2, 2012; Michael Brune, "The Sierra Club and Natural Gas," Sierra Club, February 2, 2012; personal email communication with Bob Sipchen, communications director, Sierra Club, April 21, 2014.

20. 2012 Form 990, Attachment 8, Ford Foundation, pp. 44, 48, 53.

21. In the lead-up to congressional battles over carbon trading in the U.S., philanthropists including the ClimateWorks Foundation distributed hundreds of millions of dollars to a number of environmental groups after raising money from sources such as the Hewlett Foundation and the Packard Foundation. Reportedly, this helped to create an atmosphere of pressure to focus on or to avoid distracting from the cap-and-trade fight: Petra Bartosiewicz and Marissa Miley, "The Too Polite Revolution: Why the Recent Campaign to Pass Comprehensive Climate Legislation in the United States Failed," paper presented at symposium on the Politics of America's Fight Against Global Warming, Harvard University, February 2013, p. 30; personal interview with Jigar Shah, September 9, 2013. DESIGN TO WIN: "Design to Win: Philanthropy's Role in the Fight Against Global Warming," California Environmental Associates, August 2007, pp. 14–18, 24, 42.

22. Robert Brulle, "Environmentalisms in the United States," in *Environmental Movements Around the World*, Vol. 1, ed. Timothy Doyle and Sherilyn MacGregor (Santa Barbara: Praeger, 2013), 174.

23. Global Carbon Project emissions data, 2013 Budget v2.4 (July 2014), available at http://cdiac.ornl.gov; "Caring for Climate Hosts Inaugural Business Forum to Co-Create Climate Change Solutions," United Nations Global Compact, press release, November 19, 2013; Rachel Tansey, "The COP19 Guide to Corporate Lobbying: Climate Crooks and the Polish Government's Partners in Crime," Corporate Europe Observatory and Transnational Institute, October 2013.

24. "Partners for COP19," United Nations Climate Change Conference, COP19/CMP9 Warsaw 2013, Media Centre, press release, September 17, 2013; "Who We Are," PGE Group, Investor Relations, http://www.gkpge.pl/en; "International Coal & Climate Summit 2013," World Coal Association, http://www.worldcoal.org; Adam Vaughan and John Vidal, "UN Climate Chief Says Coal Can Be Part of Global Warming Solution," *Guardian*, November 18, 2013; David Jolly, "Top U.N. Official Warns of Coal Risks," *New York Times*, November 18, 2013.

25. Pooley, *The Climate War*, 59; "25 Years After DDT Ban, Bald Eagles, Osprey Numbers Soar," Environmental Defense Fund, press release, June 13, 1997.

26. Ramachandra Guha and Joan Martínez Alier, *Varieties of Environmentalism* (Abingdon, Oxon: Earthscan, 2006), 3–21; Joan Martínez Alier, *The Environmentalism of the Poor: A Study of Ecological Conflicts and Valuation* (Cheltenham: Edward Elgar, 2002).

27. Mark Dowie, *Losing Ground: American Environmentalism at the Close of the Twentieth Century* (Cambridge, MA: MIT Press, 1996), 33, 39.

28. Lou Cannon, *Governor Reagan: His Rise to Power* (Cambridge, MA: PublicAffairs, 2003), 177–78; "Watt Says Foes Want Centralization of Power," Associated Press, January 21, 1983.

29. Riley E. Dunlap et al., "Politics and Environment in America: Partisan and Ideological Cleavages in Public Support for Environmentalism," *Environmental Politics* 10 (2001): 31; "Endangered Earth, Planet of the Year," *Time*, January 2, 1989; FOOTNOTE: Dunlap et al., "Politics and Environment in America," 31.

30. "Principles of Environmental Justice," First National People of Color Environmental Leadership Summit, October 1991, http://www.ejnet.org.

31. Gus Speth, "American Environmentalism at the Crossroads," speech, Climate Ethics and Climate Equity series, Wayne Morse Center for Law and Politics, University of Oregon, April 5, 2011.

32. "Corporations," Conservation Fund, http://www.conservationfund.org; "History," Conservation International, http://www.conservation.org, version saved by the Internet Archive Wayback Machine on December 3, 2013, http:// web.archive.org.

33. Ottaway and Stephens, "Nonprofit Land Bank Amasses Billions"; Joe Stephens and David B. Ottaway, "Nonprofit Sells Scenic Acreage to Allies at a Loss," *Washington Post*, May 6, 2003; Monte Burke, "Eco-Pragmatists; The Nature Conservancy Gets in Bed with Developers, Loggers and Oil Drillers," *Forbes*, September 3, 2001.

34. "Environmentalists Disrupt Financial Districts in NYC, San Francisco," Associated Press, April 23, 1990; Donatella Lorch, "Protesters on the Environment Tie Up Wall Street," *New York*

Times, April 24, 1990; Martin Mittelstaedt, "Protesters to Tackle Wall Street," *Globe and Mail*, April 23, 1990.

35. Elliot Diringer, "Environmental Demonstrations Take Violent Turn," *San Francisco Chronicle*, April 24, 1990; "Environmentalists Disrupt Financial Districts in NYC, San Francisco."

36. Pooley, *The Climate War*, 69.

37. Fred Krupp, "New Environmentalism Factors in Economic Needs," *Wall Street Journal*, November 20, 1986; "Partnerships: The Key to Lasting Solutions," How We Work, Environmental Defense Fund, http://www.edf.org.

38. Michael Kranish, "The Politics of Pollution," *Boston Globe Magazine*, February 8, 1998; Pooley, *The Climate War*, 74–81; personal interview with Laurie Williams and Allan Zabel, lawyers, Environmental Protection Agency, interviewed in their personal capacity, April 4, 2014.

39. "Fred Krupp," Our People, Environmental Defense Fund, http://www.edf.org; "Our Finances," About Us, Environmental Defense Fund, http://www.edf.org; Pooley, *The Climate War*, 98; FOOTNOTE: Ken Wells, "Tree-Hitter Tercek Channels Goldman at Nature Conservancy," Bloomberg, May 31, 2012.

40. $65 MILLION: "2011 Grant Report," Walton Family Foundation, http://www.waltonfamily foundation.org; "2011 Annual Report", Environmental Defense Fund, p. 31, http://www.edf.org. DONATION POLICY: "Corporate Donation Policy," How We Work, Environmental Defense Fund, http://www.edf.org; "WOULD UNDERMINE": Eric Pooley, "Viewpoint: Naomi Klein's Criticism of Environmental Groups Missed the Mark," Climate Progress, September 11, 2013; Michelle Harvey, "Working Toward Sustainability with Walmart," Environmental Defense Fund, September 18, 2013; FAMILY-CONTROLLED: 2012 Form 990, Attachment 14, Walton Family Foundation, https://www.guidestar.org; NO DIRECT DONATIONS: Stephanie Clifford, "Unexpected Ally Helps Wal-Mart Cut Waste," *New York Times*, April 13, 2012; SAM RAWLINGS WALTON: "Our Board of Trustees," About Us, Environmental Defense Fund, http://www .edf.org.

41. Stacy Mitchell, "Walmart Heirs Quietly Fund Walmart's Environmental Allies," Grist, May 10, 2012; Stacy Mitchell, "Walmart's Assault on the Climate," Institute for Local Self-Reliance, November 2013.

42. "2011 Grant Report," Walton Family Foundation, http://www.waltonfamilyfoundation.org; "Walmart Announces Goal to Eliminate 20 Million Metric Tons of Greenhouse Gas Emissions from Global Supply Chain," Environmental Defense Fund, press release, February 25, 2010; Daniel Zwerdling and Margot Williams, "Is Sustainable-Labeled Seafood Really Sustainable?," NPR, February 11, 2013; "Walmart Adds a New Facet to Its Fine Jewelry Lines: Traceability," Walmart, July 15, 2008, http://news.walmart.com; Mitchell, "Walmart Heirs Quietly Fund Walmart's Environmental Allies."

43. McIntosh, "Where Now 'Hell and High Water'?"

44. FOOTNOTE: "Universal Pictures, Illumination Entertainment and the Nature Conservancy Launch 'The Lorax Speaks' Environmental Action Campaign on Facebook," Universal Pictures, press release, February 17, 2012; Raymund Flandez, "Nature Conservancy Faces Flap Over Fundraising Deal to Promote Swimsuit Issues," Chronicle of Philanthropy, March 6, 2012; "Sports Illustrated Swimsuit Inspired Swimwear, Surfboards and Prints on Gilt.com," Inside Sports Illustrated, January 30, 2012.

45. George Marshall, "Can This Really Save the Planet?" *Guardian*, September 12, 2007.

46. Edward Roby, Untitled, UPI, June 11, 1981; Joseph Romm, "Why Natural Gas Is a Bridge to Nowhere," Energy Collective, January 24, 2012; Martha M. Hamilton, "Natural Gas, Nuclear Backers See Opportunity in 'Greenhouse' Concern," *Washington Post*, July 22, 1988.

47. "Nation's Environmental Community Offers 'Sustainable Energy Blueprint' to New Administration," Blueprint Coalition, press release, November 18, 1992; statement of Patricio Silva, project attorney, Natural Resources Defense Council, Hearing Before the Subcommittee on Energy and Air Quality, Committee on Energy and Commerce, United States House of Representatives, 107th Congress, February 28, 2001.

48. "Golden Rules for a Golden Age of Gas," World Energy Outlook Special Report, International

Energy Agency, May 29, 2012, pp. 9, 15, http://www.worldenergyoutlook.org; Nidaa Bakhsh and Brian Swint, "Fracking Spreads Worldwide," *Bloomberg Businessweek*, November 14, 2013.

49. Anthony Ingraffea, "Gangplank to a Warm Future," *New York Times*, July 28, 2013.

50. "Climate Experts Call for Moratorium on UK Shale Gas Extraction," University of Manchester, press release, January 20, 2011; Sandra Steingraber, "A New Environmentalism for an Unfractured Future," EcoWatch, June 6, 2014.

51. Personal interview with Mark Z. Jacobson, April 7, 2014.

52. "Companies We Work With: JPMorgan Chase & Co.," Nature Conservancy, http://www.nature.org; Marc Gunther, "Interview: Matthew Arnold on Steering Sustainability at JP Morgan," *Guardian*, February 18, 2013; Ann Chambers Noble, "The Jonah Field and Pinedale Anticline: A Natural-Gas Success Story," WyoHistory.org (Wyoming State Historical Society), http://www.wyohistory.org; Bryan Schutt, et al., "For Veteran Producing States, Hydraulic Fracturing Concerns Limited," *SNL Energy Gas Utility Week*, July 11, 2011; "Working with Companies: BP and Development by Design," Nature Conservancy, http://www.nature.org.

53. "Strategic Partners," Center for Sustainable Shale Development, www.sustainableshale.org; J. Mijin Cha, "Voluntary Standards Don't Make Fracking Safe," Huffington Post, March 22, 2013.

54. "Big Green Fracking Machine," Public Accountability Initiative, June 2013, p. 1; Joyce Gannon, "Heinz Endowments President's Departure Leaves Leadership Void," *Pittsburgh Post-Gazette*, January 14, 2014; Kevin Begos, "Heinz Endowments Shift on Environmental Grants," Associated Press, August 4, 2013; personal email communication with Carmen Lee, communications officer, Heinz Endowments, June 25, 2014.

55. "Environmental Defense Fund Announces Key Grant from Bloomberg Philanthropies," Environmental Defense Fund, August 24, 2012; Peter Lattman, "What It Means to Manage the Mayor's Money," *New York Times*, October 15, 2010; "Company Overview of Willett Advisors LLC," Capital Markets, *Bloomberg Businessweek*, http://investing.businessweek.com; personal email communication with Bloomberg Philanthropies representative, April 16, 2014.

56. "First Academic Study Released in EDF's Groundbreaking Methane Emissions Series," Environmental Defense Fund, press release, September 16, 2013; Michael Wines, "Gas Leaks in Fracking Disputed in Study," *New York Times*, September 16, 2013; "University of Texas at Austin Study Measures Methane Emissions Released from Natural Gas Production," Cockrell School of Engineering, press release, October 10, 2012; David T. Allen et al., "Measurements of Methane Emissions at Natural Gas Production Sites in the United States," *Proceedings of the National Academy of Sciences* 110 (2013): 17, 768–773; Robert Howarth, "Re: Allen et al. Paper in the *Proceedings of the National Academy of Sciences*," Cornell University, press release, September 11, 2013.

57. *Ibid.*; Denver Nicks, "Study: Leaks at Natural Gas Wells Less Than Previously Thought," *Time*, September 17, 2013; Seth Borenstein and Kevin Begos, "Study: Methane Leaks from Gas Drilling Not Huge," Associated Press, September 16, 2013; "Fracking Methane Fears Overdone," *The Australian*, September 19, 2013.

58. Lindsay Abrams, "Josh Fox: 'Democracy Itself Has Become Contaminated,'" Salon, August 1, 2013.

59. Pooley, *The Climate War*, 88–89.

60. William Drozdiak, "Global Warming Talks Collapse," *Washington Post*, November 26, 2000; "Special Report," *International Environment Reporter*, February 4, 1998.

61. "The EU Emissions Trading System," Policies, Climate Action, European Commission, http://ec.europa.eu; "State and Trends of the Carbon Market 2011," Environment Department, World Bank, June 2011, p. 9; personal email communication with Jacob Ipsen Hansen, energy efficiency consultant, UNEP DTU Partnership, April 15, 2014; email communication with Larry Lohmann, carbon trading expert, The Corner House.

62. Oscar Reyes, "Future Trends in the African Carbon Market," in Trusha Reddy, ed., "Carbon Trading in Africa: A Critical Review," Institute for Security Studies, Monograph No. 184, November 2011, pp. 21–28; Fidelis Allen, "Niger Delta Oil Flares, Illegal Pollution and Oppression," in Patrick Bond et al., "The CDM Cannot Deliver the Money to Africa: Why the Clean

Development Mechanism Won't Save the Planet from Climate Change, and How African Civil Society is Resisting," EJOLT Report No. 2, December 2012, pp. 57–61; "Green Projects," Carbon Limits (Nigeria), http://carbonlimitsngr.com.

63. Elisabeth Rosenthal and Andrew W. Lehren, "Profits on Carbon Credits Drive Output of a Harmful Gas," *New York Times*, August 8, 2012; John McGarrity, "India HFC-23 Emissions May Rise if CDM Boon Ends—Former Official," Reuters (Point Carbon), October 31, 2012; "Two Billion Tonne Climate Bomb: How to Defuse the HFC-23 Problem," Environmental Investigation Agency, June 2013, p. 5.

64. "CDM Panel Calls for Investigation over Carbon Market Scandal," CDM Watch and Environmental Investigation Agency, press release, July 2, 2010, http://eia-global.org.

65. "CDM Projects by Type," CDM/JI Pipeline Analysis and Database, UNEP DTU partnership, updated September 1, 2013, http://www.cdmpipeline.org.

66. Rowan Callick, "The Rush Is on for Sky Money," *The Australian*, September 5, 2009; "Voices from Madagascar's Forests: 'The Strangers, They're Selling the Wind,'" No REDD in Africa Network, http://no-redd-africa.org.

67. Ryan Jacobs, "The Forest Mafia: How Scammers Steal Millions Through Carbon Markets," *The Atlantic*, October 11, 2013; Luz Marina Herrera, "Piden Que Defensoría del Pueblo Investigue a Presunto Estafador de Nacionalidad Australiana," *La Región*, April 4, 2011; Chris Lang, "AIDESEP and COICA Condemn and Reject 'Carbon Cowboy' David Nilsson and Demand His Expulsion from Peru," REDD-Monitor, May 3, 2011; Chris Lang, "David Nilsson: Carbon Cowboy," Chris Lang.org, November 22, 2011, http://chrislang.org; "Perú: Amazónicos exigen 'REDD+ Indígena' y rechazan falsas soluciones al cambio global," Servindi, May 2, 2011; FOOTNOTE: Patrick Bodenham and Ben Cubby, "Carbon Cowboys," *Sydney Morning Herald*, July 23, 2011; "Record of Proceedings (Hansard)," 48th Parliament of Queensland, December 3, 1996, pp. 4781–83, http://www.parliament.qld.gov.au.

68. Larry Lohmann, "Carbon Trading: A Critical Conversation on Climate Change, Privatisation and Power," *Development Dialogue* no. 48, September 2006, p. 219; Deb Niemeier and Dana Rowan, "From Kiosks to Megastores: The Evolving Carbon Market," *California Agriculture* 63 (2009); Chris Lang, "How a Forestry Offset Project in Guatemala Allowed Emissions in the USA to Increase," REDD-Monitor, October 9, 2009.

69. *The Carbon Rush*, directed by Amy Miller (Kinosmith, 2012); Anjali Nayar, "How to Save a Forest," *Nature* 462 (2009): 28.

70. Mark Schapiro, "GM's Money Trees," *Mother Jones*, November/December 2009; "The Carbon Hunters" (transcript), reported by Mark Schapiro, *Frontline/World*, PBS, May 11, 2010; Chris Lang, "Uganda: Notes from a Visit to Mount Elgon," ChrisLang.org, February 28, 2007, http://chrislang.org.

71. Rosie Wong, "The Oxygen Trade: Leaving Hondurans Gasping for Air," *Foreign Policy in Focus*, June 18, 2013; Rosie Wong, "Carbon Blood Money in Honduras," *Foreign Policy in Focus*, March 9, 2012.

72. Personal email communication with Chris Lang, September 28, 2013.

73. Bram Büscher, "Nature on the Move: The Value and Circulation of Liquid Nature and the Emergence of Fictitious Conservation," *New Proposals: Journal of Marxism and Interdisciplinary Inquiry* 6, (2013): 20–36; personal email communication with Bram Büscher, April 16, 2014.

74. EUROPEAN CARBON MARKET: Stanley Reed and Mark Scott, "In Europe, Paid Permits for Pollution Are Fizzling," *New York Times*, April 21, 2013; "MEPs' Move to Fix EU Carbon Market Praised," BBC, July 4, 2013; U.K. COAL: "Digest of UK Energy Statistics 2012," United Kingdom Department of Energy and Climate Change, press release, July 26, 2012, p. 5; "Digest of UK Energy Statistics 2013," United Kingdom Department of Energy and Climate Change, press release, July 25, 2013, p. 6; UN COMMISSIONED REPORT: "Climate Change, Carbon Markets and the CDM: A Call to Action," Report of the High Level Panel on the CDM Policy Dialogue, 2012, p. 67; "99 PERCENT": personal email communication with Oscar Reyes, May 2, 2014; Alessandro Vitelli, "UN Carbon Plan Won't Reverse 99% Price Decline, New Energy Says," Bloomberg, December 12, 2013.

75. Gillian Mohney, "John Kerry Calls Climate Change a 'Weapon of Mass Destruction,'" ABC News, February 16, 2014.

76. "It Is Time the EU Scraps Its Carbon Emissions Trading System," Scrap the EU-ETS, press release, February 18, 2013; "Declaration Signatories," Scrap the EU-ETS, http://scrap-the-euets.make noise.org; "Declaration Scrap ETS," Scrap the EU-ETS, http://scrap-the-euets.makenoise.org.

77. "EU ETS Phase II—The Potential and Scale of Windfall Profits in the Power Sector," Point Carbon Advisory Services for WWF, March 2008; Suzanne Goldenberg, "Airlines 'Made Billions in Windfall Profits' from EU Carbon Tax," *Guardian*, January 24, 2013.

78. Michael H. Smith, Karlson Hargroves, Cheryl Desha, *Cents and Sustainability: Securing Our Common Future by Decoupling Economic Growth from Environmental Pressures* (London: EarthScan, 2010), 211.

79. Pooley, *The Climate War*, 371, 377.

80. Bartosiewicz and Miley, "The Too Polite Revolution," p. 26.

81. "Comparison Chart of Waxman-Markey and Kerry-Lieberman," Center for Climate and Energy Solutions, http://www.c2es.org; Bartosiewicz and Miley, "The Too Polite Revolution," p. 20.

82. Johnson, "Duke Energy Quits Scandal-Ridden American Coalition for Clean Coal Electricity"; Jane Mayer, "Covert Operations," *The New Yorker*, August 30, 2010; Ian Urbina, "Beyond Beltway, Health Debate Turns Hostile," *New York Times*, August 7, 2009; Rachel Weiner, "Obama's NH Town Hall Brings Out Birthers, Deathers, and More," Huffington Post, September 13, 2009.

83. MEMBERS DROP OUT: Steven Mufson, "ConocoPhillips, BP and Caterpillar Quit USCAP," *Washington Post*, February 17, 2010; "UNRECOVERABLE": Statement of Red Cavaney, senior vice president, government affairs, ConocoPhillips, U.S. Climate Action Partnership: Hearings Before the Committee on Energy and Commerce, United States House of Representatives, 111th Congress, 5 (2009); ConocoPhillips, 2012 Annual Report, February 19, 2013, p. 20; CONOCOPHILLIPS WEBSITE: Kate Sheppard, "ConocoPhillips Works to Undermine Climate Bill Despite Pledge to Support Climate Action," Grist, August 18, 2009; EMPLOYEES: "ConocoPhillips Intensifies Climate Focus," ConocoPhillips, press release, February 16, 2010; "LOWEST-COST OPTION": Michael Burnham, "Conoco, BP, Caterpillar Leave Climate Coalition," *New York Times* (Greenwire), February 16, 2010.

84. "Representative Barton on Energy Legislation" (video), C-SPAN, May 19, 2009; FOOTNOTE: Session 4: Public Policy Realities (video), 6th International Conference on Climate Change, The Heartland Institute, June 30, 2011; Chris Horner, "Al Gore's Inconvenient Enron," *National Review Online*, April 28, 2009.

85. John M. Broder and Clifford Krauss, "Advocates of Climate Bill Scale Down Their Goals," *New York Times*, January 26, 2010.

86. Theda Skocpol, "Naming the Problem: What It Will Take to Counter Extremism and Engage Americans in the Fight Against Global Warming," paper presented at symposium on the Politics of America's Fight Against Global Warming, Harvard University, February 2013, p. 11.

87. "Environmentalist Slams Exxon over EPA" (video), CNN Money, April 5, 2011; Colin Sullivan, "EDF Chief: 'Shrillness' of Greens Contributed to Climate Bill's Failure in Washington," *New York Times* (Greenwire), April 4, 2011.

88. "Fortune Brainstorm Green 2011," Fortune Conferences, http://fortuneconferences.com.

CHAPTER 7: NO MESSIAHS

1. This quote appeared in the first edition of Branson's book, published in 2007. All citations to the book refer to the updated 2008 version. Richard Branson, *Screw It, Let's Do It: Expanded Edition* (New York: Virgin, 2008), 114.

2. Katherine Bagley and Maria Gallucci, "Bloomberg's Hidden Legacy: Climate Change and the Future of New York City, Part 5," InsideClimate News, November 22, 2013.

3. Branson, *Screw It, Let's Do It*, 118.

4. Ibid., 122–24.

5. Ibid., 119, 127.

6. Andrew C. Revkin, "Branson Pledges Billions to Fight Global Warming," *New York Times*, Sep-

tember 21, 2006; Marius Benson, "Richard Branson Pledges $3 Billion to Tackle Global Warming," *The World Today*, ABC (Australia), September 22, 2006.

7. Bruce Falconer, "Virgin Airlines: Powered by Pond Scum?" *Mother Jones*, January 22, 2008; Branson, *Screw It, Let's Do It*, 131.

8. "Virgin Founder Richard Branson Pledges $3 Billion to Fight Global Warming," Reuters, September 22, 2006; Michael Specter, "Branson's Luck," *The New Yorker*, May 14, 2007.

9. "The Virgin Earth Challenge: Sir Richard Branson and Al Gore Announce a $25 Million Global Science and Technology Prize," The Virgin Earth Challenge, Virgin Atlantic, http://www.virgin-atlantic.com; "Branson, Gore Announce $25 Million 'Virgin Earth Challenge,'" *Environmental Leader*, February 9, 2007; Branson, *Screw It, Let's Do It*, 138; "Virgin Offers $25 Million Prize to Defeat Global Warming," Virgin Earth Prize, press release, February 9, 2007.

10. Branson, *Screw It, Let's Do It*, 140.

11. Joel Kirkland, "Branson's 'Carbon War Room' Puts Industry on Front Line of U.S. Climate Debate," ClimateWire, *New York Times*, April 22, 2010; Rowena Mason, "Sir Richard Branson: The Airline Owner on His New War," *Telegraph*, December 28, 2009.

12. Bryan Walsh, "Global Warming: Why Branson Wants to Step In," *Time*, December 31, 2009.

13. Carlo Rotella, "Can Jeremy Grantham Profit from Ecological Mayhem?" *New York Times*, August 11, 2011; Jeremy Grantham, "The Longest Quarterly Letter Ever," Quarterly Letter, GMO LLC, February 2012, http://www.capitalinstitute.org; FOOTNOTE: "Grantees," The Grantham Foundation for the Protection of the Environment," http://www.granthamfoundation.org.

14. Whitney Tilson, "Whitney Tilson's 2007 Berkshire Hathaway Annual Meeting Notes," Whitney Tilson's Value Investing Website, May 5, 2007, http://www.tilsonfunds.com.

15. "NV Energy to Join MidAmerican Energy Holdings Company," MidAmerican Energy Holdings Company, press release, May 19, 2013, http://www.midamerican.com; "Berkshire Hathaway Portfolio Tracker," CNBC, http://www.cnbc.com; Nick Zieminski, "Buffett Buying Burlington Rail in His Biggest Deal," Reuters, November 3, 2009; Alex Crippen, "CNBC Transcript: Warren Buffett Explains His Railroad 'All-In Bet' on America," CNBC, November 3, 2009.

16. Keith McCue, "Reinsurance 101," presentation, RenaissanceRe, 2011; personal interview with Eli Lehrer, August 20, 2012; Eli Lehrer, "The Beach House Bailout," *Weekly Standard*, May 10, 2010.

17. Josh Wingrove, "Meet the U.S. Billionaire Who Wants to Kill the Keystone XL Pipeline," *Globe and Mail*, April 6, 2013; FOOTNOTE: Joe Hagan, "Tom Steyer: An Inconvenient Billionaire," *Men's Journal*, March 2014; "Unprecedented Measurements Provide Better Understanding of Methane Emissions During Natural Gas Production," University of Texas at Austin, press release, September 16, 2013; Tom Steyer and John Podesta, "We Don't Need More Foreign Oil and Gas," *Wall Street Journal*, January 24, 2012.

18. "Beyond Coal Campaign," Philanthropist: Moving Beyond Coal, Mike Bloomberg, http://www.mikebloomberg.com; "Bloomberg Philanthropies Grant Awarded to Environmental Defense Fund," Bloomberg Philanthropies, press release, August 27, 2012; Katherine Bagley and Maria Gallucci, "Bloomberg's Hidden Legacy: Climate Change and the Future of New York City, Part 1," InsideClimate News, November 18, 2013; FOOTNOTE: Tom Angotti, "Is New York's Sustainability Plan Sustainable?" paper presented at Association of Collegiate Schools of Planning and Association of European Schools of Planning joint conference, July 2008; Michael R. Bloomberg and George P. Mitchell, "Fracking is Too Important to Foul Up," *Washington Post*, August 23, 2012.

19. "Introducing Our Carbon Risk Valuation Tool," Bloomberg, December 5, 2013; Dawn Lim, "Willett Advisors Eyes Real Assets for Bloomberg's Philanthropic Portfolio," Foundation & Endowment Intelligence, May 2013.

20. "Risky Business Co-Chair Michael Bloomberg" (video), Next Generation, YouTube, June 23, 2014; Robert Kopp, et al., "American Climate Prospectus: Economic Risks in the United States, prepared by Rhodium Group for the Risky Business Project, June 2014; "Secretary-General Appoints Michael Bloomberg of United States Special Envoy for Cities and Climate Change," United Nations, press release, January 31, 2014.

21. Holdings in other major oil and gas companies include Shell, ConocoPhillips, and Chevron, and the foundation has investments in a large number of other oil and gas exploration, production, services, and engineering firms, as well as coal, gas, and mining companies: Bill & Melinda Gates Foundation Trust, Form 990-PF, Return of Private Foundation, 2012, Attachment C, pp. 1–18, and Attachment D, pp. 1–15; U.S. Securities and Exchange Commission, December 31, 2013, http://www.sec.gov.

22. "ENERGY MIRACLES": Bill Gates, "Innovating to Zero!" (video), TED, February 2010; TERRAPOWER: "Chairman of the Board," http://terrapower.com; Robert A. Guth, "A Window into the Nuclear Future," Wall Street Journal, February 28, 2011; CARBON-SUCKING INVESTMENT: "About CE," Carbon Engineering, http://carbonengineering.com; MILLIONS OF HIS OWN MONEY: "Fund for Innovative Climate and Energy Research," http://dge.stanford.edu; PATENTS: U.S. Patent 8,702,982, "Water Alteration Sructure and System," filed January 3, 2008; U.S. Patent 8,685,254, "Water Alteration Structure Applications and Methods," filed January 3, 2008; U.S. Patent 8,679,331, "Water Alteration Structure Movement Method and System," filed January 3, 2008; U.S. Patent 8,348,550, "Water Alteration Structure and System Having Heat Transfer Conduit," filed May 29, 2009; "WE FOCUS": Interview with David Leonhardt (transcript), Washington Ideas Forum, November 14, 2012; "CUTE": Dave Mosher, "Gates: 'Cute' Won't Solve Planet's Energy Woes," Wired, May 3, 2011; "NON-ECONOMIC": "Conversation with Bill Gates" (transcript), Charlie Rose Show, January 30, 2013; 25 PERCENT: "Production: Gross Electricity Production in Germany from 2011 to 2013," Statistisches Bundesamt, https://www.destatis.de.

23. "Texas Oilman T. Boone Pickens Wants to Supplant Oil with Wind," USA Today, July 11, 2008; "T. Boone Pickens TV Commercial" (video), PickensPlan, YouTube, July 7, 2008.

24. Dan Reed, "An Apology to Boone Pickens: Sorry, Your Plan Never Had a Chance," Energy Viewpoints, December 9, 2013; Carl Pope, "T. Boone and Me," Huffington Post, July 3, 2008.

25. Christopher Helman, "T. Boone Reborn," Forbes, March 31, 2014; Kirsten Korosec, "T. Boone Pickens Finally Drops the 'Clean' from His 'Clean Energy' Plan," MoneyWatch, CBS, May 19, 2011; Fen Montaigne, "A New Pickens Plan: Good for the U.S. or Just for T. Boone?" Yale Environment 360, April 11, 2011; "T. Boone Pickens on Why He's for the Keystone XL Pipeline, Why the Tax Code Should Be 'Redone' and No One Person Is to Blame for Gas Prices," CNN, April 25, 2012.

26. VIRGIN GREEN FUND: Nicholas Lockley, "Eco-pragmatists," Private Equity International, November 2007, pp. 76–77; AGROFUEL INVESTMENTS: "Khosla Ventures and Virgin Fuels Invest in Gevo, Inc.," Gevo, Inc., press release, July 19, 2007; Kabir Chibber, "How Green Is Richard Branson?" Wired, August 5, 2009; NOT AN INVESTOR IN BIOFUEL PROJECTS: personal email communication with Freya Burton, director of European relations, LanzaTech, April 18, 2014; Ross Kelly, "Virgin Australia Researching Eucalyptus Leaves as Jet Fuel," Wall Street Journal, July 6, 2011; email communication, communications manager, Future Farm Industries Cooperative Research Center, April 29, 2014; "HASN'T": Branson, Screw It, Let's Do It, 132; BIOFUELS STALLED: "What Happened to Biofuels?" The Economist, September 7, 2013; "INCREASINGLY CLEAR": personal email communication with Richard Branson, May 6, 2014; FOOTNOTE: National Research Council, Renewable Fuel Standard: Potential Economic and Environmental Effects of U.S. Biofuel Policy (Washington, D.C.: National Academies Press, 2011), 130–34.

27. "Our Companies: Gevo," Virgin Green Fund, version saved by the Internet Archive Wayback Machine on September 28, 2013, http://web.archive.org; "Our Companies: Seven Seas Water," Virgin Green Fund, version saved by the Internet Archive Wayback Machine on April 4, 2014, http://web.archive.org; "Our Companies: Metrolight," Virgin Green Fund, version saved by the Internet Archive Wayback Machine on October 30, 2013, http://web.archive.org; "Our Companies: GreenRoad," Virgin Green Fund, version saved by the Internet Archive Wayback Machine on November 29, 2013, http://web.archive.org; personal interview with Evan Lovell, September 3, 2013.

28. Personal interview with Jigar Shah, September 9, 2013.
29. Chibber, "How Green Is Richard Branson?"
30. Branson took part in Solazyme's Series D round of financing, during which the company raised about $50 million from at least 10 investors. The round was led by Morgan Stanley and Braemar Energy Ventures, which says it typically invests "between $1 million and $10 million in a single round of financing," and up to $25 million total. Even if the majority of the Series D funding had come from Branson (an unlikely scenario), the value of his publicized investments would remain millions short of $300 million: "Solazyme Announces Series D Financing Round of More Than $50 Million," Solazyme Inc., press release, August 9, 2010; "Solazyme Adds Sir Richard Branson as Strategic Investor," Solazyme Inc., press release, September 8, 2010; "About Braemar Energy Ventures," Braemar Energy Ventures, http://www.braemarenergy.com. "TWO OR THREE HUNDRED MILLION": "Richard Branson on Climate Change" (video), *The Economist*, September 23, 2010; "HUNDREDS OF MILLIONS": John Vidal, "Richard Branson Pledges to Turn Caribbean Green," *Observer*, February 8, 2014; Lovell interview, September 3, 2013.
31. Branson email communication, May 6, 2014; Irene Klotz, "Profile: Sir Richard Branson, Founder, Virgin Galactic," *SpaceNews*, November 11, 2013; Vidal, "Richard Branson Pledges to Turn Caribbean Green."
32. Chibber, "How Green Is Richard Branson?"; Branson email communication, May 6, 2014.
33. Branson, *Screw It, Let's Do It*, xi; Dan Reed, "Virgin America Takes Off," *USA Today*, August 8, 2007; personal email communication with Madhu Unnikrishnan, manager of media relations, Virgin America, September 6, 2013; Victoria Stilwell, "Virgin America Cuts Airbus Order, Delays Jets to Survive," Bloomberg, November 16, 2012; Grant Robertson, "Virgin America Sets Course for Canada," *Globe and Mail*, March 19, 2010.
34. "Virgin America Orders 60 New Planes, Celebrates 'Growing Planes' with Sweet 60 Fare Sale," Virgin America, press release, January 17, 2010.
35. NUMBER OF PEOPLE: " Annual Report 2012," Virgin Australia Holdings Ltd., p. 2, http://www.virginaustralia.com; "STIFF COMPETITION": "Richard Branson Beats off Stiff Competition for Scottish Airport Links," *Courier*, April 9, 2013; NO FARES: Alastair Dalton, "Virgin's 'Zero Fares' on Scots Routes in BA Battle," *Scotsman*, March 18, 2013; "Taxi Fares," Transport for London, http://www.tfl.gov.uk; FOOTNOTE: Mark Pilling, "Size Does Matter for Virgin Boss Branson," *Flight Daily News*, July 23, 2002; Peter Pae, "New Airline Begins Service Between Los Angeles and Australia," *Los Angeles Times*, February 28, 2009; Lucy Woods, "5 Virgin Aviation Stunts by Sir Richard Branson," *Travel Magazine*, May 7, 2013.
36. Fleet expansions for Virgin Atlantic and Virgin America were confirmed by media representatives at both airlines. Virgin Australia's fleet expansion was estimated using information in the airline's 2007 annual report, as well as its half-year report for 2014, and includes chartered aircraft and other services. Additional airlines in which Virgin was temporarily invested, such as Brussels Airlines, Air Asia X, and Virgin Nigeria (now Air Nigeria) were not included: "Annual Report 2007," Virgin Blue Holdings Ltd., p. 3; "2014 Half Year Results" (presentation), Virgin Australia Holdings Ltd., February 28, 2014, p. 11. Emissions growth was estimated by comparing the combined total emissions of Virgin Atlantic and Virgin Australia in 2007 to the combined total emissions of the three major Virgin airlines in 2012 (Virgin America began its operations in mid-2007). Virgin Australia's emissions were reported for the 2006–7 and 2011–12 fiscal years: "Supply Chain 2013," Virgin Atlantic Airways Ltd., Carbon Disclosure Project, p. 8, https://www.cdp.net; "Annual Report 2007," Virgin Blue Holdings Ltd., p. 5; "Annual Report 2012," Virgin Australia Holdings Ltd., p. 29; 2008 and 2012 emissions information submitted to the Climate Registry, Virgin America Inc., p. 2, https://www.crisreport.org. For Virgin Australia in 2012–2013, see: "Annual Report 2013," Virgin Australia Holdings Ltd., p. 32. DIP: "Sustainability Report: Winter 2011/12," Virgin Atlantic Airways Ltd., p. 4, http://www.virgin-atlantic.com.
37. Mazyar Zeinali, "U.S. Domestic Airline Fuel Efficiency Ranking 2010," International Council on Clean Transportation, September 2013, http://theicct.org.
38. "Virgin and Brawn Agree Sponsorship to Confirm Branson's Entry to Formula One," *Guardian*,

March 28, 2009; Daisy Carrington, "What Does a $250,000 Ticket to Space with Virgin Galactic Actually Buy You?" CNN, August 16, 2013; Peter Elkind, "Space-Travel Startups Take Off," *Fortune*, January 16, 2013; FOOTNOTE: Salvatore Babones, "Virgin Galactic's Space Tourism Venture for the 1% Will Warm the Globe for the Rest of Us," *Truthout*, August 14, 2012.

39. Chibber, "How Green Is Richard Branson?"

40. Richard Wachman, "Virgin Brands: What Does Richard Branson Really Own?" *Observer*, January 7, 2012; David Runciman, "The Stuntman," *London Review of Books*, March 20, 2014; Heather Burke, "Bill Gates Tops Forbes List of Billionaires for the 12th Year," Bloomberg, March 9, 2006; "The World's Billionaires: #308 Richard Branson," *Forbes*, as of July 2014; Vidal, "Richard Branson Pledges to Turn Caribbean Green."

41. Chibber, "How Green is Richard Branson?"

42. James Kanter, "Cash Prize for Environmental Help Goes Unawarded," *New York Times*, November 21, 2010; Paul Smalera, "Richard Branson Has Deep-Sea Ambitions, Launches Virgin Oceanic," *Fortune*, April 5, 2011.

43. Kanter, "Cash Prize for Environmental Help Goes Unawarded."

44. Branson email communication, May 6 2014; Helen Craig, "Virgin Earth Challenge Announces Leading Organisations," Virgin Unite, November 2011.

45. *Ibid.*; "$25 Million Prize Awarded to Green Technology" (video), SWTVChannel, YouTube, November 3, 2011; "The Finalists," Virgin Earth Challenge, http://www.virginearth.com; "Biochar: A Critical Review of Science and Policy," *Biofuelwatch*, November 2011.

46. Craig, "Virgin Earth Challenge Announces Leading Organisations"; "Virgin Coming to Global Clean Energy Congress in Calgary," Calgary Economic Development, press release, September 9, 2011.

47. Knight's job as the independent Sustainable Development Advisor for the Virgin Group ended in 2012, though he is still linked to the Earth Prize: "Management Team," The Virgin Earth Challenge, http://www.virginearth.com. OTHER CLIENTS: "My Corporate Expertise," Dr. Alan Knight, http://www.dralanknight.com; "PRIVATE ACCESS": Alan Knight, "Oil Sands Revisited," Dr. Alan Knight, November 10, 2011, http://www.dralanknight.com; OSLI: "Contact," Oil Sands Leadership Initiative, http://www.osli.ca.

48. Knight, "Oil Sands Revisited"; personal interview with Alan Knight, December 12, 2011.

49. Rebecca Penty, "Calgary Firm a Finalist in Virgin's $25M Green Technology Challenge," *Calgary Herald*, September 28, 2011; Alan Knight, "Alberta Oil Sands Producers 'Distracted from Ambition and Creativity,'" *Financial Post*, November 1, 2011.

50. According to the U.S. Energy Information Administration, proved reserves of crude oil were 26.5 billion barrels in 2012. Estimates for additional, economically recoverable reserves that could be extracted using current and "next-generation" CO_2-EOR technologies have been added to the 2012 baseline: "Crude Oil Proved Reserves," International Energy Statistics, U.S. Energy Information Administration; Vello A. Kuuskraa, Tyler Van Leeuwen, and Matt Wallace, "Improving Domestic Energy Security and Lowering CO_2 Emissions with 'Next Generation' CO_2-Enhanced Oil Recovery (CO_2-EOR)," National Energy Technology Laboratory, U.S. Department of Energy, DOE/NETL-2011/1504, June 20, 2011, p. 4. "SINGLE LARGEST DETERRENT": Marc Gunther, "Rethinking Carbon Dioxide: From a Pollutant to an Asset," *Yale Environment 360*, February 23, 2012.

51. Marc Gunther, "Nations Stalled on Climate Action Could 'Suck It Up,'" Bloomberg, June 18, 2012; Marc Gunther, "The Business of Cooling the Planet," *Fortune*, October 7, 2011.

52. Penty, "Calgary Firm a Finalist in Virgin's $25M Green Technology Challenge"; Robert M. Dilmore, "An Assessment of Gate-to-Gate Environmental Life Cycle Performance of Water-Alternating-Gas CO_2-Enhanced Oil Recovery in the Permian Basin," Executive Summary, National Energy Technology Laboratory, U.S. Department of Energy, DOE/NETL-2010/1433, September 30, 2010, p. 1; Paulina Jaramillo, W. Michael Griffin, and Sean T. McCoy, "Life Cycle Inventory of CO_2 in an Enhanced Oil Recovery System," *Environmental Science & Technology* 43 (2009): 8027–8032.

53. Marc Gunther, "Direct Air Carbon Capture: Oil's Answer to Fracking?" GreenBiz.com, March 12, 2012.

54. "NRDC Calls on Major Airlines to Steer Clear of Highly Polluting New Fuel Types," Natural Resources Defense Council, press release, January 10, 2008; Liz Barratt-Brown, "NRDC Asks Airlines to Oppose Dirty Fuels and Cut Global Warming Pollution," Natural Resources Defense Council, January 10, 2008; Letter from Peter Lehner, Executive Director of the Natural Resources Defense Council, to Gerard J. Arpey, Chief Executive Officer of American Airlines, January 9, 2008, http://docs.nrdc.org.

55. Alan Knight, "Alberta Oil Sands Producers 'Distracted from Ambition and Creativity,'" Financial Post, November 1, 2011; FOOTNOTE: Brendan May, "Shell Refuses to Save the Arctic, but Its Customers Still Could," Business Green, July 24, 2013.

56. Julie Doyle, "Climate Action and Environmental Activism: The Role of Environmental NGOs and Grassroots Movements in the Global Politics of Climate Change," in Tammy Boyce and Justin Lewis, eds., Climate Change and the Media, (New York: Peter Lang, 2009), 103–116; Mark Engler, "The Climate Justice Movement Breaks Through," Yes!, December 1, 2009; "Heathrow North-west Third Runway Option Short-Listed by Airports Commission," Heathrow Airport, press release, December 17, 2013.

57. James Sturcke, "Climate Change Bill to Balance Environmental and Energy Concerns, Guardian, November 15, 2006. George Monbiot, "Preparing for Take-off," Guardian, December 19, 2006; Dan Milmo, "Brown Hikes Air Passenger Duty," Guardian, December 6, 2006; "Euro MPs Push for Air Fuel Taxes," BBC News, July 4, 2006.

58. Jean Chemnick, "Climate: Branson Calls Carbon Tax 'Completely Fair' but Dodges Question on E.U. Airline Levy," E&E News, April 26, 2012; Gwyn Topham, "Virgin Atlantic Planning Heathrow to Moscow Flights," Guardian, July 2012; Richard Branson, "Don't Run Heathrow into the Ground," Times (London), June 30, 2008; FOOTNOTE: Roland Gribben, "Sir Richard Branson's 5bn Heathrow Offer Rejected," Telegraph, March 12, 2012.

59. "Branson Criticises Carbon Tax, Backs Biofuels," PM, ABC (Australia), July 6, 2011; Rowena Mason, "Sir Richard Branson Warns Green Taxes Threaten to Kill Aviation," Telegraph, December 16, 2009; FOOTNOTE: "Behind Branson," The Economist, February 19, 1998; Juliette Garside, "Richard Branson Denies Being a Tax Exile," Guardian, October 13, 2013; Branson, Screw It, Let's Do It, 113-116.

60. Matthew Lynn, "Branson's Gesture May Not Save Aviation Industry," Bloomberg, September 26, 2006.

61. "Virgin America Selling Carbon Offsets to Passengers," Environmental Leader, December 5, 2008; John Arlidge, "I'm in a Dirty Old Business but I Try," Sunday Times (London), August 9, 2009.

62. Knight, "Alberta Oil Sands Producers 'Distracted from Ambition and Creativity.'"

63. Karl West, "Virgin Gravy Trains Rolls On," Sunday Times (London), January 16, 2011; Phillip Inman, "Privatised Rail Will Remain Gravy Train," Guardian, July 4, 2011; Richard Branson, "It's Nonsense to Suggest Virgin's Success Has Depended on State Help," Guardian, November 23, 2011.

64. Gwyn Topham, "Privatised Rail Has Meant 'Higher Fares, Older Trains and Bigger Taxpayers' Bill,'" Guardian, June 6, 2013; Adam Whitnall, "Virgin Trains Set for £3.5m Refurbishment—to Remove Smell from Corridors," Independent (London), October 6, 2013; Will Dahlgreen, "Nationalise Energy and Rail Companies, Say Public," YouGov, November 4, 2013.

65. Penty, "Calgary Firm a Finalist in Virgin's $25M Green Technology Challenge"; Gunther, "The Business of Cooling the Planet."

CHAPTER 8: DIMMING THE SUN

1. Newt Gingrich, "Stop the Green Pig: Defeat the Boxer-Warner-Lieberman Green Pork Bill Capping American Jobs and Trading America's Future," Human Events, June 3, 2008.

2. William James, The Will to Believe: And Other Essays in Popular Philosophy (New York: Longmans Green, 1907), 54.

3. "Geoengineering the Climate: Science, Governance and Uncertainty," Royal Society, September 2009, p. 62; "Solar Radiation Management: the Governance of Research," Solar Radiation Management Governance Initiative, convened by the Environmental Defense Fund, the Royal Society, and TWAS, 2011, p. 11.

4. Environmental Defense Fund, "Geoengineering: A 'Cure' Worse Than the Disease?" *Solutions* 41 (Spring 2010): 10–11.

5. EXPERIMENTS: Patrick Martin et al., "Iron Fertilization Enhanced Net Community Production but not Downward Particle Flux During the Southern Ocean Iron Fertilization Experiment LOHAFEX," *Global Biogeochemical Cycles* (2013): 871–881; "The Haida Salmon Restoration Project: The Story So Far," Haida Salmon Restoration Corporation, September 2012; PEER-REVIEWED PAPERS: GeoLibrary, Oxford Geoengineering Programme, http://www.geoengineer ing.ox.ac.uk; SHIPS AND PLANES: John Latham et al., "Marine Cloud Brightening," *Philosophical Transactions of the Royal Society* A 370 (2012): 4247–4255; HOSES: David Rotman, "A Cheap and Easy Plan to Stop Global Warming," *MIT Technology Review*, February 8, 2013; Daniel Cressey, "Cancelled Project Spurs Debate over Geoengineering Patents," *Nature* 485 (2012): 429.

6. P. J. Crutzen, "Albedo Enhancement by Stratospheric Sulfur Injections: A Contribution to Resolve a Policy Dilemma?" *Climatic Change* 77 (2006): 212; Oliver Morton, "Is This What It Takes to Save the World?" *Nature* 447 (2007): 132.

7. Ben Kravitz, Douglas G. MacMartin, and Ken Caldeira, "Geoengineering: Whiter Skies?" *Geophysical Research Letters* 39 (2012): 1, 3–5; "Geoengineering: A Whiter Sky," Carnegie Institution for Science, press release, May 30, 2012.

8. "Solar Radiation Management," p. 16.

9. Roger Revelle et al., "Atmospheric Carbon Dioxide," in *Restoring the Quality of Our Environment*, Report of the Environmental Pollution Panel, President's Science Advisory Committee, The White House, November 1965, Appendix Y4, p. 127.

10. James Rodger Fleming, *Fixing the Sky: The Checkered History of Weather and Climate Control* (New York: Columbia University Press, 2010), 165–188.

11. Crutzen, "Albedo Enhancement by Stratospheric Sulfur Injections," 216.

12. "When Patents Attack!" *Planet Money*, NPR, July 22, 2011.

13. "The Stratospheric Shield," *Intellectual Ventures*, 2009, pp. 3, 15–16; "Solving Global Warming with Nathan Myhrvold" (transcript), *Fareed Zakaria GPS*, CNN, December 20, 2009.

14. Steven D. Levitt and Stephen J. Dubner, *SuperFreakonomics* (New York: HarperCollins, 2009), 194.

15. "A Future Tense Event: Geoengineering," New America Foundation, http://www.newamerica.net.

16. Eli Kintisch, *Hack the Planet: Science's Best Hope—or Worst Nightmare—for Averting Climate Catastrophe* (Hoboken, NJ: John Wiley & Sons, 2010), 8; personal interview with James Fleming, November 5, 2010.

17. "Inventors," Intellectual Ventures, http://www.intellectualventures.com.

18. GATES AND FUND: "Fund for Innovative Climate and Energy Research," Carnegie Institution for Science, Stanford University, http://dge.stanford.edu; GATES AND CARBON ENGINEERING: "About CE," Carbon Engineering, http://carbonengineering.com; GATES AND INTELLECTUAL VENTURES: Jason Pontin, "Q&A: Bill Gates," *MIT Technology Review*, September/October 2010; PATENTS: U.S. Patent 8,702,982, "Water Alteration Structure and System," filed January 3, 2008; U.S. Patent 8,685,254, "Water Alteration Structure Applications and Methods," filed January 3, 2008; U.S. Patent 8,679,331, "Water Alteration Structure Movement Method and System," filed January 3, 2008; U.S. Patent 8,348,550, "Water Alteration Structure and System Having Heat Transfer Conduit," filed May 29, 2009; TERRAPOWER: "Nathan Myhrvold, Ph.D.," TerraPower, http://terrapower.com; BRANSON: "Stakeholder Partners," Solar Radiation Management Governance Initiative, http://www.srmgi.org.

19. Jon Taylor, "Geo-engineering—Useful Tool for Tackling Climate Change, or Dangerous Distraction?" WWF-UK, September 6, 2012, http://blogs.wwf.org.uk.

20. Alan Robock, "20 Reasons Why Geoengineering May Be a Bad Idea," *Bulletin of the Atomic Scientists* 64 (2008): 14–18; Clive Hamilton, "The Ethical Foundations of Climate Engineering," in *Climate Change Geoengineering: Philosophical Perspectives, Legal Issues, and Governance Frameworks*, ed. Wil C. G. Burns and Andrew L. Strauss (New York: Cambridge University Press, 2013), 48.

21. Francis Bacon, *Bacon's New Atlantis*, ed. A. T. Flux (London: Macmillan, 1911); John Gascoigne, *Science in the Service of Empire: Joseph Banks, the British State and the Uses of Science in the Age of Revolution* (Cambridge: Cambridge University Press, 1998), 175.

22. Personal email communication with Sallie Chisholm, October 28, 2012.

23. "PRINTING PRESS AND FIRE": Matthew Herper, "With Vaccines, Bill Gates Changes the World Again," *Forbes*, November 2, 2011; RUSS GEORGE: "Background to the Haida Salmon Restoration Project," Haida Salmon Restoration Corporation, October 19, 2012, p. 2; ONE HUNDRED TONS: "Haida Gwaii Geo-engineering, Pt 2," *As It Happens with Carol Off & Jeff Douglas*, CBC Radio, October 16, 2012; "THE CHAMPION": Mark Hume and Ian Bailey, "Businessman Russ George Defends Experiment Seeding Pacific with Iron Sulphate," *Globe and Mail*, October 19, 2012; "PANDORA'S BOX": Jonathan Gatehouse, "Plan B for Global Warming," *Maclean's*, April 22, 2009; "IRRIGATION": personal interview with David Keith, October 19, 2010.

24. Wendell Berry, *The Way of Ignorance: And Other Essays* (Emeryville, CA: Shoemaker & Hoard, 2005), 54.

25. Petra Tschakert, "Whose Hands Are Allowed at the Thermostat? Voices from Africa," presentation at "The Ethics of Geoengineering: Investigating the Moral Challenges of Solar Radiation Management," University of Montana, Missoula, October 18, 2010.

26. Alan Robock, Martin Bunzl, Ben Kravitz, and Georgiy L. Stenchikov, "A Test for Geoengineering?" *Science* 327 (2010): 530; Alan Robock, Luke Oman, and Georgiy L. Stenchikov, "Regional Climate Responses to Geoengineering with Tropical and Arctic SO_2 Injections," *Journal of Geophysical Research* 113 (2008): D16101.

27. Robock, Bunzl, Kravitz, and Stenchikov, "A Test for Geoengineering?" 530.

28. Martin Bunzl, "Geoengineering Research Reservations," presentation to the American Association for the Advancement of Science, February 20, 2010; Fleming, *Fixing the Sky*, 2.

29. Robock, Oman, and Stenchikov, "Regional Climate Responses to Geoengineering with Tropical and Arctic SO_2 Injections"; K. Niranjan Kumar et al., "On the Observed Variability of Monsoon Droughts over India," *Weather and Climate Extremes* 1 (2013): 42.

30. Numerous papers have reproduced Robock's results and found that SRM could have other potentially harmful impacts on the global water cycle and regional precipitation patterns. Notable recent examples include: Simone Tilmes et al., "The Hydrological Impact of Geoengineering in the Geoengineering Model Intercomparison Project (GeoMIP)," *Journal of Geophysical Research: Atmospheres* 118 (2013): 11,036–11,058; Angus J. Ferraro, Eleanor J. Highwood, and Andrew J. Charlton-Perez, "Weakened Tropical Circulation and Reduced Precipitation in Response to Geoengineering," *Environmental Research Letters* 9 (2014): 014001. The 2012 study is: H. Schmidt et al., "Solar Irradiance Reduction to Counteract Radiative Forcing from a Quadrupling of CO_2: Climate Responses Simulated by Four Earth System Models," *Earth System Dynamics* 3 (2012): 73. An earlier study by the U.K. Met Office Hadley Centre had found that brightening clouds off the coast of southern Africa would cause an even greater, 30 percent reduction in precipitation in the Amazon that, according to the study press release, "could accelerate die-back of the forest." See: Andy Jones, Jim Haywood, and Olivier Boucher, "Climate Impacts of Geoengineering Marine Stratocumulus Clouds," *Journal of Geophysical Research* 114 (2009): D10106; "Geoengineering Could Damage Earth's Eco-systems," UK Met Office, press release, September 8, 2009. The 2013 study is: Jim M. Haywood et al., "Asymmetric Forcing from Stratospheric Aerosols Impacts Sahelian Rainfall," *Nature Climate Change* 3 (2013): 663.

31. Climate models "appear to underestimate the magnitude of precipitation changes over the 20th century," which according to some researchers carries special relevance for the risks of SRM: Gabriele C. Hegerl and Susan Solomon, "Risks of Climate Engineering," *Science* 325 (2009):

955–956. ARCTIC SEA ICE LOSS AND GLOBAL SEA LEVEL RISE: Julienne Stroeve et al., "Arctic Sea Ice Decline: Faster than Forecast," *Geophysical Research Letters* 34 (2007): L09501; Julienne C. Stroeve et al., "Trends in Arctic Sea Ice Extent from CMIP5, CMIP3 and Observations," *Geophysical Research Letters* 39 (2012): L16502; Stefan Rahmstorf et al., "Recent Climate Observations Compared to Projections," *Science* 316 (2007): 709; Ian Allison et al., "The Copenhagen Diagnosis, 2009: Updating the World on the Latest Climate Science," University of New South Wales Climate Change Research Centre, 2009, p. 38.

32. Ken Caldeira, "Can Solar Radiation Management Be Tested?" email to the Google Group listserv "Geoengineering," September 27, 2010; Levitt and Dubner, *SuperFreakonomics*, 197.

33. *Ibid.*, 176.

34. Personal interview with Aiguo Dai, June 6, 2012; Kevin E. Trenberth and Aiguo Dai, "Effects of Mount Pinatubo Volcanic Eruption on the Hydrological Cycle as an Analog of Geoengineering," *Geophysical Research Letters* 34 (2007): L15702; "Climate Change and Variability in Southern Africa: Impacts and Adaptation Strategies in the Agricultural Sector," United Nations Environment Programme, 2006, p. 2; Donatella Lorch, "In Southern Africa, Rains' Return Averts Famine," *New York Times*, April 23, 1993; Scott Kraft, "30 Million May Feel Impact of Southern Africa Drought," *Los Angeles Times*, May 18, 1992.

35. Dai interview, June 6, 2012; Trenberth and Dai, "Effects of Mount Pinatubo Volcanic Eruption on the Hydrological Cycle as an Analog of Geoengineering."

36. Volney's full name was Constantin-François de Chasseboeuf, count de Volney. "WEAKER THAN NORMAL": Personal interview with Alan Robock, October 19, 2010; "ALL HAD PERISHED": Constantin-François Volney, *Travels Through Syria and Egypt, in the Years 1783, 1784, and 1785*, Vol. 1 (London: G. and J. Robinson, 1805), 180-181.

37. John Grattan, Sabina Michnowicz, and Roland Rabartin, "The Long Shadow: Understanding the Influence of the Laki Fissure Eruption on Human Mortality in Europe," *Living Under the Shadow: Cultural Impacts of Volcanic Eruptions*, ed. John Grattan and Robin Torrence (Walnut Creek, CA: Left Coast Press, 2010), 156; Clive Oppenheimer, *Eruptions That Shook the World* (Cambridge: Cambridge University Press, 2011), 293; Rudolf Brázdil et al., "European Floods During the Winter 1783/1784: Scenarios of an Extreme Event During the 'Little Ice Age,'" *Theoretical and Applied Climatology* 100 (2010): 179–185; Anja Schmidt et al., "Climatic Impact of the Long-lasting 1783 Laki Eruption: Inapplicability of Mass-independent Sulfur Isotopic Composition Measurements," *Journal of Geophysical Research* 117 (2012): D23116; Alexandra Witze and Jeff Kanipe, *Island on Fire: The Extraordinary Story of Laki, the Volcano That Turned Eighteenth-century Europe Dark* (London: Profile Books, 2014), 141–45.

38. Luke Oman et al., "High-Latitude Eruptions Cast Shadow over the African Monsoon and the Flow of the Nile," *Geophysical Research Letters* 33 (2006): L18711; Michael Watts, *Silent Violence: Food, Famine and Peasantry in Northern Nigeria* (Berkeley: University of California Press, 1983), 286, 289–290; Stephen Devereux, "Famine in the Twentieth Century," Institute of Development Studies, IDS Working Paper 105, 2000, pp. 6, 30–31.

39. Oman et al., "High-Latitude Eruptions Cast Shadow over the African Monsoon and the Flow of the Nile"; personal interview with Alan Robock, May 29, 2012.

40. David Keith, *A Case for Climate Engineering* (Cambridge, MA: MIT Press, 2013), 10, 54.

41. Trenberth and Dai, "Effects of Mount Pinatubo Volcanic Eruption on the Hydrological Cycle as an Analog of Geoengineering."

42. Ed King, "Scientists Warn Earth Cooling Proposals Are No Climate 'Silver Bullet,'" Responding to Climate Change, July 14, 2013; Haywood et al., "Asymmetric Forcing from Stratospheric Aerosols Impacts Sahelian Rainfall," 663–64.

43. "Why We Oppose the Copenhagen Accord," Pan African Climate Justice Alliance, June 3, 2010; "Filipina Climate Chief: 'It Feels Like We Are Negotiating on Who Is to Live and Who Is to Die,'" Democracy Now!, November 20, 2013; Rob Nixon, *Slow Violence and the Environmentalism of the Poor* (Cambridge, MA: Harvard University Press, 2011).

44. "Bill Gates: Innovating to Zero!" TED Talk, February 12, 2010, http://www.ted.com; Levitt and Dubner, *SuperFreakonomics*, 199.

45. Bruno Latour, "Love Your Monsters: Why We Must Care for Our Technologies as We Do Our Children," in *Love Your Monsters: Postenvironmentalism and the Anthropocene*, ed. Michael Shellenberger and Ted Nordhaus (Oakland: Breakthrough Institute, 2011); Mark Lynas, *The God Species: How the Planet Can Survive the Age of Humans* (London: Fourth Estate, 2011).

46. Keith, *A Case for Climate Engineering*, 111.

47. Italics in original. Ed Ayres, *God's Last Offer* (New York: Four Walls Eight Windows, 1999), 195.

48. Levitt and Dubner, *SuperFreakonomics*, 195; "About CE," Carbon Engineering, http://carbon engineering.com; Nathan Vardi, "The Most Important Billionaire In Canada," Forbes, December 10, 2012.

49. "Policy Implications of Greenhouse Warming: Mitigation, Adaptation, and the Science Base," National Academy of Sciences, National Academy of Engineering, Institute of Medicine, 1992, 458, 472.

50. Dan Fagin, "Tinkering with the Environment," *Newsday*, April 13, 1992.

51. Jason J. Blackstock et al., "Climate Engineering Responses to Climate Emergencies," Novim, 2009, pp. i–ii, 30.

52. "Factsheet: American Enterprise Institute," ExxonSecrets.org, http://www.exxonsecrets.org; Robert J. Brulle, "Institutionalizing Delay: Foundation Funding and the Creation of U.S. Climate Change Counter-Movement Organizations," *Climatic Change*, December 21, 2013, p. 8; 2008 Annual Report, American Enterprise Institute, pp. 2, 10; Lee Lane, "Plan B: Climate Engineering to Cope with Global Warming," *The Milken Institute Review*, Third Quarter 2010, p. 53.

53. Juliet Eilperin, "AEI Critiques of Warming Questioned," *Washington Post*, February 5, 2007; "Factsheet: American Enterprise Institute," ExxonSecrets.org; Kenneth Green, "Bright Idea? CFL Bulbs Have Issues of Their Own," *Journal Gazette* (Fort Wayne, Indiana), January 28, 2011.

54. Rob Hopkins, "An Interview with Kevin Anderson: 'Rapid and Deep Emissions Reductions May Not Be Easy, but 4°C to 6°C Will Be Much Worse,'" Transition Culture, November 2, 2012, http://transitionculture.org.

55. "A Debate on Geoengineering: Vandana Shiva vs. Gwynne Dyer," *Democracy Now!*, July 8, 2010.

56. Jeremy Lovell, "Branson Offers $25 mln Global Warming Prize," Reuters, February 9, 2007.

57. Barbara Ward, *Spaceship Earth* (New York: Columbia University Press, 1966), 15; FOOTNOTE: Robert Poole, *Earthrise: How Man First Saw the Earth* (New Haven: Yale University Press, 2008), 92–93; Al Reinert, "The Blue Marble Shot: Our First Complete Photograph of Earth," *The Atlantic*, April 12, 2011; Andrew Chaikin, "The Last Men on the Moon," *Popular Science*, September 1994; Eugene Cernan and Don Davis, *The Last Man on the Moon* (New York: St. Martin's, 1999), 324.

58. Kurt Vonnegut Jr., "Excelsior! We're Going to the Moon! Excelsior!" *New York Times Magazine*, July 13, 1969, SM10.

59. Poole, *Earthrise*, 144–145, 162; Peder Anker, "The Ecological Colonization of Space," *Environmental History* 10 (2005): 249–254; Andrew G. Kirk, *Counterculture Green: The Whole Earth Catalog and American Environmentalism* (Lawrence: University Press of Kansas, 2007), 170–172; Stewart Brand, *Whole Earth Discipline: Why Dense Cities, Nuclear Power, Transgenic Crops, Restored Wildlands, and Geoengineering Are Necessary* (New York: Penguin, 2009).

60. Leonard David, "People to Become Martians This Century?" NBC News, June 25, 2007.

61. "Richard Branson on Space Travel: 'I'm Determined to Start a Population on Mars,'" *CBS This Morning*, September 18, 2012; "Branson's Invasion of Mars," *New York Post*, September 20, 2012; "Branson: Armstrong 'Extraordinary Individual'" (video), *Sky News*, August 26, 2012.

62. The three Virgin-branded airlines together emitted roughly 8.8 million metric tons of CO_2 in 2011, greater than the nearly 8 million metric tons emitted by Honduras that year: "Sustainability Report: Autumn 2012," Virgin Atlantic Airways Ltd., p. 11; "Annual Report 2011," Virgin Blue Holdings Ltd., p. 28; 2011 emissions information submitted to the Climate Registry, Virgin America, Inc., p. 2, https://www.crisreport.org; "International Energy Statistics," U.S. Energy Information Administration, http://www.eia.gov.

63. Kenneth Brower, "The Danger of Cosmic Genius," *The Atlantic*, October 27, 2010.

64. Christopher Borick and Barry Rabe, "Americans Cool on Geoengineering Approaches to Addressing Climate Change," Brookings Institution, Issues in Governance Studies No. 46, May

2012, p. 3-4; Malcolm J. Wright, Damon A. H. Teagle, and Pamela M. Feetham, "A Quantitative Evaluation of the Public Response to Climate Engineering," *Nature Climate Change* 4 (2014): 106–110; "Climate Engineering—What Do the Public Think?" Massey University, press release, January 13, 2014.

PART III: STARTING ANYWAY

1. Arundhati Roy, "The Trickledown Revolution," *Outlook*, September 20, 2010.
2. Translation provided by Mitchell Anderson, field consultant at Amazon Watch. Gerald Amos, Greg Brown, and Twyla Roscovich, "Coastal First Nations from BC Travel to Witness the Gulf Oil Spill" (video), 2010.

CHAPTER 9: BLOCKADIA

1. "United Nations Conference on Environment and Development: Rio Declaration on Environment and Development," *International Legal Materials* 31 (1992): 879, http://www.un.org.
2. Harold L. Ickes, *The Secret Diary of Harold L. Ickes: The First Thousand Days, 1933–1936* (New York: Simon & Schuster, 1953), 646.
3. Scott Parkin, "Harnessing Rebel Energy: Making Green a Threat Again," *CounterPunch*, January 18–20, 2013.
4. "Greece Sees Gold Boom, but at a Price," *New York Times*, January 13, 2013; Patrick Forward, David J. F. Smith, and Antony Francis, *Skouries Cu/Au Project, Greece, NI 43-101 Technical Report*, European Goldfields, July 14, 2011, p. 96; "Skouries," Eldorado Gold Corp., http://www.eldoradogold.com; Costas Kantouris, "Greek Gold Mine Savior to Some, Curse to Others," Associated Press, January 11, 2013.
5. Personal interview with Theodoros Karyotis, Greek political activist and writer, January 16, 2014.
6. Deepa Babington, "Insight: Gold Mine Stirs Hope and Anger in Shattered Greece," Reuters, January 13, 2014; Alkman Granitsas, "Greece to Approve Gold Project," *Wall Street Journal*, February 21, 2013; Jonathan Stearns, "Mountain of Gold Sparks Battles in Greek Recovery Test," Bloomberg, April 9, 2013.
7. Karyotis interview, January 16, 2014.
8. Nick Meynen, "A Canadian Company, the Police in Greece and Democracy in the Country That Invented It," EJOLT, June 13, 2013; "A Law Unto Themselves: A Culture of Abuse and Impunity in the Greek Police," Amnesty International, 2014, p. 11; Karyotis interview, January 16, 2014.
9. Luiza Ilie, "Romanian Farmers Choose Subsistence over Shale Gas," Reuters, October 27, 2013.
10. "Romania Riot Police Clear Shale Gas Protesters," Agence France-Presse, December 2, 2013; Alex Summerchild, "Pungesti, Romania: People Versus Chevron and Riot Police," *The Ecologist*, December 12, 2013; Antoine Simon and David Heller, "From the Frontline of Anti-Shale Gas Struggles: Solidarity with Pungesti," Friends of the Earth Europe, December 7, 2013, https://www.foeeurope.org; Razvan Chiruta and Petrica Rachita, "Goal of Chevron Scandal in Vaslui County: Church Wants Land Leased to US Company Back," *Romania Libera*, October 18, 2013.
11. "First Nations Chief Issues Eviction Notice to SWN Resources," CBC News, October 1, 2013; "SWN Resources Wraps Up Shale Gas Testing in New Brunswick," CBC News, December 6, 2013; Daniel Schwartz and Mark Gollom, "N.B. Fracking Protests and the Fight for Aboriginal Rights," CBC News, October 21, 2013.
12. "Shale Gas Clash: Explosives, Firearms, Seized in Rexton," CBC News, October 18, 2013; "First Nations Clash with Police at Anti-Fracking Protest," Al Jazeera, October 17, 2013; "RCMP Says Firearms, Improvised Explosives Seized at New Brunswick Protest," Canadian Press, October 18, 2013; Gloria Galloway and Jane Taber, "Native Shale-Gas Protest Erupts in Violence," *Globe and Mail*, October 18, 2013; "Police Cars Ablaze: Social Media Captures Scene of Violent New Brunswick Protest," *Globe and Mail*, October 17, 2013.
13. William Shakespeare, *King Henry IV: Part 1*, in *The Arden Shakespeare*, ed. David Scott Kastan (London: Thompson Learning, 2002), 246; James Ball, "EDF Drops Lawsuit Against Environmental Activists After Backlash," *Guardian*, March 13, 2013.

14. John Vidal, "Russian Military Storm Greenpeace Arctic Oil Protest Ship," *Guardian*, September 19, 2013; "Greenpeace Activists Being Given Russian Exit Visas After Amnesty," UPI, December 26, 2013.

15. David Pierson, "Coal Mining in China's Inner Mongolia Fuel Tensions," *Los Angeles Times*, June 2, 2011; Jonathan Watts, "Herder's Death Deepens Tensions in Inner Mongolia," *Guardian*, May 27, 2011.

16. "About," Front Line Action on Coal, http://frontlineaction.wordpress.com; Oliver Laughland, "Maules Creek Coal Mine Divides Local Families and Communities," *Guardian*, April 9, 2014; "Maules Creek Coal Project Environmental Assessment," Section 7: Impacts, Management and Mitigation, Whitehaven Coal Limited, Hansen Bailey, July 2011, pp. 90–91; Ian Lowe, "Maules Creek Proposed Coal Mine: Greenhouse Gas Emissions," submission to the Maules Creek Community Council, 2012, http://www.maulescreek.org; "Quarterly Update of Australia's National Greenhouse Gas Inventory: December 2013," Australia's National Greenhouse Accounts, Department of the Environment, Australian Government, 2014, p. 6.

17. "Dredging, Dumping and the Great Barrier Reef," Australian Marine Conservation Society, May 2014, p. 3.

18. "Final Environmental Impact Statement for the Keystone XL Project," U.S. Department of State, August 2011, Table 3.13.1-4: Reported Incidents for Existing Keystone Oil Pipeline, section 3.13, pp. 11–14; Nathan Vanderklippe, "Oil Spills Intensify Focus on New Pipeline Proposals," *Globe and Mail*, May 9, 2011; Carrie Tait, "Pump Station Spill Shuts Keystone Pipeline," *Globe and Mail*, May 31, 2011; Art Hovey, "TransCanada Cleaning Up Spill at N.D. Pump Station," *Lincoln Journal Star* (Nebraska), May 10, 2011.

19. Jamie Henn, "40,000+ Join 'Forward on Climate' Rally in Washington, DC," Huffington Post, February 17, 2013; personal email communications with Ramsey Sprague, Tar Sands Blockade, January 22–23, 2014.

20. "Oil Sands Export Ban: BC First Nations Unite to Declare Province-Wide Opposition to Crude Oil Pipeline," Yinka Dene Alliance, press release, December 1, 2011.

21. Ian Ewing, "Pipe Piling Up," *CIM Magazine*, October 2013; Shawn McCarthy, "Keystone Pipeline Approval 'Complete No-Brainer,' Harper Says," *Globe and Mail*, September 21, 2011.

22. Ossie Michelin, "Amanda Polchies, the Woman in Iconic Photo, Says Image Represents 'Wisp of Hope,'" APTN, October 24, 2013; "Greek Granny Goads Riot Police at Gold Mining Protest with Wartime Song," (video) Keep Talking Greece, March 8, 2013; David Herron, "Government Still Ensuring Hydraulic Fracturing Happens in Pungesti, Romania, Despite Protests by Villagers," The Long Tail Pipe, January 5, 2014.

23. FOOTNOTE: Maxime Combes, "Let's Frackdown the Fracking Companies," in Leah Temper, et al., "Towards a Post-Oil Civilization: Yasunization and Other Initiatives to Leave Fossil Fuels in the Soil," EJOLT Report No. 6, May 2013, p. 92.

24. Esperanza Martínez, "The Yasuní—ITT Initiative from a Political Economy and Political Ecology Perspective," in Temper et al., "Towards a Post-Oil Civilization," p. 11; KC Golden, "The Keystone Principle," Getting a GRIP on Climate Solutions, February 15, 2013.

25. "Chop Fine: The Human Rights Impact of Local Government Corruption and Mismanagement in Rivers State, Nigeria," Human Rights Watch, January 2007, p. 16; "Niger Delta Human Development Report," United Nations Development Programme, 2006, p. 76; Adam Nossiter, "Far from Gulf, a Spill Scourge 5 Decades Old," *New York Times*, June 16, 2010; Christian Purefoy, "Nigerians Angry at Oil Pollution Double Standards," CNN, June 30, 2010.

26. Nigeria flared about 515 billion cubic feet (14.6 billion cubic meters) of natural gas in 2011, according to satellite data from the U.S. National Oceanic and Atmospheric Administration; assuming 127 kilowatt hours per Mcf of natural gas, following the U.S. Energy Information Administration, this could theoretically produce nearly three times as much electricity as Nigeria consumed in 2011 (which was about 23.1 billion kWh). Roughly half of Nigerians do not currently have electricity access. Also according to EIA data, CO_2 emissions from gas flaring in Nigeria totaled about 31.1 million metric tons in 2011, just over 40 percent of Nigeria's total emissions

from energy consumption that year. For data sources, see: "Estimated Flared Volumes from Satellite Data, 2007–2011," World Bank, Global Gas Flaring Reduction, http://web.worldbank.org; "Frequently Asked Questions: How Much Coal, Natural Gas, or Petroleum is Used to Generate a Kilowatthour of Electricity?" U.S. Energy Information, U.S. Department of Energy, http://www.eia.gov; "International Energy Statistics," U.S. Energy Information Administration, U.S. Department of Energy, http://www.eia.gov. DELTA COMMUNITIES LACK: Paul Francis, Deirdre Lapin, and Paula Rossiasco, "Niger Delta: A Social and Conflict Analysis for Change," Woodrow Wilson International Center for Scholars, 2011, p. 10; Richard Essein, "Unemployment Highest in Niger Delta," *Daily Times Nigeria*, March 30, 2011; "Communities Not Criminals: Illegal Oil Refining in the Niger Delta," Stakeholder Democracy Network, October 2013, p. 4.

27. Jedrzej George Frynas, "Political Instability and Business: Focus on Shell in Nigeria," *Third World Quarterly* 19 (1998): 463; Alan Detheridge and Noble Pepple (Shell), "A Response to Frynas," *Third World Quarterly* 3 (1998): 481-482.

28. Note that after Shell pulled out, pipelines crossing Ogoni territory from other wells remained active. Godwin Uyi Ojo, "Nigeria, Three Complementary Viewpoints on the Niger Delta," in Temper et al., "Towards a Post-Oil Civilization," pp. 39–40; "Nigeria Ogoniland Oil Clean-up 'Could Take 30 Years,'" *BBC News*, August 4, 2011; Carley Petesch, "Shell Niger Delta Oil Spill: Company to Negotiate Compensation and Cleanup with Nigerians," Associated Press, September 9, 2013; Eghosa E. Osaghae, "The Ogoni Uprising: Oil Politics, Minority Agitation and the Future of the Nigerian State," *African Affairs* 94 (1995): 325–344.

29. Osuoka interview, January 10, 2014; Ojo in Temper et al., "Towards a Post-Oil Civilization," p. 40.

30. Elisha Bala-Gbogbo, "Nigeria Says Revenue Gap May Reach as Much as $12 Billion," Bloomberg, November 1, 2013; Ed Pilkington, "14 Years After Ken Saro-Wiwa's Death, Family Points Finger at Shell in Court," *Guardian*, May 26, 2009; Frank Aigbogun, "It Took Five to Hang Saro-Wiwa," Associated Press, November 13, 1995; Andrew Rowell and Stephen Kretzmann, "The Ogoni Struggle," report, Project Underground, Berkeley, California, 1996.

31. Bronwen Manby, "The Price of Oil: Corporate Responsibility and Human Rights Violations in Nigeria's Oil Producing Communities," Human Rights Watch, HRW Index No. 1-56432-225-4, January 1999, pp. 123–26.

32. "The Kaiama Declaration," United Ijaw, 1998, http://www.unitedijaw.com.

33. *Ibid.*

34. Personal interview with Isaac Osuoka, January 10, 2014.

35. Isaac Osuoka, "Operation Climate Change," in *Climate Change: Who's Carrying the Burden? The Chilly Climates of the Global Environmental Dilemma*, ed. L. Anders Sandberg and Tor Sandberg (Ottawa: The Canadian Center for Policy Alternatives, 2010), 166.

36. Bronwen Manby, "Nigeria: Crackdown in the Niger Delta," Human Rights Watch, Vol. 11, No. 2 (A), May 1999, pp. 2, 11, 13–17.

37. Ojo in Temper et al., "Towards a Post-Oil Civilization," p. 44.

38. Paul M. Barrett, "Ecuadorian Court Cuts Chevron's Pollution Bill in Half," *Bloomberg Businessweek*, November 13, 2013; "Supreme Court will hear Chevron appeal in Ecuador environmental damages case," The Canadian Press, April 3, 2014.

39. Bob Deans, "Big Coal, Cold Cash, and the GOP," *OnEarth*, February 22, 2012.

40. Clifford Krauss, "Shale Boom in Texas Could Increase U.S. Oil Output," *New York Times*, May 27, 2011.

41. Brian Milner, "'Saudi America' Heads for Energy Independence," *Globe and Mail*, March 18, 2012; "Moving Crude Oil by Rail," Association of American Railroads, December 2013; Clifford Krauss and Jad Mouawad, "Accidents Surge as Oil Industry Takes the Train," *New York Times*, January 25, 2014; "Kim Mackrael, "How Bakken Crude Moved from North Dakota to Lac-Mégantic," *Globe and Mail*, July 8, 2014; Jim Monk, "Former Gov. Sinner Proposes National Rail Safety Discussion," KFGO (North Dakota), January 7, 2014.

42. Nathan Vanderklippe and Shawn McCarthy, "Without Keystone XL, Oil Sands Face Choke Point," *Globe and Mail*, June 8, 2011.

43. "Energy: The Pros and Cons of Shale Gas Drilling," *60 Minutes*, CBS, November 14, 2010.

44. "Glenn Beck—Bernanke Confused, a Coming Caliphate and Rick Santorum," *Glenn Beck*, June 23, 2011; Suzanne Goldenberg, "Fracking Hell: What It's Really Like to Live Next to a Shale Gas Well," *Guardian*, December 13, 2013; Russell Gold and Tom McGinty, "Energy Boom Puts Wells in America's Backyards," *Wall Street Journal*, October 25, 2013.

45. Kim Cornelissen, "Shale Gas and Quebecers: The Broken Bridge Towards Renewable Sources of Energy," in Temper et al., "Towards a Post-Oil Civilization," p. 100; Emily Gosden, "Half of Britain to Be Offered for Shale Gas Drilling as Fracking Areas Face 50 Trucks Passing Each Day," *Telegraph*, December 17, 2013; Damian Carrington, "Fracking Can Take Place in 'Desolate' North-East England, Tory Peer Says," *Guardian*, July 30, 2013.

46. David Mildenberg and Jim Efstathiou Jr., "Ranchers Tell Keystone: Not Under My Backyard," *Bloomberg Businessweek*, March 8, 2012; Goldenberg, "Fracking Hell."

47. Daniel Gilbert, "Exxon CEO Joins Suit Citing Fracking Concerns," *Wall Street Journal*, February 20, 2014; "Polis Welcomes ExxonMobil CEO into 'Exclusive' Group of People Whose Neighborhood Has Been Fracked," Congressman Jared Polis, press release, February 21, 2014.

48. Thomas Paine, *Rights of Man, Common Sense, and Other Political Writings*, ed. Mark Philp (Oxford: Oxford University Press, 1998), 25.

49. Nick Engelfried, "The Extraction Backlash—How Fossil Fuel Companies Are Aiding Their Own Demise," *Waging Nonviolence*, November 22, 2013.

50. Mark Dowie, *Losing Ground: American Environmentalism at the Close of the Twentieth Century* (Cambridge, MA: MIT Press, 1996), 125.

51. "Americans, Gulf Residents and the Oil Spill," poll, CBS News/*New York Times*, June 21, 2010; Bruce Alpert, "Obama Administration 'Cannot Support' Bill Increasing Offshore Revenue Sharing," *Times-Picayune*, July 23, 2013; Annie Snider and Nick Juliano, "Will Landrieu's Rise Give New Life to Revenue Sharing?" *E&E Daily*, February 25, 2014; "The Damage for Gulf Coast Residents: Economic, Environmental, Emotional," poll, ABC News/*Washington Post*, July 14, 2010.

52. "Current High Volume Horizontal Hydraulic Fracturing Drilling Bans and Moratoria in NY State," FracTracker.org, http://www.fractracker.org.

53. "Minisink Compressor Project: Environmental Assessment," Federal Energy Regulatory Commission, March 2012; Mary Esch, "NY Town of 9/11 Workers Wages Gas Pipeline Fight," Associated Press, February 14, 2013; "Blow-Down Events at Minisink Compressor Frighten Un-Notified Residents," Stop the Minisink Compressor Station and Minisink Residents for Environmental Preservation and Safety, March 11, 2013, http://www.stopmcs.org.

54. Maxime Combes, "Let's Frackdown the Fracking Companies," in Temper et al., "Towards a Post-Oil Civilization," p. 91, 97.

55. Vince Devlin, "Proposed Big Rigs 9 Feet Longer than Howard Hughes' Spruce Goose," *Missoulian*, November 13, 2010; "747-8: Airplane Characteristics for Airport Planning," Boeing, December 2012, p. 7; "Vertical Clearance," Federal Highway Administration, U.S. Department of Transportation, http://safety.fhwa.dot.gov.

56. Personal interview with Marty Cobenais, October 17, 2010.

57. Betsy Z. Russell, "Judge Halts Megaloads on Highway 12 in Idaho," *Spokesman-Review* (Spokane), September 13, 2013; personal interview with Alexis Bonogofsky, October 21, 2010.

58. Marc Dadigan, "Umatilla Tribe Battles Mega-Loads Headed for Alberta Oil Sands," *Indian Country Today Media Network*, December 11, 2013.

59. Lesley Fleischman et al. "Ripe for Retirement: An Economic Analysis of the U.S. Coal Fleet," *The Electricity Journal* 26 (2013): 51–63; Michael Klare, "Let Them Eat Carbon: Like Big Tobacco, Big Energy Targets the Developing World for Future Profits," TomDispatch, May 27, 2014.

60. KC Golden, "Live on Stage in the Great Northwest: King Coal's Tragic Puppet Show, Part 1," Getting a GRIP on Climate Solutions, March 4, 2013.

61. Michelle Kinman and Antonia Juhasz, ed., "The True Cost of Chevron: An Alternative Annual Report," True Cost of Chevron network, May 2011, pp. 13–14; "Contra Costa County Asthma Profile," California Breathing, http://www.californiabreathing.org; Jeremy Miller, "The Bay Area Chevron Explosion Shows Gaps in Refinery Safety," High Country News, Septem-

ber 3, 2012; Robert Rogers, "Chevron Refinery Fire One Year Later: Fallout, Impact Show No Sign of Warning," *Contra Costa Times*, August 10, 2013.

62. David R. Baker, "Judge Deals Setback to Chevron Refinery Plan," *San Francisco Chronicle*, June 9, 2009; Katherine Tam, "Court Rules Richmond Refinery Plan Is Inadequate," *Contra Costa Times*, April 26, 2010; "Chevron Refinery Expansion at Richmond, CA Halted," EarthJustice, press release, July 2, 2009.

63. Personal interview with Melina Laboucan-Massimo, July 5, 2013.

64. Hannibal Rhoades, "'We Draw the Line': Coal-Impacted Lummi Nation and Northern Cheyenne Unite in Solidarity," *IC Magazine*, October 9, 2013.

65. "Jonathan Chait, "The Keystone Fight Is a Huge Environmentalist Mistake," *New York Magazine*, October 30, 2013; Joe Nocera, "How Not to Fix Climate Change," *New York Times*, February 18, 2013; Joe Nocera, "A Scientist's Misguided Crusade," *New York Times*, March 4, 2013.

66. Jad Mouawad, "U.S. Orders Tests on Rail Shipments," *New York Times*, February 25, 2014; Jad Mouawad, "Trailing Canada, U.S. Starts a Push for Safer Oil Shipping," *New York Times*, April 24, 2014; Curtis Tate, "Regulators Take Voluntary Route on Tank Car Rules," McClatchy Newspapers, May 7, 2014.

67. There is evidence suggesting that dilbit can be more corrosive than other crudes under certain conditions, particularly at high temperatures, but the matter has been contested in recent years. There is also evidence that dilbit may be more likely to cause other kinds of pipeline failure, such as cracking. Anthony Swift, Susan Casey-Lefkowitz, and Elizabeth Shope, "Tar Sands Pipelines Safety Risks," Natural Resources Defense Council, 2011, p. 3.

68. Vivian Luk, "Diluted Bitumen Sinks When Mixed with Sediments, Federal Report Says," *Globe and Mail*, January 14, 2014; "Properties, Composition and Marine Spill Behaviour, Fate and Transport of Two Diluted Bitumen Products from the Canadian Oil Sands," Federal Government Technical Report, Government of Canada, November 30, 2013.

69. FOOTNOTE: Bob Weber, "Syncrude Guilty in Death of 1,600 Ducks in Toxic Tailings Pond," The Canadian Press, June 25, 2010; Syncrude, Suncor Cleared After Duck Death Investigation," CBC News, October 4, 2012; Colleen Cassady St. Clair, Thomas Habib, and Bryon Shore, "Spatial and Temporal Correlates of Mass Bird Mortality in Oil Sands Tailings Ponds," report prepared from Alberta Environment, November 10, 2011, pp. 17–18.

70. Although there is year-to-year fluctuation based on the size of reserves, the value of the oil sands has risen in line with the industry's expansion, from C\$19 billion in 1990 to C\$460 billion in 2010: "Energy," *Canada Year Book 2012*, Statistics Canada, http://www.statcan.gc.ca. Bill Donahue was not an author of the study he commented on: "Oilsands Study Confirms Tailings Found in Groundwater, River," CBC News, February 20, 2014; Richard A. Frank et al., "Profiling Oil Sands Mixtures from Industrial Developments and Natural Groundwaters for Source Identification," *Environmental Science & Technology* 48 (2014): 2660–70. DIFFERENT CASE: Mike De Souza, "Scientists Discouraged from Commenting on Oilsands Contaminant Study," Postmedia News, November 4, 2012.

71. Florence Loyle, "Doctor Cleared over Suggested Link Between Cancer, Oilsands," *Edmonton Journal*, November 7, 2009; Vincent McDermott, "Fort Chipewyan Cancer Study Set to Begin," *Fort McMurray Today*, February 20, 2013; Michael Toledano, "We Interviewed Dr. John O'Connor, One of the First Tar Sands Whistleblowers," *Vice*, March 3, 2014.

72. Peter Moskowitz, "Report Finds Doctors Reluctant to Link Oil Sands with Health Issues," Al Jazeera America, January 20, 2014; Mike De Souza, "Scientist Speaks Out After Finding 'Record' Ozone Hole over Canadian Arctic," Postmedia News, October 21, 2011.

73. Mike De Souza, "Federal Budget Cuts Undermine Environment Canada's Mandate to Enforce Clean Air Regulations: Emails," Postmedia News, March 17, 2013; "Silence of the Labs," *The Fifth Estate*, CBC News, January 10, 2014.

74. FOOTNOTE: Abha Parajulee and Frank Wania, "Evaluating Officially Reported Polycyclic Aromatic Hydrocarbon Emissions in the Athabasca Oil Sands Region with a Multimedia Fate Model," *Proceedings of the National Academy of Sciences* 111 (2014): 3348; "Oil Sands Pollution Two to Three Times Higher than Thought," Agence France-Presse, February 3, 2014.

75. "Regulation of Hydraulic Fracturing Under the Safe Drinking Water Act," Environmental Protection Agency, http://water.epa.gov; Mary Tiemann and Adam Vann, "Hydraulic Fracturing and Safe Drinking Water Act Regulatory Issues," Congressional Research Service, Report R41760, January 10, 2013; Lisa Song, "Secrecy Loophole Could Still Weaken BLM's Tougher Fracking Regs," InsideClimate News, February 15, 2012.

76. Robert B. Jackson et al., "Increased Stray Gas Abundance in a Subset of Drinking Water Wells Near Marcellus Shale Gas Extraction," *Proceedings of the National Academy of Sciences* 110 (2013): 11250-11255; Mark Drajem, "Duke Fracking Tests Reveal Dangers Driller's Data Missed," Bloomberg, January 9, 2014.

77. Cliff Frohlich, "Two-Year Survey Comparing Earthquake Activity and Injection Well Locations in the Barnett Shale, Texas," *Proceedings of the National Academy of Sciences* 109 (2012): 13934–13938.

78. *Ibid.*; Won-Young Kim, "Induced Seismicity Associated with Fluid Injection into a Deep Well in Youngstown, Ohio," *Journal of Geophysical Research: Solid Earth* 118 (2013): 3506–3518; Charles Q. Choi, "Fracking Practice to Blame for Ohio Earthquakes," *LiveScience*, September 4, 2013; Nicholas J. van der Elst et al., "Enhanced Remote Earthquake Triggering at Fluid-Injection Sites in the Midwestern United States," *Science* 341 (2013): 164–167; Sharon Begley, "Distant Seismic Activity Can Trigger Quakes at 'Fracking' Sites," Reuters, July 11, 2013.

79. "Report Regarding the Causes of the April 20, 2010 Macondo Well Blowout," U.S. Department of the Interior, Bureau of Ocean Energy Management, Regulation and Enforcement, September 14, 2011, p. 191; "Deep Water: The Gulf Oil Disaster and the Future of Offshore Drilling," National Commission on the BP Deepwater Horizon Oil Spill and Offshore Drilling, January 2011, p. 125; Joel Achenbach, "BP's Cost Cuts Contributed to Oil Spill Disaster, Federal Probe Finds," *Washington Post*, September 14, 2011.

80. Elizabeth McGowan and Lisa Song, "The Dilbit Disaster: Inside The Biggest Oil Spill You've Never Heard Of, Part 1," InsideClimate News, June 26, 2012.

81. *Ibid.*; Charles Rusnell, "Enbridge Staff Ignored Warnings in Kalamazoo River Spill," CBC News, June 22, 2012; "Oil Cleanup Continues on Kalamazoo River," U.S. Environmental Protection Agency, June 2013.

82. In explaining his previous denials, Daniel appears to have been trying to argue that because the diluted bitumen traveling in Enbridge's pipeline had been extracted with newer "in situ" steam injection technology, rather than mined, it would not qualify as tar sands oil: Todd Heywood, "Enbridge CEO Downplays Long-Term Effects of Spill," *Michigan Messenger*, August 12, 2010. MORE THAN A WEEK: McGowan and Song, "The Dilbit Disaster"; DANIEL: Kari Lyderson, "Michigan Oil Spill Increases Concern over Tar Sands Pipelines," *OnEarth*, August 6, 2010; Kari Lyderson, "Michigan Oil Spill: The Tar Sands Name Game (and Why It Matters)," *OnEarth*, August 12, 2010.

83. Cobenais interview, October 17, 2010.

84. Dan Joling, "Shell Oil-Drilling Ship Runs Aground on Alaska's Sitkalidak Island," Associated Press, January 1, 2013; Rachel D'Oro, "Nobel Discoverer, Shell Oil Drilling Vessel, Shows No Signs of Damage, Coast Guard Claims," Associated Press, July 15, 2012; John Ryan, "Sea Trial Leaves Shell's Arctic Oil-Spill Gear 'Crushed Like a Beer Can,'" KUOW.org, November 30, 2012.

85. Mike Soraghan, "Oil Spills: U.S. Well Sites in 2012 Discharged More Than Valdez," *EnergyWire*, Monday, July 8, 2013; Dan Frosch and Janet Roberts, "Pipeline Spills Put Safeguards Under Scrutiny," *New York Times*, September 9, 2011.

86. Jim Paulin and Carey Restino, "Shell Rig Grounds off Kodiak," *Bristol Bay Times*, January 4, 2013.

87. "SINGLE ENGINEER": Bruce Campbell, "Lac-Mégantic: Time for an Independent Inquiry," *Toronto Star*, February 27, 2014; UNTIL THE 1980s: personal interview with Ron Kaminkow, general secretary, Railroad Workers United, January 29, 2014; "CUTTING": Julian Sher, "Lac Megantic: Railway's History of Cost-Cutting," *Toronto Star*, July 11, 2013; "OFTEN DON'T TEST": Grant Robertson, "Fiery North Dakota Train Derailment Fuels Oil-Shipping Fears," *Globe and Mail*, December 30, 2013; NORTH DAKOTA: Daniella Silva, "Mile-Long Train Car-

rying Crude Oil Derails, Explodes in North Dakota," NBC News, December 30, 2013; NEW BRUNSWICK: Solarina Ho, "Train Carrying Oil Derails, Catches Fire in New Brunswick, Canada," *Reuters*, January 8, 2014; VIRGINIA: Selam Gebrekidan, "CSX Train Carrying Oil Derails in Virginia in Fiery Blast," Reuters, April 30, 2014.

88. Charlie Savage, "Sex, Drug Use and Graft Cited in Interior Department," *New York Times*, September 10, 2008.

89. "Americans Less Likely to Say 18 of 19 Industries Are Honest and Trustworthy This Year," Harris Interactive, December 12, 2013; Jeffrey Jones, "U.S. Images of Banking, Real Estate Making Comeback," Gallup, August 23, 2013; André Turcotte, Michal C. Moore, and Jennifer Winter, "Energy Literacy in Canada," School of Public Policy SPP Research Papers, Vol. 5, No. 31, October 2012; "How Companies Influence Our Society: Citizens' View," TNS Political and Social, European Commission, Flash Eurobarometer 363, April 2013, Q3, p. 25.

90. Sandra Steingraber, "It's Alive! In Defense of Underground Organisms," *Orion Magazine*, January/February 2012, p. 15.

91. Wendell E. Berry, "It All Turns on Affection," Jefferson Lecture in the Humanities, National Endowment for the Humantities, 2012, http://www.neh.gov.

CHAPTER 10: LOVE WILL SAVE THIS PLACE

1. Rachel Carson, "The Real World Around Us," speech to Theta Sigma Phi, Columbus, Ohio, 1954, in *Lost Woods: The Discovered Writing of Rachel Carson*, ed. Linda Lear (Boston: Beacon Press, 1998), 163.

2. Paige Lavender and Corbin Hiar, "Blair Mountain: Protesters March to Save Historic Battlefield," Huffington Post, June 10, 2011.

3. The largest class of tanker that Northern Gateway plans to use in BC waters has a maximum capacity of 2.2 million barrels of oil, about 74 percent more than the 1,264,155 barrels carried by the *Exxon Valdez*: "Section 3.9: Ship Specifications," TERMPOL Surveys and Studies, Northern Gateway Partnership Inc., Enbridge Northern Gateway Project, January 20, 2010, pp. 2–7; "Oil Spill Facts: Questions and Answers," Exxon Valdez Oil Spill Trustee Council, http://www.evostc .state.ak.us.

4. Jess Housty, "Transformations," Coast, April 1, 2013.

5. "Protesters Blamed for Cancelled Pipeline Hearing," CTV News Vancouver, April 2, 2012.

6. Personal email communication with Tyler McCreary, PhD candidate, York University, January 30, 2014.

7. Sheri Young, letter to the Heiltsuk Tribal Council, Heiltsuk Hereditary Chiefs and Heiltsuk Economic Development Corporation on behalf of the Enbridge Northern Gateway Project Joint Review Panel, April 2, 2012; Housty, "Transformations"; Alexis Stoymenoff, "Enbridge Northern Gateway Protest in Bella Bella Was 'Absolutely Peaceful,'" *Vancouver Observer*, April 2, 2012.

8. Housty, "Transformations"; Kai Nagata, "Enbridge Misses Heiltsuk Pipeline Hearings," *The Tyee*, July 27, 2012; FOOTNOTE: *Ibid*.

9. Jess Housty, "At the JRP Final Hearings," Coast, June 20, 2013.

10. Personal interview with Melachrini Liakou, May 31, 2013.

11. Personal interview with Alexis Bonogofsky, March 27, 2013.

12. Andrew Nikiforuk, *Tar Sands: Dirty Oil and the Future of a Continent* (Vancouver: Greystone, 2010), 44.

13. Personal interview with Jeff King, June 23, 2011.

14. Luiza Ilie, "Romanian Farmers Choose Subsistence over Shale Gas," Reuters, October 27, 2013.

15. "Oil Sands Export Ban: BC First Nations Unite to Declare Province-Wide Opposition to Crude Oil Pipeline and Tanker Expansion," Yinka Dene Alliance, press release, December 1, 2011; "First Nations Gain Powerful New Allies in Fight Against Enbridge Northern Gateway Pipeline, Tankers," Yinka Dene Alliance, press release, December 5, 2013; author's original reporting, December 1, 2011.

16. "Read the Declaration," Save the Fraser Declaration, Gathering of Nations, savethefraser.ca.

17. Sheila Leggett, Kenneth Bateman, and Hans Matthews, "Report of the Joint Review Panel for the Enbridge Northern Gateway Project," Volume 2, National Energy Board, 2013, pp. 222, 271.

18. "White House Could Cast Decisive Vote to Permit 20,000 Fracking Wells in Delaware River Basin," Democracy Now!, November 11, 2011; "Natural Gas Development Regulations," Delaware River Basin Commission, November 8, 2011, p. 19.

19. "High Plains Aquifer Water Quality Currently Acceptable but Human Activities Could Limit Future Use," U.S. Department of the Interior U.S. Geological Survey, press release, July 16, 2009; "Ogallala Aquifer Initiative," Natural Resources Conservation Service, U.S. Department of Agriculture, http://www.nrcs.usda.gov.

20. 2.3 BARRELS: "Oil Sands Water Use" (2013 data), Oil Sands Information Portal, Government of Alberta, http://osip.alberta.ca; CONVENTIONAL CRUDE: "Growth in the Canadian Oil Sands: Finding the New Balance," IHS Cambridge Energy Research Associates, 2009, pp. III–7; REQUIRES MORE WATER: Trisha A. Smrecak, "Understanding Drilling Technology," *Marcellus Shale* no. 6, Paleontological Research Institution, January 2012, p. 3; "70 TO 300 TIMES": Seth B. Shonkoff, "Public Health Dimensions of Horizontal Hydraulic Fracturing: Knowledge, Obstacles, Tactics, and Opportunities," 11th Hour Project, Schmidt Family Foundation, April 18, 2012, http://www.psr.org; 280 BILLION: Elizabeth Ridlington and John Rumpler, "Fracking by the Numbers: Key Impacts of Dirty Drilling at the State and National Level," Environment America, October 2013, p. 4. http://www.environmentamerica.org; "ENOUGH TO FLOOD": Suzanne Goldenberg, "Fracking Produces Annual Toxic Water Enough to Flood Washington DC," *Guardian*, October 4, 2013.

21. Monika Freyman, "Hydraulic Fracturing and Water Stress: Water Demand by the Numbers," *Ceres*, February 2014, pp. 49–50, 59–63; David Smith, "Proposed Fracking in South Africa Beauty Spot Blasted," *Guardian*, August 23, 2013; "Hydraulic Fracturing and the Karoo," Shell South Africa, July 2012, http://www.shell.com/zaf.html; "Tampering with the Earth's Breath" (video), Green Renaissance, Vimeo, May 11, 2011.

22. Ilie, "Romanian Farmers Choose Subsistence over Shale Gas."

23. Personal interview with Anni Vassiliou, June 1, 2013.

24. Marion W. Howard, Valeria Pizarro, and June Marie Mow, "Ethnic and Biological Diversity Within the Seaflower Biosphere Reserve," *International Journal of Island Affairs* 13 (2004): 113; "Caribbean Archipelago Spared from Oil Drilling," Rainforest Rescue, June 21, 2012, http://www.rainforest-rescue.org; FOOTNOTE: "Nicaragua Files New Claim Against Colombia over San Andres," BBC, September 16, 2013.

25. "Victories," Beyond Coal, Sierra Club, http://content.sierraclub.org; Mary Anne Hitt, "Protecting Americans from Power Plant Pollution," Sierra Club, September 17, 2013; "Proposed Coal Plant Tracker," Beyond Coal, Sierra Club, http://contentsierraclub.org.

26. James E. Casto, "Spokesmen for Coal Blast EPA Regulatory Mandates," *State Journal* (West Virginia), November 15, 2013.

27. Jeremy van Loon, "Canada's Oil-Sand Fields Need U.S. Workers, Alberta Minister Says," *Bloomberg News*, September 7, 2011; Shawn McCarthy and Richard Blackwell, "Oil Industry Rebuts 'Trash-Talking' Celebrity Critics," *Globe and Mail*, January 15, 2014.

28. T. S. Sudhir, "After Police Firing, Srikakulam Power Plants Under Review," NDTV.com, July 16, 2010.

29. Barbara Demick, "Residents of Another South China Town Protest Development Plans," *Los Angeles Times*, December 21, 2011; Gillian Wong, "Thousands Protest China Town's Planned Coal Plant," Associated Press, December 20, 2011; Gillian Wong, "Tear Gas Fired at Protesters in China Seaside Town," Associated Press, December 24, 2011.

30. Personal interview with Li Bo, January 11, 2014

31. "Beijing's Air Pollution at Dangerously High Levels," Associated Press, January 16, 2014; Ma Yue, "Alarm System to Close Schools in Severe Smog," *Shanghai Daily*, January 16, 2014; "Chinese Anger over Pollution Becomes Main Cause of Social Unrest," Bloomberg, March 6, 2013, accessed January 29, 2014.

32. Bruce Einhorn, "Why China Is Suddenly Content with 7.5 Percent Growth," *Bloomberg Businessweek*, March 5, 2012; "GDP Growth (Annual %)," World Development Indicators, World Bank, http://data.worldbank.org; James T. Areddy and Brian Spegele, "China Chases Renewable Energy as Coast Chokes on Air," *Wall Street Journal*, December 6, 2013; Justin Guay, "The Chinese Coal Bubble," Huffington Post, May 29, 2013; Katie Hunt, "China Faces Steep Climb to Exploit Its Shale Riches," *New York Times*, September 30, 2013.

33. Christian Lelong et al., "The Window for Thermal Coal Investment Is Closing," Goldman Sachs, July 24, 2013; Dave Steves, "Goldman Sachs Bails on Coal Export Terminal Investment," *Portland Tribune*, January 8, 2014.

34. "Shale Gas: Member States Need Robust Rules on Fracking, Say MEPs," European Parliament, press release, November 21, 2012.

35. Andrea Schmidt, "Heirs of Anti-Apartheid Movement Rise Up," Al Jazeera, December 15, 2013.

36. Naomi Klein, "Time for Big Green to Go Fossil Free," *The Nation*, May 1, 2013; "Commitments," Fossil Free, 350.org, http://gofossilfree.org; "Stanford to Divest from Coal Companies," Stanford University, press release, May 6, 2014.

37. "Harvard University Endowment Earns 11.3% Return for Fiscal Year," *Harvard Gazette*, September 24, 2013; Andrea Schmidt, "Heirs of Anti-Apartheid Movement Rise Up," Al Jazeera, December 15, 2013; Mark Brooks, "Banking on Divestment," *Alternatives Journal*, November 2013.

38. Mark Brownstein, "Why EDF Is Working on Natural Gas," Environmental Defense Fund, September 10, 2012.

39. FOOTNOTE: Letter to Fred Krupp from Civil Society Institute, et al., May 22, 2013, http://www.civilsocietyinstitute.org.

40. Ben Casselman, "Sierra Club's Pro-Gas Dilemma," *Wall Street Journal*, December 22, 2009; Bryan Walsh, "How the Sierra Club Took Millions from the Natural Gas Industry—and Why They Stopped," *Time*, February 2, 2012; Dave Michaels, "Natural Gas Industry Seeks Greater Role for Power Plants, Vehicles," *Dallas Morning News*, September 18, 2009; Sandra Steingraber, "Breaking Up with the Sierra Club," *Orion*, March 23, 2012.

41. Felicity Barringer, "Answering for Taking a Driller's Cash," *New York Times*, February 13, 2012; "48 Arrested at Keystone Pipeline Protest as Sierra Club Lifts 120-Year Ban on Civil Disobedience," Democracy Now!, February 14, 2013; personal email communication with Bob Sipchen, communications director, Sierra Club, April 21, 2014.

42. Robert Friedman, "Tell Your Alma Mater, Fossil Fuel Divestment Just Went Mainstream," Natural Resources Defense Council, April 30, 2014; Klein, "Time for Big Green to Go Fossil Free."

43. Andrea Vittorio, "Foundations Launch Campaign to Divest from Fossil Fuels," Bloomberg, January 31, 2014; "Philanthropy," Divest-Invest, http://divestinvest.org.

44. "Global 500," *Fortune*, 2013, http://fortune.com; Stanley Reed, "Shell Profit Rises 15% but Disappoints Investors," *New York Times*, January 31, 2013; Stanley Reed, "Shell Says Quarterly Earnings Will Fall 48%," *New York Times*, January 17, 2014.

45. *Ibid*.

46. "Notice of Arbitration Under the Arbitration Rules of the United Nations Commission on International Trade Law and Chapter Eleven of the North American Free Trade Agreement," Lone Pine Resources Inc., September 6, 2013, pp. 4, 15–18.

47. The General Agreement on Tariffs and Trade (GATT 1947), World Trade Organization, Article XI: 1, http://www.wto.org.

48. Personal interview with Ilana Solomon, August 27, 2013.

49. Sarah Anderson and Manuel Perez-Rocha, "Mining for Profits in International Tribunals: Lessons for the Trans-Pacific Partnership," Institute for Policy Studies, April 2013, p. 1; Lori Wallach, "Brewing Storm over ISDR Clouds: Trans-Pacific Partnership Talks—Part I," Kluwer Arbitration Blog, January 7, 2013.

50. Lindsay Abrams, "The Real Secret to Beating the Koch Brothers: How Our Broken Political System Can Still Be Won," *Salon*, April 29, 2014; personal interview with Marily Papanikolaou, May 29, 2013; Mark Strassman, "Texas Rancher Won't Budge for Keystone Pipeline," CBS

Evening News, February 19, 2013; Kim Murphy, "Texas Judge Deals Setback to Opponents of Keystone XL Pipeline," *Los Angeles Times*, August 23, 2012.

51. FOOTNOTE: Suzanne Goldenberg, "Terror Charges Faced by Oklahoma Fossil Fuel Protesters 'Outrageous,'" *Guardian*, January 10, 2014; Molly Redden, "A Glitter-Covered Banner Got These Protesters Arrested for Staging a Bioterror Hoax," *Mother Jones*, December 17, 2013; personal email communication with Moriah Stephenson, Great Plains Tar Sands Resistance, January 22, 2014; Will Potter, "Two Environmentalists Were Charged with 'Terrorism Hoax' for Too Much Glitter on Their Banner," *Vice*, December 18, 2013.

52. Adam Federman, "We're Being Watched: How Corporations and Law Enforcement Are Spying on Environmentalists," *Earth Island Journal*, Summer 2013; Richard Black, "EDF Fined for Spying on Greenpeace Nuclear Campaign," BBC, November 10, 2011; Matthew Millar, "Canada's Top Spy Watchdog Lobbying for Enbridge Northern Gateway Pipeline," *Vancouver Observer*, January 4, 2014; Jordan Press, "Chuck Strahl Quits Security Intelligence Review Committee," Postmedia News, January 24, 2014.

53. Greg Weston, "Other Spy Watchdogs Have Ties to Oil Business," CBC News, January 10, 2014; Press, "Chuck Strahl Quits Security Intelligence Review Committee."

54. Leggett, Bateman, and Matthews, "Report of the Joint Review Panel for the Enbridge Northern Gateway Project," Volume 2, pp. 209, 384.

55. A more recent poll found that 64 percent of British Columbians were opposed to an increase in tanker traffic, with four times as many respondents "strongly" opposed than "strongly" in favor: "Oil Tanker Traffic in B.C.: The B.C. Outlook Omnibus," Justason Market Intelligence, January 2014, p. 5. COMMUNITY HEARINGS: Larry Pynn, "Environmentalists Pledge Renewed Fight to Stop Northern Gateway Pipeline," *Vancouver Sun*, December 19, 2013; 80 PERCENT: Scott Simpson, "Massive Tankers, Crude Oil and Pristine Waters," *Vancouver Sun*, June 5, 2010; "SYSTEM IS BROKEN": Christopher Walsh, "Northern Gateway Pipeline Approved by National Energy Board," *Edmonton Beacon*, December 19, 2013,

56. Edgardo Lander, "Extractivism and Protest Against It in Latin America," presented at The Question of Power: Alternatives for the Energy Sector in Greece and Its European and Global Context, Athens, Greece, October 2013. George Monbiot, "After Rio, We Know. Governments Have Given Up on the Planet," *Guardian*, June 25, 2012.

57. "Initiative Figures," Transition Network, updated September 2013, https://www.transitionnet work.org; Transition Network, "What Is a Transition Initiative?," http://www.transitionnetwork.org.

58. David Roberts, "Climate-Proofing Cities: Not Something Conservatives Are Going to Be Good At," Grist, January 9, 2013.

59. Jesse McKinley, "Fracking Fight Focuses on a New York Town's Ban," *New York Times*, October 23, 2013.

60. "Panel Fails to Listen to British Columbians," Sierra Club BC, press release, December 19, 2013.

CHAPTER 11: YOU AND WHAT ARMY?

1. Melanie Jae Martin and Jesse Fruhwirth, "Welcome to Blockadia!" *YES!*, January 11, 2013.

2. Mary Harris Jones, *Autobiography of Mother Jones* (Mineola, NY: Dover [1925], 2004), 144.

3. Gurston Dacks, "British Columbia After the *Delgamuukw* Decision: Land Claims and Other Processes," *Canadian Public Policy* 28 (2002): 239–255.

4. "Statement of Claim between Council of the Haida Nation and Guujaaw suing on his own behalf and on behalf of all members of the Haida Nation (plaintiffs) and Her Majesty the Queen in Right of the Province of British Columbia and the Attorney General of Canada (defendants)," Action No. L020662, Vancouver Registry, November 14, 2002, http://www.haidanation.ca; *Haida Nation v. British Columbia (Minister of Forests)* 3 SCR 511 (SCC 2004); "Government Must Consult First Nations on Disputed Land, Top Court Rules," CBC News, November 18, 2004; personal interview with Arthur Manuel, August 25, 2004.

5. Personal email communication with Tyler McCreary, PhD candidate, York University, January 30, 2014.

6. *Delgamuukw v. British Columbia*, [1997], 3 SCR 1010; British Columbia Treaty Commission, "A Lay Person's Guide to Delgamuukw v. British Columbia," November 1999, http://www.bctreaty.net; Chelsea Vowel, "The Often-Ignored Facts About Elsipogtog," *Toronto Star*, November 14, 2013.

7. Melanie G. Wiber and Julia Kennedy, "Impossible Dreams: Reforming Fisheries Management in the Canadian Maritimes After the Marshall Decision," in *Law and Anthropology: International Yearbook for Legal Anthropology*, Vol. 2, ed. René Kuppe and Richard Potz (The Hague: Martinus Nijhoff Publishers, 2001), pp. 282–297; William Wicken, "Treaty of Peace and Friendship 1760," Aboriginal Affairs and Northern Development Canada, https://www.aadnc-aandc.gc.ca; *R. v. Marshall*, 3 SCR 456 (1999); "Supreme Court Decisions: R. v. Marshall," Aboriginal Affairs and Northern Development Canada.

8. "Map of Treaty-Making in Canada," Aboriginal Affairs and Northern Development Canada, https://www.aadnc-aandc.gc.ca; "Alberta Oil Sands," Alberta Geological Survey, last modified June 12, 2013, http://www.ags.gov.ab.ca; "Treaty Texts—Treaty No. 6; Copy of Treaty No. 6 Between Her Majesty the Queen and the Plain and Wood Cree Indians and Other Tribes of Indians at Fort Carlton, Fort Pitt, and Battle River with Adhesions," Aboriginal Affairs and Northern Development Canada, https://www.aadnc-aandc.gc.ca.

9. "Emergency Advisory: Mi'kmaq say, 'We Are Still Here, and SWN Will Not Be Allowed to Frack,'" press release, Halifax Media Co-op, November 3, 2013.

10. Martha Stiegman and Miles Howe, "Summer of Solidarity—A View from the Sacred Fire Encampment in Elsipogtog," (video), Halifax Media Co-op, July 3, 2013.

11. "'Crown Land Belongs to the Government, Not to F*cking Natives,'" APTN, October 17, 2013; Martin Lukacs, "New Brunswick Fracking Protests Are the Frontline of a Democratic Fight," *Guardian*, October 21, 2013; Renee Lewis, "Shale Gas Company Loses Bid to Halt Canada Protests," Al Jazeera America, October 21, 2013.

12. "FORUMe Research Results," PowerPoint, MQO Research, presented at FORUMe conference, New Brunswick, June 2012, http://www.amiando.com; Kevin Bissett, "Alward Facing Opposition from N.B. Citizens over Fracking," The Canadian Press, August 30, 2011.

13. Stiegman and Howe, "Summer of Solidarity."

14. Richard Walker, "In Washington, Demolishing Two Dams So That the Salmon May Go Home," *Indian Country Today*, September 22, 2011; "Press Release 02/26/2014," Shield the People, press release, February 26, 2014; "Keystone XL Pipeline Project Compliance Follow-up Review: The Department of State's Choice of Environmental Resources Management, Inc., To Assist in Preparing the Supplemental Environmental Impact Statement," United States Department of State and the Broadcasting Board of Governors, February 2014; Jorge Barrera, "Keystone XL 'Black Snake' Pipeline to Face 'Epic' Opposition from Native American Alliance," APTN, January 31, 2014.

15. Steve Quinn, "U.S. Appeals Court Throws Arctic Drilling into Further Doubt," Reuters, January 23, 2014; *Native Village of Point Hope v. Jewell*, 44 ELR 20016, No. 12-35287 (9th Cir., 01/22/2014); "Native and Conservation Groups Voice Opposition to Lease Sale 193 in the Chukchi Sea;" World Wildlife Fund, press release, February 6, 2008; Faith Gemmill, "Shell Cancels 2014 Arctic Drilling—Arctic Ocean and Inupiat Rights Reality Check," Platform, January 30, 2014.

16. *Native Village of Point Hope v. Jewell.*

17. Terry Macalister, "Shell's Arctic Drilling Set Back by US Court Ruling," *Guardian*, January 23, 2014; "New Shell CEO Ben van Beurden Sets Agenda for Sharper Performance and Rigorous Capital Discipline," Shell, press release, January 30, 2014.

18. Erin Parke, "Gas Hub Future Unclear After Native Title Dispute," ABC (Australia), February 7, 2013; "Environmentalists Welcome Scrapping of LNG Project," ABC (Australia), April 12, 2013; Andrew Burrell, "Gas Fracking Wars to Open Up on a New Front," *Australian*, December 30, 2013; "Native Title Challenge to Canning Gas Bill," Australian Associated Press, June 20, 2013; Vicky Validakis, "Native Title Claimants Want to Ban Mining," *Australian Mining*, May 14, 2013.

19. "Ecuador: Inter-American Court Ruling Marks Key Victory for Indigenous People," Amnesty International, press release, July 27, 2012.

20. ORIGINAL VOTE: United Nations News Centre, "United Nations Adopts Declaration on Rights of Indigenous Peoples," United Nations press release, September 13, 2007; LATER EN-DORSEMENTS: "Indigenous Rights Declaration Endorsed by States," Office of the United Nations High Commissioner for Human Rights, press release, December 23, 2010; "HAVE THE RIGHT", "REDRESS": *United Nations Declaration on the Rights of Indigenous Peoples*, G.A. Res. 61/295, U.N. Doc. A/Res/61/295 September 13, 2007, pp. 10–11, http://www.un.org; CONSTI-TUTION (ORIGINAL SPANISH): República del Bolivia, Constitución de 2009, Capítulo IV: Derechos de las Naciones y Pueblas Indígena Originario Campesinos, art. 30, sec. 2; CONSTI-TUTION (ENGLISH TRANSLATION): Leah Temper et al., "Towards a Post-Oil Civilization: Yasunization and Other Initiatives to Leave Fossil Fuels in the Soil," EJOLT Report No. 6, May 2013, p. 71.

21. Alexandra Valencia, "Ecuador Congress Approves Yasuni Basin Oil Drilling in Amazon," Reuters, October 3, 2013; Amnesty International, "Annual Report 2013: Bolivia," May 23, 2013, http://www.amnesty.org.

22. John Otis, "Chevron vs. Ecuadorean Activists," *Global Post*, May 3, 2009.

23. "Beaver Lake Cree Sue over Oil and Gas Dev't," *Edmonton Journal*, May 14, 2008; "Beaver Lake Cree Nation Draws a Line in the (Oil) Sand," Beaver Lake Cree Nation, press release, May 14, 2008.

24. *Ibid.*; Court of the Queen's Bench, Government of Alberta, 2012 ABQB 195, Memorandum of Decision of the Honourable Madam Justice B. A. Browne, March 28, 2012.

25. Bob Weber, "Athabasca Chipewyan File Lawsuit Against Shell's Jackpine Oil Sands Expansion," The Canadian Press, January 16, 2014; Chief Allan Adam, "Why I'm on Tour with Neil Young and Diana Krall," Huffington Post Canada, January 14, 2014; "Administration and Finance," Athabasca Chipewyan First Nation, http://www.acfn.com; "Shell at a Glance," Shell Global, http://www.shell.com/global.

26. Emma Gilchrist, "Countdown Is On: British Columbians Anxiously Await Enbridge Recom-mendation," DesmogCanada, December 17, 2013; personal interview with Mike Scott, Octo-ber 21, 2010.

27. Benjamin Shingler, "Fracking Protest Leads to Bigger Debate over Indigenous Rights in Canada," Al Jazeera America, December 10, 2013.

28. OMNIBUS BILLS: Bill C-38, Jobs, Growth and Long-Term Prosperity Act, 41st Parliament, 2012, S.C. 2012, c. 19, http://laws-lois.justice.gc.ca; Bill C-45, Jobs and Growth Act 2012, 41st Parliament, 2012, S.C. 2012, c. 31, http://laws-lois.justice.gc.ca; REVIEWS: Tonda MacCharles, "Tories Have Cancelled Almost 600 Environmental Assessments in Ontario," *Toronto Star*, August 29, 2012; COMMUNITY INPUT: Andrea Janus, "Activists Sue Feds over Rules That 'Block' Canadians from Taking Part in Hearings," *CTV News*, August 15, 2013; ACT: Navigable Waters Protection Act, Revised Statutes of Canada 1985, c. N-22, http://laws-lois.justice.gc.ca; FROM PRACTICALLY 100 PERCENT: "Omnibus Bill Changes Anger Water Keepers," CBC News, October 19, 2012; TO LESS THAN 1 PERCENT: "Legal Backgrounder: Bill C-45 and the Navigable Waters Protection Act" (RSC 1985, C N-22), EcoJustice, October 2012; "Hun-dreds of N.S. Waterways Taken off Protected List; Nova Scotia First Nation Joins Idle No More Protest," CBC News, December 27, 2012; PIPELINES: See amendments 349(5) and 349(9) of Bill C-45, Jobs and Growth Act 2012, 41st Parliament, 2012, S.C. 2012, c. 31; DOCUMENTS REVEALED: Heather Scoffield, "Documents Reveal Pipeline Industry Drove Changes to 'Navi-gable Waters' Act," The Canadian Press, February 20, 2013.

29. "Electoral Results by Party: 41st General Election (2011.05.02)," Parliament of Canada, http://www.parl.gc.ca; Ian Austen, "Conservatives in Canada Expand Party's Hold," *New York Times*, May 2, 2011.

30. Julie Gordon and Allison Martell, "Canada Aboriginal Movement Poses New Threat to Miners," Reuters, March 17, 2013.

31. Martin Lukacs, "Indigenous Rights Are the Best Defence Against Canada's Resource Rush," *Guardian*, April 26, 2013.

32. "Neil Young at National Farmers Union Press Conference" (video), YouTube, Thrasher Wheat, September 9, 2013; Jian Ghomeshi, "Q exclusive: Neil Young Says 'Canada Trading Integrity for Money'"(video), CBC News, January 13, 2014.

33. Personal interview with Eriel Deranger, communications manager, Athabasca Chipewyan First Nation January 30, 2014; "Poll: How Do You Feel About Neil Young Attacking the Oilsands?" *Edmonton Journal*, January 12, 2014.

34. Ghomeshi, "Q exclusive: Neil Young Says 'Canada Trading Integrity for Money'"; Adam, "Why I'm on Tour with Neil Young and Diana Krall."

35. "National Assessment of First Nations Water and Wastewater Systems," prepared by Neegan Burnside for Department of Indian and Northern Affairs, Canada, April 2011, 16, http://www.aadnc-aandc.gc.ca.

36. In 2012, Greenland's subsidy from Denmark was about 3.6 billion Danish kroner, equal to 31 percent of its GDP that year. The subsidy was also about 3.6 billion Danish kroner in 2013: "Greenland in Figures: 2014," Statistics Greenland, 2014, pp. 7–8; Jan. M. Olsen, "No Economic Independence for Greenland in Sight," Associated Press, January 24, 2014; "OUR INDEPENDENCE": McKenzie Funk, *Windfall: The Booming Business of Global Warming* (New York: Penguin, 2014), 78.

37. Angela Sterritt, "Industry and Aboriginal Leaders Examine Benefits of the Oilsands," CBC News, January 24, 2014.

38. Personal interview with Phillip Whiteman Jr., October 21, 2010.

CHAPTER 12: SHARING THE SKY

1. Leah Temper, "Sarayaku Wins Case in the Inter-American Court of Human Rights but the Struggle for Prior Consent Continues," EJOLT, August 21, 2012.

2. Sivan Kartha, Tom Athanasiou, and Paul Baer, "The North-South Divide, Equity and Development—The Need for Trust-Building for Emergency Mobilisation," *Development Dialogue* no. 61, September 2012, p. 62.

3. According to the U.S. Geological Survey, there are 162 billion short tons of technically recoverable coal in the Powder River Basin. Using the 2012 total coal consumption figure from the U.S. Energy Information Administration of 889 million short tons, this resource could last approximately 182 years: David C. Scott and James A. Luppens, "Assessment of Coal Geology, Resources, and Reserve Base in the Powder River Basin, Wyoming and Montana," U.S. Geological Survey, February 26, 2013; "International Energy Statistics," U.S. Energy Information Administration, U.S. Department of Energy, http://www.eia.gov.

4. "Many Stars CTL," Beyond Coal, Sierra Club, http://content.sierraclub.org; Homepage, Many Stars Project, http://www.manystarsctl.com/index.html.

5. Personal interview with Mike Scott, October 21, 2010; personal interview with Alexis Bonogofsky, October 21, 2010.

6. "2013 American Indian Population and Labor Force Report," U.S. Department of the Interior, Office of the Secretary, Office of the Assistant Secretary—Indian Affairs, January 2014, p. 47; "Cheyenne Warriors," *Day One*, ABC News, July 6, 1995.

7. Personal interview with Charlene Alden, October 22, 2010.

8. Personal interview with Henry Red Cloud, June 22, 2011.

9. Andreas Malm, "The Origins of Fossil Capital: From Water to Steam in the British Cotton Industry," *Historical Materialism* 21 (2013): 45.

10. Personal interview with Larry Bell, July 1, 2011.

11. Carolyn Merchant, "Environmentalism: From the Control of Nature to Partnership," Bernard Moses Lecture, University of California, Berkeley, May 2010.

12. Personal interview with Landon Means, June 24, 2011; personal interview with Jeff King, June 23, 2011.

13. Personal interview with Henry Red Cloud, June 22, 2011; personal interview with Alexis Bonogofsky, June 22, 2011.

14. Matthew Brown, "Wildfires Ravage Remote Montana Indian Reservation," Associated Press, August 31, 2012; personal interview with Vanessa Braided Hair, March 27, 2013.

15. Personal interview with Henry Red Cloud, June 24, 2011.

16. Author's original reporting, March 21, 2013; audio recording, courtesy of Alexis Bonogofsky, January 17, 2013.

17. "Our Work," Black Mesa Water Coalition, http://www.blackmesawatercoalition.org; "Black Mesa Water Coalition" (video), Black Mesa Peeps, YouTube, December 19, 2011.

18. Marc Lee, *Enbridge Pipe Dreams and Nightmares: The Economic Costs and Benefits of the Proposed Northern Gateway Pipeline*, Vancouver, BC: Canadian Centre for Policy Alternatives, March 2012, 4-7

19. *Ibid.*, p. 6.

20. Dan Apfel, "Why Investors Must Do More Than Divest from Fossil Fuels," *The Nation*, June 17, 2013

21. Diane Cardwell, "Foundations Band Together to Get Rid of Fossil-Fuel Investments," *New York Times*, January 29, 2014; Brendan Smith, Jeremy Brecher, and Kristen Sheeran, "Where Should the Divestors Invest?" Common Dreams, May 17, 2014.

22. *Ibid.*

23. Melanie Wilkinson, "Pipeline Fighters Dedicate Structure on Route," *York News-Times* (Nebraska), September 24, 2013.

24. "Our Mission," REPOWERBalcombe, http://www.repowerbalcombe.com.

25. Personal interview with Bill McKibben, November 5, 2011.

26. Personal email communication with John Jordan, January 13, 2011.

27. Patrick Quinn, "After Devastating Tornado, Town is Reborn 'Green,'" *USA Today*, April 23, 2013.

28. *Ibid.*

29. Scott Wallace, "Rain Forest for Sale," *National Geographic*, January 2013; Kevin Gallagher, "Pay to Keep Oil in the Ground," *The Guardian*, August 7, 2009.

30. Esperanza Martinez, "The Yasuní—ITT initiative from a Political Economy and Political Ecology perspective," in Leah Temper, et al., "Towards a Post-Oil Civilization: Yasunization and Other Initiatives to Leave Fossil Fuels in the Soil," EJOLT Report No. 6, May 2013, pp. 11, 27.

31. Angélica Navarro Llanos, "Climate Debt: The Basis of a Fair and Effective Solution to Climate Change," presentation to Technical Briefing on Historical Responsibility, Ad Hoc Working Group on Long-term Cooperative Action, United Nations Framework Convention on Climate Change, Bonn, Germany, June 4, 2009.

32. Susan Solomon et al., "Persistence of Climate Changes Due to a Range of Greenhouse Gases," *Proceedings of the National Academy of Sciences* 107 43 (2010): 18355.

33. "Kyoto Protocol," Kyoto Protocol, United Nations Framework Convention on Climate Change, http://unfccc.int.

34. Matthew Stilwell, "Climate Debt—A Primer," *Development Dialogue* no. 61, September 2012, p. 42; Global Carbon Project emissions data, 2013 Budget v2.4 (July 2014), available at http://cdiac.ornl.gov.

35. *Ibid.*; "Global Status of Modern Energy Access," International Energy Agency, World Energy Outlook 2012; Barbara Freese, *Coal: A Human History* (New York: Penguin, 2004), 64.

36. "Status of Ratification of the Convention," UNFCCC, http://unfccc.int; "Article 3: Principles," Full Text of the Convention, United Nations Framework Convention on Climate Change, http://unfccc.int; Kyoto Protocol, United Nations Framework Convention on Climate Change, http://unfccc.int.

37. Martínez in Temper et al., "Towards a Post-Oil Civilization," p. 32; Jonathan Watts, "Ecuador Approves Yasuni National Park Oil Drilling in Amazon Rainforest," *Guardian*, August 16, 2013.

38. Mercedes Alvaro, "Coalition to Halt Ecuador Oil-Block Development to Appeal Invalidation

of Signatures," *Wall Street Journal*, May 9, 2014; Kevin M. Koenig, "Ecuador Breaks Its Amazon Deal," *New York Times*, June 11, 2014.

39. James M. Taylor, "Cancun Climate Talks Fizzle, but U.S. Agrees to Expensive New Program," *Heartlander Magazine*, The Heartland Institute, January 3, 2011.

40. Personal interview with Alice Bows-Larkin, January 14, 2013; David Remnick, "Going the Distance: On and off the Road with Barack Obama," *The New Yorker*, January 27, 2014.

41. Sustainable Buildings and Climate Initiative, *Buildings and Climate Change: Summary for Decision Makers*, United Nations Environment Programme, 2009, http://www.unep.org; "Global Building Stock Will Expand 25 Percent by 2012, Driven by Growth in Asia Pacific, Forecasts Pike Research," *BusinessWire*, December 28, 2012; "Retail and Multi-Unit Residential Segments to Drive Global Building Space Growth through 2020," Navigant Research, press release, September 19, 2011, http://www.navigantresearch.com.

42. "Climate Change Leadership—Politics and Culture," CSD Uppsala, http://www.csduppsala.uu .se; Tariq Banuri and Niclas Hällström, "A Global Programme to Tackle Energy Access and Climate Change," *Development Dialogue* no. 61, September 2012, p. 275.

43. "'The Most Obdurate Bully in the Room': U.S. Widely Criticized for Role at Climate Talks," Democracy Now!, December 7, 2012.

44. Personal interview with Sunita Narain, director general, Centre for Science and Environment, May 6, 2013.

45. Nicole Itano, "No Unity at Racism Conference," *Christian Science Monitor*, September 7, 2001; Declaration of the World Conference Against Racism, Racial Discrimination, Xenophobia and Related Intolerance, http://www.un.org/WCAR/durban.pdf; Ben Fox, "Caribbean Nations Seeking Compensation for Slavery," Associated Press, July 25, 2013; "Statement by the Honorable Baldwin Spencer, Prime Minister of Antigua and Barbuda to 34th Regular Meeting of the Conference of Heads of Government of the Caribbean Community, July 2013—On the Issue of Reparations for Native Genocide and Slavery," Caribbean Community Secretariat, press release, July 6, 2013.

46. Ta-Nehisi Coates, "The Case for Reparations," *The Atlantic*, May 21, 2014.

47. Eric Williams, *Capitalism and Slavery* (Chapel Hill: University of North Carolina Press, [1944] 1994); "Legacies of British Slave-ownership," University College London, http://www.ucl.ac.uk

48. Sanchez Manning, "Britain's Colonial Shame: Slave-owners Given Huge Payouts After Abolition," *Independent*, February 24, 2013; "Legacies of British Slave-ownership," University College London.

49. Paul Baer, Tom Athanasiou, Sivan Kartha, and Eric Kemp-Benedict, "The Greenhouse Development Rights Framework: The Right to Development in a Climate Constrained World," revised 2nd edition, Heinrich Böll Foundation, Christian Aid, EcoEquity, and the Stockholm Environment Institute, 2008; Kartha, Athanasiou, and Paul Baer, "The North-South Divide, Equity and Development," p. 54.

50. For more information about Greenhouse Development Rights, and to explore what the framework could look like in practice, refer to the interactive equity calculators and other information available at the GDRs website: http://gdrights.org. 30 PERCENT AND CARBON TRADING: Kartha, Athanasiou, and Baer, "The North-South Divide, Equity and Development," pp. 59–60, 64; personal interview with Sivan Kartha, January 11, 2013.

CHAPTER 13: THE RIGHT TO REGENERATE

1. Personal interview with Tracie Washington, May 26, 2010.

2. Katsi Cook, "Woman Is the First Environment," speech, Live Earth, National Museum of the American Indian, Washington, D.C., July 7, 2007, http://nmai.si.edu.

3. "Global In Vitro Fertilization Market to Reach $21.6 Billion by 2020," Allied Market Research, press release, January 29, 2014; F. E. van Leeuwen et al., "Risk of Borderline and Invasive Ovarian Tumours After Ovarian Stimulation for in Vitro Fertilization in a Large Dutch Cohort," *Human Reproduction* 26 (2011): 3456–3465; L. Lerner-Geva et al., "Infertility, Ovulation Induc-

tion Treatments and the Incidence of Breast Cancer—A Historical Prospective Cohort of Israeli Women," *Breast Cancer Research Treatment* 100 (2006): 201–212; Peter Henriksson et al., "Incidence of Pulmonary and Venous Thromboembolism in Pregnancies After In Vitro Fertilisation: Cross Sectional Study," *BMJ* 346 (2013): e8632.

4. Personal interview with Jonathan Henderson, May 25, 2010.
5. Cain Burdeau and Seth Borenstein, "6 Months After Oil Spill, Scientists Say Gulf Is Sick but Not Dying," Associated Press, October 18, 2010.
6. Doug O'Harra, "Cordova on the Brink," *Anchorage Daily News*, May 1, 1994.
7. Sandra Steingraber, *Raising Elijah: Protecting Our Children in an Age of Environmental Crisis* (Philadelphia: Da Capo, 2011), 28; Sandra Steingraber, *Having Faith: An Ecologist's Journey to Motherhood* (Cambridge, MA: Perseus, 2001), 88.
8. Lisa M. McKenzie et al. "Birth Outcomes and Maternal Residential Proximity to Natural Gas Development in Rural Colorado," *Environmental Health Perspectives* 122 (2014): 412–417.
9. Mark Whitehouse, "Study Shows Fracking Is Bad for Babies," *Bloomberg View*, January 4, 2014.
10. Constanze A. Mackenzie, Ada Lockridge, and Margaret Keith, "Declining Sex Ratio in a First Nation Community," *Environmental Health Perspectives* 113 (2005): 1295–1298; Melody Petersen, "The Lost Boys of Aamjiwnaang," *Men's Health*, November 5, 2009; Nil Basu et al., "Biomarkers of Chemical Exposure at Aamjiwnaang," McGill Environmental Health Sciences Lab Occasional Report, 2013.
11. Personal email communication with Monique Harden, codirector, Advocates for Environmental and Human Rights, February 13, 2012; personal interview with Wilma Subra, chemist and environmental consultant, January 26, 2012; David S. Martin, "Toxic Towns: People of Mossville 'Are Like an Experiment,'" CNN, February 26, 2010.
12. Living on Earth, "Human Rights in Cancer Alley," April 23, 2010, http://www.loe.org; personal email communications with Monique Harden, February 13 and 15, 2012.
13. Personal interview with Debra Ramirez, May 27, 2010; Martin, "Toxic Towns"; Subra interview, January 26, 2012.
14. "Initial Exploration Plan, Mississippi Canyon Block 252," BP Exploration & Production Inc., p. 14-3.
15. Personal interview with Donny Waters, February 3, 2012.
16. Monica Hernandez, "Fishermen Angry as BP Pushes to End Payments for Future Losses," WWLTV, July 8, 2011; personal interviews with Fred Everhardt, crabber and former St. Bernard Parish Councilman, February 22, 2012, and March 7, 2014; personal interviews with George Barisich, president, United Commercial Fisherman's Association, February 22, 2012, and March 10, 2014.
17. "Scientists Find Higher Concentrations of Heavy Metals in Post–Oil Spill Oysters from Gulf of Mexico," California Academy of Sciences, press release, April 18, 2012; "Gulf of Mexico Clean-Up Makes 2010 Spill 52-Times More Toxic," Georgia Institute of Technology, press release, November 30, 2012; Roberto Rico-Martínez, Terry W. Snell, and Tonya L. Shearer, "Synergistic Toxicity of Macondo Crude Oil and Dispersant Corexit 9500A(R) to the Brachionus Plicatilis Species Complex (Rotifera)," *Environmental Pollution* 173 (2013): 5–10.
18. Personal interview with Andrew Whitehead, February 1, 2012; Andrew Whitehead et al., "Genomic and Physiological Footprint of the *Deepwater Horizon* Oil Spill on Resident Marsh Fishes," *Proceedings of the National Academy of Sciences* 109 (2012): 20298–20302; Benjamin Dubansky et al., "Multitissue Molecular, Genomic, and Developmental Effects of the Deepwater Horizon Oil Spill on Resident Gulf Killifish (*Fundulus grandis*)," *Environmental Science & Technology* 47 (2013): 5074–5082.
19. "2010–2014 Cetacean Unusual Mortality Event in Northern Gulf of Mexico," Office of Protected Resources, NOAA Fisheries, National Oceanic and Atmospheric Administration, http://www.nmfs.noaa.gov; Rob Williams et al., "Underestimating the Damage: Interpreting Cetacean Carcass Recoveries in the Context of the *Deepwater Horizon*/BP Incident," *Conservation Letters* 4 (2011): 228.

20. Harlan Kirgan, "Dead Dolphin Calves Found in Mississippi, Alabama," *Mobile Press-Register*, February 24, 2011; FOOTNOTE: "2010–2014 Cetacean Unusual Mortality Event in Northern Gulf of Mexico," Office of Protected Resources, NOAA Fisheries, National Oceanic and Atmospheric Administration, http://www.nmfs.noaa.gov.

21. Lori H. Schwacke et al., "Health of Common Bottlenose Dolphins (*Tursiops truncatus*) in Barataria Bay, Louisiana, Following the *Deepwater Horizon* Oil Spill," *Environmental Science & Technology* 48 (2014): 93–103; "Scientists Report Some Gulf Dolphins Are Gravely Ill," NOAA Fisheries, National Oceanic and Atmospheric Administration, press release, December 18, 2013.

22. "Dolphin Deaths Related to Cold Water in Gulf of Mexico, Study Says," Associated Press, July 19, 2012.

23. Moises Velasquez-Manoff, "Climate Turns Up Heat on Sea Turtles," *Christian Science Monitor*, June 21, 2007; A. P. Negri, P. A. Marshall, and A. J. Heyward, "Differing Effects of Thermal Stress on Coral Fertilization and Early Embryogenesis in Four Indo Pacific Species," *Coral Reefs* 26 (2007): 761; Andrew C. Baker, Peter W. Glynn, and Bernhard Riegl, "Climate Change and Coral Reef Bleaching: An Ecological Assessment of Long-Term Impacts, Recovery Trends and Future Outlook," *Estuarine, Coastal and Shelf Science* 80 (2008): 435–471.

24. The accelerated acidification is likely due to a combination of human emissions leading to more carbon being absorbed as well as natural upwelling of deeper, corrosive water. "MUCH MORE SENSITIVE": Personal interview with Richard Feely, November 20, 2012; SCALLOP DIE-OFF: Mark Hume, "Mystery Surrounds Massive Die-Off of Oysters and Scallops off B.C. Coast," *Globe and Mail*, February 27, 2014.

25. CARIBOU CALVES: Eric Post and Mads C. Forchhammer, "Climate Change Reduces Reproductive Success of an Arctic Herbivore Through Trophic Mismatch," *Philosophical Transactions of the Royal Society* B 363 (2008): 2369–2372; PIED FLYCATCHER: Christiaan Both, "Food Availability, Mistiming, and Climatic Change," in *Effects of Climate Change on Birds*, ed. Anders Pape Moller et al. (Oxford: Oxford University Press, 2010), 129–131; Christiaan Both et al., "Climate Change and Population Declines in a Long-Distance Migratory Birds," *Nature* 441 (2006): 81–82; ARCTIC TERN: Darryl Fears, "Biologists Worried by Migratory Birds' Starvation, Seen as Tied to Climate Change," *Washington Post*, June 19, 2013; DENS COLLAPSING, DANGEROUSLY EXPOSED: Ed Struzik, "Trouble in the Lair," Postmedia News, June 25, 2012; personal interview with Steven Amstrup, chief scientist, Polar Bears Interrnational, January 7, 2013.

26. "Arctic Rain Threatens Baby Peregrine Falcons," CBC News, December 4, 2013; Dan Joling, "Low-Profile Ring Seals Are Warming Victims," Associated Press, March 5, 2007; Jon Aars, "Variation in Detection Probability of Polar Bear Maternity Dens," *Polar Biology* 36 (2013): 1089–1096.

27. Schwake et al., "Health of Common Bottlenose Dolphins (*Tursiops truncatus*) in Barataria Bay, Louisiana, Following the *Deepwater Horizon* Oil Spill"; L. Lauria, "Reproductive Disorders and Pregnancy Outcomes Among Female Flight Attendants,"*Aviation, Space and Environmental Medicine* 77 (2006): 533–539.

28. FOOTNOTE: C.D. Lynch, et. al., "Preconception Stress Increases the Risk of Infertility: Results from a Couple-based Prospective Cohort Study—The LIFE Study," *Human Reproduction* 29 (May 2014), 1067–1075.

29. Wes Jackson, "We Can Now Solve the 10,000-Year-Old Problem of Agriculture," in Allan Eaglesham, Ken Korth, and Ralph W. F. Hardy, eds., *NABC Report 24: Water Sustainability in Agriculture*, National Agricultural Biotechnology Council, 2012, p. 41.

30. Wendell Berry, "It All Turns on Affection," Jefferson Lecture in the Humanities, Washington, D.C., April 23, 2012, http://www.neh.gov.

31. Tyrone B. Hayes et al., "Demasculinization and Feminization of Male Gonads by Atrazine: Consistent Effects Across Vertebrate Classes," *Journal of Steroid Biochemistry and Molecular Biology* 127 (2011): 65, 67; Karla Gale, "Weed Killer Atrazine May Be Linked to Birth Defect," Reuters, February 8, 2010; Kelly D. Mattix, Paul D. Winchester, and L. R. "Tres" Scherer, "Incidence of Abdominal Wall Defects Is Related to Surface Water Atrazine and Nitrate Levels," *Journal of*

Pediatric Surgery 42 (2007): 947–949; Tye E. Arbuckle et al., "An Exploratory Analysis of the Effect of Pesticide Exposure on the Risk of Spontaneous Abortion in an Ontario Farm Population," *Environmental Health Perspectives* 109 (2001): 851–857; Rachel Aviv, "A Valuable Reputation," *The New Yorker*, February 10, 2014.

32. Charles C. Mann, *1491: New Revelations of the Americas Before Columbus* (New York: Vintage, 2006), 226.

33. "Transforming Agriculture with Perennial Polycultures," The Land Institute, http://landinstitute .org.

34. Blair Fannin, "Updated 2011 Texas Agricultural Drought Losses Total $7.62 Billion," *Agrilife Today*, March 21, 2012.

35. James A. Lichatowich, *Salmon Without Rivers: A History of the Pacific Salmon Crisis* (Washington, D.C.: Island Press, 2001), 54.

36. "Leanne Simpson Speaking at Beit Zatoun Jan 23rd 2012" (video), YouTube, Dreadedstar's Channel, January 25, 2012.

37. Personal interview with Leanne Simpson, February 22, 2013.

38. John Vidal, "Bolivia Enshrines Natural World's Rights with Equal Status for Mother Earth," *Guardian*, April 10, 2011; Clare Kendall, "A New Law of Nature," *Guardian*, September 23, 2008; FOOTNOTE: Edgardo Lander, "Extractivism and Protest Against It in Latin America," presented at the Question of Power: Alternatives for the Energy Sector in Greece and Its European and Global Context, Athens, October 2013; República del Ecuador, Constitución de la República del Ecuador de 2008, Capítulo Séptimo: Derechos de la Naturaleza, art. 71; "Peoples Agreement of Cochabamba," World People's Conference on Climate Change and the Rights of Mother Earth, April 24, 2010, http://pwccc.wordpress.com.

39. Fiona Harvey, "Vivienne Westwood Backs Ecocide Law," *Guardian*, January 16, 2014; "FAQ Ecocide," End Ecocide in Europe, April 16, 2013, https://www.endecocide.eu.

40. Personal interview with Mike Scott, March 23, 2013.

41. Wes Jackson, *Consulting the Genius of the Place: An Ecological Approach to a New Agriculture* (Berkeley: Counterpoint, 2010).

42. Personal email communication with Gopal Dayaneni, March 6, 2014.

CONCLUSION: THE LEAP YEARS

1. Martin Luther King Jr., "Beyond Vietnam," speech, New York, April 4, 1967, Martin Luther King Jr. Research and Education Institute, Stanford University, http://mlk-kpp01.stanford.edu.

2. Marlene Moses, Statement on Behalf of Pacific Small Island Developing States, presented at Youth Delegates Demand Climate Justice, side event for United Nations Youth Delegates, New York October 13, 2009.

3. Personal email communication with Brad Werner, December 22, 2012.

4. "The Future of Human-Landscape Systems II" (video), American Geophysical Union (AGU), YouTube, December 5, 2012; personal interview with Brad Werner, October 2, 2013; Dave Levitan, "After Extensive Mathematical Modeling, Scientist Declares 'Earth Is F**ked,'" io9, December 7, 2012.

5. "The Future of Human-Landscape Systems II" (video), YouTube; personal email communication with Brad Werner, December 22, 2012; personal interviews with Brad Werner, February 15 and October 2, 2013.

6. "The Future of Human-Landscape Systems II" (video), YouTube.

7. John Fullerton, "The Big Choice," Capital Institute, July 19, 2011.

8. Martin Luther King Jr., *Where Do We Go from Here: Chaos or Community?* (Boston: Beacon, [1967] 2010), 5–6.

9. Johannes G. Hoogeveen and Berk Özler, "Not Separate, Not Equal: Poverty and Inequality in Post-Apartheid South Africa," Working Paper No. 739, William Davidson Institute, University of Michigan Business School, January 2005.

10. For work exploring the multi-faceted parallels between climate change, slavery, and abolitionism more broadly, see: Jean-François Mouhot, "Past Connections and Present Similarities in Slave

Ownership and Fossil Fuel Usage," *Climatic Change* 105 (2011): 329–355; Jean-François Mouhot, *Des Esclaves Énergétiques: Réflexions sur le Changement Climatique* (Seyssel: Champ Vallon, 2011); Andrew Nikiforuk, *The Energy of Slaves* (Vancouver: Greystone Books, 2012); HAYES: Christopher Hayes, "The New Abolitionism," *The Nation*, April 22, 2014.

11. Greg Grandin, "The Bleached Bones of the Dead: What the Modern World Owes Slavery (It's More Than Back Wages)," TomDispatch, February 23, 2014; Adam Hochschild, *Bury the Chains: Prophets and Rebels in the Fight to Free an Empire's Slaves* (New York: Houghton Mifflin, 2006), 13–14, 54–55.

12. Christopher Hayes, "The New Abolitionism," *The Nation*, April 22, 2014; FOOTNOTE: Seth Rockman and Sven Beckert, eds., *Slavery's Capitalism: A New History of American Economic Development* (Philadelphia: University of Pennsylvania Press, forthcoming); Sven Beckert and Seth Rockman, "Partners in Iniquity," *New York Times*, April 2, 2011; Julia Ott, "Slaves: The Capital That Made Capitalism," Public Seminar, April 9, 2014; Edward E. Baptist and Louis Hyman, "American Finance Grew on the Back of Slaves," *Chicago Sun-Times*, March 7, 2014; Katie Johnston, "The Messy Link Between Slave Owners and Modern Management," *Forbes*, January 16, 2013.

13. Lauren Dubois, *Haiti: The Aftershocks of History* (New York: Metropolitan Books, 2012), 97–100.

14. Frantz Fanon, *The Wretched of the Earth* (New York: Grove, 2004), 55.

15. Kari Marie Norgaard, *Living in Denial: Climate Change, Emotions, and Everyday Life* (Cambridge, MA: MIT Press, 2011), 61.

16. Adam Smith, *The Wealth of Nations*, Books I–III, ed. Andrew Skinner (London: Penguin, 1999), 183–84, 488–89.

17. Seymour Drescher, *The Mighty Experiment: Free Labor Versus Slavery in British Emancipation* (Oxford: Oxford University Press, 2002), 34–35, 233; Thomas Clarkson, *The History of the Rise, Progress, and Accomplishment of the Abolition of the African Slave-Trade, by the British Parliament*, Vol. 2 (London: Longman, Hurst, Rees, and Orme, 1808), 580–81.

18. Wendell Phillips, "Philosophy of the Abolition Movement: Speech Before the Massachusetts Antislavery Society (1853)," in *Speeches, Lectures, and Letters* (Boston: James Redpath, 1863), 109–10; Frederick Douglass, "The Meaning of July Fourth for the Negro," speech at Rochester, New York, July 5, 1852, in *Frederick Douglass: Selected Speeches and Writings*, ed. Philip S. Foner and Yuval Taylor (Chicago: Chicago Review Press, 2000), 196.

19. David Brion Davis, *Inhuman Bondage: The Rise and Fall of Slavery in the New World* (New York: Oxford University Press, 2006), 1.

20. Desmond Tutu, "We Need an Apartheid-Style Boycott to Save the Planet," *Guardian*, April 10, 2014.

21. Luis Hernández Navarro, "Repression and Resistance in Oaxaca," *CounterPunch*, November 21, 2006.

22. Personal interview with Sivan Kartha, January 11, 2013.

ACKNOWLEDGMENTS

One of the best decisions of my professional life was hiring Rajiv Sicora as lead researcher on this project in early 2010. Far more than a top-notch researcher, Rajiv has been a true intellectual companion on the long journey that produced this book. He has synthesized mountains of material from wildly diverse fields, sharing his own brilliant political analysis with me at every stage.

Rajiv was involved in all aspects of the book's research, but his unique mark can be felt in particular in the sections on trade, the psychology of climate denial, the history of slavery abolition, climate debt, and anything and everything related to climate science, including geoengineering. Rajiv's breadth of knowledge and command of this material is truly dazzling, as is the depth of his commitment to this project and its subject matter. I am blessed to have had him as a partner and friend throughout.

Two years ago, Rajiv and I were joined by Alexandra Tempus, another exceptional and diligent journalist and researcher. Alexandra quickly mastered her own roster of topics, from post–Superstorm Sandy disaster capitalism to financialization of nature to the opaque world of green group and foundation funding to climate impacts on fertility. She developed important new contacts, uncovered new and shocking facts, and always shared her thoughtful analysis.

Both Rajiv and Alexandra communicated with and interviewed dozens of experts. As this book went into its final stages and thousands of facts needed sourcing, checking, rechecking, and legal vetting, I have been greatly moved by their willingness to do whatever it took to get the job done, including far too many nights without sleep. To be supported by two such serious and committed colleagues is a true gift.

My next debt is to the team of tremendously tough and talented editors who pushed me to continually improve the draft. A decade and a half after we first published *No Logo* together, I am delighted to still have the honor of working with Louise Dennys, the legendary and fearless publisher of Random House of Canada. As always, my dear friend Louise knows me best, and pushes me hardest editorialy, in the most encouraging way possible. Helen Conford at Penguin U.K., a key collaborator on *The Shock Doctrine*, once again strengthened the manuscript with her thoughtful queries and insights and continues to be an inspiring publishing partner.

This is my first time publishing with Simon & Schuster in the U.S. and I would not have made the move were it not for the visionary leadership of Jonathan Karp and the editorial acumen of Bob Bender. I am so glad I did. They took what I consider to have been a very feminist risk in signing an author who was seven months pregnant, believing that the book would get written. It clearly did, but not without its share of delays, and I will always be grateful for their patience and fierce faith in this project. Bob, you have steered the editorial group with grace and improved the manuscript again and again. Thank you.

I am so fortunate to have Amanda Urban as my agent, as well as her wonderful colleagues Karolina Sutton and Helen Manders. They continue to find the perfect publishing collaborators around the world and are the most loyal friends and fighters when things get tough. I adore you guys.

And then there is Jackie Joiner: the woman who runs my life as well as Klein Lewis Productions, our little book and movie-making outfit. Only Jackie could have managed so many moving parts in a way that carved out the time and space for me to both write this book and enjoy new motherhood. As we approach launch, it is Jackie who will keep us from capsizing. Jackie, you are family and Avi and I would be lost without you.

Debra Levy, my longtime research assistant, regrettably had to leave this project in 2012. Before she did, she made enormous contributions, particularly to the sections on geoengineering, messianic billionaires, and climate debt. She also helped train Rajiv and Alexandra. She is one of the great collaborators of my career and I miss her still.

In the final months before deadline, Alleen Brown and Lauren Sutherland went above and beyond, helping enormously with the fact-check on absurdly tight deadlines. Lauren also did dynamite research for the billionaires chapter. Dave Oswald Mitchell contributed wise and comprehensive research on the growth imperative, and Mara Kardas-Nelson did the same on local power movements in Germany and Boulder.

Rajiv and I are also deeply grateful to the team of very busy climate scientists who agreed to read sections of the book relating to climate change impacts and projections. Our readers ended up being an all-star cast of scientific experts including: Kevin Anderson (Tyndall Centre for Climate Change Research), Alice Bows-Larkin (Tyndall Centre), James Hansen (Columbia University), Peter Gleick (Pacific Institute), and Sivan Kartha (Stockholm Environment Institute), all of whom vetted large sections of the book for accuracy. Michael E. Mann (Penn State University) and Olivia Serdeczny (Climate Analytics) also looked over the projections of a 4-degree world and provided helpful feedback. As a nonscientist, having this team of experts vet the accuracy of this material was critical; all political conclusions drawn from those scientific findings are mine alone and in no way reflect on these generous readers.

When Bill McKibben asked me to join the board of 350.org in 2011, I had no idea what a wild ride it would be. Through the Keystone XL campaign and the kickoff of the fossil fuel divestment movement, working with 350.org's brilliant team—particularly its imaginative executive director May Boeve—has given me a front row seat to the fast changing climate justice movement partially documented in these pages. Bill, you are one of the world's truly great people, a rock of a friend, and you wrote most of this years ago. I love being in this fight with you. All views expressed here are my own and have nothing to do with 350.org as an organization.

Other experts in their respective fields who agreed to review sections of this book for accuracy include Riley Dunlap, Aaron M. McCright, Robert Brulle, Steven Shrybman, Oscar Reyes, Larry Lohmann, Patrick Bond, Tadzio Mueller, and Tom Kruse. I am most grateful to all of them.

My dear friends Kyo Maclear, Eve Ensler, Betsy Reed, and Johann Hari all read portions of the book and shared their great skills as writers and editors. Johann, in fact, provided some of the most transformative editorial advice I received, and I am forever in his debt. This unofficial, backroom publishing team supported me in countless ways, from helping me come up with the right title to endless conversations about the book's themes.

My parents, Bonnie and Michael Klein, also provided helpful feedback, and my father, who has spent a lifetime researching the risks of obstetrical interventions and advocating for women's health, acted as a laughably overqualified research assistant in my investigations into the medical risks of fertility treatments. I am particularly grateful to my brother Seth Klein for his careful and detailed edit, and to all of his colleagues at the B.C. Canadian Centre for Policy Alternatives for their groundbreaking work on climate justice.

My husband, Avi Lewis, is always my first reader and primary collaborator. On this project we made it official: as I have been writing this book, Avi has been directing a documentary film on the same subject, a parallel process that often allowed us to research and travel together. The film work also fed into the book and though the film credits will do the real job of thanking the people involved, these acknowledgments would not be complete without some of our film collaborators, including: Joslyn Barnes, Katie McKenna, Anadil Hossain, Mary Lampson, Shane Hofeldt, Mark Ellam, Daniel Hewett, Chris Miller, Nicolas Jolliet, Martin Lukacs, Michael Premo, Alex Kelly, Daphne Wysham, Jacqueline Soohen, as well as Ellen Dorsey, Tom Kruse, Cara Mertes, and Amy Rao for their tremendous support from the earliest days.

People we met and worked with in the field shaped this work in many ways, including Theodoros Karyotis, Apostolis Fotiadis, Laura Gottesdiener, Crystal Lameman, Alexis Bonogofsky, Mike Scott, Nastaran Mohit and Sofia Gallisá Muriente, Wes Jackson, Phillip Whiteman Jr. and Lynette Two Bulls, David Hollander, and Charles Kovach, among many more.

Others who shared their expertise above and beyond include Soren Ambrose, Dan Apfel, Tom Athanasiou, Amy Bach, Diana Bronson, John Cavanagh, Stan Cox, Brendan DeMelle, Almuth Ernsting, Joss Garman, Justin Guay, Jamie Henn, Jess Housty, Steve Horn, Martin Khor, Kevin Koenig, F. Gerald Maples, Lidy Nacpil, Michael Oppenheimer, Sam Randalls, Mark Randazzo, Janet Redman, Alan Robock, Mark Schapiro, Scott Sinclair, Rachel Smolker, Ilana Solomon, Matthew Stilwell, Jesse Swanhuyser, Sean Sweeney, Jim Thomas, Kevin Trenberth, Aaron Viles, Ben West, Ivonne Yanez, and Adam Zuckerman.

Many research institutions, NGOs, and media outlets provided valuable support, and I am particularly grateful to the Climate Science Rapid

Response Team, DeSmogBlog, EJOLT (Environmental Justice Organisations, Liabilities and Trade), the Pembina Institute, Greenpeace Canada, the Carbon Dioxide Information Analysis Center, and Oil Change International. I rely heavily on Grist and Climate Progress for my climate news, and the wonderful writers at *Orion* for deeper analysis. And we would all be lost without Democracy Now!'s unflagging commitment to covering climate when no one else will, providing free transcripts for every interview.

Many books and reports are acknowledged in the text and notes, but I am particularly grateful to: Mark Dowie for *Losing Ground*, Christine MacDonald for *Green Inc.*, Petra Bartosiewicz and Marissa Miley for *The Too Polite Revolution*, and Herbert Docena for his writing on the history of carbon trading. Andreas Malm's work on the history of coal had a huge influence on me, as have the complete works of Clive Hamilton. Leanne Betasamosake Simpson helped me to understand the underlying logic of extractivism, and Renee Lertzman, Kari Marie Norgaard, Sally Weintrobe, and Rosemary Randall made me see climate change denial in a whole new light.

The political economy of the climate crisis is an incredibly dense field, and there is no way I could possibly cite all the critical thinkers who laid the foundation on which this book rests. Without hope of being exhaustive, let me mention a few whose work has been particularly important to my education and who have not been listed above: Joan Martínez Alier, Nnimmo Bassey, Robert D. Bullard, Erik M. Conway, Herman Daly, Joshua Farley, John Bellamy Foster, David Harvey, Richard Heinberg, Tim Jackson, Derrick Jensen, Van Jones, Michael T. Klare, Winona LaDuke, Edgardo Lander, Carolyn Merchant, George Monbiot, Naomi Oreskes, Christian Parenti, Ely Peredo, Andrew Ross, Juliet B. Schor, Joni Seager, Andrew Simms, Pablo Solón, James Gustave Speth, Sandra Steingraber, and Peter Victor.

Publishing is a finicky business, with more attention to detail than is at all fashionable. I am so grateful to all the people who labored over these important details, especially the stellar team at Simon & Schuster, including Johanna Li, Ruth Fecych, Fred Chase, and Phil Metcalf. At Knopf/Random House Canada, Amanda Lewis read diligently and contributed helpful editorial comments. Scott Richardson at Random House of Canada is responsible for the book's bold cover design. No one but Scott could have produced a design that would have convinced me to take my name off the cover of my own book. Thank you in advance to the three talented and dedicated publicists responsible for launching this book into the world: Julia Prosser at Simon & Schuster, Shona Cook at Random House of Canada, and Annabel Huxley at Penguin U.K. And thanks, too, to the lawyers who vetted this text: Brian MacLeod Rogers, Elisa Rivlin, and David Hirst.

Other researchers and *Nation* interns dipped in and out of the project over its five-year life, including: Jake Johnston, Dawn Paley, Michelle Chen, Kyla Neilan, Natasja Sheriff, Sarah Woolf, Eric Wuestewald, Lisa Boscov-Ellen, Saif Rahman, Diana Ruiz, Simon Davis-Cohen, Owen Davis, and Ryan Devereaux. All did excellent work. Alonzo Ríos Mira provided invaluable help with interview transcriptions, as did several others.

My writing continues to be supported by The Nation Institute, where I am a Puffin Foundation Writing Fellow, and the Institute generously provided office space to Rajiv throughout this project, while *The Nation* magazine did the same for Alexandra. I am grateful to all my colleagues in the Nation orbit, in particular my editor, Betsy Reed, as well as to Katrina vanden Heuvel, Peter Rothberg, Richard Kim, Taya Kitman, Ruth Baldwin, and Esther Kaplan. Special thanks also to the Wallace Global Fund, the Lannan Foundation, and the NoVo Foundation for their support over the years.

Rajiv extends a special thank you to Hannah Shaw and to his parents, Durga Mallampalli and Joseph Sicora. Alexandra does likewise to her parents, Robyn and Kenneth Shingler, Kent Tempus, and Denise Sheedy-Tempus, and to her grandmother Sandra Niswonger. We are all grateful for their understanding and support through this long and immersive project.

Friends with whom I have an ongoing and enriching conversation on these subjects include many of those listed above, as well as Justin Podur, Clayton Thomas-Muller, Katharine Viner, Arthur Manuel, Harsha Walia, Andréa Schmidt, Seumas Milne, Melina Laboucan-Massimo, Robert Jensen, Michael Hardt, John Jordan, Raj Patel, Brendan Martin, Emma Ruby-Sachs, Jane Saks, Tantoo Cardinal, and Jeremy Scahill. Gopal Dayaneni and the whole gang at Movement Generation provide me with an ongoing education and no end of inspiration. More personal thanks go to Misha Klein, Michele Landsberg, Stephen Lewis, Frances Coady, Nancy Friedland, David Wall, Sarah Polley, Kelly O'Brien, Cecilie Surasky and Carolyn Hunt, Sara Angel, Anthony Arnove, Brenda Coughlin, John Greyson, Stephen Andrews, Anne Biringer, Michael Sommers, Belinda Reyes, and Ofelia Whiteley.

My deepest thanks go to little Toma, for his truly heroic feats of toddler patience. He is about to learn that the world is a lot bigger than our neighborhood.

INDEX